Functional Analysis

Functional Analysis is a part of mathematics that deals with linear spaces equipped with a topology. The subject began with the work of Fredholm, Hilbert, Banach and others in the early 20th century. They developed an algebraic/topological framework which could be used to address a variety of questions in analysis. The subject immediately saw connections to abstract algebra, partial differential equations, geometry and much more.

This book is meant to introduce the reader to functional analysis. The first half of the book will cover the basic material that is taught in Masters programs across the world and prove all the major theorems in great detail. The second half of the book will focus on operators on a Hilbert space and is built around the proof of the spectral theorem – a central result in the subject that ties together traditional functional analysis with the modern theory of operator algebras.

The book aims to provide an accessible, interesting and readable introduction to the subject. It will also take the reader a little further than most courses do by introducing them to the language of operator algebras. This will help future researchers by giving them a jumping off point as they dive into deeper books on the subject.

Prahlad Vaidyanathan teaches at the Indian Institute of Science Education and Research, Bhopal. He has published several research papers on C*-algebras and their K-theory.

Functional Analysis

Prahlad Vaidyanathan

CAMBRIDGE
UNIVERSITY PRESS

Shaftesbury Road, Cambridge CB2 8EA, United Kingdom

One Liberty Plaza, 20th Floor, New York, NY 10006, USA

477 Williamstown Road, Port Melbourne, VIC 3207, Australia

314–321, 3rd Floor, Plot 3, Splendor Forum, Jasola District Centre, New Delhi – 110025, India

103 Penang Road, #05-06/07, Visioncrest Commercial, Singapore 238467

Cambridge University Press is part of Cambridge University Press & Assessment, a department of the University of Cambridge.

We share the University's mission to contribute to society through the pursuit of education, learning and research at the highest international levels of excellence.

www.cambridge.org
Information on this title: www.cambridge.org/9781009243902

First published 2023

Printed in India by Nutech Print Services, New Delhi 110020

A catalogue record for this publication is available from the British Library

ISBN 978-1-009-24390-2 Paperback

Contents

vi   *Contents*

Preface

This book is an attempt to emulate the classroom learning experience. It seems appropriate in a world where online education has become par for the course and the student does not always have access to a teacher who can help fill in the blanks. As a result, the book is thorough (sometimes to a fault) and somewhat more conversational than most others of its ilk.

The classroom is a place where one often engages in free-wheeling discussions that cut across disciplines. The subject of Functional Analysis, which lies at the confluence of modern analysis, algebra and topology, seems well-placed to transfer such discussions to the written word. It seamlessly mixes ideas from these different subjects, is widely applicable, and is therefore appealing to a broad spectrum of people. My hope is to present an introduction to the subject that is useful to everyone, regardless of their tastes.

The book is intended to be used for a year-long course in Functional Analysis aimed at Master's or Ph.D. students. After a short review in Chapter 1, Chapters 2–6 constitute the core of the subject. Here, one proves the Hahn-Banach theorems, the consequences of the Baire Category theorem, and the Banach-Alaoglu and Krein-Milman Theorems. Barring a few specialized topics, these chapters may be taught in a single semester.

The second half of the book (Chapter 7–10) is a little more advanced, and is meant to be taught in the second semester as an introduction to the theory of Operator Algebras. Ostensibly, the goal is to prove the Spectral Theorem for Normal Operators on a Hilbert space. However, I have chosen to take the scenic route, introducing as much operator algebra theory as I can given the time constraints. Perhaps the most egregious detour is in Chapter 9, where one encounters a proof of the Riesz-Markov-Kakutani theorem (due to Garling) that uses C*-algebra theory. I hope that such discussions will encourage students to look further into this fascinating subject.

A word on the exercises: there are plenty of them at the end of each chapter. Many are there to complement the results proved in the text, while others are there to allow students to practice using these results. At the end of each chapter, there is a list of problems that are somewhat tangential to the topic at hand (for example, Reproducing Kernel Hilbert spaces, Amenable Groups, etc.). These are meant to introduce students to interesting questions and avenues of research. As such, they may be used as jumping-off points for short projects. I have also mentioned some books and articles one might use for further investigation.

Finally, I should say that this book is far from comprehensive. Indeed, it would be impossible to do justice to a subject that is as vast and varied as Functional Analysis. I hope, though, that the book will give the reader enough tools to understand more advanced texts with confidence.

The book was conceived and written during the interminable lockdowns necessitated by the COVID-19 pandemic. During this time, Namrata and Nayan have had to put up with me, my temper, and my all-pervading supply of paper. I want to thank them for their patience and good humour through all of it. I also want to thank Vidya, Viswa, Kartik, Vidya (Sr.) and all the others in the extended family whose affection and support over the years means so much. Finally, I owe the greatest debt of gratitude to my parents. It is thanks to their sacrifices that I am able to live a comfortable and happy life today. For this (and so much more), this book is fondly dedicated to them.

Notation

Throughout the book, I plan to use some notational conventions which I have described below. Apart from these, a variety of symbols are used *locally*, which will be defined as and when they appear.

Notation	What it represents
$\mathbb{C}$	The set of complex numbers
$\mathbb{N}$	The set of natural numbers
$\mathbb{Q}$	The set of rational numbers
$\mathbb{R}$	The set of real numbers
$\mathbb{Z}$	The set of integers
μ, ν, λ	Measures
$\mathfrak{M}, \mathfrak{N}$	σ-algebras
$\mathfrak{L}$	The Lebesgue σ-algebra
$\mathfrak{B}_X$	The Borel σ-algebra on X
$\mathfrak{A}_X$	The Baire σ-algebra on X
$\mathbb{K}$	The base field for a vector space
$\alpha, \beta, \ldots$	Scalars in a field
$x, y, z, \ldots$	Elements of a vector space
$f, g, h, \ldots$	Scalar-valued functions, and elements of function spaces
$\mathbf{A}, \mathbf{B}, \mathbf{C}$	Banach or C*-algebras
$\mathbf{I}, \mathbf{J}$	Ideals in a Banach or C*-algebra
$\mathbf{E}, \mathbf{F}, \mathbf{W}, \mathbf{N}$	Normed linear spaces
$\mathbf{H}, \mathbf{K}, \mathbf{M}$	Hilbert spaces
φ, ψ	(Bounded) linear functionals
$T, S, \ldots$	(Bounded) linear operators
Φ, Ψ	Homomorphisms between Banach or C*-algebras

Notation	What it represents
$\mathcal{B}(\mathbf{E})$	The set of bounded linear operators on **E**
$\mathcal{K}(\mathbf{E})$	The set of compact linear operators on **E**
$\mathcal{F}(\mathbf{E})$	The set of bounded finite rank operators on **E**
$\lhd$	Ideal (used as $\mathbf{I} \lhd \mathbf{A}$)
$\mathbf{1}$	Constant function 1
$\xrightarrow{w}$	Weak convergence; also used for WOT-convergence
$\xrightarrow{s}$	Strong (norm) convergence; also used for SOT-convergence
$\xrightarrow{w^*}$	Weak-$*$ convergence
$\xrightarrow{bp}$	Bounded pointwise convergence

Among the exercises, those marked with the (⚒) symbol may be treated as an extension of the text, and must be solved by any serious student of the subject.

Chapter 1

Preliminaries

1.1 Review of Linear Algebra

Linear algebra is primarily the study of finite dimensional vector spaces and linear transformations between them. In a first course, one encounters the fact that every such vector space has a basis, that linear transformations may be associated to matrices, and one finally understands why matrix multiplication looks so clumsy. This section sets out to relive those glory days.

1.1.1 DEFINITION A *vector space* over a field $\mathbb{K}$ is a set $\mathbf{E}$ together with two operations $a\colon \mathbf{E} \times \mathbf{E} \to \mathbf{E}$ (vector addition) and $s\colon \mathbb{K} \times \mathbf{E} \to \mathbf{E}$ (scalar multiplication) written as $a(x, y) = x + y$ and $s(\alpha, x) = \alpha x$, which satisfy the following properties.

(a) $(\mathbf{E}, +)$ is an abelian group, whose identity is denoted by 0. If $x \in \mathbf{E}$, we write $(-x)$ for the inverse of x.

(b) If $1 \in \mathbb{K}$ denotes the multiplicative identity, then $1x = x$ for all $x \in \mathbf{E}$.

(c) $\alpha(\beta x) = (\alpha \beta)x$ for all $\alpha, \beta \in \mathbb{K}$ and $x \in \mathbf{E}$.

(d) $\alpha(x + y) = \alpha x + \alpha y$ for all $\alpha \in \mathbb{K}$ and all $x, y \in \mathbf{E}$.

(e) $(\alpha + \beta)x = \alpha x + \beta x$ for all $\alpha, \beta \in \mathbb{K}$ and $x \in \mathbf{E}$.

1

Note. For reasons we will discuss later, all vector spaces in this book will be over $\mathbb{R}$ or $\mathbb{C}$ (denoted by $\mathbb{K}$ when we do not wish to specify which). Furthermore, for the sake of sanity, we will always assume that our vector spaces are *non-zero*.

1.1.2 DEFINITION   Let **E** be a vector space over $\mathbb{K}$ and let $S = \{x_1, x_2, \ldots, x_n\}$ be a finite set of vectors in **E**.

(i) A *linear combination* of these vectors is an expression of the form $\alpha_1 x_1 + \alpha_2 x_2 + \ldots + \alpha_n x_n$, where $\alpha_i \in \mathbb{K}$ for all $1 \leq i \leq n$.

(ii) The set S is said to be *linearly dependent* if there is some vector in S which can be expressed as a linear combination of the remaining vectors. Equivalently, S is linearly dependent if there are scalars $\alpha_1, \alpha_2, \ldots, \alpha_n$ in $\mathbb{K}$, not all of which are zero, such that

$$\alpha_1 x_1 + \alpha_2 x_2 + \ldots + \alpha_n x_n = 0. \tag{1.1}$$

(iii) The set S is said to be *linearly independent* if it is not linearly dependent. Equivalently, S is linearly independent if, whenever $\alpha_1, \alpha_2, \ldots, \alpha_n$ are scalars such that Equation 1.1 holds, then each α_i is forced to be zero.

1.1.3 DEFINITION   A *Hamel basis* for a vector space **E** is a set $\Lambda \subset \mathbf{E}$ such that every element of **E** can be expressed uniquely as a finite linear combination of elements in Λ.

The word 'finite' is crucial in the above definition. An infinite sum is necessarily defined as a *limit* of partial sums and therefore only makes sense in a vector space that is equipped with a topology. There will come a time when we do equip vector spaces with topologies, and we will discuss series in that context. However, a linear combination will always mean a finite sum.

The next few results ought to be familiar, so we omit the proofs. Hoffman and Kunze [27] and Halmos [24] are good references for all this and more.

1.1.4 PROPOSITION   *For a subset $\Lambda \subset \mathbf{E}$, the following are equivalent:*

(i) Λ *is a Hamel basis for* **E**.

(ii) Λ *is a maximal linearly independent set.*

(iii) Λ *is a minimal spanning set.*

1.1.5 THEOREM (ZORN'S LEMMA) *Let $(\mathcal{F}, \leq)$ be a partially ordered set such that every totally ordered subset has an upper bound. Then $\mathcal{F}$ has a maximal element.*

Aside. If $\mathcal{C}$ is a subset of $\mathcal{F}$, an upper bound for $\mathcal{C}$ is an element $x_0 \in \mathcal{F}$ (not necessarily in $\mathcal{C}$), with the property that $x \leq x_0$ for all $x \in \mathcal{C}$. A maximal element in $\mathcal{F}$ is an element m with the property that, if $x \in \mathcal{F}$ and $m \leq x$, then $m = x$. An important point that is often confusing is this: A maximal element of $\mathcal{F}$ need not be an upper bound for $\mathcal{F}$!

1.1.6 THEOREM *Every vector space has a Hamel basis. In fact, if $\Lambda_0 \subset \mathbf{E}$ is any linearly independent set, then there exists a Hamel basis Λ of $\mathbf{E}$ such that $\Lambda_0 \subset \Lambda$.*

1.1.7 EXAMPLE

(i) For $\mathbf{E} = \mathbb{K}^n$, we write $e_i := (0, 0, \ldots, 0, 1, 0, \ldots, 0)$ (with 1 in the ith position). The set $\{e_i : 1 \leq i \leq n\}$ is called the **standard basis** for $\mathbb{K}^n$.

(ii) Define

$$c_{00} := \{(x_n)_{n=1}^\infty : x_i \in \mathbb{K}, \text{ and there exists } N \in \mathbb{N} \text{ such that } x_i = 0 \text{ for all } i \geq N\}.$$

Members of c_{00} are sequences that are *eventually* zero (or equivalently, sequences with finite support). It is a vector space over $\mathbb{K}$ where the vector space operators are defined componentwise. Write e_i for the sequence

$$(e_i)_j = \delta_{i,j} = \begin{cases} 1 & : \text{ if } i = j, \\ 0 & : \text{ otherwise.} \end{cases}$$

Then $\{e_i : i \in \mathbb{N}\}$ is a basis for c_{00}.

(iii) Define

$$c_0 = \{(x_n)_{n=1}^\infty : x_i \in \mathbb{K}, \text{ and } \lim_{i \to \infty} x_i = 0\}.$$

Note that $\{e_i : i \in \mathbb{N}\}$ as above is a linearly independent set but *not* a basis for c_0. We will prove later that any basis of c_0 must be uncountable (see 5.1.4 Corollary). For now, though, give an example of an element in c_0 that cannot be expressed as a linear combination of the $\{e_i\}$.

(iv) Let $a, b \in \mathbb{R}$ with $a < b$, and define

$$C[a, b] := \{f : [a, b] \to \mathbb{K} \text{ continuous}\}.$$

This is a vector space over $\mathbb{K}$ under pointwise addition and scalar multiplication. For $n \geq 0$, let $e_n(x) := x^n$, then $\{e_n : n \geq 0\}$ is a linearly independent set, but it is not a basis for $C[a, b]$ (once again, do verify this).

More generally, if X is a compact, Hausdorff space, we set $C(X)$ to denote the space of continuous, $\mathbb{K}$-valued functions on X. This is a vector space under pointwise operations as well.

We will continue to develop this collection of examples as we go through the book. For the most part, all examples will fall into three 'types': finite dimensional vector spaces, sequence spaces and function spaces. While all of them may be profitably thought of as function spaces, it is more intuitive to think of them as different objects.

1.1.8 THEOREM *If* **E** *is a vector space, then any two Hamel bases of* **E** *have the same cardinality. This common number is called the* **dimension** *of* **E**.

We omit the proof of this result. In the finite dimensional case, this is proved in Hoffman and Kunze [27, Section 2.3] while the proof in the infinite dimensional case is similar to that of Lemma 3.4.1 in Chapter 3.

1.1.9 DEFINITION Let **E** and **F** be two vector spaces.

(i) A function $T : \mathbf{E} \to \mathbf{F}$ is said to be a *linear transformation* or an *operator* if

$$T(\alpha x + y) = \alpha T(x) + T(y)$$

for all $x, y \in \mathbf{E}$ and $\alpha \in \mathbb{K}$.

(ii) We write $L(\mathbf{E}, \mathbf{F})$ for the set of all linear operators from **E** to **F**. If $S, T \in L(\mathbf{E}, \mathbf{F})$ and $\alpha \in \mathbb{K}$, we define the operators $(S + T)$ and αS by

$$(S + T)(x) := S(x) + T(x), \text{ and } (\alpha S)(x) = \alpha S(x).$$

Clearly, this makes $L(\mathbf{E}, \mathbf{F})$ a $\mathbb{K}$-vector space.

(iii) If $\mathbf{F} = \mathbb{K}$, then a linear transformation $T : \mathbf{E} \to \mathbb{K}$ is called a *linear functional*.

(iv) Given a linear transformation $T : \mathbf{E} \to \mathbf{F}$, there are two sets associated to T that we will often refer to:

$$\ker(T) := \{x \in \mathbf{E} : T(x) = 0\} \text{ and } \text{Range}(T) := \{T(x) : x \in \mathbf{E}\}.$$

It is easy to check that $\ker(T)$ and $\text{Range}(T)$ are subspaces of $\mathbf{E}$ and $\mathbf{F}$, respectively.

(v) A linear transformation $T : \mathbf{E} \to \mathbf{F}$ is said to be an *isomorphism* if T is bijective. If such a map exists, we write $\mathbf{E} \cong \mathbf{F}$.

1.1.10 EXAMPLE

(i) Let $\mathbf{E} = \mathbb{K}^n, \mathbf{F} = \mathbb{K}^m$; then any $m \times n$ matrix A with entries in $\mathbb{K}$ defines a linear transformation $T_A : \mathbf{E} \to \mathbf{F}$ given by $x \mapsto A(x)$. Conversely, if $T \in L(\mathbf{E}, \mathbf{F})$, then the matrix whose columns are $\{T(e_i) : 1 \leq i \leq n\}$ defines an $m \times n$ matrix A such that $T = T_A$. If $M_{m \times n}(\mathbb{K})$ denotes the vector space of all such matrices, then there is an isomorphism of vector spaces

$$L(\mathbf{E}, \mathbf{F}) \cong M_{m \times n}(\mathbb{K})$$

given by $T_A \mapsto A$. If we replace the standard basis $\{e_1, e_2, \ldots, e_n\}$ by another basis Λ of $\mathbf{E}$, we get another isomorphism from $L(\mathbf{E}, \mathbf{F}) \to M_{m \times n}(\mathbb{K})$. Thus, the isomorphism is not canonical (it depends on the choice of basis).

(ii) Let $\mathbf{E} = c_{00}$ and define $\varphi : \mathbf{E} \to \mathbb{K}$ by

$$\varphi((x_j)) := \sum_{n=1}^{\infty} x_n.$$

Note that φ is well-defined and linear. Thus, $\varphi \in L(c_{00}, \mathbb{K})$.

(iii) Let $\mathbf{E} = C[a, b]$ and define $\varphi : \mathbf{E} \to \mathbb{K}$ by

$$\varphi(f) := \int_a^b f(t)dt.$$

Then $\varphi \in L(C[a, b], \mathbb{K})$.

(iv) Let $\mathbf{E} = \mathbf{F} = C[0, 1]$. Define $T : \mathbf{E} \to \mathbf{F}$ by

$$T(f)(x) := \int_0^x f(t)dt.$$

Note that T is well-defined (from Calculus) and linear. Thus $T \in L(\mathbf{E}, \mathbf{F})$.

1.1.11 DEFINITION Let **E** be a vector space and **F** be a subspace of **E**.

(i) The *quotient space*, denoted by **E/F**, is the quotient group, viewing **E** as an abelian group under addition and **F** as a (normal) subgroup. Note that **E/F** has a natural vector space structure, with addition given by

$$(x + \mathbf{F}) + (y + \mathbf{F}) := (x + y) + \mathbf{F},$$

and scalar multiplication given by $\alpha(x + \mathbf{F}) := \alpha x + \mathbf{F}$ for $\alpha \in \mathbb{K}$ and $x, y \in \mathbf{E}$.

(ii) The *quotient map*, denoted by $\pi : \mathbf{E} \to \mathbf{E/F}$, is given by $x \mapsto x + \mathbf{F}$. It is a surjective linear transformation such that $\ker(\pi) = \mathbf{F}$.

(iii) Furthermore, we define the *codimension* of **F** by $\mathrm{codim}(\mathbf{F}) := \dim(\mathbf{E/F})$.

(iv) If $\mathrm{codim}(\mathbf{F}) = 1$, then we say that **F** is a *hyperplane* of **E**.

Given a non-zero linear functional $\varphi : \mathbf{E} \to \mathbb{K}$, the subspace $\ker(\varphi)$ is a hyperplane in **E**. Conversely, every hyperplane is of this form.

The next result is a simple consequence of Theorem 1.1.6, and we will omit its proof. Henceforth, we will write '**F** < **E**' to indicate that **F** is a subspace of **E**.

1.1.12 PROPOSITION *Let* **E** *be a finite dimensional vector space and* **F** < **E**. *Then* $codim(\mathbf{F}) = \dim(\mathbf{E}) - \dim(\mathbf{F})$.

One rarely mentions the First Isomorphism Theorem in the context of vector spaces but that is perhaps because the Rank–Nullity Theorem hogs the limelight. Also, the proof is completely analogous to the case of groups.

1.1.13 THEOREM (FIRST ISOMORPHISM THEOREM) *Let* $T : \mathbf{E} \to \mathbf{F}$ *be a linear transformation. Then*

(i) $\ker(T) < \mathbf{E}$ *and* $\mathrm{Range}(T) < \mathbf{F}$.

(ii) *Furthermore, the map* $\widehat{T} : \mathbf{E}/\ker(T) \to \mathrm{Range}(T)$ *given by*

$$x + \ker(T) \mapsto T(x)$$

is an isomorphism.

Let us now put the Rank–Nullity Theorem in its place. It is a direct consequence of the First Isomorphism Theorem, with a touch of Proposition 1.1.12.

1.1.14 THEOREM (RANK–NULLITY THEOREM) *If* $T : \mathbf{E} \to \mathbf{F}$ *is a linear transformation and* **E** *is finite dimensional, then* $\dim(\ker(T)) + \dim(\mathrm{Range}(T)) = \dim(\mathbf{E})$.

Recall that the **nullity** of T is $\dim(\ker(T))$ and the **rank** of T is $\dim(\text{Range}(T))$. Note that the dimension of the co-domain of the linear transformation plays no role in the Rank–Nullity Theorem; one merely needs the domain to be finite dimensional.

1.2   Review of Measure Theory

Historically, Lebesgue's theory of measure and integration provided great impetus to the then fledgling subject of Functional Analysis. In fact, it can be argued that Functional Analysis grew out of a need to understand the L^p spaces and operators between them. Therefore measure theory will be used liberally throughout this book.

However, to start with, we do not assume that the reader is necessarily familiar with all the nuances of measure theory. Instead, we would like to take a middle path. We will assume some familiarity with the notion of a measure, measurable functions, and basic integration theory, such as those available in the first few chapters of Royden [49] or Rudin [51]. As we go along, we will need more and more, and we hope that the reader will pick those things up as and when needed. For now, though, let us refresh our collective memories with the basic notions.

1.2.1 DEFINITION   Let X be a set. A σ-**algebra** on X is a collection $\mathfrak{M}$ of subsets of X satisfying the following axioms:

(a) $\varnothing \in \mathfrak{M}$.

(b) If $E \in \mathfrak{M}$, then $E^c := X \setminus E \in \mathfrak{M}$.

(c) If $\{E_1, E_2, \ldots\}$ is a sequence of sets in $\mathfrak{M}$, then $\bigcup_{n=1}^{\infty} E_n \in \mathfrak{M}$.

The pair $(X, \mathfrak{M})$ is called a **measurable space** and the members of $\mathfrak{M}$ are called **measurable sets**.

It is a useful fact (and one that is easy to prove) that if $\{\mathfrak{M}_\alpha : \alpha \in J\}$ is a family of σ-algebras on a set X, then the intersection $\bigcap_{\alpha \in J} \mathfrak{M}_\alpha$ is also a σ-algebra. In particular, if $\mathcal{S}$ is a collection of subsets of X, then there is a unique smallest σ-algebra on X that contains $\mathcal{S}$. This is called the σ-algebra **generated by** $\mathcal{S}$. The most important example of this phenomenon is the following.

1.2.2 DEFINITION   Let X be a topological space. The σ-algebra generated by the topology on X is called the **Borel** σ-**algebra** on X and is denoted by $\mathfrak{B}_X$. The members of this σ-algebra are called **Borel sets**.

Important examples of Borel sets are the following: A countable union of closed sets is called an F_σ-set and the countable intersection of open sets is called a G_δ-set.

1.2.3 DEFINITION Let $(X, \mathfrak{M})$ be a measurable space and Y be a topological space. A function $f : X \to Y$ is said to be *measurable* if $f^{-1}(U) \in \mathfrak{M}$ for every open set $U \subset Y$.

For the most part, measurable functions in this book will take values in $\mathbb{K}$ ($= \mathbb{R}$ or $\mathbb{C}$), where the latter is equipped with the usual topology. When it is important to make a distinction, we will refer to such functions as *real-measurable* or *complex-measurable*, as the case may be.

1.2.4 EXAMPLE

(i) Given a subset $E \subset X$, the *characteristic function* of E is the map $\chi_E : X \to \mathbb{R}$ given by

$$\chi_E(x) = \begin{cases} 1 & : \text{ if } x \in E, \\ 0 & : \text{ otherwise.} \end{cases}$$

Clearly, χ_E is a measurable function if and only if E is a measurable set.

(ii) More generally, a linear combination of characteristic functions of measurable sets is measurable. Such a function is called a *simple function*. Alternatively, a simple function is a measurable function whose range is a finite set.

(iii) If X and Y are both topological spaces and we take $\mathfrak{M} = \mathfrak{B}_X$, then any measurable function $f : X \to Y$ is said to be *Borel measurable*. Notice that every continuous function is Borel measurable (however, there are Borel measurable functions that are not continuous).

The class of measurable functions is closed under a number of operations, which we list below.

1.2.5 PROPOSITION *Let $(X, \mathfrak{M})$ be a measurable space.*

(i) *If $f : X \to \mathbb{K}$, $g : X \to \mathbb{K}$ are measurable functions and $\alpha \in \mathbb{K}$, then $\alpha f + g$ is also measurable. So is the pointwise product $fg : X \to \mathbb{K}$, which is given by $x \mapsto f(x)g(x)$.*

(ii) *If $u : X \to \mathbb{R}$ and $v : X \to \mathbb{R}$ are real-measurable functions, then $f := u + iv$ is complex-measurable. Conversely, if $f : X \to \mathbb{C}$ is complex-measurable, then its real and imaginary parts are real-measurable functions.*

(iii) *If $f, g : X \to \mathbb{R}$ are measurable, then so are $\max\{f, g\}$ and $\min\{f, g\}$, which are defined by $\max\{f, g\}(x) := \max\{f(x), g(x)\}$, and $\min\{f, g\}(x) := \min\{f(x), g(x)\}$. In particular,*

$$f^+ := \max\{f, 0\}, \text{ and } f^- := -\min\{f, 0\}$$

are both measurable.

(iv) *If $f : X \to \mathbb{R}$ is measurable, then so is $|f| = f^+ + f^-$.*

(v) *If $\{f_n\}$ are a sequence of $\mathbb{K}$-valued measurable functions, then $\limsup_{n \to \infty} f_n$ and $\liminf_{n \to \infty} f_n$ are both measurable. In particular, the pointwise limit of measurable functions (if it exists) is measurable.*

One important result that allows us to prove theorems about arbitrary measurable functions by first proving them for characteristic functions is the following.

1.2.6 Theorem *Let $f : X \to \mathbb{R}_+$ be a non-negative measurable function. Then there is a sequence (s_n) of simple functions such that for each $x \in X$, $(s_n(x))$ is an increasing sequence of non-negative real numbers with $\lim_{n \to \infty} s_n(x) = f(x)$.*

Let us now turn to the notion of measure. This is a vast generalization of the idea of the 'volume' of a set and, unlike geometric notions of volume, it turns out to be flexible enough that we can prove interesting theorems about it.

1.2.7 Definition Let $(X, \mathfrak{M})$ be a measurable space. A *positive measure* on $(X, \mathfrak{M})$ is a function $\mu : \mathfrak{M} \to [0, \infty]$ satisfying the following axioms.

(a) $\mu(\varnothing) = 0$.

(b) μ is *countably additive*: If $\{E_1, E_2, \ldots\}$ is a sequence of mutually disjoint sets in $\mathfrak{M}$, then

$$\mu\left(\bigcup_{n=1}^{\infty} E_n\right) = \sum_{n=1}^{\infty} \mu(E_n).$$

The triple $(X, \mathfrak{M}, \mu)$ is called a *measure space*.

We will encounter both *real* and *complex* measures later on in the book, but the notion of a positive measure is the most basic. Therefore a positive measure will simply be referred to as a measure (without any qualification).

1.2.8 EXAMPLE

(i) Let X be any set and $x_0 \in X$ be a fixed point. Let $\mathfrak{M} := 2^X$ be the set of all subsets of X and let $\mu : \mathfrak{M} \to \mathbb{R}$ be the function

$$\mu(E) := \begin{cases} 1 & : \text{ if } x_0 \in E, \\ 0 & : \text{ if } x_0 \notin E. \end{cases}$$

This is called the **Dirac measure** at x_0 and is denoted by δ_{x_0}.

(ii) Let X be any set and $\mathfrak{M} := 2^X$ as above. Define $\mu : \mathfrak{M} \to [0, \infty]$ by

$$\mu(E) := \begin{cases} |E| & : \text{ if } E \text{ is finite}, \\ \infty & : \text{ otherwise}. \end{cases}$$

(where $|\cdot|$ denotes the cardinality function). It is clear that this is a measure on $(X, \mathfrak{M})$, and is called the **counting measure**.

(iii) If X is a topological space, a measure on X is called a **Borel measure** if its domain contains $\mathfrak{B}_X$. Note that the domain of the measure may be larger than $\mathfrak{B}_X$ as well.

(iv) A measure μ on a measurable space $(X, \mathfrak{M})$ is said to be a **finite measure** if $\mu(X) < \infty$ and it is said to be σ-**finite** if X can be expressed as a countable union of sets of finite measure.

The most important measure is the **Lebesgue measure** on $\mathbb{R}$. No matter how you do it, the construction of the measure is long and complicated. However, we will describe it in enough detail so as to have a working understanding of it.

Consider $\mathbb{R}$, equipped with the usual topology. Then there is a σ-algebra $\mathfrak{L}$ which contains $\mathfrak{B}_{\mathbb{R}}$, and a positive measure $m : \mathfrak{L} \to [0, \infty]$ satisfying the following properties:

(a) If $a, b \in \mathbb{R}$ with $a < b$, then $m([a, b)) = (b - a)$.

(b) If $E \in \mathfrak{L}$ and $x \in \mathbb{R}$, then $E + x \in \mathfrak{L}$ and $m(E + x) = m(E)$. This property is called **translation invariance** of the measure m (here, $E + x$ is the set $\{y + x : y \in E\}$).

(c) If $E \in \mathfrak{L}$, then

$$m(E) = \inf\{m(U) : U \text{ open}, E \subset U\} = \sup\{m(K) : K \text{ compact}, K \subset E\}.$$

This property is called **regularity** of the measure m.

(d) If $E \in \mathfrak{L}$ is such that $m(E) = 0$, and $F \subset E$, then $F \in \mathfrak{L}$ (and hence, $m(F) = 0$).
   This property is called **completeness** of the measure m.

The members of $\mathfrak{L}$ are called **Lebesgue measurable sets**. By construction, every Borel set is Lebesgue measurable. It is important to point out that there do exist subsets of $\mathbb{R}$ which are not Lebesgue measurable (although constructing such a set is itself non-constructive!).

Having done this, we may do the same for $\mathbb{R}^n$ when $n \geq 2$. There is a σ-algebra $\mathfrak{L}_n$ on $\mathbb{R}^n$, which contains the Borel σ-algebra $\mathfrak{B}_{\mathbb{R}^n}$ and a measure $m = m_n$ on $\mathfrak{L}_n$ with the property that

$$m\left(\prod_{i=1}^{n}[a_i, b_i)\right) = \prod_{i=1}^{n}(b_i - a_i)$$

and satisfying properties (b)–(d) exactly as above. Since any such rectangle has finite measure, it follows that m is a σ-finite measure on $\mathbb{R}^n$.

Finally, if $X \subset \mathbb{R}$ is a measurable set, then we may define a σ-algebra on E by

$$\mathfrak{L}_X := \{E \cap X : E \in \mathfrak{L}\}.$$

Then $\mathfrak{L}_X \subset \mathfrak{L}$ and therefore we may restrict the Lebesgue measure to $\mathfrak{L}_X$ to obtain a measure on X. We will simply refer to this as the Lebesgue measure on X. In the first few chapters, we will almost entirely focus on the case when $X = [a, b]$ is a compact interval in $\mathbb{R}$ so as to have a finite measure to work with.

Having defined measurable functions and measures, we now turn our attention to integration; this is, after all, the main reason this language was invented! Given a measure space $(X, \mathfrak{M}, \mu)$ and a measurable function $f : X \to \mathbb{K}$, we would like to make sense of the symbol

$$\int_X f d\mu = \int_X f(x) d\mu(x).$$

Now, if $f = \chi_E$ is a characteristic function, then it makes sense to define

$$\int_X \chi_E d\mu := \mu(E).$$

More generally, if $s = \sum_{i=1}^{n} \alpha_i \chi_{E_i}$ is a non-negative simple function, and the sets $\{E_1, E_2, \ldots, E_n\}$ are mutually disjoint, then we may define

$$\int_X s d\mu := \sum_{i=1}^{n} \alpha_i \mu(E_i)$$

(we require s to be non-negative, because μ is allowed to take the value ∞ and we want to avoid potential landmines such as '$\infty - \infty$'). If $f : X \to \mathbb{R}_+$ is a non-negative measurable function, we lean on Theorem 1.2.6 and define

$$\int_X f d\mu := \sup \left\{ \int_X s d\mu : s \text{ is simple, and } 0 \le s \le f \right\}.$$

This definition has the important property that it is monotone: if $0 \le g \le f$ are both measurable functions, then $\int_X g d\mu \le \int_X f d\mu$. In order to make a definition for an *arbitrary* $\mathbb{K}$-valued measurable function, we need to jump through some hoops, unfortunately. Suppose, to begin with, that $f : X \to \mathbb{R}$ is real-measurable, so that we may define $f^+ := \max\{f, 0\}$ and $f^- := -\min\{f, 0\}$ as we did above. Then $|f| = f^+ + f^-$ is a non-negative function, which allows us to make the following definition.

1.2.9 DEFINITION   A function $f : X \to \mathbb{R}$ is said to be *integrable* if

$$\int_X |f| d\mu < \infty.$$

By the monotonicity of the integral, it follows that if f is integrable, then $\int_X f^+ d\mu < \infty$ and $\int_X f^- d\mu < \infty$. Therefore, we may define

$$\int_X f d\mu := \int_X f^+ d\mu - \int_X f^- d\mu.$$

Finally, if f is a complex-valued measurable function, then we may define integrability exactly as in Definition 1.2.9. Furthermore, if we write $f = u + iv$, where u and v are real-measurable functions, then integrability of f implies the integrability of u and v. Hence, we may define

$$\int_X f d\mu := \int_X u d\mu + i \int_X v d\mu = \int_X u^+ d\mu - \int_X u^- d\mu + i \int_X v^+ d\mu - i \int_X v^- d\mu.$$

This (admittedly unwieldy) definition does have the virtue of being linear. If $f : X \to \mathbb{C}$ and $g : X \to \mathbb{C}$ are integrable functions and $\alpha \in \mathbb{C}$ is a scalar, then $(\alpha f + g)$ is integrable and

$$\int_X (\alpha f + g) d\mu = \alpha \int_X f d\mu + \int_X g d\mu.$$

1.2.10 EXAMPLE

(i) Let X be any set, $x_0 \in X$, and $\mu = \delta_{x_0}$ be the Dirac measure at x_0 as in Example 1.2.8. Then any function $f : X \to \mathbb{C}$ is measurable, integrable and

$$\int_X f d\delta_{x_0} = f(x_0).$$

(ii) If $X = \mathbb{N}$ and μ denotes the counting measure as in Example 1.2.8, then a function $f : X \to \mathbb{C}$ corresponds to a sequence $(f(n))_{n=1}^\infty$ of complex numbers. Also, f is integrable if and only if the corresponding series is absolutely convergent and, in that case,

$$\int_{\mathbb{N}} f d\mu = \sum_{n=1}^\infty f(n).$$

(iii) If $f : [a,b] \to \mathbb{K}$ is a bounded, Riemann integrable function, then f is Lebesgue measurable and its Lebesgue integral coincides with its Riemann integral (see [49, Section 4.2]). Therefore, whenever $f : [a,b] \to \mathbb{K}$ is a measurable function, we have the liberty to write

$$\int_a^b f = \int_a^b f(t)dt := \int_{[a,b]} f dm.$$

Now, the main reason that Lebesgue's theory of integration is so successful is that it behaves well with respect to *pointwise* limits of functions (Riemann integration does not). The most important theorems in this context are the following results.

1.2.11 THEOREM (FATOU'S LEMMA) *Let* $(X, \mathfrak{M}, \mu)$ *be a measure space and* (f_n) *be a sequence of non-negative measurable functions. Then*

$$\int_X \liminf_{n\to\infty} f_n d\mu \le \liminf_{n\to\infty} \int_X f_n d\mu.$$

1.2.12 THEOREM (MONOTONE CONVERGENCE THEOREM) *Let* $(X, \mathfrak{M}, \mu)$ *be a measure space and* (f_n) *be a sequence of non-negative measurable functions such that*

(i) $0 \le f_1(x) \le f_2(x) \le \ldots$ *for all* $x \in X$.

(ii) $\lim_{n\to\infty} f_n(x) = f(x)$ *for all* $x \in X$.

Then f *is measurable and* $\int_X f d\mu = \lim_{n\to\infty} \int_X f_n d\mu$.

An immediate consequence of the Monotone Convergence Theorem is the following result which will be used liberally throughout the book.

1.2.13 PROPOSITION *Let* $(X, \mathfrak{M}, \mu)$ *be a measure space and* (f_n) *be a sequence of non-negative measurable functions. Then* $\sum_{n=1}^{\infty} f_n$ *is measurable and*

$$\int_X \left(\sum_{n=1}^{\infty} f_n \right) d\mu = \sum_{n=1}^{\infty} \int_X f_n d\mu.$$

The previous three results are concerned with sequences of non-negative functions. The last of these theorems, and the most widely used, speaks of *any* sequence of functions.

1.2.14 THEOREM (DOMINATED CONVERGENCE THEOREM) *Let* $(X, \mathfrak{M}, \mu)$ *be a measure space and* (f_n) *be a sequence of* $\mathbb{K}$*-valued measurable functions. Suppose that*

(i) $\lim_{n \to \infty} f_n(x) = f(x)$ *for all* $x \in X$, *and*

(ii) *there is an integrable functions* $g : X \to [0, \infty]$ *such that* $|f_n(x)| \leq g(x)$ *for all* $x \in X$ *and* $n \in \mathbb{N}$.

Then f *is integrable and* $\int_X f d\mu = \lim_{n \to \infty} \int_X f_n d\mu$.

Now, a property is said to hold **almost everywhere** (in symbols, we write 'a.e.' or 'a.e. [μ]') if the set on which it fails to hold is contained in a set of measure zero. For instance, we would write "$f = \lim_{n \to \infty} f_n$ a.e." if the set $\{x \in X : \lim_{n \to \infty} f_n(x) \neq f(x)\}$ is contained in a set of measure zero. It is a fact that the convergence theorems mentioned above hold if we assume that the sequence converges almost everywhere (in other words, it need not converge *everywhere* for the conclusions to hold).

At various points in the book, we will often need to integrate functions of two variables as 'iterated integrals'. The fundamental theorem in this context is the Fubini–Tonelli Theorem, which is what we turn to next.

Let $(X, \mathfrak{M}, \mu)$ and $(Y, \mathfrak{N}, \nu)$ be two measure spaces. We wish to construct a σ-algebra and a measure on $X \times Y$. A measurable rectangle is a set of the form $E \times F$, where $E \in \mathfrak{M}$ and $F \in \mathfrak{N}$. The σ-algebra on $X \times Y$ generated by all such measurable rectangles is called the product σ-algebra and is denoted by $\mathfrak{M} \otimes \mathfrak{N}$. Now, there is a measure λ on $\mathfrak{M} \otimes \mathfrak{N}$ such that

$$\lambda(E \times F) = \mu(E)\nu(F)$$

for any $E \in \mathfrak{M}$ and $F \in \mathfrak{N}$. Furthermore, if both μ and ν are σ-finite measures, then there is exactly one measure on $\mathfrak{M} \otimes \mathfrak{N}$ satisfying this property. This is called the

product measure on $\tilde{X} \times Y$ and is denoted by $\mu \times \nu$. The Fubini–Tonelli Theorem now tells us how one may integrate functions on $X \times Y$ with respect to this product measure.

1.2.15 THEOREM (FUBINI–TONELLI THEOREM) *Let $(X, \mathfrak{M}, \mu)$ and $(Y, \mathfrak{N}, \nu)$ be two σ-finite measure spaces and let $f : X \times Y \to \mathbb{C}$ be a measurable function.*

(i) *(Tonelli, 1909) If f is a non-negative function, then the functions*

$$g(x) := \int_Y f(x, y) d\nu(y), \text{ and } h(y) := \int_X f(x, y) d\mu(x)$$

are both non-negative measurable functions on $(X, \mathfrak{M})$ and $(Y, \mathfrak{N})$, respectively, and

$$\begin{aligned}
\int_{X \times Y} f d(\mu \times \nu) &= \int_X \left(\int_Y f(x, y) d\nu(y) \right) d\mu(x) \\
&= \int_Y \left(\int_X f(x, y) d\mu(x) \right) d\nu(y).
\end{aligned} \tag{1.2}$$

(ii) *(Fubini, 1907) If f is integrable over $X \times Y$, then the functions g and h as above are defined almost everywhere, both g and h are integrable over X and Y, respectively, and Equation 1.2 holds.*

The most important example of this phenomenon is, once again, the Lebesgue measure. If $\mathfrak{B}_{\mathbb{R}^n}$ denotes the Borel σ-algebra on $\mathbb{R}^n$, then $\mathfrak{B}_{\mathbb{R}^n} \otimes \mathfrak{B}_{\mathbb{R}^k} = \mathfrak{B}_{\mathbb{R}^{n+k}}$. If we restrict the Lebesgue measure m_n to $\mathfrak{B}_{\mathbb{R}^n}$, the product measure $m_n \times m_k$ is precisely the restriction of m_{n+k} to $\mathfrak{B}_{\mathbb{R}^{n+k}}$. The measure m_{n+k} is thus an extension of the product measure $m_n \times m_k$.

With this, we end our rapid refresher course on measure theory. If you have not seen any measure theory, it is probably best if you read chapters 3 and 4 of Royden [49] or chapter 1 of Rudin [51]. The Fubini–Tonelli Theorem may be postponed until it is needed later in the book or you could take a look at chapter 6 of Wheeden and Zygmund [61].

Chapter 2

Normed Linear Spaces

2.1 Definitions and Examples

We now introduce the fundamental object in our investigations. A vector space, while very useful, is somewhat unwieldy when it is infinite dimensional. Equipping it with a metric, especially one that understands the linear structure of the underlying set, is a simple and effective way to alleviate this problem.

2.1.1 DEFINITION A *norm* on a $\mathbb{K}$-vector space $\mathbf{E}$ is a function

$$\|\cdot\| : \mathbf{E} \to \mathbb{R}_+$$

which satisfies the following properties for all $x, y \in \mathbf{E}$ and all $\alpha \in \mathbb{K}$.

(a) $\|x\| \geq 0$, and $\|x\| = 0$ if and only if $x = 0$.

(b) $\|\alpha x\| = |\alpha| \|x\|$.

(c) (Triangle inequality) $\|x + y\| \leq \|x\| + \|y\|$.

The pair $(\mathbf{E}, \|\cdot\|)$ is called a *normed linear space*.

2.1.2 REMARK Let $(\mathbf{E}, \|\cdot\|)$ be a normed linear space.

(i) The function $d(x, y) := \|x - y\|$ defines a metric on $\mathbf{E}$, called the metric *induced by the norm*. This makes $\mathbf{E}$ a topological space.

(ii) A sequence $(x_n) \subset \mathbf{E}$ converges to $x \in \mathbf{E}$ if and only if $\lim_{n \to \infty} \|x_n - x\| = 0$. When this happens, we will write $x_n \to x$.

(iii) By the triangle inequality, vector space addition is a continuous map from $\mathbf{E} \times \mathbf{E} \to \mathbf{E}$. Therefore if $x_n \to x$ and $y_n \to y$, then $(x_n + y_n) \to (x + y)$.

Similarly, scalar multiplication is also a continuous map from $\mathbb{K} \times \mathbf{E} \to \mathbf{E}$. (Here, $\mathbf{E} \times \mathbf{E}$ and $\mathbb{K} \times \mathbf{E}$ may both be equipped with a product metric).

(iv) For any $x, y \in \mathbf{E}$, the triangle inequality implies that

$$\big| \|x\| - \|y\| \big| \leq \|x - y\|.$$

Thus, the norm function $\mathbf{E} \to \mathbb{R}_+$ is continuous. Hence if $x_n \to x$ in $\mathbf{E}$ then $\|x_n\| \to \|x\|$ in $\mathbb{R}$.

Now let us revisit the examples from Example 1.1.7 and equip those spaces with norms. The normed linear spaces described below (and their cousins) will play a vital role throughout the rest of the book. Understanding these objects will lead to more general theorems and, conversely, applying the big theorems to these objects will help clarify and illuminate the theorems themselves.

2.1.3 EXAMPLE $\mathbb{K}(= \mathbb{R} \text{ or } \mathbb{C})$ is a normed linear space with the absolute value norm.

2.1.4 EXAMPLE $\mathbb{K}^n$ may be equipped with many different norms. We give two such norms and we will give more later on.

(i) The B_∞ $B_\infty 1$-*norm* is given by $\|(x_1, x_2, \ldots, x_n)\|_1 := \sum_{i=1}^n |x_i|$.

(ii) The *supremum norm* is given by $\|(x_1, x_2, \ldots, x_n)\|_\infty := \sup_{1 \leq i \leq n} |x_i|$.

2.1.5 EXAMPLE c_{00} is a normed linear space with a variety of norms. In fact, the 1-norm and supremum norm may be defined exactly as above (except that we need to take an infinite sum, in principle).

2.1.6 EXAMPLE c_0 (the space of sequences 'vanishing at infinity') is a normed linear space with the supremum norm. Note that the 1-norm no longer makes sense on c_0.

2.1.7 EXAMPLE $C[a, b]$ may also be equipped with many norms. The definitions of the 1-norm and supremum norm as similar to the case of $\mathbb{K}^n$ and c_{00} above.

(i) The 1-norm is given by

$$\|f\|_1 := \int_a^b |f(t)| dt.$$

Note that it is a norm because if $\|f\|_1 = 0$ and f is continuous, then $f \equiv 0$ (this is no longer true if we replace $C[a, b]$ by the larger class of Riemann-integrable functions).

(ii) The supremum norm given by

$$\|f\|_\infty := \sup_{x \in [a,b]} |f(x)|.$$

Once again this only makes sense because continuous functions on a compact set are necessarily bounded. If we replace the closed interval $[a,b]$ by the real line, this norm ceases to be well-defined (see Exercise 2.45 for a work-around in that case).

2.1.8 EXAMPLE More generally, let X be a compact, Hausdorff space and $C(X)$ be the space of continuous, $\mathbb{K}$-valued functions on X. Since every such function is bounded, we may equip $C(X)$ with the supremum norm (however, the analog of the 1-norm only makes sense if X is equipped with a Borel measure with full support).

One fundamental way to construct a norm is by means of an inner product. Historically, inner product spaces pre-date the more general notion of a normed linear space and for good reason. Much of the analysis one does in Euclidean space translates quite naturally over to the setting of inner product spaces. We will delve deeper into this in Chapter 3.

2.1.9 DEFINITION An **inner product** on a vector space **E** is a function

$$\langle \cdot, \cdot \rangle : \mathbf{E} \times \mathbf{E} \to \mathbb{K}$$

satisfying the following properties for all $x, y, z \in \mathbf{E}$ and $\alpha, \beta \in \mathbb{K}$.

(a) $\langle \alpha x + \beta y, z \rangle = \alpha \langle x, z \rangle + \beta \langle y, z \rangle$,

(b) $\langle x, y \rangle = \overline{\langle y, x \rangle}$,

(c) $\langle x, x \rangle \geq 0$, and $\langle x, x \rangle = 0$ if and only if $x = 0$.

The pair $(\mathbf{E}, \langle \cdot, \cdot \rangle)$ is called an **inner product space**.

Note that some books define an inner product to be linear in the second variable in axiom (a), making it conjugate linear in the first variable. This does cause some unnecessary confusion but does not change anything conceptual.

Now it is not clear how to construct a norm from such a function (although condition (c) gives you an idea), but the next inequality helps us get there.

2.1.10 LEMMA (CAUCHY–SCHWARZ INEQUALITY (CAUCHY, 1821, BUNYAKOVSKY, 1859, AND SCHWARZ, 1888)) *If* **E** *is an inner product space and* $x, y \in \mathbf{E}$, *then*

$$|\langle x, y \rangle|^2 \leq \langle x, x \rangle \langle y, y \rangle.$$

Moreover, equality holds if and only if the set $\{x, y\}$ *is linearly dependent.*

Proof. The inequality is clearly true if $x = 0$, so we may assume that $x \neq 0$. Now set $z := y - \frac{\langle y, x \rangle}{\langle x, x \rangle} x$ and compute

$$
\begin{aligned}
\langle z, y \rangle &= \left\langle y - \frac{\langle y, x \rangle}{\langle x, x \rangle} x, y \right\rangle \\
&= \langle y, y \rangle - \frac{\langle y, x \rangle \langle x, y \rangle}{\langle x, x \rangle} \\
&= \langle y, y \rangle - \frac{|\langle x, y \rangle|^2}{\langle x, x \rangle}.
\end{aligned}
$$

Since $\langle z, x \rangle = 0$, we see that $\langle z, y \rangle = \langle z, z \rangle \geq 0$. This gives us the required inequality. The second statement concerning equality is left as an exercise. $\square$

2.1.11 COROLLARY *If* **E** *is an inner product space, then the function*

$$\|x\| := \sqrt{\langle x, x \rangle}$$

defines a norm on **E**. *This is called the norm* **induced by the inner product**.

Proof. We only check the triangle inequality since the other axioms are obvious. For $x, y \in \mathbf{E}$, consider

$$
\begin{aligned}
\|x + y\|^2 &= \langle x + y, x + y \rangle \\
&= \|x\|^2 + \langle x, y \rangle + \langle y, x \rangle + \|y\|^2 \\
&= \|x\|^2 + 2\,\mathrm{Re}(\langle x, y \rangle) + \|y\|^2 \\
&\leq \|x\|^2 + 2|\langle x, y \rangle| + \|y\|^2 \\
&\leq \|x\|^2 + 2\|x\|\|y\| + \|y\|^2 \\
&= (\|x\| + \|y\|)^2.
\end{aligned}
$$

Note that the penultimate step follows from the Cauchy–Schwarz Inequality. Taking square roots now gives us the triangle inequality. $\square$

One important consequence of the Cauchy–Schwarz Inequality is the fact that the inner product is a continuous map from $\mathbf{E} \times \mathbf{E}$ to $\mathbb{K}$ when $\mathbf{E}$ is equipped with this norm. In other words, if $x_n \to x$ and $y_n \to y$, then $\langle x_n, y_n \rangle \to \langle x, y \rangle$ in $\mathbb{K}$.

Let us now look at the most basic examples of inner product spaces.

2.1.12 EXAMPLE $\mathbb{K}^n$ with the Euclidean inner product given by

$$\langle (x_j), (y_j) \rangle = \sum_{i=1}^{n} x_i \overline{y_i}.$$

The induced norm is denoted by $\| \cdot \|_2$.

2.1.13 EXAMPLE c_{00} with the Euclidean inner product given by

$$\langle (x_j), (y_j) \rangle = \sum_{i=1}^{\infty} x_i \overline{y_i}.$$

Note that this is a finite sum for any two vectors in c_{00} and is thus well-defined.

2.1.14 EXAMPLE The most important inner product space is the sequence space ℓ^2.

$$\ell^2 := \left\{ (x_n)_{n=1}^{\infty} : x_i \in \mathbb{K} \text{ for all } i \geq 1, \text{ and } \sum_{i=1}^{\infty} |x_i|^2 < \infty \right\}.$$

This is the space of *square-summable* sequences and is equipped with the Euclidean inner product

$$\langle (x_j), (y_j) \rangle = \sum_{i=1}^{\infty} x_i \overline{y_i}.$$

It is not obvious that this series converges, so let us prove that. Fix $(x_n), (y_n) \in \ell^2$ and define $s_m := \sum_{i=1}^{m} x_i \overline{y_i}$. Fix integers $n > m$ and observe that

$$|s_n - s_m| = \left| \sum_{i=m+1}^{n} x_i \overline{y_i} \right| \leq \left(\sum_{i=m+1}^{n} |x_i|^2 \right)^{1/2} \left(\sum_{i=m+1}^{n} |y_i|^2 \right)^{1/2}.$$

by the Cauchy–Schwarz Inequality in $\mathbb{K}^n$. For a fixed $\epsilon > 0$, we may choose $M \in \mathbb{N}$ such that $\sum_{i=M+1}^{\infty} |x_i|^2 < \epsilon$ and $\sum_{i=M+1}^{\infty} |y_i|^2 < \epsilon$. If $n > m \geq M$, it follows that $|s_n - s_m| < \epsilon$. Thus, the sequence (s_n) is a Cauchy sequence in $\mathbb{K}$ and the series $\sum_{i=1}^{\infty} x_i \overline{y_i}$ converges. Once the inner product is well-defined, the other axioms are trivial to check.

Notice that we crucially used the fact that $\mathbb{K}$ (be it either $\mathbb{R}$ or $\mathbb{C}$) is *complete* (every Cauchy sequence in $\mathbb{K}$ converges to a point in $\mathbb{K}$). This is a recurrent theme and the reason we restricted ourselves to vector spaces over these two fields.

We now introduce the most important class of normed linear spaces: the L^p spaces. For the purposes of this discussion, we will fix a compact interval $[a, b] \subset \mathbb{R}$ and equip it with the Lebesgue measure.

2.1.15 DEFINITION Fix $0 < p < \infty$.

(i) A **p-integrable** function $f : [a, b] \to \mathbb{K}$ is a Lebesgue measurable function such that

$$\int_a^b |f(t)|^p dt < \infty.$$

(ii) Let $\mathcal{L}^p[a, b]$ be the set of all p-integrable measurable functions. Observe that if $f, g \in \mathcal{L}^p[a, b]$, then

$$|f + g|^p \leq [2 \max\{|f|, |g|\}]^p \leq 2^p [|f|^p + |g|^p]. \tag{2.1}$$

Hence, $f + g \in \mathcal{L}^p[a, b]$, making $\mathcal{L}^p[a, b]$ a vector space.

(iii) Define $\mu_p : \mathcal{L}^p[a, b] \to \mathbb{R}_+$ by

$$\mu_p(f) := \left(\int_a^b |f|^p(t) dt \right)^{1/p}.$$

Then μ_p enjoys the following properties:

(a) $\mu_p(f) \geq 0$ for all $f \in \mathcal{L}^p[a, b]$.

(b) $\mu_p(\alpha f) = |\alpha| \mu_p(f)$ for all $\alpha \in \mathbb{K}$.

(c) Note that $\mu_p(f) = 0$ implies that $f \equiv 0$ a.e. (not that $f = 0$).

We do not know, as yet, if μ_p satisfies the triangle inequality.

(iv) Define $\mathbf{N} := \{f \in \mathcal{L}^p[a, b] : \mu_p(f) = 0\} = \{f \in \mathcal{L}^p[a, b] : f \equiv 0 \text{ a.e.}\}$. Then $\mathbf{N}$ is a subspace of $\mathcal{L}^p[a, b]$ by Equation 2.1, so we may define

$$L^p[a, b] := \mathcal{L}^p[a, b] / \mathbf{N}.$$

Then $L^p[a, b]$ is a vector space.

(v) For $f + \mathbf{N} \in L^p[a, b]$, we write

$$\|f + \mathbf{N}\|_p := \mu_p(f).$$

Note that if $f + \mathbf{N} = g + \mathbf{N}$, then $f \equiv g$ a.e., and so $\mu_p(f) = \mu_p(g)$. Hence, $\| \cdot \|_p : L^p[a, b] \to \mathbb{R}_+$ is well-defined and clearly satisfies the first two axioms of a norm.

Henceforth, we identify two functions that are equal a.e. and merely write $\|f\|_p$ for $\|f + \mathbf{N}\|_p$. This is a standard, and quite understandable, abuse of notation. If you are uncomfortable with such things, then you are going to have a very long day, so do get used to it. That said, it is instructive to verify that all the definitions that follow are *well-defined* in the sense that they do not change if you alter a function's values on a set of measure zero.

2.1.16 EXAMPLE For $0 < p < 1$, the triangle inequality fails. Take $f = \chi_{(0,1/2)}, g = \chi_{(1/2,1)} \in \mathcal{L}^p[0, 1]$, then $\|f\|_p = \|g\|_p = 2^{-1/p}, \|f + g\|_p = 1$ and

$$\|f\|_p + \|g\|_p = 2^{-1/p} + 2^{-1/p} = 2^{1-1/p} < 1.$$

That the triangle inequality holds when $p \geq 1$ is a beautiful fact and gives us a rich supply of normed linear spaces to play with. The next lemma is too clever for its own good, but it does have the virtue of giving us a direct path to Hölder's Inequality.

2.1.17 LEMMA *If $a, b \geq 0$ and $0 < \lambda < 1$, then $a^\lambda b^{1-\lambda} \leq \lambda a + (1 - \lambda)b$. Moreover, equality holds if and only if $a = b$.*

Proof. If $b = 0$, there is nothing to prove, so assume $b \neq 0$ and set $t = a/b$. Then we wish to prove that

$$t^\lambda \leq \lambda t + (1 - \lambda)$$

with equality if and only if $t = 1$. The function $f : [0, \infty) \to \mathbb{R}$ given by $t \mapsto t^\lambda - \lambda t$ satisfies $f'(t) = \lambda t^{\lambda-1} - \lambda$. Since $0 < \lambda < 1$, f is increasing for $t < 1$ and decreasing for $t > 1$. Hence, the global maximum of f occurs at $t = 1$. The result now follows from the fact that $f(1) = 1 - \lambda$. $\square$

2.1.18 THEOREM (HÖLDER'S INEQUALITY (ROGERS, 1888, AND HÖLDER, 1889))
Let $1 < p < \infty$ and $q \in \mathbb{R}$ such that $1/p + 1/q = 1$. If $f, g : [a, b] \to \mathbb{K}$ are measurable

functions, then

$$\int_a^b |fg| \le \|f\|_p \|g\|_q.$$

Furthermore, equality holds if and only if there exist constants $\alpha, \beta \in \mathbb{K}$ such that $\alpha\beta \ne 0$ and $\alpha|f|^p \equiv \beta|g|^q$ a.e.

Proof.   If either term on the right hand side is 0 or ∞, there is nothing to prove. Furthermore, if the inequality holds for any pair f, g, then it also holds for all pairs $\alpha f, \beta g$ for $\alpha, \beta \in \mathbb{K}$. Therefore replacing f by $f / \|f\|_p$ and g by $g/\|g\|_q$, it suffices to assume that

$$\|f\|_p = \|g\|_q = 1.$$

Aside. What just happened here? Well, we have effectively assumed that f and g are both functions of norm 1. This is possible because the required inequality *scales*. That is, if it works for f, then it works for αf as well. This is a convenience that appears frequently. If a property scales, then we may as well prove it for elements of norm 1. One has to avoid the zero vector in this argument, but that case is usually trivial.

Take a moment to reflect on these last few lines. This kind of logic will be applied many times in this book. Some textbooks go so far as to say, 'Assume without loss of generality that $\|f\|_p = \|g\|_q = 1$'. Once you parse the comments above, such statements will become more palatable.

So fix $x \in [a, b]$ and let $a = |f(x)|^p, b = |g(x)|^q$ and $\lambda = 1/p$ in Lemma 2.1.17 so that

$$|f(x)g(x)| \le \frac{|f(x)|^p}{p} + \frac{|g(x)|^q}{q}.$$

Integrating both sides, we get

$$\int_a^b |fg| \le \frac{1}{p}\int_a^b |f|^p + \frac{1}{q}\int_a^b |g|^q = \frac{1}{p} + \frac{1}{q} = 1 = \|f\|_p\|g\|_q.$$

Furthermore, equality holds if and only if $|f(x)|^p = |g(x)|^q$ a.e. $\square$

Given $1 \le p \le \infty$, the real number q satisfying the relation $\frac{1}{p} + \frac{1}{q} = 1$ is called the *conjugate exponent* of p. Note that if $p = 1$, then $q = \infty$, and vice-versa. The pair (p, q) will often occur together and they share a deep relationship which we will explore in Chapter 4.

2.1.19 THEOREM (MINKOWSKI'S INEQUALITY) *If* $1 \leq p < \infty$ *and* $f, g \in L^p[a,b]$, *then*

$$\|f + g\|_p \leq \|f\|_p + \|g\|_q.$$

Thus $(L^p[a,b], \| \cdot \|_p)$ *is a normed linear space.*

Proof.   The result is obvious if $p = 1$ or $f + g = 0$ a.e. Otherwise

$$|f + g|^p \leq (|f| + |g|)|f + g|^{p-1}.$$

Let q be the conjugate exponent of p. By Hölder's Inequality,

$$\int_a^b |f + g|^p \leq \|f\|_p \||f + g|^{p-1}\|_q + \|g\|_p \||f + g|^{p-1}\|_q$$

$$= [\|f\|_p + \|g\|_p] \left[\int_a^b |f + g|^{(p-1)q} \right]^{1/q}.$$

Now $(p - 1)q = p$ and Equation 2.1 tells us that $\int_a^b |f + g|^p < \infty$. Thus we may divide by $\left[\int_a^b |f + g|^p \right]^{1/q}$ on both sides to obtain

$$\|f + g\|_p = \left[\int_a^b |f + g|^p \right]^{1-1/q} \leq \|f\|_p + \|g\|_p.$$

$\square$

We have one more space in this family to introduce. This one, $L^\infty[a, b]$, is the only one of the L^p spaces whose norm is *not* defined in terms of an integral. Therefore many arguments concerning convergence that apply for the other L^p spaces (for instance, using the Monotone or Dominated Convergence Theorems) do not apply to $L^\infty[a, b]$. This makes $L^\infty[a, b]$ different in almost every way from the other L^p spaces.

2.1.20 DEFINITION   In what follows, we will write 'm' to denote the Lebesgue measure.

(i) A function $f : [a, b] \rightarrow \mathbb{K}$ is said to be ***essentially bounded*** if there exists $M \in \mathbb{R}$ such that

$$m(\{x \in [a, b] : |f(x)| > M\}) = 0.$$

A number M satisfying this condition is called an **essential bound** of f.

(ii) Let $\mathcal{L}^{\infty}[a,b]$ be the set of all essentially bounded measurable functions. If $f, g \in \mathcal{L}^{\infty}[a,b]$, then $f + g \in \mathcal{L}^{\infty}[a,b]$ (this is because the union of two sets of measure zero has measure zero). Therefore $\mathcal{L}^{\infty}[a,b]$ is a vector space.

(iii) For $f \in \mathcal{L}^{\infty}[a,b]$, define

$$\mu_{\infty}(f) := \inf\{M > 0 : M \text{ is an essential bound for } f\}.$$

Predictably, the quantity $\mu_{\infty}(f)$ is called the **essential supremum** of f.

The next lemma has a simple consequence: If $f \in \mathcal{L}^{\infty}[a,b]$ is such that $\mu_{\infty}(f) = 0$, then $f = 0$ almost everywhere. This is the only hurdle to clear before we are able to define $L^{\infty}[a,b]$, which we do immediately after.

2.1.21 LEMMA   *If $f \in \mathcal{L}^{\infty}[a,b]$, then $|f| \leq \mu_{\infty}(f)$ a.e.*

Proof.   We wish to prove that $\mu_{\infty}(f)$ is an essential bound for f. If $n \in \mathbb{N}$, then $\mu_{\infty}(f) + 1/n$ is not a lower bound for the set $A_f := \{M > 0 : M \text{ is an essential bound for } f\}$. Hence there exists $M_n \in A_f$ such that $M_n \leq \mu_{\infty}(f) + 1/n$. Now

$$\{x \in [a,b] : |f(x)| > \mu_{\infty}(f)\} = \bigcup_{n=1}^{\infty} \{x \in [a,b] : |f(x)| > \mu_{\infty}(f) + 1/n\}$$

$$\subset \bigcup_{n=1}^{\infty} \{x \in [a,b] : |f(x)| > M_n\}.$$

But each set $\{x \in [a,b] : |f(x)| > M_n\}$ has measure zero and thus $m(\{x \in [a,b] : |f(x)| > \mu_{\infty}(f)\}) = 0$. Therefore $|f| \leq \mu_{\infty}(f)$ a.e. □

2.1.22 DEFINITION   Consider $\mathbf{N} := \{f \in \mathcal{L}^{\infty}[a,b] : \mu_{\infty}(f) = 0\}$. Then $f \in \mathbf{N}$ if and only if $f = 0$ a.e. Therefore $\mathbf{N}$ is a subspace of $\mathcal{L}^{\infty}[a,b]$. We define

$$L^{\infty}[a,b] := \mathcal{L}^{\infty}[a,b]/\mathbf{N}$$

and for any $f + \mathbf{N} \in L^{\infty}[a,b]$, we write

$$\|f + \mathbf{N}\|_{\infty} := \mu_{\infty}(f).$$

Then $\|\cdot\|_{\infty}$ is well-defined and a norm on $L^{\infty}[a,b]$.

Of course, all this needs a proof, so do try Exercise 2.9. It is interesting to note that these proofs are considerably simpler than the corresponding ones for the L^p spaces for $p < \infty$. As in the case of $L^p[a,b]$, we will henceforth think of elements of $L^\infty[a,b]$ as functions rather than cosets.

Having done all this work, one would hope that we get something for free. Well, if $(X, \mathfrak{M}, \mu)$ is any measure space, then the definitions of the L^p spaces go through *mutatis mutandis*. The two most important cases are described below.

2.1.23 DEFINITION

(i) If $X = \{1, 2, \ldots, n\}$ is a finite set equipped with the counting measure, then a function $f : X \to \mathbb{K}$ is determined by a tuple $(f(1), f(2), \ldots, f(n))$ of scalars. Therefore we may identify

$$\mathbb{K}^n = L^p(X, \mu),$$

equipped with the norm

$$\|(x_i)\|_p := \begin{cases} \left(\sum_{i=1}^n |x_i|^p\right)^{1/p} & : \text{if } 1 \le p < \infty, \\ \max_{1 \le i \le n} |x_i| & : \text{if } p = \infty. \end{cases}$$

(ii) If $X = \mathbb{N}$, equipped with the counting measure, then a function $f : X \to \mathbb{K}$ is determined by a sequence $(f(n))_{n=1}^\infty$. Therefore elements in $L^p(X, \mu)$ are thought of as sequences. For $1 \le p < \infty$, we define the *little* ℓ^p spaces by

$$\ell^p := L^p(X, \mu) = \left\{ (x_n)_{n=1}^\infty : x_i \in \mathbb{K} \text{ for all } i \in \mathbb{N}, \text{ and } \sum_{i=1}^\infty |x_i|^p < \infty \right\}$$

equipped with the norm given by

$$\|(x_n)\|_p := \left(\sum_{n=1}^\infty |x_n|^p\right)^{1/p}.$$

For $p = \infty$, we define

$$\ell^\infty := L^\infty(X, \mu) = \{(x_n)_{n=1}^\infty : x_i \in \mathbb{K} \text{ for all } i \in \mathbb{N}, \text{ and } (x_n) \text{ is bounded}\}$$

equipped with the supremum norm $\|(x_n)\|_\infty := \sup_{n \in \mathbb{N}} |x_n|$.

The subject of **Functional Analysis** has its roots in the second half of the 19th century. Bernhard Riemann (1826–1866), Karl Weierstrass (1815–1897) and many others had rigorously developed the notion of a function, understanding questions of uniform convergence, series of functions, discontinuities of a function, etc. Georg Cantor (1845–1918) and others created modern set theory, which was to play a crucial role in the genesis of the subject. Guiseppe Peano (1858–1932) wrote a book in 1888, where he laid down the axioms of a vector space. Meanwhile, the subject of Point–Set Topology was starting to reach a mature form around 1900 although it would continue to see tremendous growth alongside Functional Analysis.

All these strands of knowledge were synthesized in four fundamental investigations, which were to lay the foundation for the subject in the 20th century. Erik Ivar Fredholm (1866–1927) studied integral equations in 1900; Henri Lebesgue (1875–1941) published his thesis in 1902, which dealt with integration; David Hilbert (1862–1943) published a series of six papers on spectral theory between 1904–1910; and Maurice Fréchet (1878–1973) studied metric spaces in his thesis of 1906.

Through the work of many hands, the subject grew rapidly in the next three decades and culminated in the publication of *Théorie des Opérations Linéaires* in 1932 by Stefan Banach (1892–1945). The book is an important landmark in the history of the field and it exerted a tremendous influence on the development of the subject in the years to come.

2.2   Bounded Linear Operators

While studying a collection of mathematical objects, one is often interested in understanding the natural maps (or *morphisms*) between these objects. These are typically functions between two such objects that preserve structure. In our case, we have vector spaces, so we would naturally consider linear maps. Moreover, these vector spaces are also equipped with a topology, so we consider *continuous* linear maps.

Let $\mathbf{E}$ be a normed linear space. For $x \in \mathbf{E}$ and $r > 0$, we write $B(x,r) := \{y \in \mathbf{E} : \|y - x\| < r\}$, and $B[x,r] := \{y \in \mathbf{E} : \|y - x\| \leq r\}$. Note that $B(x,r)$ is open and $B[x, r]$ is closed. When it is necessary to do so, we will write $B_\mathbf{E}(x, r)$ instead of $B(x, r)$ to emphasize that this is a subset of $\mathbf{E}$. The **closed unit ball** is the set $B[0, 1]$ (which we will also denote by $B_\mathbf{E}$) and the **open unit ball** is $B(0, 1)$. The **unit sphere** is the set $S_\mathbf{E} = \{x \in \mathbf{E} : \|x\| = 1\}$.

2.2.1 DEFINITION Let $\mathbf{E}$ and $\mathbf{F}$ be normed linear spaces. A linear operator $T : \mathbf{E} \to \mathbf{F}$ is said to be

(i) **continuous** if it is continuous with respect to the norm topologies on $\mathbf{E}$ and $\mathbf{F}$.

(ii) **bounded** if there exists $M \geq 0$ such that $\|T(x)\| \leq M\|x\|$ for all $x \in \mathbf{E}$.

> **Aside.** Typically, a bounded function means one whose range is a bounded subset of the co-domain. In our case, the word *bounded* refers to the fact that the operator maps the unit ball $B[0, 1]$ in $\mathbf{E}$ into a bounded subset of $\mathbf{F}$. This may be confusing at first until you realize that the only linear map that has bounded range is the zero map.

When one first encounters continuous functions, it is drilled into our heads that continuity is a local property, that a function that is continuous at one point need not be continuous at another. One of the thoroughly satisfying parts of studying linear maps is that one no longer has to worry about this problem. A linear map that is continuous at one point is continuous everywhere!

2.2.2 THEOREM *For a linear operator $T : \mathbf{E} \to \mathbf{F}$ between normed linear spaces, the following are equivalent:*

(i) *T is continuous.*

(ii) *T is continuous at one point of $\mathbf{E}$.*

(iii) *T is continuous at $0 \in \mathbf{E}$.*

(iv) *T is bounded.*

(v) *T is uniformly continuous.*

Proof. Observe that (i) $\Rightarrow$ (ii) and (v) $\Rightarrow$ (i) hold by definition.

(ii) $\Rightarrow$ (iii): If T is continuous at a point $x_0 \in \mathbf{E}$, then for any $\epsilon > 0$, choose $\delta > 0$ such that

$$\|x - x_0\| < \delta \Rightarrow \|T(x) - T(x_0)\| < \epsilon.$$

So if $\|x\| < \delta$, then let $z := x + x_0$, so that $\|z - x_0\| < \delta$. Then $\|T(z) - T(x_0)\| < \epsilon$, which implies that $\|T(x)\| < \epsilon$. Therefore T is continuous at 0.

(iii) $\Rightarrow$ (iv): Suppose T is continuous at 0, then for $\epsilon = 1$ there exists $\delta > 0$ such that

$$\|x\| < \delta \Rightarrow \|T(x)\| < 1.$$

So for any non-zero vector $y \in \mathbf{E}$, let $x := \frac{\delta}{2}\frac{y}{\|y\|}$. Then $\|x\| < \delta$ and so $\|T(x)\| < 1$. Therefore

$$\|T(y)\| < \frac{2}{\delta}\|y\|.$$

Since this holds for any $y \in \mathbf{E}$, T is bounded.

(iv) $\Rightarrow$ (v): Suppose that there exists $M > 0$ such that $\|T(x)\| \leq M\|x\|$ for all $x \in \mathbf{E}$, then for any $\epsilon > 0$, choose $\delta := \frac{\epsilon}{2M}$. If $\|x - y\| < \delta$, then

$$\|T(x) - T(y)\| = \|T(x - y)\| \leq M\|x - y\| \leq \frac{\epsilon}{2} < \epsilon.$$

Therefore, T is continuous on $\mathbf{E}$.   □

Aside. Buried in this proof is a trick that is worth pointing out. Look at the implication (iii) $\Rightarrow$ (iv). Here, we knew something about vectors $x \in B(0, \delta)$. To deduce something analogous for other vectors, we applied the *scaling trick*. For any non-zero vector $y \in \mathbf{E}$, the vector

$$x := \frac{\delta}{2}\frac{y}{\|y\|}$$

lies in the ball $B(0, \delta)$. The conclusion then follows from the linearity of T. This idea of exploiting scaling (together with linearity) is used everywhere. In fact, you have already seen a version of it in Hölder's Inequality.

Now, it is time to sink our teeth into some examples. Do understand these now or else they will come back to haunt you later.

2.2.3 EXAMPLE Let $\mathbf{E}$ be any inner product space and $y \in \mathbf{E}$ be fixed. Define $\varphi : \mathbf{E} \to \mathbb{K}$ by

$$\varphi(x) := \langle x, y \rangle.$$

Then $|\varphi(x)| \leq \|x\|\|y\|$ by the Cauchy–Schwarz Inequality and so φ is bounded.

2.2.4 EXAMPLE Let $T : \mathbb{K}^n \to \mathbf{E}$ be any operator, where $\mathbb{K}^n$ is endowed with the supremum norm and $\mathbf{E}$ is any normed linear space. Then for any $x = (x_1, x_2, \ldots, x_n) \in \mathbb{K}^n$, $x = \sum_{i=1}^n x_i e_i$, where $\{e_1, e_2, \ldots, e_n\}$ is the standard basis for $\mathbb{K}^n$. Then

$$\|T(x)\| = \|\sum_{i=1}^n x_i T(e_i)\| \leq \sum_{i=1}^n |x_i| \|T(e_i)\| \leq \|x\| \left(\sum_{i=1}^n \|T(e_i)\|\right).$$

If $M := \sum_{i=1}^n \|T(e_i)\|$, then $\|T(x)\| \leq M\|x\|$. We have thus proved that any linear operator $T : \mathbb{K}^n \to \mathbf{E}$ is continuous.

2.2.5 EXAMPLE Let $\mathbf{E} = c_{00}$ and $\varphi : \mathbf{E} \to \mathbb{K}$ be given by

$$\varphi((x_j)) = \sum_{n=1}^\infty x_n.$$

(i) If $\mathbf{E}$ has the 1-norm, then φ is continuous since $|\varphi(x)| \leq \|x\|_1$.

(ii) If $\mathbf{E}$ has the supremum norm, let $x^k = (1, 1, \ldots, 1, 0, 0, \ldots)$, where the 1 appears k times. Then $\|x^k\|_\infty = 1$ for all $k \in \mathbb{N}$, but $|\varphi(x^k)| = k$. Hence there is no $M > 0$ such that $|\varphi(x)| \leq M\|x\|_\infty$ for all $x \in \mathbf{E}$, and so φ cannot be continuous.

2.2.6 EXAMPLE Let $\mathbf{E} = \mathbf{F} = (C[0,1], \|\cdot\|_\infty)$, and $T : \mathbf{E} \to \mathbf{F}$ be the operator

$$T(f)(x) = \int_0^x f(t)dt.$$

Then for any $x \in [0,1]$,

$$|T(f)(x)| \leq \int_0^x |f(t)|dt \leq x\|f\|_\infty \leq \|f\|_\infty.$$

Hence $\|T(f)\|_\infty \leq \|f\|_\infty$, and T is continuous.

2.2.7 EXAMPLE Let $\mathbf{E} = C[0,1]$ and define $\varphi : \mathbf{E} \to \mathbb{K}$ by $\varphi(f) := f(0)$.

(i) If $\mathbf{E}$ has the supremum norm, then $|\varphi(f)| \leq \|f\|_\infty$, so φ is continuous.

(ii) If $\mathbf{E}$ has the 1-norm, then consider a sequence (f_k) of non-negative continuous functions such that $f_k(0) = k$, and $\int_0^1 f_k(t)dt = 1$ (triangles of large height but area 1). Thus $\|f_k\| = 1$ for all $k \in \mathbb{N}$ but $|\varphi(f_k)| = k$. As before, this implies that φ is not continuous.

2.2.8 DEFINITION   Let **E** and **F** be two normed linear spaces

(i) Write $\mathcal{B}(\mathbf{E}, \mathbf{F})$ for the set of all bounded linear operators from **E** to **F**. Note that $\mathcal{B}(\mathbf{E}, \mathbf{F})$ is a subset of $L(\mathbf{E}, \mathbf{F})$. Furthermore, if $S, T \in \mathcal{B}(\mathbf{E}, \mathbf{F})$, then $S + T \in \mathcal{B}(\mathbf{E}, \mathbf{F})$ because addition is a continuous operation on **F**. Similarly, $\alpha T \in \mathcal{B}(\mathbf{E}, \mathbf{F})$ for any $\alpha \in \mathbb{K}$. Thus, $\mathcal{B}(\mathbf{E}, \mathbf{F})$ is a vector space.

(ii) Write $\mathcal{B}(\mathbf{E})$ for the space $\mathcal{B}(\mathbf{E}, \mathbf{E})$.

(iii) Write $\mathbf{E}^*$ for the space $\mathcal{B}(\mathbf{E}, \mathbb{K})$. This is called the (continuous) *dual space* of **E**.

We now wish to define a norm on $\mathcal{B}(\mathbf{E}, \mathbf{F})$, the so-called *operator norm*. In other words, to each $T \in \mathcal{B}(\mathbf{E}, \mathbf{F})$, we wish to associate a non-negative real number $\nu(T)$ with the understanding that $\nu(T)$ measures the distance of T from the zero operator. For any $T \in \mathcal{B}(\mathbf{E}, \mathbf{F})$, we write

$$\nu(T) := \inf\{M > 0 : \|T(x)\| \le M\|x\| \text{ for all } x \in \mathbf{E}\}.$$

Note that the set on the right hand side is not empty and thus $\nu(T) < \infty$.

2.2.9 LEMMA   *For any $T \in \mathcal{B}(\mathbf{E}, \mathbf{F})$ and any $x \in \mathbf{E}$, $\|T(x)\| \le \nu(T)\|x\|$.*

Proof.   For each $n \in \mathbb{N}$, $\nu(T) + 1/n$ is not a lower bound for the set $A_T := \{M > 0 : \|T(x)\| \le M\|x\| \text{ for all } x \in \mathbf{E}\}$. Hence there exists $M_n \in A_T$ such that $\nu(T) \le M_n < \nu(T) + 1/n$ and

$$\|T(x)\| \le M_n\|x\|$$

for all $x \in \mathbf{E}$ and $n \in \mathbb{N}$. Fixing $x \in \mathbf{E}$, we let $n \to \infty$ to obtain $\|T(x)\| \le \nu(T)\|x\|$.   □

2.2.10 PROPOSITION   *The function $\nu : \mathcal{B}(\mathbf{E}, \mathbf{F}) \to \mathbb{R}_+$ defined above is a norm on $\mathcal{B}(\mathbf{E}, \mathbf{F})$.*

Proof.

(i) Clearly, $\nu(T) \ge 0$ and $\nu(0) = 0$. If $\nu(T) = 0$, then the fact that $T = 0$ follows from Lemma 2.2.9.

(ii) Fix $T \in \mathcal{B}(\mathbf{E}, \mathbf{F})$ and $0 \ne \lambda \in \mathbb{K}$ and define $A_1 := \{M > 0 : \|T(x)\| \le M\|x\| \text{ for all } x \in \mathbf{E}\}$ and $A_2 := \{K > 0 : \|(\lambda T)(y)\| \le K\|y\| \text{ for all } y \in \mathbf{E}\}$.

For any $M \in A_1$,

$$\|\lambda T(x)\| = \|T(\lambda x)\| \leq M\|\lambda x\| = M|\lambda|\|x\|$$

for all $x \in \mathbf{E}$. Therefore, $M|\lambda| \in A_2$, and $\nu(\lambda T) = \inf A_2 \leq M|\lambda|$. This is true for any $M \in A_1$, so

$$\nu(\lambda T) \leq |\lambda| \inf A_1 = |\lambda|\nu(T).$$

Replacing λ with $1/\lambda$ and T by λT, we conclude that

$$\nu(T) = \nu\left(\frac{1}{\lambda}\lambda T\right) \leq \frac{1}{|\lambda|}\nu(\lambda T).$$

Hence, $\nu(\lambda T) \geq |\lambda|\nu(T)$ as well.

(iii) Fix $S, T \in \mathcal{B}(\mathbf{E}, \mathbf{F})$, then for any $x \in \mathbf{E}$, we have

$$\|(S + T)(x)\| = \|S(x) + T(x)\| \leq \|S(x)\| + \|T(x)\|$$
$$\leq \nu(S)\|x\| + \nu(T)\|x\| = (\nu(S) + \nu(T))\|x\|.$$

By definition, this implies that $\nu(S + T) \leq \nu(S) + \nu(T)$.

$\square$

As is customary, we write $\|T\| := \nu(T)$ from here on. An important inequality that merits attention here is that for any $T \in \mathcal{B}(\mathbf{E}, \mathbf{F})$ and any $x \in \mathbf{E}$, one has

$$\|T(x)\| \leq \|T\|\|x\|.$$

Note that the symbol $\|\cdot\|$ is being abused here. Each time it is used in the above inequality, it refers to a different norm!

2.2.11 PROPOSITION *If $T \in \mathcal{B}(\mathbf{E}, \mathbf{F})$, then*

$$\|T\| = \sup\{\|T(x)\| : x \in \mathbf{E}, \|x\| = 1\}$$
$$= \sup\{\|T(x)\| : x \in \mathbf{E}, \|x\| \leq 1\}.$$

Proof. Let $\alpha := \sup\{\|T(x)\| : x \in \mathbf{E}, \|x\| = 1\}$ and $\beta := \sup\{\|T(x)\| : x \in \mathbf{E}, \|x\| \leq 1\}$, then clearly $\alpha \leq \beta$.

For any $x \in \mathbf{E}$ with $\|x\| \le 1$, $\|T(x)\| \le \|T\|\|x\| \le \|T\|$. Hence $\beta \le \|T\|$. To complete the proof, it suffices to show that $\alpha \ge \|T\|$. To that end, set

$$A := \{M > 0 : \|T(x)\| \le M\|x\| \text{ for all } x \in \mathbf{E}\}.$$

For any $n \in \mathbb{N}$, $\|T\| - 1/n \notin A$, so there exists $x_n \in \mathbf{E}$ such that

$$\|T(x_n)\| > (\|T\| - 1/n)\|x_n\|.$$

In particular, $x_n \ne 0$, so if $y_n := x_n/\|x_n\|$, then $\|y_n\| = 1$ and $\|T(y_n)\| > \|T\| - 1/n$. Thus $\alpha > \|T\| - 1/n$ for each $n \in \mathbb{N}$. Hence $\alpha \ge \|T\|$ as well. $\qquad\square$

It is now instructive to compute the norms of the operators we saw earlier. A typical way to do this is as follows. Given $T \in \mathcal{B}(\mathbf{E}, \mathbf{F})$, one takes an educated guess to find $M > 0$ such that $\|T(x)\| \le M\|x\|$ for all $x \in \mathbf{E}$. This ensures that $\|T\| \le M$ by definition. To prove equality, one hopes to find a vector $x \in \mathbf{E}$ such that $\|x\| = 1$ and $\|T(x)\| = M$. This would be ideal and works in some cases.

Unfortunately, this is not always possible (see Example 2.2.15 below). Instead, one may try to find a sequence $(x_n) \subset \mathbf{E}$ such that $\|x_n\| = 1$ for all $n \in \mathbb{N}$ and $\|T(x_n)\| \to M$ as $n \to \infty$. Together with the previous inequality, this would prove that $\|T\| = M$.

2.2.12 EXAMPLE Let $\mathbf{E}$ be an inner product space, $y \in \mathbf{E}$ and define $\varphi : \mathbf{E} \to \mathbb{K}$ by $x \mapsto \langle x, y \rangle$. Then by the Cauchy–Schwarz Inequality, $|\varphi(x)| \le \|x\|\|y\|$ for all $x \in \mathbf{E}$. Hence $\|\varphi\| \le \|y\|$. Furthermore, if $x = y$, then $\|y\|^2 = |\varphi(y)| \le \|\varphi\|\|y\|$ and so $\|\varphi\| = \|y\|$.

2.2.13 EXAMPLE Let $\mathbf{E} = \mathbb{K}^n$ with the 1-norm and let $\{e_1, e_2, \dots, e_n\}$ be the standard basis for $\mathbf{E}$. Let $\mathbf{F}$ be a normed linear space and $T : \mathbf{E} \to \mathbf{F}$ be a linear operator. Then for $x = \sum_{i=1}^n x_i e_i$, we have

$$\|T(x)\| \le \sum_{i=1}^n |x_i|\|T(e_i)\|$$

and so T is continuous with $\|T\| \le \max_{1 \le i \le n} \|T(e_i)\|$. If $x = e_i$, then $\|x\|_1 = 1$ and $\|T(x)\| = \|T(e_i)\|$ and so by Proposition 2.2.11, $\|T\| \ge \|T(e_i)\|$. This is true for all $1 \le i \le n$, so $\|T\| = \max_{1 \le i \le n} \|T(e_i)\|$.

2.2.14 EXAMPLE Let $\mathbf{E} = c_{00}$ with the 1-norm and $\varphi : \mathbf{E} \to \mathbb{K}$ be given by

$$\varphi((x_j)) := \sum_{n=1}^\infty x_n,$$

then $\|\varphi\| \le 1$. Also, for $x = e_1$, we have $\|x\| = 1$ and $|\varphi(x)| = 1$ so that $\|\varphi\| = 1$.

2.2.15 EXAMPLE Define $T : L^1[0,1] \to L^1[0,1]$ by

$$T(f)(x) = \int_0^x f(t)dt.$$

Note that T is well-defined because

$$\int_0^1 |T(f)(x)|dx = \int_0^1 \left| \int_0^x f(t)dt \right| dx$$

$$\leq \int_0^1 \int_0^x |f(t)|dtdx \qquad (2.2)$$

$$\leq \int_0^1 \int_0^1 |f(t)|dtdx = \|f\|_1.$$

This also proves that T is bounded with $\|T\| \leq 1$. To prove that $\|T\| = 1$, we set $f_n = n\chi_{[0,1/n]}$. Then $\|f_n\|_1 = 1$ and

$$T(f_n)(x) = \int_0^x n\chi_{[0,1/n]}(t)dt = \begin{cases} 1 & : \text{if } x \geq 1/n, \\ nx & : \text{if } x < 1/n. \end{cases}$$

Hence

$$\|T(f_n)\|_1 = \int_0^{1/n} ntdt + \int_{1/n}^1 dt = n\frac{1}{2n^2} + 1 - \frac{1}{n} = 1 - \frac{1}{2n}.$$

Thus $\|T(f_n)\|_1 \to 1$ as $n \to \infty$ and therefore $\|T\| = 1$. To see that there is no non-zero function $f \in L^1[0,1]$ such that $\|T(f)\|_1 = \|f\|_1$, one has to examine Equation 2.2 a little closely (see Exercise 2.17).

We know that any linear operator between finite dimensional vector spaces may be associated to a matrix. The next example (and its brethren) show that, under suitable conditions, one may use matrices to define linear operators between sequence spaces.

2.2.16 EXAMPLE Let $M := (a_{i,j})_{i,j \in \mathbb{N}}$ be an infinite matrix whose entries are in $\mathbb{K}$. Suppose that

$$\alpha := \sup_{j \in \mathbb{N}} \left\{ \sum_{i=1}^\infty |a_{i,j}| \right\} < \infty.$$

For $x = (x_1, x_2, \dots) \in \ell^1$, define a sequence $T(x) = (y_1, y_2, \dots)$ by

$$y_i := \sum_{j=1}^\infty a_{i,j}x_j.$$

Then

$$\sum_{i=1}^{\infty} |y_i| \le \sum_{i=1}^{\infty} \sum_{j=1}^{\infty} |a_{i,j}||x_j| = \sum_{j=1}^{\infty} \sum_{i=1}^{\infty} |a_{i,j}||x_j| \le \alpha \|x\|_1.$$

Note that the second equality holds by Proposition 1.2.13. Therefore $T : \ell^1 \to \ell^1$ is a well-defined operator. Furthermore, T is bounded and $\|T\| \le \alpha$.

In fact, we claim that $\|T\| = \alpha$. To see this, fix $j \in \mathbb{N}$ and let $x = e_j \in \ell^1$ denote the jth standard basis vector. Then $\|x\|_1 = 1$ and

$$\|T(x)\|_1 = \sum_{i=1}^{\infty} |T(x)_i| = \sum_{i=1}^{\infty} |a_{i,j}|.$$

Hence $\sum_{i=1}^{\infty} |a_{i,j}| \le \|T\|\|x\|_1 = \|T\|$. This is true for each $j \in \mathbb{N}$ and therefore $\alpha \le \|T\|$ holds as well.

The analogous operator on ℓ^{∞} is described in Exercise 2.22. The next example may be thought of as a continuous analogue of the previous one. These operators are important for applications and their study was one of the motivating reasons to develop the subject of Functional Analysis in the early 20th century.

2.2.17 EXAMPLE Let $\mathbf{H} := L^2[0,1]$ and $K \in L^2([0,1]^2)$ be a fixed function. Since $\int_0^1 \left(\int_0^1 |K(s,t)|^2 dt \right) ds < \infty$, it follows that $\int_0^1 |K(s,t)|^2 dt < \infty$ for almost every $s \in [0,1]$. In other words, $K(s,\cdot) \in L^2[0,1]$ for almost every $s \in [0,1]$. Therefore we may define $T : \mathbf{H} \to \mathbf{H}$ by

$$T(f)(s) := \int_0^1 K(s,t)f(t)dt = \langle f, \overline{K(s,\cdot)} \rangle.$$

We claim that T is a bounded operator. To see this, fix $f \in \mathbf{H}$ and observe that $|T(f)(s)| \le \|f\|_2 \|K(s,\cdot)\|_2$ by the Cauchy–Schwarz Inequality. Therefore

$$|T(f)(s)|^2 \le \left(\int_0^1 |K(s,t)|^2 dt \right) \|f\|_2^2.$$

Integrating over s, we see that T is bounded and that $\|T\| \le \|K\|_2$. This operator is called an ***integral operator with kernel K***.

From the calculation above, it may be tempting to say that $\|T\| = \|K\|_2$. However, this is *not* true in general. While one can construct examples of this phenomenon by brute force, it seems best to leave this for later in the book, where we will understand the norm of the operator through its *spectrum* (see Example 8.2.18).

2.3 Banach Spaces

A metric space is said to be complete if every Cauchy sequence converges to a point in the space. This is a notion one first encounters while studying the real line, where it is intimately tied to the least upper bound property. As a result, it is ever present in a variety of different proofs in analysis. However, it is in the study of normed linear spaces that it really steals the show. It not only governs the structure of the underlying space but also of the operator spaces defined on them.

2.3.1 DEFINITION

(i) A normed linear space that is complete with respect to the induced metric is called a **Banach Space**.

(ii) An inner product space that is complete with respect to the norm induced by the inner product is called a **Hilbert space**.

Stefan Banach (1892–1945) was a Polish mathematician, whose working life coincided with the turbulent times in Poland between the two World Wars. Together with Steinhaus, Nikodym and others, Banach established the Polish Mathematical Society and a number of important mathematical publications. It was in his dissertation (published in 1922) that he introduced what is today called a Banach space and his name is perhaps the one that appears most often in this book! He was an energetic and sociable person, who enjoyed doing his mathematics in crowded, noisy places such as the Scottish Cafe in Lvov.

2.3.2 PROPOSITION *For $1 \leq p \leq \infty$, $(\mathbb{K}^n, \| \cdot \|_p)$ is a Banach space.*

Proof. Assume $p = \infty$ as the case when $p < \infty$ is similar. Suppose (x^m) is a Cauchy sequence in $(\mathbb{K}^n, \| \cdot \|_p)$ with $x^m = (x_1^m, x_2^m, \ldots, x_n^m)$. Then for any $1 \leq i \leq n$, the sequence (x_i^m) is Cauchy in $\mathbb{K}$. Since $\mathbb{K}$ is complete, there exists $y_i \in \mathbb{K}$ such that $\lim_{m \to \infty} x_i^m = y_i$. Thus for any $\epsilon > 0$ there exists $N_i \in \mathbb{N}$ such that

$$|x_i^m - y_i| < \epsilon$$

for all $m \geq N_i$. Let $N_0 = \max\{N_1, N_2, \ldots, N_n\}$. Then for all $m \geq N_0$,

$$\|x^m - y\|_\infty = \sup_{1 \leq i \leq n} |x_i^m - y_i| < \epsilon.$$

Thus $x^m \to y$ in norm. $\square$

Notice that the completeness of $\mathbb{K}$ plays a crucial role in this proof. In fact, the only other fact we needed was that the projection maps $\pi_i : \mathbb{K}^n \to \mathbb{K}, (1 \le i \le n)$ are all bounded linear operators. The next proof is also similar in spirit.

2.3.3 PROPOSITION *For $1 \le p \le \infty$, ℓ^p is a Banach space.*

Proof.   We prove this for if $p < \infty$ as the $p = \infty$ case is similar. Suppose (x^k) is a Cauchy sequence in ℓ^p with $x^k = (x_1^k, x_2^k, \ldots, x_n^k, \ldots)$. We prove that (x^k) converges in the following steps.

(i) For any $n \in \mathbb{N}$, $|x_n^k - x_n^m| \le \|x^k - x^m\|_p$, so (x_n^k) is Cauchy in $\mathbb{K}$. Since $\mathbb{K}$ is complete, there exists $y_n \in \mathbb{K}$ such that $\lim_{k \to \infty} x_n^k = y_n$.

(ii) We wish to prove that $y = (y_n) \in \ell^p$. Since (x^k) is Cauchy, it is bounded, so there exists $R > 0$ such that $\|x^k\|_p \le R$ for all $k \in \mathbb{N}$. For any fixed $j \in \mathbb{N}$, this implies

$$\left(\sum_{n=1}^{j} |x_n^m|^p \right)^{1/p} \le R.$$

Now let $m \to \infty$ in the finite sum to conclude that $\left(\sum_{n=1}^{j} |y_n|^p \right)^{1/p} \le R$. This is true for all $j \in \mathbb{N}$, so

$$\left(\sum_{n=1}^{\infty} |y_n|^p \right)^{1/p} \le R.$$

Hence $y \in \ell^p$ as required.

(iii) We wish to prove that $x^k \to y$ in $\| \cdot \|_p$. For any $\epsilon > 0$, there exists $N_0 \in \mathbb{N}$ such that $\|x^k - x^m\|_p < \epsilon$ for all $k, m \ge N_0$. Now if $j \in \mathbb{N}$ is fixed, consider

$$\left(\sum_{n=1}^{j} |x_n^k - x_n^m|^p \right)^{1/p} \le \|x^k - x^m\|_p < \epsilon.$$

Let $m \to \infty$ in the finite sum to obtain $\left(\sum_{n=1}^{j} |x_n^k - y_n|^p \right)^{1/p} \le \epsilon$. This is true for all $j \in \mathbb{N}$, so

$$\left(\sum_{n=1}^{\infty} |x_n^k - y_n|^p \right)^{1/p} \le \epsilon$$

for all $k \ge N_0$. Thus $\|x^k - y\|_p \to 0$ as desired. $\qquad\square$

Aside. The proofs of the previous two theorems are remarkably similar. They both rely on the fact that the underlying field is complete, to be sure. However, there are two crucial differences. Firstly, in Proposition 2.3.2, there was no need to prove that the limit point was in our space since $\mathbb{K}^n$ is all encompassing.

The more important difference, however, is in step (iii) of Proposition 2.3.3. In $\mathbb{K}^n$ (equipped with any p-norm), a sequence converges to a limit point if and only if it converges in each component. This statement is no longer true in ℓ^p (you can construct simple counter-examples in ℓ^2, for instance). Therefore we are forced to resort to other methods to ensure that the sequence converges to the desired limit.

2.3.4 PROPOSITION *For $1 \le p < \infty$, c_{00} is dense in ℓ^p. In particular, $(c_{00}, \|\cdot\|_p)$ is not complete.*

Proof. Fix $x = (x_n) \in \ell^p$ and $\epsilon > 0$. Then there exists $N_0 \in \mathbb{N}$ such that $\sum_{n=N_0}^{\infty} |x_n|^p < \epsilon$. Hence $y := (x_1, x_2, \ldots, x_{N_0}, 0, 0, \ldots) \in c_{00}$ and $\|x - y\|_p^p < \epsilon$. $\qquad\square$

Note that, in contrast to Proposition 2.3.4, c_{00} is *not* dense in ℓ^∞ (do you see why?). Even so, $(c_{00}, \|\cdot\|_\infty)$ is not complete (See Exercise 2.26).

2.3.5 PROPOSITION $L^\infty[a, b]$ *is a Banach space.*

Proof. Let $(f_n) \subset L^\infty[a, b]$ be a Cauchy sequence and define

$$A_k := \{x \in [a, b] : |f_k(x)| > \|f_k\|_\infty\}, \text{ and}$$
$$B_{k,m} := \{x \in [a, b] : |f_k(x) - f_m(x)| > \|f_k - f_m\|_\infty\}.$$

By Lemma 2.1.21, each of these sets has measure zero, so

$$C := \left(\bigcup_{k=1}^{\infty} A_k \right) \cup \left(\bigcup_{k,m=1}^{\infty} B_{k,m} \right)$$

also has measure zero. Furthermore, on $D := [a, b] \setminus C$, each f_k is bounded and uniformly Cauchy. So for any $x \in D$, the inequality

$$|f_k(x) - f_m(x)| \le \|f_k - f_m\|_\infty$$

implies that $(f_m(x))_{m=1}^\infty$ is a Cauchy sequence in $\mathbb{K}$. Hence we may define $f : D \to \mathbb{K}$ by

$$f(x) = \lim_{m \to \infty} f_m(x).$$

We may extend f to all of $[a, b]$ by defining $f \equiv 0$ on C. Furthermore, for $\epsilon > 0$, there exists $N_0 \in \mathbb{N}$ such that $\|f_k - f_m\|_\infty < \epsilon$ for all $k, m \geq N_0$. For any fixed $x \in D$ and $k \geq N_0$, $|f_k(x) - f_m(x)| < \epsilon$ so that

$$|f_k(x) - f(x)| \leq \epsilon.$$

Hence $f - f_k$ is bounded on D and so $f - f_k \in L^\infty[a, b]$. Therefore $f = (f - f_k) + f_k \in L^\infty[a, b]$. Finally, the above inequality also proves that $\|f_k - f\|_\infty \leq \epsilon$ for all $k \geq N_0$. Therefore $f_k \to f$ in $L^\infty[a, b]$.   □

Notice that the essential supremum of a continuous function is the same as its supremum. Therefore we may think of $(C[a, b], \|\cdot\|_\infty)$ as a subspace of $L^\infty[a, b]$. The next proposition is a simple consequence of the fact that the uniform limit of continuous functions is continuous, which would have been proved in an earlier course in Real Analysis. The proof is a straight-forward use of the triangle inequality in $\mathbb{K}$, so we omit it.

2.3.6 PROPOSITION $(C[a, b], \|\cdot\|_\infty)$ *is closed in* $L^\infty[a, b]$. *In particular,* $(C[a, b], \|\cdot\|_\infty)$ *is a Banach space.*

Our next target is an important theorem in analysis: All the L^p spaces are complete. In order to prove this, we need a little set-up. It is also high time we started talking about series.

2.3.7 DEFINITION Let $\mathbf{E}$ be a normed linear space and $(x_n) \subset \mathbf{E}$ be a sequence.

(i) We say that the series $\sum_{n=1}^\infty x_n$ is ***convergent*** if the sequence (s_n) of partial sums defined by $s_n := \sum_{k=1}^n x_k$ converges to a point in $\mathbf{E}$. In other words, there exists $s \in \mathbf{E}$ such that for any $\epsilon > 0$ there exists $N_0 \in \mathbb{N}$ such that

$$\left\| \left(\sum_{k=1}^n x_k \right) - s \right\| < \epsilon$$

for all $n \geq N_0$.

(ii) We say that the series $\sum_{n=1}^\infty x_n$ is ***absolutely convergent*** if $\sum_{n=1}^\infty \|x_n\| < \infty$. Note that this is a series of non-negative real numbers, so to say that the series $\sum_{n=1}^\infty x_n$ is absolutely convergent is the same as saying that there is a real number $M \geq 0$ such that $\sum_{i=1}^n \|x_i\| \leq M$ for all $n \in \mathbb{N}$.

The next proposition is derived from a fact that we have known for some time now: An absolutely convergent series of complex numbers is convergent. If you think through the proof of this fact (also reproduced below), it relies on the completeness of $\mathbb{C}$. An interesting bit of mathematical gymnastics shows that this property *characterizes* completeness.

2.3.8 PROPOSITION *A normed linear space* $\mathbf{E}$ *is a Banach space if and only if every absolutely convergent series is convergent in* $\mathbf{E}$.

Proof. Let $\mathbf{E}$ be a Banach space and $(x_n) \subset \mathbf{E}$ such that $\sum_{n=1}^{\infty} \|x_n\| < \infty$. Let $s_n := \sum_{j=1}^{n} x_j$. Then it suffices to show that (s_n) is a Cauchy sequence. Now if $\epsilon > 0$, there exists $N_0 \in \mathbb{N}$ such that $\sum_{n=N_0}^{\infty} \|x_n\| < \epsilon$. Hence if $n, m \geq N_0$ with $n > m$, then

$$\|s_n - s_m\| = \left\| \sum_{k=m+1}^{n} x_k \right\| \leq \sum_{k=m+1}^{n} \|x_n\| \leq \sum_{k=N_0}^{\infty} \|x_n\| < \epsilon.$$

Thus the series $\sum_{n=1}^{\infty} x_n$ is convergent.

To prove the converse, suppose every absolutely convergent series is convergent in $\mathbf{E}$, choose a Cauchy sequence $(x_n) \subset \mathbf{E}$. Since (x_n) is Cauchy, it suffices (does it?) to prove that (x_n) has a convergent subsequence. Now for each $j \in \mathbb{N}$ there exists $N_j \in \mathbb{N}$ such that $\|x_k - x_l\| < 2^{-j}$ for all $k, l \geq N_j$. By induction, we may choose $N_1 < N_2 < \ldots$ and so we obtain a subsequence (x_{N_j}) such that $\|x_{N_{j+1}} - x_{N_j}\| < 2^{-j}$ for all $j \in \mathbb{N}$. Thus

$$\sum_{j=1}^{\infty} \|x_{N_{j+1}} - x_{N_j}\| < \infty.$$

By hypothesis, the series $\sum_{j=1}^{\infty} (x_{N_{j+1}} - x_{N_j})$ converges in $\mathbf{E}$. But this is a telescoping series. Consider the partial sum and it collapses as $\sum_{j=1}^{n} (x_{N_{j+1}} - x_{N_j}) = x_{N_{n+1}} - x_{N_1}$. So if the partial sums converge, so does (x_{N_j}). $\qquad\square$

2.3.9 THEOREM (F. RIESZ AND FISCHER, 1907) *For* $1 \leq p < \infty, L^p[a,b]$ *is a Banach space.*

Proof. If $(f_n) \in L^p[a,b]$ is such that $M := \sum_{n=1}^{\infty} \|f_n\|_p < \infty$. Define $g_k := \sum_{n=1}^{k} |f_n|$ and $g := \sum_{n=1}^{\infty} |f_n|$. By Minkowski's Inequality, $\|g_k\|_p \leq M$ and so by Fatou's

Lemma,

$$\int_a^b g^p \leq \liminf \int_a^b g_k^p \leq M^p.$$

In particular, $g(x) < \infty$ a.e. Hence we may define

$$f := \sum_{n=1}^{\infty} f_n$$

and this converges a.e. (we may define $f \equiv 0$ on the set of measure zero where the series does not converge). Then f is measurable and $|f| \leq g$, so $f \in L^p[a, b]$ by the above inequality. Now define $s_k := \sum_{n=1}^k f_n$, then $s_k \in L^p[a, b]$ and we want to prove that $\|s_k - f\|_p \to 0$. Observe that $s_k \to f$ pointwise and $|f - s_k|^p \leq (2g)^p \in L^1[a, b]$. Therefore by the Dominated Convergence Theorem,

$$\|f - s_k\|_p^p = \int_a^b |f - s_k|^p \to 0.$$

Thus we have verified that any absolutely convergent series in $L^p[a, b]$ converges. By Proposition 2.3.8, $L^p[a, b]$ is complete.   □

We now prove that continuous functions are dense in L^p (provided $1 \leq p < \infty$). This is in stark contrast to the fact that $C[a, b]$ is a (proper) closed subspace of $L^\infty[a, b]$ (Proposition 2.3.6). The proof rests on Urysohn's Lemma, which we describe for metric spaces.

2.3.10 REMARK   Let (X, d) be a metric space.

(i) If $A \subset X$ is a set, then for any $x \in X$, define $d(x, A) := \inf\{d(x, y) : y \in A\}$. The function $x \mapsto d(x, A)$ is continuous, and $d(x, A) = 0$ if and only if $x \in \overline{A}$.

(ii) If $A, B \subset X$ are two disjoint closed sets, then define

$$f(x) := \frac{d(x, A)}{d(x, A) + d(x, B)}.$$

Then, $f : X \to [0, 1]$ is continuous, $f \equiv 0$ on A, and $f \equiv 1$ on B.

We will have need for Urysohn's Lemma for arbitrary (normal) topological spaces, but we will defer that discussion until we need it. For now we will apply it to compact subsets of $\mathbb{R}$.

2.3.11 LEMMA   *Let $K \subset [a, b]$ be a compact set. Then there exists $g \in C[a, b]$ such that $g \equiv 1$ on K and $g < 1$ on $[a, b] \setminus K$.*

Proof. Let d denote the usual metric on $[a, b]$. For each $n \in \mathbb{N}$, consider $G_n = \{x \in [a, b] : d(x, K) \geq 1/n\}$. Then G_n is closed and $G_n \cap K = \emptyset$. Hence by Remark 2.3.10 there exists $g_n \in C[a, b]$ such that $0 \leq g_n \leq 1$, $g_n \equiv 1$ on K and $g_n \equiv 0$ on G_n. Now the series

$$g := \sum_{n=1}^{\infty} \frac{1}{2^n} g_n$$

converges in $C[a, b]$ (since $C[a, b]$ is a Banach space and the series is absolutely convergent). The function g satisfies the required properties. $\square$

The proof of the next result uses a technique that should be familiar to you from your experience with the measure theory. To prove something for a large class of measurable functions, one usually begins with proving it for characteristic functions. That done, one can use linearity to prove it for simple functions and then apply the fact that every measurable function is a pointwise limit of simple functions (usually this is applied together with the convergence theorems).

Here we have one additional simplification: Every measurable set can be approximated by compact sets from the inside (this is the *regularity* of the Lebesgue measure mentioned in Chapter 1).

2.3.12 PROPOSITION *If $1 \leq p < \infty$, then $C[a, b]$ is dense in $L^p[a, b]$. In particular, $(C[a, b], \| \cdot \|_p)$ is not complete.*

Proof. Given $f \in L^p[a, b]$, and $\epsilon > 0$, we want to prove that there exists $g \in C[a, b]$ such that $\|f - g\|_p < \epsilon$.

(i) Suppose $f = \chi_K$ where $K \subset [a, b]$ is compact: Let $g \in C[a, b]$ be as in Lemma 2.3.11. For $n \in \mathbb{N}$, define $g^n \in C[a, b]$ by the pointwise product $g^n(x) = g(x)^n$. Then $g^n \to f$ pointwise. Furthermore, $|g^n - f|^p \leq 2^p \in L^1[a, b]$ for all $n \in \mathbb{N}$. By the Dominated Convergence Theorem, $\|g^n - f\|_p^p = \int_a^b |g^n - f|^p \to 0$. Hence there exists $N \in \mathbb{N}$ such that $\|g^N - f\|_p < \epsilon$.

(ii) If $f = \chi_E$ where $E \subset [a, b]$ measurable, then for $\epsilon > 0$, there exists a compact set $K \subset E$ such that $m(E \setminus K) < \epsilon$. Hence, $\|\chi_K - \chi_E\|_p^p < \epsilon$. Now apply part (i).

(iii) If $f = \sum_{i=1}^{n} \alpha_i \chi_{E_i} \in L^1[a, b]$ is a simple function, then apply part (ii) to each E_i and take a linear combination.

(iv) If $f \in L^p[a, b]$ is non-negative, then choose a sequence of simple functions (s_n) such that $0 \leq s_n \leq s_{n+1} \to f$ pointwise (from Theorem 1.2.6). Since $|s_n - f|^p \leq (2f)^p \in L^1[a, b]$, the Dominated Convergence Theorem implies that $\|s_n - f\|_p \to 0$. Now apply part (iii) to s_N for N large enough.

(v) If $f \in L^p[a,b]$ is real-valued, then write it as $f = f^+ - f^-$ and apply part (iv) to each of f^+ and f^-.

(vi) If $f \in L^p[a,b]$ is complex-valued, then apply part (v) to the real and imaginary parts of f. $\qquad\square$

As we go through this book, operators will start playing a more important role. One of the central reasons that operator theory works so beautifully is the fact that, for any Banach space $\mathbf{E}$, the space $\mathcal{B}(\mathbf{E})$ of bounded operators on $\mathbf{E}$ is itself a Banach space. The proof is analogous to that of Proposition 2.3.5.

2.3.13 THEOREM *Let $\mathbf{E}$ be a normed linear space.*

(i) *If $\mathbf{F}$ is complete, then $\mathcal{B}(\mathbf{E}, \mathbf{F})$ is complete.*

(ii) *In particular, $\mathbf{E}^*$ is a Banach space.*

Proof. Suppose $(T_n) \subset \mathcal{B}(\mathbf{E}, \mathbf{F})$ is a Cauchy sequence. Then for any $x \in \mathbf{E}$, the inequality $\|T_n(x) - T_m(x)\| \le \|T_n - T_m\|\|x\|$ implies that $(T_n(x)) \subset \mathbf{F}$ is a Cauchy sequence. Since $\mathbf{F}$ is complete, this sequence converges in $\mathbf{F}$. Hence we may define $T : \mathbf{E} \to \mathbf{F}$ by

$$T(x) = \lim_{n \to \infty} T_n(x).$$

It is clear that T is linear. Furthermore, since (T_n) is Cauchy, there exists $M > 0$ such that $\|T_n\| \le M$ for all $n \in \mathbb{N}$. Hence $\|T(x)\| \le M\|x\|$ for all $x \in \mathbf{E}$, so $T \in \mathcal{B}(\mathbf{E}, \mathbf{F})$. We now want to prove that $\|T_n - T\| \to 0$. To this end choose $\epsilon > 0$ and $N_0 \in \mathbb{N}$ such that $\|T_n - T_m\| < \epsilon$ for all $n, m \ge N_0$. Then for any $x \in \mathbf{E}$ and $n \ge N_0$ fixed,

$$\|T(x) - T_n(x)\| = \lim_{m \to \infty} \|T_m(x) - T_n(x)\| \le \lim_{m \to \infty} \|T_m - T_n\|\|x\| \le \epsilon\|x\|.$$

Therefore $\|T - T_n\| \le \epsilon$ for all $n \ge N_0$. Thus, $T_n \to T$ in $\mathcal{B}(\mathbf{E}, \mathbf{F})$ as required. $\qquad\square$

To wrap up this rather long section, we study an important property for a normed linear space to possess.

2.3.14 DEFINITION A topological space is said to be *separable* if it has a countable dense subset.

Once again, you would have encountered the term before. However, it is when dealing with function spaces that separability truly shines. As in the case of completeness, the reason is the same: approximation techniques are the heart and soul of functional analysis. We will see this in action as we go along. But for now we will simply collect a toolbox of examples.

2.3.15 REMARK Let $\mathbf{E}$ be a normed linear space.

(i) If $\mathbf{E}$ has a dense, separable subspace, then $\mathbf{E}$ is separable.

(ii) If $\mathbf{E}$ contains an uncountable family of disjoint open sets, then $\mathbf{E}$ is not separable.

2.3.16 EXAMPLE

(i) Depending on whether $\mathbb{K} = \mathbb{R}$ or $\mathbb{K} = \mathbb{C}$, we write $\mathbb{K}_0 := \mathbb{Q}$ or $\mathbb{K}_0 = \mathbb{Q} \times \mathbb{Q}$. In either case, $\mathbb{K}_0$ is countable and dense in $\mathbb{K}$, which makes $\mathbb{K}$ separable. Furthermore, it allows us to prove separability for a number of other spaces.

(ii) $(\mathbb{K}^n, \| \cdot \|_p)$ is separable (for any $1 \leq p \leq \infty$) since $\mathbb{K}_0^n$ is dense in $\mathbb{K}^n$. In fact, we will soon see (Theorem 2.4.8) that the norm is irrelevant; $\mathbb{K}^n$ is separable with respect to any norm.

(iii) $(c_{00}, \| \cdot \|_p)$ is separable (for any $1 \leq p \leq \infty$) since we may choose sequences with entries from $\mathbb{K}_0$. This would give a subset which has the same cardinality as

$$\bigcup_{n=1}^{\infty} \mathbb{K}_0^n$$

which is countable and dense in c_{00}.

(iv) By Proposition 2.3.4 and Example (iii), ℓ^p is separable if $1 \leq p < \infty$.

(v) ℓ^∞ is not separable.

Proof. For each subset $A \subset \mathbb{N}$, choose $\chi_A \in \ell^\infty$. Then if $A \neq B$, then $\|\chi_A - \chi_B\|_\infty = 1$. Thus, $\{B(\chi_A; 1/3) : A \subset \mathbb{N}\}$ forms an uncountable family of disjoint open sets (because the power set of an infinite set is necessarily uncountable). $\qquad \square$

(vi) By the Weierstrass Approximation Theorem, polynomials with coefficients in $\mathbb{K}_0$ form a dense subset of $(C[a,b], \| \cdot \|_\infty)$. For any $1 \leq p < \infty$ and any $f \in C[a,b]$,

$$\|f\|_p \leq \|f\|_\infty (b-a)^{1/p}.$$

Therefore, this set is also dense in $C[a,b]$ with respect to $\| \cdot \|_p$. Hence, $(C[a,b], \| \cdot \|_p)$ is separable for all $1 \leq p \leq \infty$.

(vii) By Proposition 2.3.12 and Example (vi), $L^p[a,b]$ is separable for $1 \leq p < \infty$.

(viii) $L^\infty[a,b]$ is not separable.

Proof. For each $t \in [a,b]$, consider $f_t = \chi_{[a,t]}$. Then if $s \neq t$, $\|f_s - f_t\|_\infty = 1$. Once again we obtain an uncountable family of disjoint open sets. $\qquad \square$

2.4   Finite Dimensional Spaces

The goal of this section is to describe finite dimensional normed linear spaces completely. As we will see later, this is useful even when studying infinite dimensional spaces because we may choose finite dimensional subspaces with certain desired properties and exploit the results of this section.

2.4.1 DEFINITION Let $\mathbf{E}$ be a vector space and $\| \cdot \|_1$ and $\| \cdot \|_2$ be two norms on $\mathbf{E}$. We say that these norms are *equivalent* (in symbols, $\| \cdot \|_1 \sim \| \cdot \|_2$) if they induce the same metric topologies on $\mathbf{E}$ (see Remark 2.1.2).

Note that this is an equivalence relation on the class of norms on $\mathbf{E}$.

2.4.2 EXAMPLE Let us take a moment to understand what this equivalence relation means by examining two norms on $\mathbf{E} = \mathbb{R}^2$, namely the supremum norm $\| \cdot \|_\infty$ and the Euclidean norm $\| \cdot \|_2$. Consider the open unit ball in both these norms, which we denote by B_∞ and B_2, respectively. We write τ_∞ and τ_2 for the topologies generated by $\| \cdot \|_\infty$ and $\| \cdot \|_2$, respectively. In $(\mathbf{E}, \tau_2)$ every open ball is a homeomorphic image of B_2. Specifically, any other ball is obtained by translating and dilating (or shrinking) B_2.

Now, given a point $x \in B_\infty$, then simple geometry tells us that there is a disc centered at x that is fully contained in B_∞. This is true for every point in B_∞ so B_∞ is open in $(\mathbf{E}, \tau_2)$. As in the case of $(\mathbf{E}, \tau_2)$, every open ball in $(\mathbf{E}, \tau_\infty)$ is obtained by translating and dilating B_∞. Therefore every such open ball belongs to τ_2. Since every open set is merely a union of open balls, it follows that $\tau_\infty \subset \tau_2$.

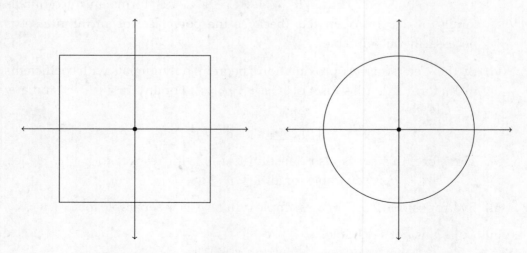

FIGURE 2.1   The open unit balls B_∞ and B_2, respectively

Conversely, given a point $y \in B_2$, there is an open square centered at y that is contained in B_2. Thus the above argument may be reversed to conclude that $\tau_2 = \tau_\infty$. Hence $\| \cdot \|_2 \sim \| \cdot \|_\infty$.

Take a moment to understand the geometry behind this example before diving into the next theorem.

2.4.3 PROPOSITION *Two norms $\| \cdot \|_1$ and $\| \cdot \|_2$ are equivalent if and only if there exist two constants $\alpha, \beta > 0$ such that*

$$\alpha \|x\|_1 \leq \|x\|_2 \leq \beta \|x\|_1 \tag{2.3}$$

for all $x \in \mathbf{E}$.

Proof. Let τ_1 and τ_2 be the topologies induced by $\| \cdot \|_1$ and $\| \cdot \|_2$, respectively.

(i) Suppose $\| \cdot \|_1 \sim \| \cdot \|_2$, then consider $B_1 := \{x \in \mathbf{E} : \|x\|_1 < 1\}$. By hypothesis, $B_1 \in \tau_2$. In particular, since $0 \in B_1$, there exists $\delta > 0$ such that

$$D_2 := \{x \in \mathbf{E} : \|x\|_2 < \delta\} \subset B_1.$$

Now we apply the scaling trick. For any $0 \neq y \in \mathbf{E}$, consider $z := \frac{\delta}{2} \frac{y}{\|y\|_2}$. Then $\|z\|_2 < \delta$, so $\|z\|_1 < 1$ and hence

$$\frac{\delta}{2} \|y\|_1 < \|y\|_2$$

for all $y \in \mathbf{E}$. Thus $\alpha := \delta/2$ satisfies the first inequality of Equation 2.3. By symmetry there exists $\beta > 0$ such that $\|y\|_2 \leq \beta \|y\|_1$ for all $y \in \mathbf{E}$.

(ii) Now suppose that there exist constants $\alpha, \beta > 0$ such that Equation 2.3 holds. Then choose $U \in \tau_1$. We want to prove that $U \in \tau_2$. For any $x \in U$ there exists $r > 0$ such that

$$U_1 := \{y \in \mathbf{E} : \|y - x\|_1 < r\} \subset U.$$

Let $U_2 := \{y \in \mathbf{E} : \|y - x\|_2 < \alpha r\}$. Then for any $y \in U_2$, $\|y - x\|_1 \leq \frac{\|y-x\|_2}{\alpha} < r$. Therefore $U_2 \subset U_1 \subset U$. This is true for every $x \in U$ and so $U \in \tau_2$. Hence $\tau_1 \subset \tau_2$. By symmetry, $\tau_2 \subset \tau_1$ as well.

$\square$

2.4.4 EXAMPLE Let $\mathbf{E} = \mathbb{K}^n$ and consider $\| \cdot \|_2$ and $\| \cdot \|_\infty$ on $\mathbf{E}$. Then for any $x \in \mathbb{K}^n$,

$$\|x\|_\infty \leq \|x\|_2 \leq \sqrt{n}\|x\|_\infty$$

and so $\| \cdot \|_2 \sim \| \cdot \|_\infty$ (as proved in Example 2.4.2).

2.4.5 EXAMPLE Let $\mathbf{E} = c_{00}$ and consider $\| \cdot \|_1$ and $\| \cdot \|_\infty$ on $\mathbf{E}$. Then $\|x\|_\infty \leq \|x\|_1$ for all $x \in \mathbf{E}$, but if $x^k = (1, 1, \ldots, 1, 0, 0, \ldots)$, then $\|x^k\|_1 = k$ and $\|x^k\|_\infty = 1$. Hence there is no constant $\beta > 0$ satisfying $\|x\|_1 \leq \beta\|x\|_\infty$ for all $x \in \mathbf{E}$. Thus $\| \cdot \|_1 \nsim \| \cdot \|_\infty$. Notice that we knew this fact already from a different point of view: In Example 2.2.5, we had constructed a linear functional $\varphi : \mathbf{E} \to \mathbb{K}$ which is continuous with respect to $\| \cdot \|_1$ but not with respect to $\| \cdot \|_\infty$. Therefore the two topologies could not have been the same!

Now suppose $\mathbf{E}$ is a vector space with two equivalent norms $\| \cdot \|_1$ and $\| \cdot \|_2$. If $\mathbf{E}$ is complete with respect to $\| \cdot \|_1$, then it is complete with respect to $\| \cdot \|_2$ (do check this fact!). Hence,

2.4.6 EXAMPLE If $\mathbf{E} = C[a, b]$ then for any $1 \leq p < \infty$, $\| \cdot \|_p \nsim \| \cdot \|_\infty$ (by Proposition 2.3.12 and Proposition 2.3.6).

The next lemma is the famous Heine–Borel Theorem, which we do not prove here. The theorem has a curious bit of history (see [48]). It was proved by Borel in his thesis of 1894. He used an approach that was similar to one that had been used by Heine in 1872 to prove that a continuous function on a closed interval was uniformly continuous. It was Schönflies who noticed the connection and named this theorem after both the authors even though Heine never proved this result!

2.4.7 LEMMA (HEINE–BOREL THEOREM (BOREL, 1894)) *Every closed and bounded subset of $(\mathbb{K}^n, \| \cdot \|_1)$ is compact.*

Now we come to the first main result of this section. Apart from the Heine–Borel Theorem, we use the fact that a continuous, real-valued function on a compact set attains its minimum.

2.4.8 THEOREM *Any two norms on a finite dimensional vector space are equivalent.*

Proof. Let $(\mathbf{E}, \|\cdot\|_\mathbf{E})$ be a finite dimensional normed linear space with basis $\{e_1, e_2, \ldots, e_n\}$. For any $x = \sum_{i=1}^n x_i e_i \in \mathbf{E}$, define

$$\|x\|_1 := \sum_{i=1}^n |x_i|$$

and note that $\|\cdot\|_1$ is a norm on $\mathbf{E}$. Since the equivalence of norms is an equivalence relation, it suffices to show that $\|\cdot\|_\mathbf{E} \sim \|\cdot\|_1$.

If $D := \max\{\|e_j\|_\mathbf{E} : 1 \leq j \leq n\}$, then $\|x\|_\mathbf{E} \leq D\|x\|_1$. This implies that, for any $x, y \in \mathbf{E}$,

$$|\|x\|_\mathbf{E} - \|y\|_\mathbf{E}| \leq \|x - y\|_\mathbf{E} \leq D\|x - y\|_1.$$

Hence the function $f : (\mathbf{E}, \|\cdot\|_1) \to \mathbb{R}_+$ given by $x \mapsto \|x\|_\mathbf{E}$ is continuous. Note that the unit sphere $S = \{x \in \mathbf{E} : \|x\|_1 = 1\}$ is a closed and bounded set and hence compact by Lemma 2.4.7. Thus $f : S \to \mathbb{R}_+$ attains its minimum on S. Therefore there exists $x_0 \in S$ and $C \in \mathbb{R}_+$ such that

$$C = \|x_0\|_\mathbf{E} \leq \|x\|_\mathbf{E}$$

for all $x \in S$. If $C = 0$, then $x_0 = 0$, contradicting the fact that $x_0 \in S$. Hence $C > 0$. Furthermore, for any non-zero $x \in \mathbf{E}$, $y := x/\|x\|_1 \in S$, so $C \leq \|y\|_\mathbf{E}$. Unwrapping this, we see that

$$C\|x\|_1 \leq \|x\|_\mathbf{E}$$

for all $x \in \mathbf{E}$. We conclude that $\|\cdot\|_\mathbf{E} \sim \|\cdot\|_1$. $\square$

2.4.9 COROLLARY *Let $\mathbf{E}$ be finite dimensional and $\mathbf{F}$ be any normed linear space. Then any linear operator $T : \mathbf{E} \to \mathbf{F}$ is continuous.*

Proof. Define a norm on $\mathbf{E}$ by $\|x\|_G := \|x\| + \|T(x)\|$. It is easy to see that this is a norm. By Theorem 2.4.8, there exists $M > 0$ such that $\|T(x)\| \leq \|x\|_G \leq M\|x\|$ for all $x \in \mathbf{E}$. Therefore T is continuous by Theorem 2.2.2. $\square$

2.4.10 DEFINITION A linear operator $T : \mathbf{E} \to \mathbf{F}$ is said to be a

(i) **topological isomorphism** if T is an isomorphism of vector spaces and a homeomorphism (i.e., T and T^{-1} are both continuous).

(ii) **isometric isomorphism** if it is a topological isomorphism that is isometric.

We write $\mathbf{E} \cong \mathbf{F}$ if they are isometrically isomorphic.

2.4.11 COROLLARY *Let* $\mathbf{E}$ *be a finite dimensional normed linear space with* $\dim(\mathbf{E}) = n$. *Then there is a topological isomorphism* $T : (\mathbb{K}^n, \|\cdot\|_1) \to \mathbf{E}$.

Proof. Choose any linear bijection $T : \mathbb{K}^n \to \mathbf{E}$ and apply Corollary 2.4.9 to both T and T^{-1}.   □

The next result follows directly from Corollary 2.4.11 and the fact that $(\mathbb{K}^n, \|\cdot\|_1)$ is complete.

2.4.12 COROLLARY *Every finite dimensional normed linear space is a Banach space.*

Since a complete subset of a metric space is necessarily closed, the next result also follows.

2.4.13 COROLLARY *Let* $\mathbf{E}$ *be an normed linear space and* $\mathbf{F} < \mathbf{E}$ *be a finite dimensional subspace. Then* $\mathbf{F}$ *is closed in* $\mathbf{E}$.

The next lemma is an important one and is a generalization of a fact that we know from Euclidean geometry. Suppose you are given a proper vector subspace $\mathbf{F} \subset \mathbb{R}^3$, where $\mathbb{R}^3$ is equipped with the Euclidean norm. Then $\mathbf{F}$ must be contained in a plane. From geometric intuition, you know that you can choose a line that is perpendicular to $\mathbf{F}$. There must be a vector x_0 on this perpendicular with $\|x_0\| = 1$. Then consider the distance

$$d(x_0, \mathbf{F}) = \inf\{\|x_0 - y\|_2 : y \in \mathbf{F}\}.$$

Since $0 \in \mathbf{F}$, $d(x_0, \mathbf{F}) \leq \|x_0\| = 1$. Furthermore, if y is any other vector in $\mathbf{F}$, then Pythagoras' Theorem tells you that $\|x_0 - y\| > \|x_0\| = 1$. Hence it follows that $d(x_0, \mathbf{F}) = 1$. The next lemma says that one can *almost* make this argument work in any normed linear space. However, one cannot really apply Pythagoras' Theorem, so one does not obtain as sharp a result as one would like.

We get the next best thing though (and, of course, this 'best approximation' theorem holds for Hilbert spaces exactly the way you imagine it should – see Theorem 3.1.6).

2.4.14 LEMMA (F. RIESZ, 1918) *Let* $\mathbf{E}$ *be an normed linear space and* $\mathbf{F} < \mathbf{E}$ *be a proper, closed subspace. Then for any* $0 < t < 1$, *there is a vector* $x_t \in \mathbf{E}$ *such that* $\|x_t\| = 1$ *and* $d(x_t, \mathbf{F}) \geq t$.

Proof. Since $\mathbf{F} < \mathbf{E}$ is a proper closed subspace, there exists $x \in \mathbf{E} \setminus \mathbf{F}$, whence $d := d(x, \mathbf{F}) > 0$. If $0 < t < 1$, then $d/t > d$ so there exists $y \in \mathbf{F}$ such that

$\|x - y\| \leq d/t$. Now $x_t := \frac{x-y}{\|x-y\|}$ satisfies $\|x_t\| = 1$ and

$$d(x_t, \mathbf{F}) = \frac{1}{\|x-y\|} d(x, F) \geq t.$$

$\square$

We are now ready for the next big result of this section. You may remember locally compact spaces from your course in topology – these are spaces with well-behaved neighbourhoods and are amenable to a lot of interesting analysis.

2.4.15 Definition A topological space is said to be **locally compact** if every point has an open neighbourhood with compact closure.

Unfortunately, *linear* spaces tend not to be locally compact (unless they are finite dimensional). Let us look at a simple example: Consider the standard basis vectors $\{e_1, e_2, \ldots\} \subset \ell^2$ described in Example 1.1.7. Then

$$\|e_i - e_j\| = \sqrt{2}$$

if $i \neq j$. Hence the sequence (e_i) does not have a convergent subsequence so the closed unit ball in ℓ^2 cannot be compact. Now any other open set must contain a closed ball, which is homeomorphic to $B[0, 1]$. Thus if $B[0, 1]$ is not compact, then that open set cannot have compact closure.

In other words, ℓ^2 is *not* locally compact. The reason is that the vectors $\{e_i : i \in \mathbb{N}\}$ determine infinitely many 'directions' and thus could not possibly have a convergent subsequence. After all a limit point of this set would not know which way to point! This heuristic understanding is surprisingly accurate: No infinite dimensional normed linear space can be locally compact.

2.4.16 Theorem *For a normed linear space* **E**, *the following are equivalent:*

(i) **E** *is finite dimensional.*

(ii) *Every closed and bounded subset of* **E** *is compact.*

(iii) **E** *is locally compact.*

(iv) *The closed unit ball* $B[0, 1]$ *is compact.*

(v) *The unit sphere* $S_{\mathbf{E}} = \{x \in \mathbf{E} : \|x\| = 1\}$ *is compact.*

Proof.

(i) $\Rightarrow$ (ii): If $\dim(\mathbf{E}) = n$, let $T : (\mathbb{K}^n, \|\cdot\|_1) \to \mathbf{E}$ be a topological isomorphism (which exists by Corollary 2.4.11). Let $B \subset \mathbf{E}$ be a closed bounded set.

Then $T^{-1}(B)$ is closed and bounded in $\mathbb{K}^n$. Hence $T^{-1}(B)$ is compact by the Heine–Borel Theorem. Since T is continuous, $B = T(T^{-1}(B))$ is compact.

(ii) $\Rightarrow$ (iii): If every closed and bounded subset of $\mathbf{E}$ is compact, then $B[x, 1]$ is compact for all $x \in \mathbf{E}$. Since $B[x, 1] = \overline{B(x, 1)}$, it follows that $\mathbf{E}$ is locally compact.

(iii) $\Rightarrow$ (iv): If $\mathbf{E}$ is locally compact, then there exists an open set U such that $0 \in U$ and $\overline{U}$ is compact. Hence there exists $r > 0$ such that $B(0, r) \subset U$. Therefore $B[0, r] \subset \overline{U}$ and so $B[0, r]$ is compact. But $B[0, r] = rB[0, 1]$ and the map $x \mapsto rx$ is a homeomorphism. We conclude that $B[0, 1]$ is compact.

(iv) $\Rightarrow$ (v): This is obvious because $S_{\mathbf{E}}$ is closed subset of $B[0, 1]$.

(v) $\Rightarrow$ (i): Suppose $S_{\mathbf{E}}$ is compact and $\mathbf{E}$ is infinite dimensional, then we repeatedly apply Riesz' Lemma to arrive at a contradiction. Choose $0 \neq x_1 \in \mathbf{E}$ with $\|x_1\| = 1$ and let $\mathbf{F}_1 := \mathrm{span}\{x_1\}$. Then $\mathbf{F}_1 < \mathbf{E}$ is a closed proper subspace (by Corollary 2.4.13) and hence there exists $x_2 \in \mathbf{E}$ such that $\|x_2\| = 1$, and $d(x_2, \mathbf{F}_1) \geq 1/2$. Let $\mathbf{F}_2 := \mathrm{span}\{x_1, x_2\}$, which is again a proper, closed subspace of $\mathbf{E}$. Once again there exists $x_3 \in \mathbf{E}$ such that $\|x_3\| = 1$, and $d(x_3, \mathbf{F}_2) \geq 1/2$. Thus proceeding, we get a sequence (x_n) such that $\|x_n\| = 1$ and

$$d(x_n, \mathbf{F}_{n-1}) \geq 1/2,$$

where $\mathbf{F}_k := \mathrm{span}\{x_1, x_2, \ldots, x_k\}$. In particular,

$$\|x_n - x_m\| \geq 1/2$$

for all $n > m$. Hence $(x_n) \subset S_{\mathbf{E}}$ cannot have a convergent subsequence. This contradicts the assumption that $S_{\mathbf{E}}$ is compact. Hence $\mathbf{E}$ must have been finite dimensional to begin with. $\qquad \square$

2.5 Quotient Spaces

Quotients of normed linear spaces play a role in Functional Analysis that is analogous to the role that quotient groups and quotient rings play in Abstract Algebra. Here, we develop the basic ideas involved and prove some useful (if unexciting) theorems. The highlight of the section is Theorem 2.5.8, which proves that quotients of Banach spaces are complete.

Let us fix some notation we will use throughout this section: **E** will denote a fixed normed linear space, **F** < **E** will be a subspace, and $\pi : \mathbf{E} \to \mathbf{E}/\mathbf{F}$ is the natural quotient map $x \mapsto x + \mathbf{F}$. The first result tells us how to equip **E**/**F** with a norm; it turns out, we need **F** to be closed for this to work.

2.5.1 DEFINITION Let **E** be a vector space. A *seminorm* on **E** is a function $p : \mathbf{E} \to \mathbb{R}_+$ such that

$$p(\alpha x) = |\alpha| p(x), \text{ and } p(x + y) \le p(x) + p(y)$$

for all $x, y \in \mathbf{E}, \alpha \in \mathbb{K}$.

2.5.2 PROPOSITION *Define $p : \mathbf{E} \to \mathbb{R}_+$ by*

$$p(x) := d(x, \mathbf{F}) = \inf\{\|x - y\| : y \in \mathbf{F}\}.$$

Then

(i) *p defines a seminorm on **E**.*

(ii) *If **F** is closed, p induces a norm on **E**/**F** given by $\|x + \mathbf{F}\| := p(x)$.*

Proof. By Exercise 2.8, it suffices to prove (i).

(a) Clearly, $p(x) \ge 0$, and $p(x) = 0$ if and only if $x \in \overline{\mathbf{F}}$.

(b) Now if $\alpha \in \mathbb{K}$, then consider $d(\alpha x, \mathbf{F}) = \inf\{\|\alpha x - y\| : y \in \mathbf{F}\}$. If $\alpha = 0$, then $\alpha x = 0 \in \mathbf{F}$, so $p(\alpha x) = 0$. If $\alpha \ne 0$, then the map $y \mapsto \alpha y$ is a bijection on **F** and hence

$$p(\alpha x) = \inf\{\|\alpha x - \alpha y\| : y \in \mathbf{F}\} = |\alpha| p(x).$$

(c) Finally, if $x_1, x_2 \in \mathbf{E}$, then for any $y_1, y_2 \in \mathbf{F}$, we have $\|(x_1 + x_2) - (y_1 + y_2)\| \le \|x_1 - y_1\| + \|x_2 - y_2\|$. Since $y_1 + y_2 \in \mathbf{F}$, it follows that

$$p(x_1 + x_2) \le \|x_1 - y_1\| + \|x_2 - y_2\|.$$

Taking infima independently, we conclude that $p(x_1 + x_2) \le p(x_1) + p(x_2)$.

This proves that p is a seminorm. $\qquad\square$

Now given a closed subspace $\mathbf{F}$ of a normed linear space $\mathbf{E}$, we wish to examine the norm topology of $\mathbf{E}/\mathbf{F}$. From your experience with abstract algebra, you would know that the group structure of a quotient group is intimately linked to the quotient *map*. Therefore we now analyze the quotient map $\pi : \mathbf{E} \to \mathbf{E}/\mathbf{F}$ carefully, in the hope that it will reveal the topology of $\mathbf{E}/\mathbf{F}$.

2.5.3 PROPOSITION   *If $\mathbf{F}$ is a proper, closed subspace of $\mathbf{E}$, then the quotient map $\pi :$ $\mathbf{E} \to \mathbf{E}/\mathbf{F}$ is continuous and $\|\pi\| = 1$.*

Proof.   If $x \in \mathbf{E}$, we have

$$\|\pi(x)\| = \|x + \mathbf{F}\| = d(x, \mathbf{F}) \leq \|x\|$$

since $0 \in \mathbf{F}$. Hence π is continuous and $\|\pi\| \leq 1$. Furthermore, by Riesz' Lemma (Lemma 2.4.14), for each $0 < t < 1$ there exists $x_t \in \mathbf{E}$ such that $\|x_t\| = 1$ and $\|\pi(x_t)\| \geq t$. Hence $\|\pi\| \geq 1$.   □

The next lemma is interesting in and of itself and will be useful when we study operators on Banach spaces as well: It simply says that the image of an open ball in $\mathbf{E}$ under π is an open ball in $\mathbf{E}/\mathbf{F}$. Naturally, this will imply that π is an open map.

2.5.4 LEMMA   *Let $\mathbf{F}$ be a closed subspace of a normed linear space $\mathbf{E}$. Then for any $x \in \mathbf{E}$ and $r > 0$, $\pi(B_{\mathbf{E}}(x, r)) = B_{\mathbf{E}/\mathbf{F}}(\pi(x), r)$.*

Proof.   We prove containment both ways. If $z \in B_{\mathbf{E}}(x, r)$ then $\|x - z\| < r$. Since $0 \in \mathbf{F}$, this implies that

$$\|\pi(x) - \pi(z)\| = \|\pi(x - z)\| \leq \|x - z\| < r$$

so that $\pi(z) \in B_{\mathbf{E}/\mathbf{F}}(\pi(x), r)$. Conversely, suppose $\pi(z) \in B_{\mathbf{E}/\mathbf{F}}(\pi(x), r)$, then $\|\pi(z - x)\| = \|\pi(z) - \pi(x)\| < r$. So there exists $y \in \mathbf{F}$ such that $\|z - x - y\| < r$. So if $w := z - y$, then $w \in B_{\mathbf{E}}(x, r)$ and

$$\pi(z) = \pi(z - y) = \pi(w) \in \pi(B_{\mathbf{E}}(x, r)).$$

                                    □

2.5.5 PROPOSITION   *Let $\mathbf{F}$ be a closed subspace of a normed linear space $\mathbf{E}$.*

   (i) a set $U \subset \mathbf{E}/\mathbf{F}$ is open if and only if $\pi^{-1}(U)$ is open in $\mathbf{E}$.

   (ii) π is an open map.

Proof.

(i) If $U \subset \mathbf{E}/\mathbf{F}$ is open, then $\pi^{-1}(U) \subset \mathbf{E}$ is open since π is continuous by Proposition 2.5.3. Conversely, suppose $\pi^{-1}(U) \subset \mathbf{E}$ is open. Choose $\pi(x) \in U$. Then $x \in \pi^{-1}(U)$ and there exists $r > 0$ such that $B_{\mathbf{E}}(x, r) \subset \pi^{-1}(U)$. By Lemma 2.5.4,

$$B_{\mathbf{E}/\mathbf{F}}(\pi(x), r) \subset \pi(\pi^{-1}(U)) = U$$

since π is surjective. Therefore U is open.

(ii) If $U \subset \mathbf{E}$ is open, then consider $V = \pi(U)$ and note that

$$\pi^{-1}(V) = U + \mathbf{F} = \{u + y : u \in U, y \in \mathbf{F}\}.$$

For each $y \in \mathbf{F}$, the map $u \mapsto u + y$ is a homeomorphism and hence

$$\pi^{-1}(V) = \bigcup_{y \in \mathbf{F}} (U + y)$$

is an open set. So by part (i), V is open in $\mathbf{E}/\mathbf{F}$. Hence π is an open map. $\square$

The previous theorem should remind you of the construction of quotient spaces in topology. There one takes a topological space X and an equivalence relation $\sim$ to get the quotient space $Y := X/ \sim$. The natural map $X \to Y$ is an open map and has the same properties as in Proposition 2.5.5. This, of course, ensures that we understand the topology of Y completely. Maps arising out of Y are continuous if and only if they are continuous when composed with the quotient map. The same is true here as well; the topology and the linear structure of $\mathbf{E}/\mathbf{F}$ are both completely understood by the quotient map. We record this useful fact.

2.5.6 COROLLARY *A linear map $T : \mathbf{E}/\mathbf{F} \to \mathbf{W}$ is bounded if and only if $T \circ \pi : \mathbf{E} \to \mathbf{W}$ is bounded.*

The next result is an interesting fact because of what it does *not* prove. Given two subsets of a U and V of a normed linear space $\mathbf{E}$, we may obviously define their sum

$$U + V := \{x + y : x \in U, y \in V\}.$$

It is interesting to determine the properties of $U + V$ given the corresponding properties of U and V individually. In this context, if U and V are both *subspaces*, then $U + V$ is clearly a subspace. As we have established in this section, closed subspaces are of particular interest to us. However, if U and V are closed subspaces, it is not necessarily true that $U + V$ is closed (see Exercise 2.43). What we do know is the following.

2.5.7 COROLLARY *If $\mathbf{F}$ is closed and $\mathbf{W} < \mathbf{E}$ finite dimensional, then $\mathbf{W} + \mathbf{F}$ is closed in $\mathbf{E}$.*

Proof. Since $\dim(\pi(\mathbf{W})) \leq \dim(\mathbf{W}) < \infty$, it follows from Corollary 2.4.13 that $\pi(\mathbf{W}) < \mathbf{E}/\mathbf{F}$ is closed. Since π is continuous, $\mathbf{W} + \mathbf{F} = \pi^{-1}(\pi(\mathbf{W}))$ is closed in $\mathbf{E}$. $\qquad\square$

The next result will be of use to us much later in the book. One might go so far as to say that this entire section was written solely to get to this point (and one would not be too far off the mark). Also, it gives us another reason to show off Proposition 2.3.8.

2.5.8 THEOREM *If $\mathbf{E}$ is a Banach space, then $\mathbf{E}/\mathbf{F}$ is a Banach space.*

Proof. By Proposition 2.3.8, it suffices to prove that every absolutely convergent series in $\mathbf{E}/F$ is convergent. So suppose $(x_n) \subset \mathbf{E}$ is such that

$$M := \sum_{n=1}^{\infty} \|\pi(x_n)\| < \infty.$$

We wish to prove that the sequence (s_k) defined by $s_k := \sum_{n=1}^{k} \pi(x_n)$ is convergent in $\mathbf{E}/\mathbf{F}$. By the definition of the quotient norm (as an infimum), for each $n \in \mathbb{N}$ there exists $y_n \in \mathbf{F}$ such that

$$\|\pi(x_n)\| + 1/2^n > \|x_n + y_n\|.$$

Hence $\sum_{n=1}^{\infty} \|x_n + y_n\| < \infty$. By Proposition 2.3.8, the sequence (t_k) defined by $t_k := \sum_{n=1}^{k}(x_n + y_n)$ converges to a point $t \in \mathbf{E}$. But since π is continuous, $s_k = \pi(t_k) \to \pi(t) \in \mathbf{E}/\mathbf{F}$. Thus the series $\sum_{n=1}^{\infty} \pi(x_n)$ converges in $\mathbf{E}/\mathbf{F}$. $\qquad\square$

Having said all this about quotient spaces, it would be unfair not to mention the First Isomorphism Theorem. It is, of course, entirely expected. Perhaps the second part of the theorem carries a nugget of a surprise, though: The map induced on the quotient has the same norm as the original map.

2.5.9 THEOREM (FIRST ISOMORPHISM THEOREM) *Let $T : \mathbf{E} \to \mathbf{W}$ be a bounded linear operator and $\mathbf{F} := \ker(T)$. Then*

(i) **F** *is a closed subspace of* **E**.

(ii) *There exists a unique injective bounded operator* $\widehat{T} : \mathbf{E}/\mathbf{F} \to \mathbf{W}$ *such that* $\widehat{T} \circ \pi = T$ *and* $\|\widehat{T}\| = \|T\|$.

Proof. Since $\mathbf{F} = T^{-1}(\{0\})$, $\mathbf{F}$ is closed. Let $\widehat{T} : \mathbf{E}/\mathbf{F} \to \mathbf{W}$ be the injective map given by

$$\widehat{T}(x + \mathbf{F}) := T(x).$$

Then $\widehat{T}$ is well-defined and linear. Furthermore, for any $x \in \mathbf{E}$ and $y \in \mathbf{F}$,

$$\|\widehat{T}(x + \mathbf{F})\| = \|T(x)\| = \|T(x - y)\| \leq \|T\|\|x - y\|.$$

This is true for all $y \in \mathbf{F}$ and hence $\|\widehat{T}(x+\mathbf{F})\| \leq \|T\|\|x+\mathbf{F}\|$, whence $\widehat{T}$ is bounded and $\|\widehat{T}\| \leq \|T\|$. Moreover,

$$\|T(x)\| = \|\widehat{T}(x+\mathbf{F})\| \leq \|\widehat{T}\|\|x+\mathbf{F}\| \leq \|\widehat{T}\|\|x\|,$$

and so $\|T\| \leq \|\widehat{T}\|$. $\qquad\square$

One word of caution: In Theorem 2.5.9, the map $\widehat{T}$ does not establish a topological isomorphism between $\mathbf{E}/\ker(T)$ and $\mathrm{Range}(T)$. It is merely a bijective, continuous linear map (see Exercise 2.24 for a simple counterexample). We will see later that it *does* become a topological isomorphism if $\mathbf{E}$ and $\mathrm{Range}(T)$ are both complete (Corollary 5.3.6).

And with this, we end our initial foray into the theory of normed linear spaces and it is time to test our mettle.

2.6 Exercises

Definitions and Examples

EXERCISE 2.1 Prove that any non-zero vector space has at least one norm.

EXERCISE 2.2 Let **E** be any vector space.

(i) Construct a metric on **E** which cannot be induced by any norm.

(ii) Let $d : \mathbf{E} \times \mathbf{E} \to \mathbb{R}_+$ be a metric on **E** such that

$$d(x,y) = d(x+z, y+z), \text{ and } d(\alpha x, \alpha y) = |\alpha| d(x,y)$$

for all $x, y, z \in \mathbf{E}$ and all $\alpha \in \mathbb{K}$. Prove that there is a norm $\|\cdot\|_d$ on **E** such that $d(x,y) = \|x - y\|_d$ for all $x, y \in \mathbf{E}$.

EXERCISE 2.3 (⚒) Complete the proof of Lemma 2.1.10.

EXERCISE 2.4 (⚒) For each of the following norms on $\mathbb{R}^2$, draw a picture of the closed unit ball $B = \{x \in \mathbb{R}^2 : \|x\| \leq 1\}$.

(i) The 1-norm $\|(x_1, x_2)\|_1 := |x_1| + |x_2|$.

(ii) The 2-norm $\|(x_1, x_2)\|_2 := \sqrt{|x_1|^2 + |x_2|^2}$.

(iii) The 3-norm $\|(x_1, x_2)\|_3 := \left(|x_1|^3 + |x_2|^3\right)^{1/3}$.

(iv) The sup-norm $\|(x_1, x_2)\|_\infty := \max\{|x_1|, |x_2|\}$.

EXERCISE 2.5 (⚒) Let $(\mathbf{E}, \|\cdot\|_{\mathbf{E}})$ and $(\mathbf{F}, \|\cdot\|_{\mathbf{F}})$ be two normed linear spaces. On the vector space $\mathbf{E} \times \mathbf{F}$, define a function $\|\cdot\| : \mathbf{E} \times \mathbf{F} \to \mathbb{R}_+$ by

$$\|(x, y)\| := \|x\|_{\mathbf{E}} + \|y\|_{\mathbf{F}}.$$

(i) Prove that this is a norm on $\mathbf{E} \times \mathbf{F}$.

(ii) If $\mathbf{E}$ and $\mathbf{F}$ are both Banach spaces, prove that $(\mathbf{E} \times \mathbf{F}, \|\cdot\|)$ is a Banach space.

This normed linear space is denoted by $\mathbf{E} \oplus_1 \mathbf{F}$ (the subscript is meant to emphasize the fact that this norm is the analog of the 1-norm on $\mathbb{R}^2$).

EXERCISE 2.6 (⚒) Let $\mathbf{E}$ be a normed linear space and $U, V \subset \mathbf{E}$. Recall that $U + V := \{x + y : x \in U, y \in V\}$.

(i) If U is open, then prove that $U + V$ is open.

(ii) If U is closed and V is compact, then prove that $U + V$ is closed.

EXERCISE 2.7

(i) Let $\mathbf{E} := (\mathbb{R}^2, \|\cdot\|_1)$, and $\mathbf{F} := (\mathbb{R}^2, \|\cdot\|_\infty)$. Prove that the map $T : \mathbf{E} \to \mathbf{F}$ given by $T(x, y) := (x + y, x - y)$ is an isometric isomorphism.

(ii) Does there exist an isometry $T : (\mathbb{R}^2, \|\cdot\|_1) \to (\mathbb{R}^2, \|\cdot\|_2)$?

Hint: Look at the *shape* of the unit ball in each space.

EXERCISE 2.8 (⚒) Let $\mathbf{E}$ be a vector space and $p : \mathbf{E} \to \mathbb{R}_+$ be a seminorm on $\mathbf{E}$ (Definition 2.5.1). Define $\mathbf{F} := \{x \in \mathbf{E} : p(x) = 0\}$.

(i) Prove that $\mathbf{F}$ is a subspace of $\mathbf{E}$.

(ii) Define $\|\cdot\| : \mathbf{E}/\mathbf{F} \to \mathbb{R}_+$ by $\|x + \mathbf{F}\| := p(x)$. Prove that $\|\cdot\|$ is well-defined and a norm on $\mathbf{E}/\mathbf{F}$.

EXERCISE 2.9 (⚒) Prove that μ_∞ is a seminorm on $\mathcal{L}^\infty[a.b]$ (Definition 2.1.20). Conclude that $L^\infty[a,b]$ is a normed linear space.

EXERCISE 2.10 Let $\mathfrak{S}$ denote the set of all real sequences and define a function $d : \mathfrak{S} \times \mathfrak{S} \to \mathbb{R}_+$ by

$$d((x_n),(y_n)) := \sum_{n=1}^{\infty} 2^{-n} \frac{|x_n - y_n|}{1 + |x_n - y_n|}.$$

(i) Prove that this defines a complete metric on $\mathfrak{S}$.

(ii) For each $n \in \mathbb{N}$, let $\varphi_n : \mathfrak{S} \to \mathbb{R}$ denote the evaluation functional given by $\varphi_n((x_j)) := x_n$. Prove that each φ_n is continuous with respect to this metric.

(iii) Prove that the topology induced by this metric coincides with the product topology on $\mathfrak{S}$ (thought of as $\prod_{n=1}^{\infty} \mathbb{R}$).

(iv) Prove that there is no norm on $\mathfrak{S}$ whose induced topology is this metric topology.

Hint: Given a norm $\|\cdot\|$, the set $V = \{x \in \mathfrak{S} : \|x\| < 1\}$ is an open set containing the origin and must therefore contain a basic open set.

EXERCISE 2.11 (⚒) If $0 < p < 1$, we may define ℓ^p exactly as in Definition 2.1.23 and define $d_p : \ell^p \times \ell^p \to \mathbb{R}_+$ by $d_p(x,y) := \|x-y\|_p^p$.

(i) Show that this defines a complete metric on ℓ^p.
Hint: If $x, y \geq 0$, then prove that $(x+y)^p \leq x^p + y^p$.

(ii) Prove that there is no norm on ℓ^p whose induced topology coincides with the metric topology.

Hint: For $\epsilon > 0$ and $n \in \mathbb{N}$, construct vectors $x_1, x_2, \ldots, x_n \in c_{00} \subset \ell^p$ such that $d_p(x_i, 0) = \epsilon$ for all $1 \leq i \leq n$. If $y := (x_1 + x_2 + \ldots + x_n)/n$, then prove that $d_p(y, 0) = n^{1-p}\epsilon$.

EXERCISE 2.12 (⚒) Let $1 \leq p < r < \infty$.

(i) Prove that $\ell^p \subset \ell^r$.

(ii) Prove that $L^r[a,b] \subset L^p[a,b]$.

In fact, in both the above cases, prove that the inclusion maps are bounded.

EXERCISE 2.13 (⚒) Let $\mathbf{E}$ be a normed linear space and $\mathbf{F} < \mathbf{E}$ be a proper subspace. Prove that $\mathbf{F}$ cannot contain a non-empty open set.

Bounded Linear Operators

EXERCISE 2.14 ($\maltese$) Let $E = (c_0, \| \cdot \|_\infty)$, fix a sequence $\alpha := (\alpha_n) \in \ell^1$ and consider $\varphi : E \to \mathbb{K}$ given by

$$\varphi((x_j)) := \sum_{n=1}^{\infty} \alpha_n x_n.$$

Prove that φ is well-defined and continuous and that $\|\varphi\| = \|\alpha\|_1$.

EXERCISE 2.15 ($\maltese$) Fix a bounded sequence $\alpha = (\alpha_n) \in \ell^\infty$ and consider the map $T : \ell^2 \to \ell^2$ given by

$$T((x_n)) := (\alpha_1 x_1, \alpha_2 x_2, \ldots).$$

Prove that T is well-defined and continuous and that $\|T\| = \|\alpha\|_\infty$.

EXERCISE 2.16 Let $\{t_1, t_2, \ldots, t_n\}$ be a finite collection of points in $[0, 1]$ and let $\{\alpha_1, \alpha_2, \ldots, \alpha_n\}$ be a finite set of complex numbers. Define $\varphi : C[0, 1] \to \mathbb{C}$ by

$$\varphi(f) := \sum_{i=1}^{n} \alpha_i f(t_i).$$

Prove that φ is continuous and determine $\|\varphi\|$.

EXERCISE 2.17 In Example 2.2.15, prove that there is no non-zero function $f \in L^1[0, 1]$ such that $\|T(f)\|_1 = \|f\|_1$.

EXERCISE 2.18 ($\maltese$) Let E be a normed linear space and let $\varphi : E \to \mathbb{K}$ be a linear functional.

   (i) Prove that φ is continuous if and only if $\ker(\varphi)$ is closed in E.

  (ii) If φ is discontinuous, then show that $\ker(\varphi)$ is dense in E.

EXERCISE 2.19 Prove that every infinite dimensional normed linear space has a discontinuous linear functional.

EXERCISE 2.20 Prove that there is a discontinuous function $f : \mathbb{R} \to \mathbb{R}$ such that $f(x + y) = f(x) + f(y)$ for all $x, y \in \mathbb{R}$.

EXERCISE 2.21 Give an example of a linear operator $T : E \to F$ between two normed linear spaces such that $\ker(T)$ is closed, but T is discontinuous.

EXERCISE 2.22 (✖) Let $M := (a_{i,j})_{i,j \in \mathbb{N}}$ be an infinite matrix whose entries are in $\mathbb{K}$. Suppose that

$$\beta := \sup_{i \in \mathbb{N}} \left\{ \sum_{j=1}^{\infty} |a_{i,j}| \right\} < \infty.$$

For $x = (x_1, x_2, \dots) \in \ell^{\infty}$, define a sequence $T(x) = (y_1, y_2, \dots)$ by setting

$$y_i := \sum_{j=1}^{\infty} a_{i,j} x_j.$$

Show that this defines a linear operator $T : \ell^{\infty} \to \ell^{\infty}$ and that $\|T\| = \beta$.

EXERCISE 2.23 Let $\mathbf{E} = \mathbf{F} = C[0,1]$, equipped with the supremum norm. Fix $K \in C([0,1]^2)$ and consider $T : \mathbf{E} \to \mathbf{F}$ given by

$$T(f)(x) = \int_0^1 K(x,t) f(t) dt.$$

(i) Prove that T is well-defined.

(ii) Prove that T is a bounded operator and that $\|T\| = \sup_{x \in [0,1]} \left\{ \int_0^1 |K(x,t)| dt \right\}$.

Hint: For $\epsilon > 0$ and $x_0 \in [0,1]$, define $f_\epsilon \in C[0,1]$ by

$$f_\epsilon(t) = \frac{\overline{K(x_0,t)}}{|K(x_0,t)| + \epsilon}.$$

EXERCISE 2.24 (✖) Let $T : \ell^2 \to \ell^2$ be the operator given by

$$T((x_n)) := \left(\frac{x_1}{1}, \frac{x_2}{2}, \frac{x_3}{3}, \dots \right).$$

(i) Prove that T is a bounded linear operator, which is injective.

(ii) Prove that $T : \ell^2 \to \mathrm{Range}(T)$ is not a homeomorphism.

Banach Spaces

This next problem needs a little knowledge of measure theory. You may skip it for now. Alternatively, you can read ahead to Definition 4.1.8 (and the ideas surrounding it) before attempting it.

EXERCISE 2.25 Let $1 \leq p < \infty$ and let $W_p^1[0,1]$ be the set of all functions $f : [0,1] \to \mathbb{C}$ such that f is absolutely continuous (and hence differentiable a.e.) and

$f' \in L^p[0,1]$. For $f \in W_p^1[0,1]$, define a norm by $\|f\| := \|f\|_p + \|f'\|_p$. Prove that $(W_p^1[0,1], \|\cdot\|)$ is a Banach space.

Hint: If a sequence converges in $L^p[0,1]$, then there is a subsequence which converges pointwise a.e.

EXERCISE 2.26 (✗) Prove that

   (i) $(c_0, \|\cdot\|_\infty)$ is a Banach space.

  (ii) $(c_{00}, \|\cdot\|_\infty)$ is dense in $(c_0, \|\cdot\|_\infty)$. Conclude that $(c_{00}, \|\cdot\|_\infty)$ is not complete.

EXERCISE 2.27 (✗) Let c be the space of all convergent sequences in $\mathbb{K}$, equipped with the supremum norm. Show that c is a closed subspace of ℓ^∞ (and hence a Banach space).

EXERCISE 2.28 Prove Proposition 2.3.6.

EXERCISE 2.29 (✗) Let $\mathbf{E}$ be a normed linear space and let $\widetilde{\mathbf{E}}$ denote the set of all Cauchy sequences in $\mathbf{E}$.

   (i) Define addition and scalar multiplication on $\widetilde{\mathbf{E}}$ by

$$(x_n) + (y_n) := (x_n + y_n), \text{ and } \alpha(x_n) := (\alpha x_n).$$

      Show that $\widetilde{\mathbf{E}}$ is a vector space and that $\mathbf{N} := \{(x_n) \in \widetilde{\mathbf{E}} : \lim_{n\to\infty} \|x_n\| = 0\}$ is a subspace of $\widetilde{\mathbf{E}}$.

  (ii) Set $\widehat{\mathbf{E}} := \widetilde{\mathbf{E}}/\mathbf{N}$. Define a map $\|\cdot\|_{\widehat{\mathbf{E}}} : \widehat{\mathbf{E}} \to \mathbb{R}_+$ by

$$\|(x_n) + \mathbf{N}\|_{\widehat{\mathbf{E}}} := \lim_{n\to\infty} \|x_n\|.$$

      Show that this limit exists and that this is a well-defined norm on $\widehat{\mathbf{E}}$.

 (iii) Show that $\widehat{\mathbf{E}}$, equipped with this norm, is a Banach space.

 (iv) Define $T : \mathbf{E} \to \widehat{\mathbf{E}}$ by $T(x) := (x, x, x, \dots) + \mathbf{N}$. Show that T is an isometry whose range is dense in $\widehat{\mathbf{E}}$.

  (v) Suppose $\mathbf{F}$ is a Banach space and $S : \mathbf{E} \to \mathbf{F}$ is an isometry with dense range, then show that there is a unique isometric isomorphism $U : \widehat{\mathbf{E}} \to \mathbf{F}$ such that $U \circ T = S$. This proves that the pair $(\widehat{\mathbf{E}}, T)$ is unique upto isomorphism.

The pair $(\widehat{\mathbf{E}}, T)$ is called the ***completion*** of $\mathbf{E}$. With a little abuse of notation, we sometimes say that $\widehat{\mathbf{E}}$ is the completion of $\mathbf{E}$ (the operator T is implicit).

Finite Dimensional Spaces and Equivalent Norms

EXERCISE 2.30 Let **E** be an infinite dimensional normed linear space. Construct two norms on **E** that are not equivalent to each other.

EXERCISE 2.31 Let $\mathbf{E} := c_{00}$, and $(a_n)_{n=1}^{\infty}$ be a sequence in $\mathbb{K}$.

(i) Prove that the function

$$\|(x_j)\|_a := \sum_{n=1}^{\infty} |a_n x_n|$$

is a norm on **E** if and only if $a_n \neq 0$ for all $n \in \mathbb{N}$.

(ii) Let $(b_n)_{n=1}^{\infty}$ be another sequence in $\mathbb{K}$ such that $b_n \neq 0$ for all $n \in \mathbb{N}$ and $\|\cdot\|_b$ be the associated norm as in part (i). Prove that $\|\cdot\|_a$ and $\|\cdot\|_b$ are equivalent if and only if

$$0 < \inf_{n \in \mathbb{N}} \left\{ \frac{|a_n|}{|b_n|} \right\} \leq \sup_{n \in \mathbb{N}} \left\{ \frac{|a_n|}{|b_n|} \right\} < \infty.$$

EXERCISE 2.32 Let $\mathbf{E} = C^1[0,1]$ denote the space of all continuously differentiable, $\mathbb{K}$-valued functions on $[0,1]$. Define two norms on **E** by

$$\|f\|_{\infty} := \sup_{x \in [0,1]} |f(x)|, \text{ and } \|f\|_d := \|f\|_{\infty} + \|f'\|_{\infty},$$

where f' denotes the derivative of f. Determine whether these two norms are equivalent to each other.

EXERCISE 2.33 Let **E** be the vector space of all polynomials in one variable with coefficients in $\mathbb{K}$. For a polynomial $p(t) = a_0 + a_1 t + \ldots + a_n t^n$, we define

$$\|p\|_1 := \sup\{|p(t)| : t \in [0,1]\}, \text{ and } \|p\|_2 := \max\{|a_0|, |a_1|, \ldots, |a_n|\}.$$

(i) Prove that there is no constant $\beta \geq 0$ such that $\|p\|_2 \leq \beta\|p\|_1$ for all $p \in \mathbf{E}$.

(ii) Prove that there is no constant $\alpha \geq 0$ such that $\|p\|_1 \leq \alpha\|p\|_2$ for all $p \in \mathbf{E}$.

For the next problem, let m denote the Lebesgue measure on $\mathbb{R}^n$, and for $x \in \mathbb{R}^n$ and $r > 0$, set $B(x,r) := \{y \in \mathbb{R}^n : \|y - x\|_2 < r\}$.

EXERCISE 2.34 For every polynomial p on $\mathbb{R}^n$, prove that there is a constant $C = C(p) > 0$ such that for any $x \in \mathbb{R}^n$ and any $r > 0$,

$$\sup_{y \in B(x,r)} |p(y)| \leq \frac{C}{m(B(x,r))} \int_{B(x,r)} |p(y)| dm(y).$$

Hint: Fix $x = 0$ and $r = 1$ and apply Theorem 2.4.8 to an appropriate space of polynomials.

Note: The constant $C > 0$ obtained in this exercise is independent of the ball, but it does depend on the polynomial.

EXERCISE 2.35 Let $\mathbf{E} := \{f \in C[0,1] : f(0) = 0\}$ (equipped with the supremum norm) and let $\mathbf{F}$ be the subspace given by

$$\mathbf{F} = \left\{ f \in \mathbf{E} : \int_0^1 f(t)dt = 0 \right\}.$$

Then prove that there is no vector $f \in \mathbf{E}$ such that $\|f\| = 1$, and $d(f, \mathbf{F}) = 1$. In other words, the conclusion of Riesz' Lemma does not hold for $t = 1$.

EXERCISE 2.36 Prove that a linear map $T : \mathbf{E} \to \mathbf{F}$ is a topological isomorphism if and only if it is surjective and there exists constants $\alpha, \beta > 0$ such that $\alpha\|x\| \leq \|T(x)\| \leq \beta\|x\|$ for all $x \in \mathbf{E}$.

EXERCISE 2.37 Let $T : c_0 \to c$ be given by $T((x_n)) := (x_1 + x_2, x_1 + x_3, x_1 + x_4, \ldots)$. Prove that T is well-defined and a topological isomorphism.

2.6.1 DEFINITION

(i) A subset S of a vector space $\mathbf{E}$ is said to be a *convex set* if for any pair of points $x, y \in S$, and any $t \in [0,1]$, the vector $tx + (1-t)y$ belongs to S.

(ii) A point $x \in S$ is said to be an ***extreme point*** of S if whenever $x_1, x_2 \in S$ and $t \in [0,1]$ are such that $x = tx_1 + (1-t)x_2$, then it must happen that either $x = x_1$ or $x = x_2$.

EXERCISE 2.38 (⚒)

(i) Prove that the unit ball in c_0 does not have any extreme points.

Hint: If $x = (x_n) \in c_0$, then there exists $N \in \mathbb{N}$ such that $|x_N| < 1$.

(ii) Prove the the unit ball in c does have extreme points.

(iii) Conclude that c_0 and c are *not* isometrically isomorphic.

EXERCISE 2.39 Let $\mathbf{E}$ be a normed linear space and $\mathbf{F} < \mathbf{E}$ a proper finite dimensional subspace. Prove that there exists $x \in \mathbf{E}$ such that $\|x\| = 1$ and $d(x, \mathbf{F}) = 1$.

Hint: First assume $\mathbf{E}$ is finite dimensional and consider $f : S_{\mathbf{E}} \to \mathbb{R}$ given by $x \mapsto d(x, \mathbf{F})$.

EXERCISE 2.40 Let $\mathbf{E}$ be a normed linear space, $\mathbf{F} < \mathbf{E}$ a finite dimensional subspace and $x \in \mathbf{E}$.

(i) Prove that there exists $x_0 \in \mathbf{F}$ such that $\|x - x_0\| = d(x, \mathbf{F})$.

Hint: Consider $S = \{y \in \mathbf{F} : \|x - y\| \leq \|x\|\}$ and $f : S \to \mathbb{R}$ given by $y \mapsto \|x - y\|$.

(ii) Is this point x_0 unique? Give a proof or a counterexample.

Quotient Spaces

EXERCISE 2.41 Let $\mathbf{E}$ be a normed linear space and $\mathbf{F} < \mathbf{E}$ a closed subspace. If $\mathbf{F}$ and $\mathbf{E}/\mathbf{F}$ are both Banach spaces, then prove that $\mathbf{E}$ is complete.

EXERCISE 2.42 Consider c_0 as a closed subspace of ℓ^∞. Then for any $x = (x_n) \in \ell^\infty$, prove that the quotient norm of $(x + c_0) \in \ell^\infty / c_0$ is given by the formula

$$\|x + c_0\| = \limsup_{n \to \infty} |x_n|.$$

EXERCISE 2.43 Give an example of a normed linear space $\mathbf{E}$ and two closed subspaces $\mathbf{F}_1, \mathbf{F}_2 < \mathbf{E}$ such that $\mathbf{F}_1 + \mathbf{F}_2$ is not closed.

Spaces of Continuous Functions

EXERCISE 2.44 Let X be a locally compact, Hausdorff space and let $C_b(X)$ be the space of all continuous, $\mathbb{K}$-valued functions on X that are bounded. That is, $f \in$

$C_b(X)$ if f is continuous and $\sup_{x \in X} |f(x)| < \infty$. Prove that $C_b(X)$, equipped with the supremum norm, is a Banach space.

EXERCISE 2.45 (⚒) Let X be a locally compact, Hausdorff space and let $C_0(X)$ denote the space of all continuous, $\mathbb{K}$-valued functions on X that *vanish at infinity*. That is, $f \in C_0(X)$ if f is continuous and for each $\epsilon > 0$ there exists a compact set $F \subset X$ such that $|f(x)| < \epsilon$ for all $x \in X \setminus F$. Note that elements of $C_0(X)$ are bounded, so we may equip $C_0(X)$ with the supremum norm.

$$\|f\|_\infty := \sup_{x \in X} |f(x)|.$$

Show that $(C_0(X), \|\cdot\|_\infty)$ is a Banach space.

EXERCISE 2.46 Let X be a non-compact, locally compact, Hausdorff space and let $X^+ = X \cup \{p\}$ be the one-point compactification of X. Show that

$$C_0(X) \cong \{f \in C(X^+) : f(p) = 0\}.$$

EXERCISE 2.47 Let X be a locally compact, Hausdorff space and let $C_c(X)$ denote the space of all continuous, $\mathbb{K}$-valued functions on X that are *compactly supported*. That is, $f \in C_c(X)$ if f is continuous and there exists a compact set $F \subset X$ such that $f(x) = 0$ for all $x \in X \setminus F$. Note that $C_c(X) \subset C_0(X)$, so we may equip $C_c(X)$ with the supremum norm.

Prove that $C_c(X)$ is dense in $C_0(X)$.

EXERCISE 2.48 Let $X = \mathbb{N}$, equipped with the discrete topology. Identify $C_b(X)$, $C_0(X)$, $C_c(X)$ and $C(X^+)$ with sequence spaces that you have encountered in the chapter.

EXERCISE 2.49 Define $\varphi : C_c(\mathbb{R}) \to \mathbb{K}$ by

$$\varphi(f) = \int_{-\infty}^{\infty} f(t)dt.$$

Determine whether φ is continuous.

EXERCISE 2.50 (⚒) Let X and Y be two locally compact, Hausdorff spaces and let $\eta : X \to Y$ be a continuous, proper map (a map is said to be *proper* if the inverse image of compact sets are compact). Define $T : C_0(Y) \to C_0(X)$ by $T(f) := f \circ \eta$.

(i) Show that T is a well-defined, bounded linear map.

(ii) If η is a homeomorphism, then show that T is an isometric isomorphism.

2.6.2 DEFINITION Let X be a compact, Hausdorff space. A function $f \in C(X)$ is said to be *positive* if $f(x) \in \mathbb{R}_+$ for all $x \in X$. A linear functional $\varphi : C(X) \to \mathbb{K}$ is said to be a *positive linear functional* if $\varphi(f) \in \mathbb{R}_+$ for all positive functions $f \in C(X)$.

EXERCISE 2.51 Let X be a compact, Hausdorff space and let $C(X)$ be equipped with the supremum norm. Prove that a positive linear functional on $C(X)$ is bounded.

Hint: If $f \in C(X)$ is real-valued, then $-\|f\| \leq f(x) \leq \|f\|$ for all $x \in X$.

EXERCISE 2.52 Let X be a compact, Hausdorff space and $C(X)$ be equipped with the supremum norm. Let $\mathbf{E} \subset C(X)$ be a finite dimensional subspace and let $\{f_1, f_2, \ldots, f_n\}$ be a basis for $\mathbf{E}$.

(i) Choose $x_1 \in X$ such that $f_1(x_1) \neq 0$. Define $g_2 : X \to \mathbb{K}$ by

$$g_2(x) = \det \begin{pmatrix} f_1(x_1) & f_2(x_1) \\ f_1(x) & f_2(x) \end{pmatrix}.$$

Prove that there exists $x_2 \in X$ such that $g_2(x_2) \neq 0$.

(ii) Prove, by induction, that there are points $\{x_1, x_2, \ldots, x_n\} \subset X$ such that

$$\det \begin{pmatrix} f_1(x_1) & f_2(x_1) & \cdots & f_n(x_1) \\ f_1(x_2) & f_2(x_2) & \cdots & f_n(x_2) \\ \vdots & \vdots & \cdots & \vdots \\ f_1(x_n) & f_2(x_n) & \cdots & f_n(x_n) \end{pmatrix} \neq 0.$$

(iii) Prove that the map $T : \mathbf{E} \to \mathbb{K}^n$ given by $T(f) := (f(x_1), f(x_2), \ldots, f(x_n))$ is a topological isomorphism. Conclude that there is a constant $c > 0$ such that

$$\|f\|_\infty \leq c \max\{|f(x_1)|, |f(x_2)|, \ldots, |f(x_n)|\}$$

for all $f \in \mathbf{E}$.

EXERCISE 2.53 Let X be a compact, Hausdorff space and $U \subset X$ be an open set.

(i) Prove that $C_0(U)$ is a closed subspace of $C(X)$.

(ii) Prove that $C(X)/C_0(U) \cong C(X \setminus U)$.

Hint: Use Tietze's Extension Theorem.

The Banach Contraction Principle

2.6.3 DEFINITION Let (X, d) be a metric space. A function $F : X \to X$ is said to be a *contraction* if there is a constant $0 < c < 1$ such that $d(F(x), F(y)) \leq c d(x, y)$ for all $x, y \in X$.

EXERCISE 2.54 Prove the *Banach Contraction Principle*: Let (X, d) be a non-empty, complete metric space and $F : X \to X$ be a contraction. Then prove that F has a unique fixed point (a point $x_* \in X$ such that $F(x_*) = x_*$). In fact, for any point $x_0 \in X$, prove that the sequence (x_n) defined inductively by $x_{n+1} := F(x_n)$ converges to the fixed point.

Hint: If (x_n) as above and $n < m$, then prove that $d(x_n, x_m) \leq \left(\sum_{i=n}^{m-1} c^i \right) d(x_1, x_0)$.

EXERCISE 2.55 Let X be a non-empty, complete metric space and $F : X \to X$ be a function. Suppose there exists $n \in \mathbb{N}$ such that

$$F^n := \underbrace{F \circ F \circ \ldots \circ F}_{n \text{ times}}$$

is a contraction. Then prove that F has a unique fixed point.

EXERCISE 2.56 Let $\mathbf{E} := C[0, 1]$, equipped with the supremum norm, and fix $K \in C([0, 1]^2)$. Let $T : \mathbf{E} \to \mathbf{E}$ be the operator given by

$$T(f)(s) := \int_0^s K(s, t) f(t) dt.$$

  (i) Prove that T is well-defined.

 (ii) Prove that T is a bounded linear operator and that $\|T\| \leq M := \sup\{|K(s, t)| : (s, t) \in [0, 1]^2\}$.

(iii) Fix $g \in C[0, 1]$ and $\lambda \in \mathbb{R}$. Define $F : \mathbf{E} \to \mathbf{E}$ by

$$F(f) = g + \lambda T(f).$$

For any $f_1, f_2 \in \mathbf{E}$ and any $n \geq 1$, prove that

$$\|F^n(f_1) - F^n(f_2)\| \leq \frac{|\lambda|^n M^n}{n!} \|f_1 - f_2\|.$$

(iv) Prove (using Exercise 2.55) that there is a unique function $f \in C[0, 1]$ such that

$$f(s) = g(s) + \lambda \int_0^s K(s, t) f(t) dt.$$

This is called a *Volterra integral equation of the second kind*.

EXERCISE 2.57 Let $f_0, f_1, f_2 \in C[0,1]$ be fixed functions and $\alpha, \beta \in \mathbb{R}$ be fixed scalars. Consider the ordinary differential equation

$$x''(s) + f_0(s)x'(s) + f_1(s)x(s) = f_2(s) \tag{2.4}$$

for $s \in (0,1)$, subject to the initial conditions $x(0) = \alpha$, and $x'(0) = \beta$.

(i) Set $f(s) := x''(s)$. Then prove that

$$x'(s) = \beta + \int_0^s f(\tau)d\tau, \text{ and}$$

$$x(s) = \alpha + \beta s + \int_0^s f(\tau)(s - \tau)d\tau.$$

(ii) Prove that Equation 2.4 can be expressed as a Volterra integral equation of the second kind. Conclude that it has a unique solution.

Additional Reading

- The space $W_p^1[0,1]$ introduced in Exercise 2.25 is an example of a *Sobolev space*. These are important Banach spaces for those who study partial differential equations. A good introduction to this fascinating subject is given in this book by Brezis.

  Brezis, Haim (2011). *Functional Analysis, Sobolev Spaces and Partial Differential Equations*. New York: Universitext. Springer. ISBN: 978-0- 387-70913-0.

- The use of functional analytic techniques to solve integral equations is intimately tied to the birth of the subject, and we have briefly described one such technique in Exercise 2.54 and the problems after it. The basic tools used in the subject are all well laid out in this book by Hochstadt.

  Hochstadt, Harry (1989). *Integral Equations*. Wiley Classics Library. Reprint of the 1973 original, A Wiley-Interscience Publication. New York: John Wiley & Sons, Inc. ISBN: 0-471-50404-1.

Chapter 3

Hilbert Spaces

3.1 Orthogonality

Hilbert space theory is a blend of Euclidean geometry and modern analysis. The inner product between vectors affords us the notion of orthogonality. This gives us access to geometric ideas such as Pythagoras' Theorem (yes, really) and a best approximation property which formalizes the idea of 'dropping a perpendicular'. This property, together with the notion of an orthonormal basis, will lead us to discover the Fourier series of an L^2 function. These ideas were originally studied in the context of integral equations and are the historical roots of all of Functional Analysis.

David Hilbert (1862–1943) is perhaps best known for the twenty-three problems he posed in the International Congress of Mathematicians (ICM) in Paris in 1900. These problems speak to the incredible range of knowledge that Hilbert possessed and have been important signposts for mathematics in the 20th century. The Paris address was sandwiched between his immense work on axiomatizing geometry (published in 1899) and his work on integral equations (1904–1910) which laid the foundations for modern analysis. His mentorship of extraordinary students, including Hermann Weyl, Ernst Zermelo, Max Dehn, and many others, has only enriched his mathematical legacy.

We begin with some definitions and notation. Throughout this section, we will use the letter **H** to denote a Hilbert space and $\langle \cdot, \cdot \rangle$ to denote the inner product on **H**.

3.1.1 DEFINITION

(i) We say that two elements $x, y \in \mathbf{H}$ are **orthogonal** if $\langle x, y \rangle = 0$. If this happens, we write $x \perp y$.

(ii) For two subsets $A, B \subset \mathbf{H}$, we write $A \perp B$ if $x \perp y$ for all $x \in A$ and $y \in B$.

(iii) For any set $A \subset \mathbf{H}$, write $A^{\perp} := \{x \in \mathbf{H} : x \perp y \text{ for all } y \in A\}$. If $A = \{x\}$, then we simply write $x^{\perp}$ instead of $\{x\}^{\perp}$.

3.1.2 REMARK Let A be any subset of **H**.

(i) For each $y \in \mathbf{H}$, the linear functional $\varphi_y : \mathbf{H} \to \mathbb{K}$ given by $\varphi_y(x) := \langle x, y \rangle$ is continuous (see Example 2.2.3) and

$$A^{\perp} = \bigcap_{y \in A} \ker(\varphi_y).$$

Thus regardless of what A is, $A^{\perp}$ is always closed and a *subspace* of **H**.

(ii) For any $A \subset \mathbf{H}$, it is easy to check that $A \cap A^{\perp} \subset \{0\}$ and that $A \subset (A^{\perp})^{\perp}$.

(iii) Note that $(A^{\perp})^{\perp}$ is always a closed subspace of **H**, so one cannot expect the equality $A = (A^{\perp})^{\perp}$ unless A were also a closed subspace (see Proposition 3.1.10).

The first collection of results do not require a proof. They simply follow from the axioms of the inner product. Give it a try!

3.1.3 PROPOSITION *Let $x, y \in \mathbf{H}$, then*

(i) *(Polarization Identity):* $\|x + y\|^2 = \|x\|^2 + 2\operatorname{Re}\langle x, y \rangle + \|y\|^2$.

(ii) *(Pythagoras' Theorem):* If $x \perp y$, then $\|x + y\|^2 = \|x\|^2 + \|y\|^2$.

(iii) *(Parallelogram law):* $\|x + y\|^2 + \|x - y\|^2 = 2(\|x\|^2 + \|y\|^2)$.

The parallelogram law gives us a quick way to determine which of our familiar normed linear spaces are Hilbert spaces.

3.1.4 EXAMPLE Fix $1 \le p \le \infty$.

(i) Let $\mathbf{E} = \ell^p$ and set $x := e_1 + e_2$ and $y := e_1 - e_2$. Then $\|x+y\|_p = \|2e_1\|_p = 2$ and $\|x - y\|_p = \|2e_2\|_p = 2$ and

$$\|x\|_p = \|y\|_p = \begin{cases} 2^{1/p} & : \text{if } 1 \le p < \infty, \\ 1 & : \text{if } p = \infty. \end{cases}$$

Hence the parallelogram law holds if and only if $8 = 4 \times 2^{2/p}$, which happens only when $p = 2$.

(ii) Let $\mathbf{E} = L^p[0,1]$ and $f(t) := t$ and $g(t) := 1 - t$. Then $\|f + g\|_p = 1$ and

$$\|f - g\|_p = \|f\|_p = \|g\|_p = \begin{cases} \frac{1}{(1+p)^{1/p}} & : \text{if } 1 \le p < \infty, \\ 1 & : \text{if } p = \infty. \end{cases}$$

Once again the parallelogram law holds if and only if $p = 2$.

We have constructed these examples by showing that a space is not a Hilbert space if the parallelogram law is violated. It is an interesting fact that this is the only obstruction: A Banach space is a Hilbert space *if and only if* the norm satisfies the parallelogram law. A proof of this lies buried in Proposition 3.1.3. Specifically, one can recover the inner product from the norm by rewriting the Polarization Identity. Have a look at Exercise 3.5.

3.1.5 DEFINITION Let $\mathbf{E}$ be a vector space. A subset $A \subset \mathbf{E}$ is said to be *convex* if, for any $x, y \in A$, the set $[x, y] := \{tx + (1 - t)y : 0 \le t \le 1\}$ is contained in A.

There are two kinds of convex sets we will be most interested in: The closed (or open) unit ball in a normed linear space $\mathbf{E}$ and any vector subspace of $\mathbf{E}$. Also, if A is a convex set, then $A + x$ is also a convex set for any $x \in \mathbf{E}$. Therefore when dealing with (non-empty) convex sets, we may often assume that $0 \in A$ by translating.

As discussed earlier, the next result is central to the subject. Notice the liberal use of the parallelogram law in the proof.

3.1.6 THEOREM (BEST APPROXIMATION PROPERTY) *Let $A \subset \mathbf{H}$ be a non-empty, closed, convex set and $x \in \mathbf{H}$. Then there exists a unique vector $x_0 \in A$ such that*

$$\|x - x_0\| = d(x, A) = \inf\{\|x - y\| : y \in A\}.$$

*This vector x_0 is called the **best approximation of x in A**.*

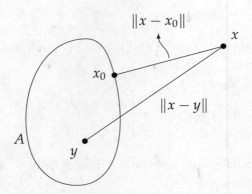

FIGURE 3.1   Best approximation property

Proof.   Since $d(x, A) = d(0, A - x)$, replacing A by $A - x$ (which is also convex), we may assume without loss of generality that $x = 0$.

(i) Existence: By definition there exists a sequence $(y_n) \subset A$ such that $d := d(0, A) = \lim_{n \to \infty} \|y_n\|$. We wish to prove that (y_n) is Cauchy. By the parallelogram law,

$$\left\| \frac{y_n - y_m}{2} \right\|^2 = \frac{1}{2}(\|y_n\|^2 + \|y_m\|^2) - \left\| \frac{y_n + y_m}{2} \right\|^2.$$

Since A is convex, $(y_n + y_m)/2 \in A$, so $\|(y_n + y_m)/2\|^2 \geq d^2$. For $\epsilon > 0$, choose $N_0 \in \mathbb{N}$ such that $\|y_n\|^2 < d^2 + \epsilon$ for all $n \geq N_0$. Then for $n, m \geq N_0$, we have

$$\left\| \frac{y_n - y_m}{2} \right\|^2 < \frac{1}{2}(2d^2 + 2\epsilon) - d^2 = \epsilon,$$

and so $\|y_n - y_m\| < 2\sqrt{\epsilon}$ for all $n, m \geq N_0$. Thus (y_n) is Cauchy and hence convergent in $\mathbf{H}$. Since A is closed, there exists $x_0 \in A$ such that $y_n \to x_0$. By the continuity of the norm (Remark 2.1.2), $d = \lim_{n \to \infty} \|y_n\| = \|x_0\|$.

(ii) Uniqueness: Suppose $x_0, x_1 \in A$ are such that $\|x_0\| = \|x_1\| = d$. Then $(x_0 + x_1)/2 \in A$ and hence

$$d \leq \left\| \frac{1}{2}(x_0 + x_1) \right\| \leq \frac{1}{2}(\|x_0\| + \|x_1\|) \leq d$$

and so $\|\frac{1}{2}(x_0 + x_1)\| = d$. The parallelogram law then implies that

$$d^2 = \left\| \frac{x_0 + x_1}{2} \right\|^2 = d^2 - \left\| \frac{x_0 - x_1}{2} \right\|^2.$$

Therefore $x_0 = x_1$.   □

The main application of the best approximation property is when the convex set in question is a closed subspace. Here the best approximation to a vector is geometrically visualized as the point obtained by 'dropping a perpendicular' from the point onto the subspace.

3.1.7 PROPOSITION *Let* $\mathbf{M} < \mathbf{H}$ *be a closed subspace and* $x \in \mathbf{H}$. *Then* $x_0 \in \mathbf{M}$ *is the best approximation of* x *in* $\mathbf{M}$ *if and only if* $(x - x_0) \perp \mathbf{M}$.

Proof. If x_0 is the best approximation of x in $\mathbf{M}$, then we wish to prove that $\langle x - x_0, y \rangle = 0$ for all $y \in \mathbf{M}$. It suffices to prove this when $\|y\| = 1$, so fix $y \in \mathbf{M}$ with $\|y\| = 1$ and let $\alpha := \langle x - x_0, y \rangle$. Then $z := x_0 + \alpha y \in \mathbf{M}$, so

$$\begin{aligned}
\|x - x_0\|^2 &\leq \|x - z\|^2 = \|(x - x_0) - \alpha y\|^2 \\
&= \|x - x_0\|^2 + \|\alpha y\|^2 - 2\operatorname{Re}\langle x - x_0, \alpha y \rangle \\
&= \|x - x_0\|^2 + |\alpha|^2 \|y\|^2 - 2|\alpha|^2 \\
&= \|x - x_0\|^2 - |\alpha|^2.
\end{aligned}$$

Hence $|\alpha|^2 = 0$, which implies that $x - x_0 \perp y$.

Conversely, suppose $(x - x_0) \perp \mathbf{M}$, then we wish to prove that $\|x - x_0\| = d(x, \mathbf{M})$. In other words, we wish to prove that $\|x - x_0\| \leq \|x - y\|$ for all $y \in \mathbf{M}$. For any $y \in \mathbf{M}$, we have $(x_0 - y) \in \mathbf{M}$, so $(x - x_0) \perp (x_0 - y)$. By Pythagoras' Theorem,

$$\|x - y\|^2 = \|(x - x_0) + (x_0 - y)\|^2 = \|x - x_0\|^2 + \|x_0 - y\|^2 \geq \|x - x_0\|^2.$$

Therefore $\|x - x_0\| \leq \|x - y\|$ for all $y \in \mathbf{M}$. $\qquad\square$

3.1.8 DEFINITION Let $\mathbf{M} < \mathbf{H}$ be a closed subspace. For $x \in \mathbf{H}$, let $P_{\mathbf{M}}(x) \in \mathbf{M}$ denote the best approximation of x in $\mathbf{M}$. In other words, $P_{\mathbf{M}}(x)$ is the unique vector in $\mathbf{M}$ such that $\|x - P_{\mathbf{M}}(x)\| = d(x, \mathbf{M})$. By Proposition 3.1.7, this is equivalent to requiring that $x - P_{\mathbf{M}}(x) \in \mathbf{M}^{\perp}$. The map $P_{\mathbf{M}} : \mathbf{H} \to \mathbf{M}$ is called the ***orthogonal projection*** of $\mathbf{H}$ onto $\mathbf{M}$ (and we will often think of it as a map from $\mathbf{H}$ to itself and denote it by P when the subspace is implicit).

That this projection map is linear is perhaps the most remarkable part of the next result.

3.1.9 PROPOSITION *Let* $P : \mathbf{H} \to \mathbf{M}$ *be the orthogonal projection onto a closed subspace* $\mathbf{M} < \mathbf{H}$. *Then*

  (i) P is a linear transformation.

(ii) *P is bounded and* $\|P\| \leq 1$. *If* $\mathbf{M} \neq \{0\}$, *then* $\|P\| = 1$.

(iii) $P \circ P = P$.

(iv) $\ker(P) = \mathbf{M}^{\perp}$ *and* $\mathrm{Range}(P) = \mathbf{M}$.

Proof.

(i) Let $x_1, x_2 \in \mathbf{H}$ and $\alpha \in \mathbb{K}$. Set $z = x_1 + \alpha x_2$ and $z_0 = P(x_1) + \alpha P(x_2)$. We wish to prove that $P(z) = z_0$. For any $y \in \mathbf{M}$,

$$\langle z - z_0, y \rangle = \langle x_1 - P(x_1), y \rangle + \alpha \langle x_2 - P(x_2), y \rangle = 0.$$

Hence $z - z_0 \in \mathbf{M}^{\perp}$. Since $z_0 \in \mathbf{M}$, it follows from Proposition 3.1.7 that $P(z) = z_0$.

(ii) For any $x \in \mathbf{H}$, $x = (x - P(x)) + P(x)$ and $(x - P(x)) \perp P(x)$. Therefore by Pythagoras' Theorem,

$$\|x\|^2 = \|x - P(x)\|^2 + \|P(x)\|^2 \geq \|P(x)\|^2.$$

Hence P is continuous and $\|P\| \leq 1$. Moreover, if $\mathbf{M} \neq \{0\}$, then choose a non-zero vector $y \in \mathbf{M}$. Then $P(y) = y$ (by definition of the best approximation) and thus $\|P(y)\| = \|y\|$. This proves that $\|P\| = 1$.

(iii) Now if $y \in \mathbf{M}$ then $P(y) = y$ (as above). If $x \in \mathbf{M}$ then $y = P(x) \in \mathbf{M}$, so $P(P(x)) = P(x)$.

(iv) If $P(x) = 0$, then $x = x - P(x) \in \mathbf{M}^{\perp}$. Hence $\ker(P) \subset \mathbf{M}^{\perp}$. Conversely, if $x \in \mathbf{M}^{\perp}$, then $0 \in \mathbf{M}$ satisfies the conditions of Proposition 3.1.7. By uniqueness of the best approximation, it follows that $P(x) = 0$. Hence $\ker(P) = \mathbf{M}^{\perp}$. That $\mathrm{Range}(P) = \mathbf{M}$ is evident from part (iii) and the definition. $\qquad\square$

3.1.10 PROPOSITION *Let* $\mathbf{M} < \mathbf{H}$ *be a subspace, then* $(\mathbf{M}^{\perp})^{\perp} = \overline{\mathbf{M}}$.

Proof. If $x \in \mathbf{M}$, then for any $y \in \mathbf{M}^{\perp}, \langle x, y \rangle = 0$. Hence $x \in (\mathbf{M}^{\perp})^{\perp}$. Thus $\mathbf{M} \subset (\mathbf{M}^{\perp})^{\perp}$. However, $(\mathbf{M}^{\perp})^{\perp}$ is closed, so $\overline{\mathbf{M}} \subset (\mathbf{M}^{\perp})^{\perp}$.

Conversely, if $x \in (\mathbf{M}^{\perp})^{\perp}$, then let $x_0 = P_{\overline{\mathbf{M}}}(x)$ denote the best approximation of x in $\overline{\mathbf{M}}$. Then $x_0 \in \overline{\mathbf{M}}$ and $x - x_0 \in \overline{\mathbf{M}}^{\perp}$. However, $\mathbf{M}^{\perp} = \overline{\mathbf{M}}^{\perp}$ (by continuity of the inner product), so it follows that $\langle x, x - x_0 \rangle = 0$ and $\langle x_0, x - x_0 \rangle = 0$. Thus $\|x - x_0\|^2 = \langle x - x_0, x - x_0 \rangle = 0$, whence $x = x_0 \in \overline{\mathbf{M}}$. $\qquad\square$

The next theorem is the first hint that the relationship between closed subspaces may be recovered from the operator-theoretic algebra of the corresponding projections. From now on, when S and T are two operators, 'ST' will be used to denote the composition $S \circ T$.

3.1.11 PROPOSITION *Let* $\mathbf{M} < \mathbf{H}$ *be a closed subspace, and* $P = P_{\mathbf{M}}$. *Since* $\mathbf{M}^{\perp}$ *is a closed subspace, we may set* $Q = P_{\mathbf{M}^{\perp}}$. *Then*

(i) $PQ = QP = 0$.

(ii) $P + Q = I$ *where* $I : \mathbf{H} \to \mathbf{H}$ *denotes the identity map.*

Proof.

(i) If $x \in \mathbf{H}$, then $Q(x) \in \mathbf{M}^{\perp}$. By Proposition 3.1.9, $\ker(P) = \mathbf{M}^{\perp}$ and hence $PQ(x) = 0$. Since $(\mathbf{M}^{\perp})^{\perp} = \mathbf{M}$ (by Proposition 3.1.10), the fact that $QP = 0$ follows by symmetry.

(ii) If $x \in \mathbf{H}$, then $(x - P(x)) \in \mathbf{M}^{\perp}$ and $Q(x) \in \mathbf{M}^{\perp}$, so

$$x - P(x) - Q(x) \in \mathbf{M}^{\perp}.$$

However, $x - Q(x) \in (\mathbf{M}^{\perp})^{\perp} = \mathbf{M}$ by Proposition 3.1.10 and $P(x) \in \mathbf{M}$, so $x - Q(x) - P(x) \in \mathbf{M}$. Since $\mathbf{M} \cap \mathbf{M}^{\perp} = \{0\}$, it follows that $x - P(x) - Q(x) = 0$. Hence, $x = (P + Q)(x)$. This is true for any $x \in \mathbf{H}$ and thus $(P + Q) = I$. $\square$

3.1.12 DEFINITION Given two Hilbert spaces $(\mathbf{H}_1, \langle \cdot, \cdot \rangle_{\mathbf{H}_1})$ and $(\mathbf{H}_2, \langle \cdot, \cdot \rangle_{\mathbf{H}_2})$, there is a natural inner product on the direct product $\mathbf{H}_1 \times \mathbf{H}_2$, given by

$$\langle (x_1, y_1), (x_2, y_2) \rangle := \langle x_1, x_2 \rangle_{\mathbf{H}_1} + \langle x_2, y_2 \rangle_{\mathbf{H}_2}.$$

Under this inner product, one can prove that $\mathbf{H}_1 \times \mathbf{H}_2$ is a Hilbert space (see Exercise 3.12). This is called the *Hilbert space direct sum* of $\mathbf{H}_1$ and $\mathbf{H}_2$ and is denoted by $\mathbf{H}_1 \oplus \mathbf{H}_2$.

The next corollary is an extremely important one. Given a closed subspace of a Hilbert space $\mathbf{H}$, it says that we can decompose $\mathbf{H}$ as a direct sum in a very natural way. This allows us to 'build up' the Hilbert space from closed subspaces and is particularly useful when analyzing operators on Hilbert spaces. This is something we will do later in the book. Also, it is worth mentioning that such a theorem does not exist for arbitrary Banach spaces. We will have more to say on this matter in Exercise 5.48 and the problems following it.

3.1.13 COROLLARY *Let* $\mathbf{M} < \mathbf{H}$ *be a closed subspace. Then the map* $T : \mathbf{M} \oplus \mathbf{M}^{\perp} \to \mathbf{H}$ *given by* $T(x,y) := x + y$ *is an isometric isomorphism.*

Proof.   Let $P : \mathbf{H} \to \mathbf{M}$ and $Q : \mathbf{H} \to \mathbf{M}^{\perp}$ be the orthogonal projections onto $\mathbf{M}$ and $\mathbf{M}^{\perp}$, respectively. Then by Proposition 3.1.11,

$$\mathbf{H} = (P + Q)(\mathbf{H}) = P(\mathbf{H}) + Q(\mathbf{H}) = \mathbf{M} + \mathbf{M}^{\perp}.$$

Furthermore, $\mathbf{M} \cap \mathbf{M}^{\perp} = \{0\}$. From this it is clear that T is a linear bijection. Finally, the parallelogram law shows that for any $x \in \mathbf{M}$ and $y \in \mathbf{M}^{\perp}$,

$$\|T(x,y)\|^2 = \|x + y\|^2 = \|x\|^2 + \|y\|^2 = \|(x,y)\|^2_{\mathbf{M} \oplus \mathbf{M}^{\perp}}.$$

Thus T is isometric as well.   □

The proof of the next corollary is an easy consequence and is relegated to the exercises.

3.1.14 COROLLARY   *If* $\mathbf{M} < \mathbf{H}$ *is any subspace, then* $\mathbf{M}$ *is dense in* $\mathbf{H}$ *if and only if* $\mathbf{M}^{\perp} = \{0\}$.

3.2   The Riesz Representation Theorem

For a finite dimensional vector space, the notion of a dual basis ensures that the dual space is isomorphic to the original vector space. The analogous statement fails quite spectacularly for arbitrary Banach spaces. In fact, the quest to understand the dual space leads to one of the most important results in the subject, the Hahn–Banach Theorem, which we will discuss in Chapter 4. For now, though, we will contend with the case of a Hilbert space. Here the Riesz Representation Theorem completely answers the question: The dual of a Hilbert space is isomorphic to itself.

Instead of relying on bases and their dual bases, the proof constructs a rather natural isomorphism. This map (denoted by Δ below) is not only useful here, but will also provide a template for answering similar questions about other Banach spaces.

3.2.1 DEFINITION If $y \in \mathbf{H}$, define $\varphi_y : \mathbf{H} \to \mathbb{K}$ by $x \mapsto \langle x, y \rangle$. Then $\varphi_y \in \mathbf{H}^*$ and $\|\varphi_y\| = \|y\|$ (see Example 2.2.12). Hence we get a map $\Delta : \mathbf{H} \to \mathbf{H}^*$ given by

$$\Delta(y) := \varphi_y.$$

Note that $\Delta(\alpha y) = \bar{\alpha}\Delta(y)$ and $\Delta(y_1 + y_2) = \Delta(y_1) + \Delta(y_2)$. In other words, Δ is a *conjugate–linear isometry*.

3.2.2 THEOREM (RIESZ REPRESENTATION THEOREM (F. RIESZ AND M. FRÉCHET, 1907)) *For any $\varphi \in \mathbf{H}^*$ there exists a unique vector $y \in \mathbf{H}$ such that*

$$\varphi(x) = \langle x, y \rangle$$

for all $x \in \mathbf{H}$. Hence, $\Delta : \mathbf{H} \to \mathbf{H}^*$ is a conjugate–linear isometric isomorphism.

Proof. Fix $0 \neq \varphi \in \mathbf{H}^*$, then $\mathbf{M} := \ker(\varphi)$ is a closed subspace of $\mathbf{H}$. Since $\varphi \neq 0$, $\mathbf{M} \neq \mathbf{H}$ and thus $\mathbf{M}^\perp \neq \{0\}$ by Corollary 3.1.14. Choose $y_0 \in \mathbf{M}^\perp$ such that $\varphi(y_0) = 1$. Now for any $x \in \mathbf{H}$, if we set $\alpha := \varphi(x)$, then $x - \alpha y_0 \in \mathbf{M}$. Hence

$$0 = \langle x - \alpha y_0, y_0 \rangle = \langle x, y_0 \rangle - \varphi(x)\|y_0\|^2.$$

Note that $y_0 \neq 0$ since $\varphi(y_0) = 1$. Hence if $y = y_0/\|y_0\|^2$, then for any $x \in \mathbf{H}$, $\varphi(x) = \langle x, y \rangle$. This completes the proof of existence. As for uniqueness, this follows from the fact that Δ is both linear and isometric (do check this!). $\qquad \square$

Note that the Riesz Representation Theorem fails for arbitrary inner product spaces (see Exercise 3.18).

3.2.3 REMARK If $\mathbf{H}$ is a Hilbert space, then $\mathbf{H}^*$ may be equipped with an inner product

$$(\varphi_y, \varphi_z) := \langle z, y \rangle.$$

Since $\mathbf{H}^*$ is complete, it is a Hilbert space.

Frigyes Riesz (1880–1956) was one of the early pioneers of Functional analysis, and his contributions to the subject include the Riesz Representation Theorem, the study of the L^p spaces for $p \neq 2$, dual spaces and the Riesz–Fischer Theorem (Theorem 2.3.9). Much of this was done with a view to connect the work of Fréchet and Lebesgue (on real functions) to the work of Hilbert and Schmidt (on integral equations). These ideas were later incorporated into the work of Banach and formed the foundation upon which the subject was built. His long career culminated in the 1952 book *Leçons d'analyse fonctionelle* (co-authored with B.Sz-Nagy), which remains an important and lucid introduction to the subject even today.

3.2.4 PROPOSITION   *Let* $\mathbf{E}$ *be a normed linear space,* $\mathbf{F}$ *a Banach space and* $\mathbf{E}_0 < \mathbf{E}$ *a dense subspace. If* $T_0 \in \mathcal{B}(\mathbf{E}_0, \mathbf{F})$, *then there exists a unique linear operator* $T \in \mathcal{B}(\mathbf{E}, \mathbf{F})$ *such that*

$$T|_{\mathbf{E}_0} = T_0.$$

Moreover, $\|T\| = \|T_0\|$, *so* T *is called the* **norm-preserving extension** *of* T_0.

Proof.   For any $x \in \mathbf{E}$, choose a sequence $(x_n) \subset \mathbf{E}_0$ such that $x_n \to x$. Then (x_n) is Cauchy. Since T_0 is bounded, it follows that $(T_0(x_n))$ is Cauchy in $\mathbf{F}$. Since $\mathbf{F}$ is complete, there exists $y \in \mathbf{F}$ such that $T_0(x_n) \to y$. Define $T : \mathbf{E} \to \mathbf{F}$ by

$$T(x) := \lim_{n \to \infty} T_0(x_n).$$

(i) We have to prove that T is well-defined: Suppose $(z_n) \subset \mathbf{E}$ is another sequence such that $z_n \to x$, then $\|T_0(z_n) - T_0(x_n)\| \leq \|T_0\| \|z_n - x_n\| \to 0$. Hence, $\lim_{n \to \infty} T_0(z_n) = \lim_{n \to \infty} T_0(x_n)$.

(ii) T is linear: If $x_n \to x$ and $y_n \to y$, then $x_n + y_n \to x + y$ (see Remark 2.1.2). Since T_0 is linear, it follows that

$$T(x + y) = \lim_{n \to \infty} T_0(x_n + y_n) = \lim_{n \to \infty} T_0(x_n) + \lim_{n \to \infty} T_0(y_n) = T(x) + T(y).$$

Similarly, $T(\alpha x) = \alpha T(x)$ for all $\alpha \in \mathbb{K}$ and $x \in \mathbf{E}$.

(iii) T is bounded: If $x_n \to x$, then $\|x_n\| \to \|x\|$ and so

$$\|T(x)\| = \lim_{n \to \infty} \|T_0(x_n)\| \leq \|T_0\| \lim_{n \to \infty} \|x_n\| = \|T_0\| \|x\|.$$

Therefore T is bounded with $\|T\| \leq \|T_0\|$. However, T is an extension of T_0 and so $\|T\| \geq \|T_0\|$ holds trivially. Thus $\|T\| = \|T_0\|$.

(iv) Uniqueness: Let $T_1, T_2 \in \mathcal{B}(\mathbf{E}, \mathbf{F})$ be two bounded linear operators such that $T_1|_{\mathbf{E}_0} = T_0 = T_2|_{\mathbf{E}_0}$. Then for any $x \in \mathbf{E}$, choose $(x_n) \subset \mathbf{E}_0$ such that $x_n \to x$, so that

$$T_1(x) = \lim_{n \to \infty} T_1(x_n) = \lim_{n \to \infty} T_0(x_n) = \lim_{n \to \infty} T_2(x_n) = T_2(x).$$

Therefore $T_1 = T_2$.

$\square$

3.2.5 COROLLARY *Let* $\mathbf{E}$ *be a normed linear space and* $\mathbf{E}_0 < \mathbf{E}$ *be a dense subspace of* $\mathbf{E}$. *Then the map* $\mathbf{E}^* \to \mathbf{E}_0^*$ *given by*

$$\varphi \mapsto \varphi|_{\mathbf{E}_0}$$

is an isometric isomorphism of Banach spaces.

Proof. The map $S : \varphi \mapsto \varphi|_{\mathbf{E}_0}$ is clearly well-defined and linear. Furthermore, by Proposition 3.2.4, for any $\psi \in \mathbf{E}_0^*$ there exists a unique element $\varphi \in \mathbf{E}^*$ such that

$$\varphi|_{\mathbf{E}_0} = \psi.$$

Hence S is surjective. Also, since $\|\varphi\| = \|\psi\|$, it follows that S is isometric and thus injective.

$\square$

The next result may not seem like much at the moment. After all it is a fairly direct application of the Riesz Representation Theorem. What is remarkable is that this result is true for arbitrary normed linear spaces and that fact is one of the cornerstones of the subject.

3.2.6 COROLLARY *Let* $\mathbf{M} < \mathbf{H}$ *be a subspace of* $\mathbf{H}$ *and let* $\varphi : \mathbf{M} \to \mathbb{K}$ *be a bounded linear functional. Then there exists* $\psi \in \mathbf{H}^*$ *such that*

$$\psi|_{\mathbf{M}} = \varphi \text{ and } \|\psi\| = \|\varphi\|.$$

We say that ψ is a ***norm-preserving extension*** of φ. It may not be unique (see Exercise 3.17).

Proof. Let $\varphi \in \mathbf{M}^*$. Then by Corollary 3.2.5 there exists $\psi_0 : \overline{\mathbf{M}} \to \mathbb{K}$ linear such that $\psi_0|_{\mathbf{M}} = \varphi$ and $\|\psi_0\| = \|\varphi\|$. Since $\overline{\mathbf{M}} < \mathbf{H}$, it is a Hilbert space. By the Riesz Representation Theorem there exists $y \in \overline{\mathbf{M}}$ such that $\psi_0(x) = \langle x, y \rangle$ for all $x \in \overline{\mathbf{M}}$. Now simply define $\psi : \mathbf{H} \to \mathbb{K}$ by

$$\psi(x) := \langle x, y \rangle.$$

Then ψ is clearly an extension of ψ_0 and hence of φ. Furthermore, $\|\psi\| = \|y\| = \|\psi_0\| = \|\varphi\|$.

$\square$

Aside. There is *another* theorem that goes by the name of 'Riesz Representation Theorem'. That theorem is substantially deeper than this one and describes bounded linear functionals on the space $C_c(X)$ of continuous, compactly supported functions on a locally compact, Hausdorff space X. We will have reason to discuss this theorem later in the book (see Chapter 9). For now, though, do keep this in mind in case you wish to look the theorem up online.

3.3 Orthonormal Bases

At the heart of analytic geometry lies the principle of cartesian coordinates. Given a vector in Euclidean space, one can assign to it a tuple of numbers that uniquely determines the vector. This simple idea acts as an important visual aid, while also providing numerous computational advantages. One of the chief reasons to develop Hilbert space theory was to study function spaces using 'coordinates'. We describe this process in this section.

3.3.1 Definition Let $\mathbf{H}$ be a Hilbert space, and Λ be a subset of $\mathbf{H}$.

(i) Λ is said to be *orthogonal* if $x \perp y$ for all distinct $x, y \in \Lambda$.

(ii) Λ is said to be *orthonormal* if it is orthogonal and $\|x\| = 1$ for all $x \in \Lambda$.

(iii) A maximal orthonormal set is called an *orthonormal basis* of $\mathbf{H}$.

Warning: An orthonormal basis may not be a Hamel basis for $\mathbf{H}$ (see Exercise 3.24).

3.3.2 Lemma *Every orthonormal set is linearly independent.*

Proof. Let Λ be an orthonormal set and $\{x_1, x_2, \ldots, x_n\} \subset \Lambda$ satisfy $\sum_{i=1}^n \alpha_i x_i = 0$. Then for any fixed $j \in \{1, 2, \ldots, n\}$, $\alpha_j = \langle \sum_{i=1}^n \alpha_i x_i, x_j \rangle = 0$. Hence $\{x_1, x_2, \ldots, x_n\}$ is linearly independent. $\square$

3.3.3 Lemma *Let $\Lambda \subset \mathbf{H}$ be an orthonormal set. Then the following are equivalent:*

(i) Λ *is an orthonormal basis of* $\mathbf{H}$.

(ii) $\Lambda^\perp = \{0\}$.

(iii) $\mathrm{span}(\Lambda)$ *is dense in* $\mathbf{H}$.

Proof.   Let $\Lambda \subset \mathbf{H}$ be an orthonormal set.

(i) $\Rightarrow$ (ii): Suppose Λ is an orthonormal basis and $\Lambda^{\perp} \neq \{0\}$, then choose $x \in \Lambda^{\perp}$ such that $\|x\| = 1$. Then $\Lambda \cup \{x\}$ is an orthonormal set. This contradicts the maximality of Λ. Therefore it must happen that $\Lambda^{\perp} = \{0\}$.

(ii) $\Rightarrow$ (iii): Suppose $\Lambda^{\perp} = \{0\}$. Since $\mathrm{span}(\Lambda)^{\perp} \subset \Lambda^{\perp}$, it follows that $\mathrm{span}(\Lambda)^{\perp} = \{0\}$. By Corollary 3.1.14, $\mathrm{span}(\Lambda)$ is dense in $\mathbf{H}$.

(iii) $\Rightarrow$ (i): Suppose $\overline{\mathrm{span}(\Lambda)} = \mathbf{H}$, and let Λ' be an orthonormal set that contains Λ. We wish to prove that $\Lambda' = \Lambda$. Suppose not, then there is a vector $x \in \Lambda' \setminus \Lambda$. Since Λ' is orthonormal, $\|x\| = 1$, and $x \perp \Lambda$. This implies that $x \perp \mathrm{span}(\Lambda)$. By continuity of the inner product (Example 2.2.3), it follows that
$$x \perp \overline{\mathrm{span}(\Lambda)}.$$
Hence $x \perp \mathbf{H}$ and so $x = 0$. This contradicts the assumption that $\|x\| = 1$. We must conclude that $\Lambda' = \Lambda$. $\qquad \square$

The next result follows directly as in Theorem 1.1.6 by the use of Zorn's Lemma.

3.3.4 THEOREM *If $\Lambda_0 \subset \mathbf{H}$ is any orthonormal set, then there is an orthonormal basis $\Lambda \subset \mathbf{H}$ that contains Λ_0.*

3.3.5 EXAMPLE

(i) Let $\mathbf{H} = (\mathbb{K}^n, \|\cdot\|_2)$. Then the standard basis is an orthonormal basis.

(ii) Let $\mathbf{H} = \ell^2$. Then $\{e_1, e_2, \ldots\}$ is an orthonormal basis.

   Proof.   Clearly, $\Lambda = \{e_1, e_2, \ldots\}$ is orthonormal. Furthermore, if $x = (x_n) \perp \Lambda$, then $x_n = \langle x, e_n \rangle = 0$ for all n and hence $\Lambda^{\perp} = \{0\}$. $\qquad \square$

(iii) Let $\mathbf{H} = L^2[-\pi, \pi]$ and $\mathbb{K} = \mathbb{C}$. For $n \in \mathbb{Z}$, define
$$e_n(t) := \frac{1}{\sqrt{2\pi}} e^{int}.$$

Then
$$\langle e_n, e_m \rangle = \frac{1}{2\pi} \int_{-\pi}^{\pi} e^{i(n-m)t} dt = \begin{cases} 1 & : \text{if } n = m, \\ 0 & : \text{otherwise.} \end{cases}$$

We will prove later (in Section 3.5) that $\{e_n : n \in \mathbb{Z}\}$ forms an orthonormal basis.

The next theorem would be familiar to you from linear algebra. The orthogonalization process described below was used by Jørgen Gram in 1883, when he tried to resolve a problem of approximating a function by a finite linear combination of a given set of functions. The form of the theorem most familiar to us comes from Erhard Schmidt, who proved it in 1907 in the context of L^2 functions. However, the theorem was known to Laplace as early as 1820 and was probably known to other mathematicians at that time as well [36].

3.3.6 THEOREM (GRAM–SCHMIDT ORTHOGONALIZATION (GRAM, 1883, AND SCHMIDT, 1907)) *Let* $\{x_1, x_2, \ldots, x_n\} \subset \mathbf{H}$ *be linearly independent. Define* $\{u_1, u_2, \ldots, u_n\}$ *inductively by* $u_1 := x_1$ *and*

$$u_j = x_j - \sum_{i=1}^{j-1} \frac{\langle x_j, u_i \rangle}{\langle u_i, u_i \rangle} u_i \tag{3.1}$$

for $j \geq 2$. *Then* $\{u_1, u_2, \ldots, u_n\}$ *is an orthogonal set and* $\mathrm{span}(\{u_1, u_2, \ldots, u_n\}) = \mathrm{span}(\{x_1, x_2, \ldots, x_n\})$.

Proof. We proceed by induction on n since this is clearly true if $n = 1$. If $n > 1$, suppose $\{u_1, u_2, \ldots, u_{n-1}\}$ is an orthogonal set such that $\mathrm{span}(\{u_1, u_2, \ldots, u_{n-1}\}) = \mathrm{span}(\{x_1, x_2, \ldots, x_{n-1}\})$. Then if u_n is given by Equation 3.1, then clearly $u_n \in \mathrm{span}(\{x_1, x_2, \ldots, x_n\})$ and $u_n \neq 0$ since $\{x_1, x_2, \ldots, x_n\}$ is linearly independent. Also, $\langle u_n, u_j \rangle = 0$ for all $j < n$, so $\{u_1, u_2, \ldots, u_n\}$ is orthogonal. In particular, $\{u_1, u_2, \ldots, u_n\}$ is linearly independent (by Lemma 3.3.2). Hence

$$\mathrm{span}(\{u_1, u_2, \ldots, u_n\}) = \mathrm{span}(\{x_1, x_2, \ldots, x_n\})$$

since both spaces have the same dimension.   □

3.3.7 COROLLARY *If* $\mathbf{H}$ *is a Hilbert space and* $\{x_1, x_2, \ldots\}$ *is a linearly independent set, then there exists an orthonormal set* $\{e_1, e_2, \ldots\}$ *such that* $\mathrm{span}(\{e_1, e_2, \ldots, e_n\}) = \mathrm{span}(\{x_1, x_2, \ldots, x_n\})$ *for all* $n \in \mathbb{N}$.

3.3.8 EXAMPLE Let $\mathbf{H} = L^2[-1, 1]$ and $x_n(t) = t^n$. Then $\{x_0, x_1, \ldots\}$ is a linearly independent subset of $\mathbf{H}$. The Gram–Schmidt process gives us an orthogonal set

$\{P_0, P_1, \ldots\}$ which is given explicitly as

$$P_n(t) = \frac{1}{2^n n!} \left(\frac{d}{dt}\right)^n (t^2 - 1)^n.$$

These polynomials are called **Legendre polynomials**. Since $\mathrm{span}(\{P_n\}) = \mathrm{span}(\{x_n\})$, it follows by the Weierstrass Approximation Theorem and Proposition 2.3.12 that $\overline{\mathrm{span}}(\{P_1, P_2, \ldots\}) = \mathbf{H}$. Therefore if we divide by their norms, the set

$$\left\{ \sqrt{\frac{2n+1}{2}} P_n : n = 0, 1, \ldots \right\}$$

forms an orthonormal basis for $\mathbf{H}$ (see Exercise 3.27 for more on these polynomials).

3.3.9 PROPOSITION *Let* $\{e_1, e_2, \ldots, e_n\}$ *be an orthonormal set in* $\mathbf{H}$, *and let* $\mathbf{M} := \mathrm{span}\{e_1, e_2, \ldots, e_n\}$. *If* $P : \mathbf{H} \to \mathbf{M}$ *denotes the orthogonal projection onto* $\mathbf{M}$, *then*

$$P(x) = \sum_{k=1}^{n} \langle x, e_k \rangle e_k$$

for any $x \in \mathbf{H}$.

Proof. Let $x \in \mathbf{H}$ and set $x_0 := \sum_{k=1}^{n} \langle x, e_k \rangle e_k$. Then $x_0 \in \mathbf{M}$, and $\langle x - x_0, e_j \rangle = 0$ for any $1 \leq j \leq n$. Hence $x - x_0 \in \mathbf{M}^\perp$. Proposition 3.1.7 then implies that $P(x) = x_0\square$

The previous result would, of course, be familiar to you from the study of finite dimensional inner product spaces. The next theorem, however, is only of relevance in the infinite dimensional setting. This innocuous looking inequality is fundamental to the theory and is the basis of all the theorems that follow. It was originally proved for the trigonometric system (part (iii) of Example 3.3.5) by Bessel in 1828 and was proved for orthonormal systems in $L^2[0, 1]$ by Erhard Schmidt in his PhD dissertation of 1905.

3.3.10 THEOREM (BESSEL'S INEQUALITY) *If* $\{e_1, e_2, \ldots\}$ *is an orthonormal set and* $x \in \mathbf{H}$, *then* $\sum_{n=1}^{\infty} |\langle x, e_n \rangle|^2 \leq \|x\|^2$.

Proof. For each $n \in \mathbb{N}$, write $x_n := x - \sum_{i=1}^{n} \langle x, e_i \rangle e_i$. Then $x_n \perp e_i$ for all $1 \leq i \leq n$, so $x_n \perp \sum_{i=1}^{n} \langle x, e_i \rangle e_i$. Thus by Pythagoras' Theorem,

$$\|x\|^2 = \|x_n\|^2 + \left\| \sum_{i=1}^{n} \langle x, e_i \rangle e_i \right\|^2 = \|x_n\|^2 + \sum_{i=1}^{n} |\langle x, e_i \rangle|^2 \geq \sum_{i=1}^{n} |\langle x, e_i \rangle|^2.$$

This is true for all $n \in \mathbb{N}$, which implies the result. $\square$

If a series of complex numbers converge, then the nth term goes to zero as $n \to \infty$. This observation gives us an important consequence of Bessel's Inequality.

3.3.11 COROLLARY (RIEMANN–LEBESGUE LEMMA) *Let* $\{e_1, e_2, \ldots\}$ *be an orthonormal set, and* $x \in \mathbf{H}$. *Then* $\lim_{n \to \infty} \langle x, e_n \rangle = 0$.

The Riemann–Lebesgue Lemma is originally a statement about the Fourier series. Specifically, a version of it states that the Fourier coefficients of an L^1 function tend to zero. As we will see later, the term $\langle x, e_n \rangle$ may be thought of as the nth Fourier coefficient of x. Therefore the version of the Riemann–Lebesgue Lemma stated here proves the original version for L^2 functions. We will discuss all this later in the chapter and discuss the Fourier series of L^1 functions in Chapter 5.

3.3.12 COROLLARY *Let* Λ *be an orthonormal set in* $\mathbf{H}$ *and* $x \in \mathbf{H}$. *Then* $\{e \in \Lambda : \langle x, e \rangle \neq 0\}$ *is a countable set.*

Proof. For each $n \in \mathbb{N}$, define $\Lambda_n := \{e \in \Lambda : |\langle x, e \rangle| \geq 1/n\}$. If $\{e_1, e_2, \ldots, e_N\} \subset \Lambda_n$, then by Bessel's Inequality,

$$\frac{N}{n^2} = \sum_{k=1}^{N} \frac{1}{n^2} \leq \sum_{k=1}^{N} |\langle x, e_k \rangle|^2 \leq \|x\|^2.$$

Thus Λ_n must be finite and $\{e \in \Lambda : \langle x, e \rangle \neq 0\} = \bigcup_{n=1}^{\infty} \Lambda_n$ must be countable. □

3.3.13 REMARK Let $\{x_\alpha : \alpha \in J\}$ be a (possibly uncountable) set in an normed linear space $\mathbf{E}$. We want to make sense of an expression of the form

$$\sum_{\alpha \in J} x_\alpha. \tag{3.2}$$

If J is countable, we can do this by enumerating $J = \mathbb{N}$ and treating Equation 3.2 as a series in the usual sense. However, if J is uncountable, we need to make some changes. Define $\mathcal{F} := \{F \subset J : F \text{ is finite}\}$. For each $F \in \mathcal{F}$, we define a partial sum

$$s_F := \sum_{\alpha \in F} x_\alpha.$$

We say that the expression in Equation 3.2 exists if there is a vector $s \in \mathbf{E}$ with the property that for all $\epsilon > 0$ there exists a finite set $F_0 \in \mathcal{F}$ such that for any $F \in \mathcal{F}$ containing F_0, one has $\|s_F - s\| < \epsilon$.

In other words, $\mathcal{F}$ is a (directed) partially ordered set under inclusion and $\{s_F : F \in \mathcal{F}\}$ forms a *net*. Our requirement is that this net be convergent in $\mathbf{E}$. We will not have much reason to discuss nets in this book. The interested reader will find more on the subject in Munkres [42].

3.3.14 COROLLARY *Let* Λ *be an orthonormal set in* **H** *and* $x \in$ **H**. *Then* $\sum_{e \in \Lambda} |\langle x, e \rangle|^2 \le \|x\|^2$.

Proof. Clearly, the sum is the same as the sum over the set $\{e \in \Lambda : \langle x, e \rangle \neq 0\}$. By Corollary 3.3.12, this set is countable and so it reduces to a countable sum. The result now follows from Bessel's Inequality. $\qquad \square$

3.3.15 LEMMA *Let* Λ *be an orthonormal set in* **H**. *Then for any* $x \in$ **H**, *the series*

$$\sum_{e \in \Lambda} \langle x, e \rangle e$$

converges in **H**.

Proof. By Corollary 3.3.12, the set $\{e \in \Lambda : \langle x, e \rangle \neq 0\}$ is countable. Denote this set by $\{e_n : n \in \mathbb{N}\}$. Then

$$\sum_{e \in \Lambda} \langle x, e \rangle e = \sum_{n=1}^{\infty} \langle x, e_n \rangle e_n.$$

Define $x_n := \sum_{i=1}^{n} \langle x, e_i \rangle e_i$ and it now suffices to prove that (x_n) is Cauchy. By Bessel's Inequality, $\sum_{i=1}^{\infty} |\langle x, e_i \rangle|^2 \le \|x\|^2 < \infty$. Hence if $\epsilon > 0$, there exists $N_0 \in \mathbb{N}$ such that $\sum_{i=N_0}^{\infty} |\langle x, e_i \rangle|^2 < \epsilon$. If $n > m \ge N_0$, Pythagoras' Theorem tells us that

$$\|x_n - x_m\|^2 = \left\| \sum_{i=m+1}^{n} \langle x, e_i \rangle e_i \right\|^2 = \sum_{i=m+1}^{n} |\langle x, e_i \rangle|^2 \le \sum_{i=N_0}^{\infty} |\langle x, e_i \rangle|^2 < \epsilon.$$

Hence (x_n) is Cauchy and converges in **H**. $\qquad \square$

The next result is the main theorem of this section. What it says is the following: Given an orthonormal basis Λ of a Hilbert space **H**, any vector $x \in$ **H** may be assigned coordinates

$$x \mapsto (\langle x, e \rangle)_{e \in \Lambda}.$$

Furthermore, the vector (and its norm) may be recovered from this tuple of scalars. These ideas originated in the study of trigonometric series and thus bear the names of Fourier and Parseval.

3.3.16 THEOREM *Let Λ be an orthonormal basis of $\mathbf{H}$. Then for each $x, y \in \mathbf{H}$, we have*

(i) *Fourier Expansion:* $x = \sum_{e \in \Lambda} \langle x, e \rangle e$.

(ii) $\langle x, y \rangle = \sum_{e \in \Lambda} \langle x, e \rangle \overline{\langle y, e \rangle}$.

(iii) *Parseval's identity:* $\|x\|^2 = \sum_{e \in \Lambda} |\langle x, e \rangle|^2$.

Proof.   Let $x, y \in \mathbf{H}$ be fixed. Then by Corollary 3.3.12, the set $\{e \in \Lambda : \langle x, e \rangle \neq 0\} \cup \{e \in \Lambda : \langle y, e \rangle \neq 0\}$ is countable. Replacing Λ by this set, we may assume that $\Lambda = \{e_n : n \in \mathbb{N}\}$ is countable.

(i) Write $z := \sum_{n=1}^{\infty} \langle x, e_n \rangle e_n$, which exists by Lemma 3.3.15. By continuity and linearity of the inner product, it follows that $\langle z, e_j \rangle = \langle x, e_j \rangle$ for any $j \in \mathbb{N}$. Hence $x - z \in \Lambda^{\perp}$. Since $\Lambda^{\perp} = \{0\}$, $x = z$.

(ii) By part (i), write $x = \sum_{n=1}^{\infty} \langle x, e_n \rangle e_n$ and $y = \sum_{m=1}^{\infty} \langle y, e_m \rangle e_m$. Then by the continuity of the inner product in both variables, we see that

$$\langle x, y \rangle = \sum_{n,m=1}^{\infty} \langle \langle x, e_n \rangle e_n, \langle y, e_m \rangle e_m \rangle .$$

Since $e_n \perp e_m$ if $n \neq m$, we get $\langle x, y \rangle = \sum_{n=1}^{\infty} \langle x, e_n \rangle \overline{\langle y, e_n \rangle}$.

(iii) Follows directly from part (ii).   □

3.4   Isomorphisms of Hilbert Spaces

Before we apply the results of the earlier section to the study of the function space L^2, we need to describe the Hilbert space that will be the target of the Fourier transform. Then we will describe the rather self-evident notion of an isomorphism between Hilbert spaces.

3.4.1 LEMMA   *Any two orthonormal bases of a Hilbert space have the same cardinality.*

This common number is called the ***dimension*** of the Hilbert space.

Proof.   If $\mathbf{H}$ is finite dimensional, then this statement is familiar from linear algebra (see [24, Section I.8], for instance). Therefore we will assume $\mathbf{H}$ is infinite dimensional. Let Λ_1 and Λ_2 be two orthonormal bases of $\mathbf{H}$. If $f \in \Lambda_1$, then there exists $e \in \Lambda_2$ such that $\langle f, e \rangle \neq 0$. Furthermore, $\Lambda_e := \{f \in \Lambda_1 : \langle f, e \rangle \neq 0\}$ is a countable set by Corollary 3.3.12. Since

$$\Lambda_1 \subset \bigcup_{e \in \Lambda_2} \Lambda_e,$$

we conclude that $|\Lambda_1| \leq |\Lambda_2 \times \mathbb{N}| = |\Lambda_2|$. By symmetry, $|\Lambda_2| \leq |\Lambda_1|$ holds as well.

$\square$

There is a clash of terminology here that you need to be aware of. The Hilbert space dimension of $\mathbf{H}$ is *not necessarily* the same as the dimension of $\mathbf{H}$ as a vector space. For instance, $\dim(\ell^2) = \aleph_0$ (countable infinity), but its vector space dimension is uncountable. Proceed with caution!

3.4.2 DEFINITION Let I be any set, and $1 \leq p < \infty$

(i) A function $f : I \to \mathbb{K}$ is said to be *p-summable* if $\operatorname{supp}(f) := \{i \in I : f(i) \neq 0\}$ is countable and $\sum_{i \in I} |f(i)|^p < \infty$ (where the series is defined in the sense of Remark 3.3.13).

(ii) Let $\ell^p(I)$ denote the set of all p-summable functions on I. Then the inequality

$$|f + g|^p \leq [2\max\{|f|, |g|\}]^p \leq 2^p[|f|^p + |g|^p]$$

shows that $\ell^p(I)$ is a vector space. If $f \in \ell^p(I)$, we define

$$\|f\|_p := \left(\sum_{i \in I} |f(i)|^p \right)^{1/p},$$

and this satisfies Minkowski's Inequality since the verification only requires a countable sum. Hence $\ell^p(I)$ is a normed linear space. Furthermore, $\ell^p(I)$ is a Banach space as before (Exercise 3.35).

(iii) Also, $\ell^2(I)$ has an inner product given by

$$\langle f, g \rangle = \sum_{i \in I} f(i)\overline{g(i)}.$$

Once again this is well-defined since $\operatorname{supp}(f) \cup \operatorname{supp}(g)$ is countable. Hence $\ell^2(I)$ is a Hilbert space.

As mentioned before, these spaces $\ell^p(I)$ are themselves special cases of L^p spaces. You simply take the set I, equip it with the counting measure and voila!

3.4.3 LEMMA *Let I be any set. Then* $\dim(\ell^2(I)) = |I|$.

Proof. For each $i \in I$, define $e_i : I \to \mathbb{K}$ by

$$e_i(j) = \delta_{i,j} = \begin{cases} 1 & : \text{if } i = j, \\ 0 & : \text{if } i \neq j. \end{cases}$$

Then the set $\Lambda := \{e_i : i \in I\}$ forms an orthonormal set in $\ell^2(I)$. If $f \in \ell^2(I)$ satisfies $f \perp \Lambda$, then for any $i \in I$,

$$f(i) = \langle f, e_i \rangle = 0$$

and so $f \equiv 0$. Hence $\Lambda^\perp = \{0\}$ and we conclude that Λ is an orthonormal basis for $\ell^2(I)$. In particular, $\dim(\ell^2(I)) = |\Lambda| = |I|$. $\square$

The basis Λ constructed as above is called the ***standard orthonormal basis*** for $\ell^2(I)$. We now examine the notion of isomorphisms of Hilbert spaces. Obviously, any isomorphism must be an isometry. Let us isolate one useful fact about isometries which we will need later on, the proof of which we will leave to the exercises.

3.4.4 LEMMA   *Let* **H** *and* **K** *be Hilbert spaces. A linear map* $T : \mathbf{H} \to \mathbf{K}$ *is an isometry if and only if* $\langle Tx, Ty \rangle_{\mathbf{K}} = \langle x, y \rangle_{\mathbf{H}}$ *for all* $x, y \in \mathbf{H}$.

A simple example of an isometry that is not an isomorphism is the ***right shift operator*** $S : \ell^2 \to \ell^2$ given by

$$S(x_1, x_2, x_3, \ldots) := (0, x_1, x_2, x_3, \ldots).$$

Clearly, the only thing that prevents an isometry from being an isomorphism is surjectivity.

3.4.5 DEFINITION   An operator $U : \mathbf{H} \to \mathbf{K}$ is called a ***unitary*** operator if U is a surjective isometry. **H** is said to be isomorphic to **K** if there is a unitary map $U : \mathbf{H} \to \mathbf{K}$. If such a map exists, we write $\mathbf{H} \cong \mathbf{K}$.

Clearly, If U is a unitary, then U^{-1} is also a unitary. Also, if U and V are unitaries (with appropriate range and domain), then UV is a unitary. Therefore the notion of isomorphism is indeed an equivalence relation on the collection of Hilbert spaces.

The proof of the next result is more important than its statement. The isomorphism described below is the 'Fourier transform' afforded by a fixed orthonormal basis of the Hilbert space.

3.4.6 THEOREM   *Let* **H** *be a Hilbert space and* Λ *be an orthonormal basis of* **H**. *Then* $\mathbf{H} \cong \ell^2(\Lambda)$.

Proof. For any $x \in \mathbf{H}$, define $\hat{x} : \Lambda \to \mathbb{K}$ by

$$\hat{x}(e) := \langle x, e \rangle.$$

By Corollary 3.3.12, supp$(\hat{x})$ is countable and by Bessel's Inequality, $\hat{x} \in \ell^2(\Lambda)$. Thus we define $U : \mathbf{H} \to \ell^2(\Lambda)$ given by $x \mapsto \hat{x}$. Note that U is linear by the axioms of the inner product. Furthermore, by Theorem 3.3.16, we see that

$$\langle U(x), U(y) \rangle = \sum_{e \in \Lambda} \langle x, e \rangle \overline{\langle y, e \rangle} = \langle x, y \rangle$$

and so U is an isometry by Lemma 3.4.4. Finally, for each $f \in \Lambda$, $\hat{f}(e) = \langle f, e \rangle = \delta_{e,f}$. Therefore $\{U(f) : f \in \Lambda\}$ is the standard orthonormal basis for $\ell^2(\Lambda)$ and hence Range$(U)^{\perp} = \{0\}$. By Corollary 3.1.14, Range(U) is dense in $\ell^2(\Lambda)$. However, U is an isometry and so Range(U) is a complete subspace of $\ell^2(\Lambda)$. Thus Range(U) is closed and U is surjective. $\square$

The next (rather surprising) fact makes the study of Hilbert spaces line up particularly well with what we know of finite dimensional vector spaces. In particular, it implies that separable infinite dimensional Hilbert spaces are all isomorphic to one another!

3.4.7 COROLLARY *Two Hilbert spaces are isomorphic if and only if they have the same dimension.*

Proof. Suppose $U : \mathbf{H} \to \mathbf{K}$ is an isomorphism and $\Lambda_{\mathbf{H}} \subset \mathbf{H}$ is any orthonormal basis for $\mathbf{H}$, then $U(\Lambda_{\mathbf{H}}) \subset \mathbf{K}$ is an orthonormal set. Hence by Theorem 3.3.4, there is an orthonormal basis $\Lambda_{\mathbf{K}}$ of $\mathbf{K}$ such that $U(\Lambda_{\mathbf{H}}) \subset \Lambda_{\mathbf{K}}$. In particular,

$$\dim(\mathbf{H}) = |\Lambda_{\mathbf{H}}| = |U(\Lambda_{\mathbf{H}})| \leq |\Lambda_{\mathbf{K}}| = \dim(\mathbf{K}).$$

By symmetry, $\dim(\mathbf{K}) \leq \dim(\mathbf{H})$ as well.

Conversely, suppose $\dim(\mathbf{H}) = \dim(\mathbf{K})$, let $\Lambda_{\mathbf{H}}$ and $\Lambda_{\mathbf{K}}$ denote two orthonormal bases of $\mathbf{H}$ and $\mathbf{K}$, respectively. Then $|\Lambda_{\mathbf{H}}| = |\Lambda_{\mathbf{K}}|$ and hence there is a natural isomorphism $\ell^2(\Lambda_{\mathbf{H}}) \overset{\cong}{\to} \ell^2(\Lambda_{\mathbf{K}})$. Now apply Theorem 3.4.6 to conclude that $\mathbf{H} \cong \mathbf{K}$. $\square$

3.4.8 THEOREM *A Hilbert space $\mathbf{H}$ is separable if and only if $\dim(\mathbf{H})$ is countable.*

Proof. If $\mathbf{H}$ is separable and $\Lambda \subset \mathbf{H}$ is an orthonormal set, then for any pair of distinct vectors $e, f \in \Lambda$, we have

$$\|e - f\| = \sqrt{2}.$$

Hence the balls $\{B(e; \sqrt{2}/2) : e \in \Lambda\}$ form a family of disjoint open sets in $\mathbf{H}$. Since $\mathbf{H}$ is separable, this family must be countable and hence Λ must be countable.

Conversely, if **H** has a countable orthonormal basis $\Lambda := \{e_n : n \in \mathbb{N}\}$, then span($\Lambda$) is dense in **H**. Now set $\mathbb{K}_0 = \mathbb{Q}$ or $\mathbb{K}_0 = \mathbb{Q} \times \mathbb{Q}$, according as $\mathbb{K} = \mathbb{R}$ or $\mathbb{K} = \mathbb{C}$. Then $A := \text{span}_{\mathbb{K}_0}(\Lambda)$ is countable and $\overline{A} = \mathbf{H}$. Therefore **H** is separable. $\square$

3.4.9 COROLLARY *Any two separable, infinite dimensional Hilbert spaces are isomorphic.*

In particular, there is an isomorphism

$$L^2[-\pi, \pi] \to \ell^2(\mathbb{Z}).$$

Our goal in the next section is to the describe one such isomorphism explicitly.

3.5   Fourier Series of L^2 Functions

In what follows, we will perforce be working with complex-valued functions. In other words, we will tacitly assume that $\mathbb{K} = \mathbb{C}$ throughout this section. Back in Example 3.3.5, we had defined an orthonormal set $\{e_n : n \in \mathbb{Z}\}$ in $L^2[-\pi, \pi]$ by

$$e_n(t) = \frac{1}{\sqrt{2\pi}} e^{int}.$$

Now we show that this forms an orthonormal basis for $L^2[-\pi, \pi]$. The proof is relatively straightforward: We show that the span of this set is dense in $C[-\pi, \pi]$ and then appeal to the fact that $C[-\pi, \pi]$ is dense in $L^2[-\pi, \pi]$.

For the first step of the proof, we need the following fact. It is a consequence of the Stone–Weierstrass Theorem, whose proof we have described in the appendix (Section A.1). If you have not seen a proof of this theorem, do go ahead and take a look at that section. The Stone–Weierstrass Theorem is an important one in modern analysis and will play a crucial role for us in this book.

3.5.1 PROPOSITION *Let $S^1 := \{z \in \mathbb{C} : |z| = 1\}$ and let $C(S^1)$ be equipped with the supremum norm. Let **F** denote the space of all polynomials in z and $\overline{z}$, thought of as a subspace of $C(S^1)$. Then **F** is dense in $C(S^1)$.*

Fix $\mathbf{E} := \{f \in C[-\pi, \pi] : f(-\pi) = f(\pi)\}$. For $n \in \mathbb{Z}$, define $e_n \in \mathbf{E}$ as above and set $\mathcal{A} := \text{span}(\{e_n\}) \subset \mathbf{E}$. Note that

$$\mathcal{A} = \text{span}(\{\cos(nt), \sin(nt) : n \in \mathbb{Z}\}).$$

Hence an element of $\mathcal{A}$ is called a ***trigonometric polynomial***.

Now if $f \in C[-\pi, \pi]$ and $1 \le p < \infty$, then $\|f\|_p \le (2\pi)^{1/p}\|f\|_\infty$. Hence if a subset of $C[-\pi, \pi]$ is dense with respect to the supremum norm, then it is dense with respect to any p-norms. We wish to prove that the set $\mathcal{A}$ defined above is dense in $C[-\pi, \pi]$ with respect to $\|\cdot\|_2$. In order to do this, we need one fact about S^1. It is the quotient space of the interval $[-\pi, \pi]$ via the identification $-\pi \sim \pi$. Specifically, the map $q : [-\pi, \pi] \to S^1$ given by

$$t \mapsto e^{it}$$

is the quotient map. Therefore a function $F : S^1 \to \mathbb{C}$ is continuous if and only if $F \circ q : [-\pi, \pi] \to \mathbb{C}$ is continuous. We use this fact below.

3.5.2 PROPOSITION *$\mathcal{A}$ is dense in $\mathbf{E}$ with respect to the supremum norm. Therefore it is dense with respect to $\|\cdot\|_p$ for all $1 \le p \le \infty$.*

Proof. Define $\theta : C(S^1) \to \mathbf{E}$ by

$$\theta(f)(t) := f(e^{it}).$$

Then θ is linear and since the function $q : [-\pi, \pi] \to S^1$ is surjective, it follows that $\|\theta(f)\|_\infty = \|f\|_\infty$. Therefore θ is isometric and hence injective.

Given $f \in \mathbf{E}$, define $F : S^1 \to \mathbb{C}$ by $F(e^{it}) := f(t)$. Since $f(-\pi) = f(\pi)$, F is well-defined and continuous (by the discussion preceding this proof). Hence $F \in C(S^1)$ and $\theta(F) = f$. Therefore θ is an isometric isomorphism.

If $\zeta \in C(S^1)$ denotes the identity function $\zeta(z) = z$, we have $\theta(\zeta)(t) = e^{it}$ and $\theta(\overline{\zeta})(t) = e^{-it}$. Therefore if $\mathbf{F}$ is the subspace of $C(S^1)$ defined in Proposition 3.5.1, then $\theta(\mathbf{F}) = \mathcal{A}$. Since $\mathbf{F}$ is dense in $C(S^1)$, $\mathcal{A}$ is dense in $\mathbf{E}$. $\qquad\square$

3.5.3 LEMMA *$\mathbf{E}$ is dense in $L^p[-\pi, \pi]$ with respect to $\|\cdot\|_p$ for all $1 \le p < \infty$.*

Proof. For $n \in \mathbb{N}$, define

$$f_n(t) = \begin{cases} 1 & : \text{if } -\pi + 1/n \le t \le \pi - 1/n, \\ 0 & : \text{if } t \in \{-\pi, \pi\}, \\ \text{linear} & : \text{otherwise.} \end{cases}$$

Then $\|f_n - 1\|_p \le \frac{2}{n}$. For any $g \in C[-\pi, \pi]$, note that $f_n g \in \mathbf{E}$ and $\|f_n g - g\|_p \le \frac{2}{n}\|g\|_\infty \to 0$. Hence $C[-\pi, \pi] \subset \overline{\mathbf{E}}^{\|\cdot\|_p}$. Now the result follows from the fact that $C[-\pi, \pi]$ is dense in $L^p[-\pi, \pi]$ (Proposition 2.3.12). $\qquad\square$

3.5.4 Theorem *If*

$$e_n(t) := \frac{1}{\sqrt{2\pi}} e^{int},$$

then the set $\{e_n : n \in \mathbb{Z}\}$ *forms an orthonormal basis of* $L^2[-\pi, \pi]$.

Proof. By Example 3.3.5, we know that it is an orthonormal set. By Proposition 3.5.2 and Lemma 3.5.3, $\overline{\text{span}(\{e_n\})} = L^2[-\pi, \pi]$. By Lemma 3.3.3, it is an orthonormal basis. $\square$

This result is, of course, what we have been after this whole time. And now that we have an orthonormal basis for $L^2[-\pi, \pi]$, we are free to apply the results of Section 3.3. In what follows, it is customary to normalize the Lebesgue measure on $[-\pi, \pi]$ so that the interval has measure 1. This is done by simply dividing the measure by 2π. In that case, we set $e_n(t) := e^{int}$ and observe that $\{e_n : n \in \mathbb{Z}\}$ is now an orthonormal basis for $L^2[-\pi, \pi]$ (with this normalized measure).

3.5.5 Theorem *Let* $f \in L^2[-\pi, \pi]$. *For* $n \in \mathbb{Z}$, *the* n^{th} *Fourier coefficient of* f *is defined as*

$$\widehat{f}(n) := \langle f, e_n \rangle = \frac{1}{2\pi} \int_{-\pi}^{\pi} f(t) e^{-int} dt.$$

Then we have

(i) *Fourier Expansion:* $f(t) \sim \sum_{n=-\infty}^{\infty} \widehat{f}(n) e^{int}$.

   Note: It is traditional to use the symbol $\sim$ *here to indicate that the convergence is in the* L^2 *norm but not necessarily pointwise.*

(ii) *Fourier Series: The map* $U : L^2[-\pi, \pi] \to \ell^2(\mathbb{Z})$ *given by* $f \mapsto \widehat{f}$ *is an isomorphism of Hilbert spaces.*

(iii) *Parseval's identity:* $\|f\|_2^2 = \sum_{n=-\infty}^{\infty} |\widehat{f}(n)|^2$.

(iv) *Riemann–Lebesgue Lemma:*

$$\lim_{n \to \pm\infty} \int_{-\pi}^{\pi} f(t) e^{int} dt = \lim_{n \to \pm\infty} \int_{-\pi}^{\pi} f(t) \cos(nt) dt = \lim_{n \to \pm\infty} \int_{-\pi}^{\pi} f(t) \sin(nt) dt = 0.$$

(v) *Riesz–Fischer Theorem: For any* $(c_n) \in \ell^2(\mathbb{Z})$ *there exists* $f \in L^2[-\pi, \pi]$ *such that* $\widehat{f}(n) = c_n$ *for all* $n \in \mathbb{Z}$.

This concludes our discussion on Hilbert spaces and the Fourier series for the moment. We will return to these ideas repeatedly as we go along. For now it is time to consolidate our understanding with a few problems.

3.6 Exercises

Preliminaries

Exercise 3.1 Let $(\mathbf{E}, \langle \cdot, \cdot \rangle)$ be an inner product space and let $T : \mathbf{E} \to \mathbf{E}$ be an injective, linear map. Define a bilinear map $(\cdot, \cdot) : \mathbf{E} \times \mathbf{E} \to \mathbb{K}$ by $(x, y) := \langle T(x), T(y) \rangle$.

(i) Show that $(\cdot, \cdot)$ is an inner product on $\mathbf{E}$.

(ii) Let $a, b, c \in \mathbb{R}$ with $b^2 \leq 4ac$ and consider the ellipse $\Gamma := \{(x, y) \in \mathbb{R}^2 : ax^2 + bxy + cy^2 = 1\}$. Prove that there is an inner product $(\cdot, \cdot)$ on $\mathbb{R}^2$ such that $\Gamma = \{v \in \mathbb{R}^2 : (v, v) = 1\}$.

Exercise 3.2 For $n \in \mathbb{N}$, let $\mathcal{P}_n$ denote the space of all polynomials in one variable over $\mathbb{K}$ with degree atmost n. Let $A := \{x_1, x_2, \ldots, x_k\}$ be a finite subset of $\mathbb{K}$ and define a bilinear map $\langle \cdot, \cdot \rangle_A : \mathcal{P}_n \times \mathcal{P}_n \to \mathbb{K}$ by

$$\langle f, g \rangle_A := \sum_{i=1}^{k} f(x_i)\overline{g(x_i)}.$$

Determine conditions on the set A under which this an inner product on $\mathcal{P}_n$.

3.6.1 Definition A ***semi-inner product*** on a vector space $\mathbf{E}$ is a function

$$(\cdot, \cdot) : \mathbf{E} \times \mathbf{E} \to \mathbb{K}$$

such that for all $x, y, z \in \mathbf{E}, \alpha, \beta \in \mathbb{K}$,

(i) $(\alpha x + \beta y, z) = \alpha(x, z) + \beta(y, z)$,

(ii) $(x, y) = \overline{(y, x)}$,

(iii) $(x, x) \geq 0$.

Note: It is possible that $(x, x) = 0$ for non-zero $x \in \mathbf{E}$.

Exercise 3.3 (⚒) Let $(\cdot, \cdot)$ be a semi-inner product on a vector space $\mathbf{E}$.

(i) If $x \in \mathbf{E}$ is such that $(x, x) = 0$, then prove that $(x, y) = 0$ for all $y \in \mathbf{E}$.

Hint: Consider $z := t(y, x)x - y$, where $t \in \mathbb{R}$ is arbitrary.

(ii) Prove the Cauchy–Schwarz Inequality: For any $x, y \in \mathbf{E}$, prove that

$$|(x, y)|^2 \leq (x, x)(y, y).$$

(iii) Let $p : \mathbf{E} \to \mathbb{R}_+$ denote the function $p(x) := \sqrt{(x,x)}$. Prove that p is a seminorm on $\mathbf{E}$.

(iv) Let $\mathbf{N} := \{x \in \mathbf{E} : p(x) = 0\}$ and $\mathbf{H} := \mathbf{E}/\mathbf{N}$. Prove that the map $\langle \cdot, \cdot \rangle : \mathbf{H} \times \mathbf{H} \to \mathbb{K}$ given by

$$\langle x + \mathbf{N}, y + \mathbf{N} \rangle := (x, y)$$

is a well-defined inner product on $\mathbf{H}$.

Orthogonality

EXERCISE 3.4 (⚒) Prove Proposition 3.1.3.

EXERCISE 3.5 (⚒) Assume that $(\mathbf{E}, \|\cdot\|)$ is a a Banach space such that the norm satisfies the parallelogram law. If $\mathbf{E}$ is a real Banach space, define

$$\langle x, y \rangle := \frac{1}{4}(\|x+y\|^2 - \|x-y\|^2).$$

If $\mathbf{E}$ is a complex Banach space, define

$$\langle x, y \rangle := \frac{1}{4}(\|x+y\|^2 - \|x-y\|^2 + i\|x+iy\|^2 - i\|x-iy\|^2).$$

In either case, show that this formula defines an inner product on $\mathbf{E}$ which induces the given norm.

EXERCISE 3.6 Let $\mathbf{H}$ be a Hilbert space and $A \subset \mathbf{H}$ be any set.

(i) Prove that $(A^\perp)^\perp = \overline{\text{span}(A)}$.

(ii) If $\mathbf{M}_1$ and $\mathbf{M}_2$ are two closed subspaces of $\mathbf{H}$, then show that $(\mathbf{M}_1 \cap \mathbf{M}_2)^\perp = \overline{\mathbf{M}_1^\perp + \mathbf{M}_2^\perp}$.

3.6.2 DEFINITION Let $\{x_1, x_2, \ldots, x_n\}$ be a finite set of vectors in an inner product space $\mathbf{H}$. The ***Gram determinant*** of these vectors is defined as the determinant

$$G(x_1, x_2, \ldots, x_n) = \begin{vmatrix} \langle x_1, x_1 \rangle & \langle x_1, x_2 \rangle & \ldots & \langle x_1, x_n \rangle \\ \langle x_2, x_1 \rangle & \langle x_2, x_2 \rangle & \ldots & \langle x_2, x_n \rangle \\ \vdots & \vdots & \ddots & \vdots \\ \langle x_n, x_1 \rangle & \langle x_n, x_2 \rangle & \ldots & \langle x_n, x_n \rangle \end{vmatrix}.$$

EXERCISE 3.7 For $1 \leq k \leq n$, define $\mathbf{H}_k := \text{span}\{x_1, x_2, \ldots, x_k\}$.

(i) Prove that $G(x_1, x_2, \ldots, x_{n-1}) d(x_n, \mathbf{H}_{n-1})^2 = G(x_1, x_2, \ldots, x_n)$.

Hint: Let $z = \sum_{i=1}^{n-1} \alpha_i x_i$ denote the orthogonal projection of x_n onto $\mathbf{H}_{n-1}$. From the last row of this matrix, subtract the first $(n-1)$ rows multiplied by $\alpha_1, \alpha_2, \ldots, \alpha_{n-1}$, respectively. This does not change the determinant. Now do something similar with the columns.

(ii) Prove that $G(x_1, x_2, \ldots, x_n) \geq 0$ and that equality holds if and only if the vectors are linearly dependent.

(iii) If $\{x_1, x_2, \ldots, x_n\}$ are linearly independent, then prove that

$$d(x_n, \mathbf{H}_{n-1}) = \sqrt{\frac{G(x_1, x_2, \ldots, x_n)}{G(x_1, x_2, \ldots, x_{n-1})}}.$$

The quantity $\sqrt{G(x_1, x_2, \ldots, x_n)}$ represents the volume of the parallelotope spanned by the vectors $\{x_1, x_2, \ldots, x_n\}$.

EXERCISE 3.8 (✖) Let $\mathbf{M}_1, \mathbf{M}_2 < \mathbf{H}$ be two closed subspaces such that $\mathbf{M}_1 \perp \mathbf{M}_2$, and let P_1 and P_2 denote the orthogonal projections onto $\mathbf{M}_1$ and $\mathbf{M}_2$, respectively. Prove that

(i) $\mathbf{M}_1 + \mathbf{M}_2$ is a closed subspace of $\mathbf{H}$.

(ii) $P_1 P_2 = P_2 P_1 = 0$.

(iii) $P_1 + P_2$ is the orthogonal projection onto $\mathbf{M}_1 + \mathbf{M}_2$.

EXERCISE 3.9 Let $\mathbf{H}$ be a Hilbert space over $\mathbb{R}$. Show that there is a Hilbert space $\mathbf{K}$ over $\mathbb{C}$ and an $\mathbb{R}$-linear map $T : \mathbf{H} \to \mathbf{K}$ satisfying the following properties:

(a) $\langle T(x), T(y) \rangle = \langle x, y \rangle$ for all $x, y \in \mathbf{H}$.

(b) For any $z \in \mathbf{K}$ there exist unique $x, y \in \mathbf{H}$ such that $z = T(x) + iT(y)$.

(c) If $\widetilde{\mathbf{K}}$ is a complex Hilbert space and $\widetilde{T} : \mathbf{H} \to \widetilde{\mathbf{K}}$ is an $\mathbb{R}$-linear map satisfying conditions as in parts (a) and (b), then there is a $\mathbb{C}$-linear unitary $U : \mathbf{K} \to \widetilde{\mathbf{K}}$ such that $U \circ T = \widetilde{T}$.

This proves that the pair $(\mathbf{K}, T)$ is unique upto isomorphism. The Hilbert space $\mathbf{K}$ is called the ***complexification*** of $\mathbf{H}$.

Hint: Take $\mathbf{K} = \mathbf{H} \times \mathbf{H}$ (as a set) and define addition and scalar multiplication over $\mathbb{C}$ appropriately.

The next two problems are meant to indicate that the Best Approximation Theorem (Theorem 3.1.6) can fail quite miserably in arbitrary Banach spaces.

EXERCISE 3.10 Let $\mathbf{E} := C[0,1]$, equipped with the supremum norm, and set

$$A := \left\{ f \in \mathbf{E} : \int_0^{1/2} f(t)dt - \int_{1/2}^1 f(t)dt = 1 \right\}.$$

(i) Prove that A is a closed, convex subset of $\mathbf{E}$.

(ii) Prove that $\inf\{\|f\| : f \in A\} = 1$.

(iii) Prove that there does not exist any $f \in A$ such that $\|f\| = 1$.

EXERCISE 3.11 Let A and B be the subsets of ℓ^1 given by

$$A := \left\{ (x_n) \in \ell^1 : \sum_{i=1}^{\infty} x_i = 1 \right\} \text{ and}$$

$$B := \left\{ (x_n) \in \ell^1 : \sum_{i=1}^{\infty} \frac{i}{i+1} x_i = 1 \right\}.$$

(i) Prove that A and B are both closed, convex subsets of ℓ^1.

(ii) Prove that $\inf\{\|x\|_1 : x \in A\} = \inf\{\|x\|_1 : x \in B\} = 1$.

(iii) Prove that there exist infinitely many vectors in A with norm 1.

(iv) Prove that there does not exist any vector $x \in B$ with $\|x\|_1 = 1$.

EXERCISE 3.12 (✖) For each $n \geq 1$, let $(\mathbf{H}_n, \langle \cdot, \cdot \rangle_{\mathbf{H}_n})$ be a Hilbert space. Set

$$\mathbf{H} := \left\{ (x_n) : x_i \in \mathbf{H}_i \text{ for each } i \geq 1, \text{ and } \sum_{i=1}^{\infty} \|x_i\|_{\mathbf{H}_i}^2 < \infty \right\}.$$

Define addition and scalar multiplication pointwise. Define an inner product on $\mathbf{H}$ by

$$\langle (x_n), (y_n) \rangle := \sum_{i=1}^{\infty} \langle x_i, y_i \rangle_{\mathbf{H}_i}.$$

Show that this inner product is well-defined and that $(\mathbf{H}, \langle \cdot, \cdot \rangle)$ is a Hilbert space. This Hilbert space is called the ***direct sum*** of the $\{\mathbf{H}_n : n \geq 1\}$ and is denoted by $\bigoplus_{n=1}^{\infty} \mathbf{H}_n$.

EXERCISE 3.13 (⚒) Let $(\mathbf{E}, \langle\cdot,\cdot\rangle_{\mathbf{E}})$ be an inner product space and let $\widehat{\mathbf{E}}$ denote the completion of $\mathbf{E}$ as a normed linear space (following the notation of Exercise 2.29).

(i) Define a bilinear map on $\widehat{\mathbf{E}}$ by

$$\langle (x_n) + \mathbf{N}, (y_n) + \mathbf{N} \rangle := \lim_{i\to\infty} \langle x_i, y_i \rangle_{\mathbf{E}}.$$

Show that this limit exists and defines an inner product on $\widehat{\mathbf{E}}$.

(ii) Show that $\widehat{\mathbf{E}}$, equipped with this inner product, is a Hilbert space.

(iii) Define $T : \mathbf{E} \to \widehat{\mathbf{E}}$ by

$$T(x) = (x, x, x, \ldots) + \mathbf{N}.$$

Show that T is an isometry whose range is dense in $\widehat{\mathbf{E}}$.

(iv) Suppose $\mathbf{K}$ is a Hilbert space and $S : \mathbf{E} \to \mathbf{K}$ is an isometry with dense range, then show that there is a unique unitary map $U : \widehat{\mathbf{E}} \to \mathbf{K}$ such that $U \circ T = S$. This proves that the pair $(\widehat{\mathbf{E}}, T)$ is unique upto isomorphism.

As in Exercise 2.29, the space $\widehat{\mathbf{E}}$ is called the Hilbert space ***completion*** of $\mathbf{E}$.

EXERCISE 3.14 (⚒) For each $n \geq 1$, let $(\mathbf{H}_n, \langle\cdot,\cdot\rangle_{\mathbf{H}_n})$ be a Hilbert space. Set

$$\mathbf{E} := \{(x_n) : x_i \in \mathbf{H}_i \text{ for each } i \geq 1, \text{ and there is } N \in \mathbb{N} \text{ such that } x_i = 0 \text{ for all } i \geq N\}$$

Define addition and scalar multiplication pointwise. Define an inner product on $\mathbf{E}$ by

$$\langle (x_n), (y_n) \rangle := \sum_{i=1}^{\infty} \langle x_i, y_i \rangle_{\mathbf{H}_i}.$$

Note that this is well-defined because it is a finite sum. Show that the completion of $\mathbf{E}$ with respect to this inner product is the Hilbert space direct sum $\bigoplus_{n=1}^{\infty} \mathbf{H}_n$.

EXERCISE 3.15 (⚒) Prove Corollary 3.1.14.

EXERCISE 3.16 Let $\mathbf{E}$ denote the space of all functions $f : [-1,1] \to \mathbb{R}$ such that $f(-1) = f(1) = 0$, and f is infinitely differentiable on $(-1,1)$ (the space $\mathbf{E}$ is usually denoted by $C_0^\infty(-1,1)$). Note that we may think of functions in $\mathbf{E}$ as defined on all of $\mathbb{R}$ by extending them to be zero outside $[-1,1]$.

(i) If p is a polynomial such that $\int_{\mathbb{R}} f(t)p(t)dt = 0$ for all $f \in \mathbf{E}$, then prove that $p = 0$.

Hint: Follow the argument of Step (i) of Proposition 2.3.12.

(ii) For $m \in \mathbb{N}$ fixed, define $T : \mathbf{E} \to \mathbb{R}^{m+1}$ by $T(f) = (y_0, y_1, \ldots, y_m)$, where

$$y_i := \int_{\mathbb{R}} f(t) t^i dt.$$

Prove that T is a surjective, linear operator. In particular there exists $f \in \mathbf{E}$ such that $T(f) = (1, 0, 0, \ldots 0)$.

Hint: Use Corollary 3.1.14 on $\mathbf{M} = \text{Range}(T) < \mathbb{R}^n$.

(iii) Conclude that for each $m \in \mathbb{N}$ there exists $f \in \mathbf{E}$ such that

$$f_\epsilon * q = q$$

for every polynomial q of degree $\leq m$ and every $\epsilon > 0$ (here f_ϵ denotes the function $f_\epsilon(x) = \epsilon^{-1} f(x/\epsilon)$ and $*$ denotes the convolution operation $u * v(x) := \int_{\mathbb{R}} u(y) v(x - y) dy$).

Note: The results of Exercise 3.16 also hold for functions defined on $\mathbb{R}^n$ for $n \geq 2$ with appropriate changes to the spaces concerned.

Riesz Representation Theorem

EXERCISE 3.17 (✕) Let $\mathbf{H}$ be a Hilbert space and $\mathbf{M} < \mathbf{H}$ be any subspace. The map $\mathbf{H}^* \to \mathbf{M}^*$ given by $\varphi \mapsto \varphi|_{\mathbf{M}}$ is surjective by Corollary 3.2.6. Prove that it is injective if and only if $\overline{\mathbf{M}} = \mathbf{H}$.

EXERCISE 3.18 (✕) Let $\mathbf{E} = c_{00}$ equipped with the inner product inherited from ℓ^2. Define $\varphi : \mathbf{E} \to \mathbb{K}$ by

$$\varphi((x_j)) = \sum_{n=1}^{\infty} \frac{x_n}{n}.$$

(i) Prove that φ is a bounded linear functional.

(ii) Prove that there does not exist any $y \in \mathbf{E}$ such that $\varphi(x) = \langle x, y \rangle$ for all $x \in \mathbf{E}$.

(iii) If $\mathbf{M} = \ker(\varphi)$, then prove that $\mathbf{M}^{\perp} = \{0\}$. Is $\mathbf{M}$ dense in $\mathbf{E}$?

EXERCISE 3.19 Let $\mathbf{E}$ be an inner product space such that for any $\varphi \in \mathbf{E}^*$ there exists $y \in \mathbf{E}$ such that $\varphi(x) = \langle x, y \rangle$ for all $x \in \mathbf{E}$. Prove that $\mathbf{E}$ is a Hilbert space.

Orthonormal Bases

EXERCISE 3.20 Prove Theorem 3.3.4.

EXERCISE 3.21 Let $\mathbf{M}$ be the plane $\mathbf{M} = \{(x,y,z) \in \mathbb{R}^3 : x+y+2z = 0\}$. Determine the orthogonal projection $P_\mathbf{M} : \mathbb{R}^3 \to \mathbf{M}$ (in other words, given a vector $(x,y,z) \in \mathbb{R}^3$, determine the vector $P_\mathbf{M}(x,y,z) \in \mathbf{M}$).

EXERCISE 3.22 Consider the Sobolev space $W_2^1[0,1]$ introduced in Exercise 2.25 (once again read ahead to Definition 4.1.8 before you attempt this problem). Let $\mathbf{H}$ denote the subspace of $W_2^1[0,1]$ consisting of those functions $f \in W_2^1[0,1]$ such that $f(0) = f(1) = 0$. Observe that the map

$$\langle f, g \rangle := \int_0^1 f'(t)g'(t)dt$$

defines an inner product on $\mathbf{H}$.

   (i) Prove that $\mathbf{H}$ is a Hilbert space.

   (ii) For $n \in \mathbb{Z} \setminus \{0\}$, define $e_n \in \mathbf{H}$ by

$$e_n(t) := \frac{e^{2\pi i n t} - 1}{2\pi n}.$$

   Prove that $\{e_n : n \neq 0\}$ forms an orthonormal basis for $\mathbf{H}$.

Hint: You will need the fact that $\{t \mapsto e^{2\pi i n t} : n \in \mathbb{Z}\}$ spans a dense subspace of $L^2[0,1]$ (as in Theorem 3.5.4).

EXERCISE 3.23 Let Λ be an orthonormal set in a Hilbert space $\mathbf{H}$ and let $\mathbf{M} = \overline{\operatorname{span}(\Lambda)}$. If $P_\mathbf{M} : \mathbf{H} \to \mathbf{M}$ denotes the orthogonal projection onto $\mathbf{M}$, then prove that

$$P_\mathbf{M}(x) = \sum_{e \in \Lambda} \langle x, e \rangle e$$

for all $x \in \mathbf{H}$, where the sum is defined as in Remark 3.3.13.

EXERCISE 3.24 ($\times$) Let Λ be an orthonormal basis for a Hilbert space $\mathbf{H}$.

   (i) For any $x \in \mathbf{H}$, prove that x can be expressed *uniquely* as $x = \sum_{e \in \Lambda} \alpha_e e$, where $\alpha_e \in \mathbb{K}$ for all $e \in \Lambda$.

   (ii) Conclude that Λ is a Hamel basis if and only if $\mathbf{H}$ is finite dimensional.

EXERCISE 3.25 (⚒) Let $\mathbf{H}$ be an infinite dimensional Hilbert space and $\Lambda = \{e_n : n \in \mathbb{N}\}$ be a countably infinite orthonormal set.

(i) Prove that the series $x := \sum_{n=1}^{\infty} \frac{e_n}{n}$ converges in $\mathbf{H}$ and that $x \notin \text{span}(\Lambda)$.

(ii) Conclude that $\mathbf{H}$ cannot have a countable Hamel basis.

EXERCISE 3.26 Let $\mathbf{H} := L^2[0,1]$. Define

$$A := \{1, \sqrt{2}\cos(\pi t), \sqrt{2}\cos(2\pi t), \ldots\} \text{ and}$$
$$B := \{1, \sqrt{2}\sin(\pi t), \sqrt{2}\sin(2\pi t), \ldots\}.$$

Show that A and B are both orthonormal bases for $\mathbf{H}$.

EXERCISE 3.27 (⚒) Let $\mathbf{H} := L^2[-1,1]$. For $n \geq 0$, define the Legendre polynomials $P_n \in \mathbf{H}$ by

$$P_n(t) = \frac{1}{2^n n!}\left(\frac{d}{dt}\right)^n (t^2-1)^n.$$

(i) Compute the first three Legendre polynomials P_0, P_1 and P_2.

(ii) If $x_n(t) = t^n$, then show that $x_m \perp P_n$ for any $m < n$. Conclude that the set $\{P_n : n \geq 0\}$ is orthogonal.

Hint: Use integration by parts multiple times.

(iii) Show that the Legendre polynomials are obtained by Gram–Schmidt orthogonalization of the monomials $\{x_n : n \geq 0\}$ in $\mathbf{H}$.

(iv) Show that $\|P_n\|_2^2 = \frac{2}{2n+1}$.

(v) Show that P_n satisfies the differential equation

$$\frac{d}{dt}(1-t^2)\frac{d}{dt}P_n + n(n+1)P_n = 0.$$

(vi) Let $\mathbf{M} := \text{span}\{x_0, x_1, x_2\}$ be the subspace of $\mathbf{H}$ consisting of polynomials of degree ≤ 2 and let $P : \mathbf{H} \to \mathbf{M}$ denote the corresponding orthogonal projection. Let $f \in \mathbf{H}$ be the function $f(t) := e^t$. Determine the best approximation $P(f) \in \mathbf{M}$.

EXERCISE 3.28 Define

$$\widetilde{\mathbf{H}} := \left\{ f : \mathbb{R} \to \mathbb{R} : \int_{-\infty}^{\infty} e^{-x^2}|f(x)|^2 dx < \infty \right\}.$$

(i) Show that $\widetilde{\mathbf{H}}$ is a real vector space and that $(\cdot,\cdot) : \widetilde{\mathbf{H}} \times \widetilde{\mathbf{H}} \to \mathbb{R}$ given by

$$(f,g) := \int_{-\infty}^{\infty} e^{-x^2} f(x)g(x)\,dx$$

is a semi-inner product on $\widetilde{\mathbf{H}}$.

(ii) Let $(\mathbf{H}, \langle\cdot,\cdot\rangle)$ be the inner product space constructed as in Exercise 3.3. Prove that $\mathbf{H}$ is a Hilbert space.

EXERCISE 3.29 Let $\mathbf{H}$ be the Hilbert space defined in Exercise 3.28. Define the *Hermite polynomials* by

$$H_n(t) = (-1)^n e^{t^2} \left(\frac{d}{dt}\right)^n e^{-t^2}.$$

(i) Show that $H_0(t) = 1$ and that

$$H_{n+1}(t) = 2tH_n(t) - \frac{d}{dt}H_n(t).$$

Conclude that each H_n is a polynomial of degree n.

(ii) Show that $\{H_n : n \geq 0\}$ is an orthogonal set in $\mathbf{H}$.

Hint: Use integration by parts to evaluate

$$\int_{-\infty}^{\infty} H_m(t) \left(\frac{d}{dt}\right)^n e^{-t^2}\,dt.$$

(iii) Show that $\|H_n\|^2 = 2^n n! \sqrt{\pi}$.

It takes some work, but one can show (with the help of the Fourier transform) that the collection $\{H_n : n \geq 0\}$ spans a dense subspace of $\mathbf{H}$. Thus after normalization, one obtains an orthonormal basis of $\mathbf{H}$. The Hermite polynomials (like the Legendre polynomials from Exercise 3.27) are useful functions that are studied by both physicists and mathematicians alike.

EXERCISE 3.30 Let $[a,b] \subset \mathbb{R}$ be a closed and bounded interval and let $\{e_n : n \in \mathbb{N}\}$ be an orthonormal basis for $L^2[a,b]$. Define $f_{i,j} \in L^2([a,b]^2)$ by $f_{i,j}(s,t) = e_i(s)\overline{e_j(t)}$. Suppose $f \in L^2([a,b]^2)$ is orthogonal to $\{f_{i,j} : i,j \in \mathbb{N}\}$.

(i) Set $f_s(t) := f(s,t)$. As in Example 2.2.17, $f_s \in L^2[a,b]$ for almost every $s \in [a,b]$. Now show that

$$\|f\|_2^2 = \sum_{j=1}^{\infty} \int_a^b \left| \int_a^b e_j(t) f_s(t)\,dt \right|^2 ds$$

(ii) Set $g_j(s) := \int_a^b e_j(t) f_s(t) dt$ and show that $\langle f, f_{i,j}\rangle = \int_a^b g_j(s)\overline{e_i(s)} ds$.

(iii) Prove that $f = 0$. Conclude that the set $\{f_{i,j} : i,j \in \mathbb{N}\}$ forms an orthonormal basis for $L^2([a,b]^2)$.

Hint: Use the Fubini–Tonelli Theorem liberally.

EXERCISE 3.31 (⚒) Let **H** and **K** be Hilbert spaces and $T : \mathbf{H} \to \mathbf{K}$ be a bounded linear operator of finite rank (in other words, Range(T) is finite dimensional). Prove that there exist vectors $x_1, x_2, \dots, x_n \in \mathbf{H}$ and $y_1, y_2, \dots, y_n \in \mathbf{K}$ such that

$$T(z) = \sum_{i=1}^{n} \langle z, x_i\rangle y_i$$

for all $z \in \mathbf{H}$.

Hint: Start with a Fourier expansion in Range(T).

EXERCISE 3.32 Let **H** be a Hilbert space and $\varphi : \mathbf{H} \to \mathbb{K}$ be a bounded linear functional on **H**. Let $\Lambda := \{e_\alpha : \alpha \in J\}$ be an orthonormal basis for **H**.

(i) If $y \in \mathbf{H}$ is such that $\varphi(x) = \langle x, y\rangle$ for all $x \in \mathbf{H}$, then determine the Fourier coefficients of y with respect to the basis Λ.

(ii) Use this to construct a *proof* of the Riesz representation that uses the Fourier expansion.

Isomorphisms of Hilbert Spaces

EXERCISE 3.33 (⚒) Prove Lemma 3.4.4.

EXERCISE 3.34 Let **H** and **K** be Hilbert spaces and $T : \mathbf{H} \to \mathbf{K}$ be a surjective function such that $\langle Tx, Ty\rangle = \langle x, y\rangle$ for all $x, y \in \mathbf{H}$. Prove that T is linear.

EXERCISE 3.35 (⚒) Let I be any set and let $1 \leq p < \infty$. Prove that $\ell^p(I)$ is a Banach space.

EXERCISE 3.36 (⚒) Let **H** be a separable Hilbert space. Prove that there is a sequence $(P_n) \subset \mathcal{B}(\mathbf{H})$ satisfying the following conditions.

(a) For each $x \in \mathbf{H}$, $\lim_{n\to\infty} P_n(x) = x$.

(b) $\sup_{n\in\mathbb{N}} \|P_n\| = 1$.

Fourier Series of L^2 Functions

EXERCISE 3.37 Let $f \in L^2[-\pi, \pi]$ be a function such that $\sum_{n=-\infty}^{\infty} |\widehat{f}(n)| < \infty$. Prove that the Fourier series of f converges uniformly (in the supremum norm).

EXERCISE 3.38 Let $f \in L^2[-\pi, \pi]$ be a function whose Fourier series converges uniformly (in the supremum norm). Prove that the sum of the series must be f.

In order to attempt the next problem, you will need to know Lebesgue's Fundamental Theorem of Calculus (see Theorem 4.1.10 and the surrounding ideas).

EXERCISE 3.39 Let $f \in C[-\pi, \pi]$ be an absolutely continuous function, with derivative f'. Prove that the Fourier series of f' takes the form

$$f'(t) \sim \sum_{n=-\infty}^{\infty} in\widehat{f}(n)e^{int}.$$

In other words, the Fourier series of f' is obtained from the Fourier series of f by termwise differentiation.

EXERCISE 3.40 Let $f \in L^2[-\pi, \pi]$ and let $F : [-\pi, \pi] \to \mathbb{C}$ be defined by

$$F(x) = \int_a^x f(t)dt$$

for some $a \in [-\pi, \pi]$ (note that this is well-defined because $f \in L^1[-\pi, \pi]$ by Exercise 2.12). Then prove that there are constants $\alpha, \beta \in \mathbb{C}$ such that the Fourier series of F takes the form

$$F(t) - \alpha t \sim \beta + \sum_{n \neq 0} \frac{\widehat{f}(n)}{in} e^{int}.$$

EXERCISE 3.41 Suppose $f \in C[-\pi, \pi]$ is a continuously differentiable function.

(i) Use Exercise 3.40 to prove that $\sum_{n=-\infty}^{\infty} |\widehat{f}(n)| < \infty$.

(ii) Conclude from Exercise 3.37 and Exercise 3.38 that the Fourier series of f converges uniformly to f.

Note: The question of uniform (indeed pointwise) convergence of the Fourier series will be taken up more seriously in Chapter 5. Specifically, have a look at Remark 5.2.3 and the ensuing discussion.

EXERCISE 3.42 Consider $f : [-\pi, \pi] \to \mathbb{C}$ given by $f(x) = x^2$. By Exercise 3.41, the Fourier series of f converges uniformly to f. Use this to compute the sum of the series $\sum_{n=1}^{\infty} \frac{1}{n^2}$ and $\sum_{n=1}^{\infty} \frac{1}{n^4}$.

EXERCISE 3.43 Prove that the set

$$\Lambda := \{1, \cos(t), \sin(t), \ldots, \cos(kt), \sin(kt), \ldots\}$$

forms an orthonormal basis of $L^2[-\pi, \pi]$ when $[-\pi, \pi]$ is equipped with the normalized Lebesgue measure. Conclude that every $f \in L^2[-\pi, \pi]$ can be expressed as a series of the form

$$f(t) \sim \frac{1}{2}a_0 + \sum_{k=1}^{\infty} (a_k \cos(kt) + b_k \sin(kt))$$

where the convergence is in the L^2 norm (as in Theorem 3.5.5).

Note: The numbers a_k and b_k appearing above are called the **Fourier cosine** and the **Fourier sine** coefficients of f, respectively.

Reproducing Kernel Hilbert Spaces

EXERCISE 3.44 (✖) Let $\mathbb{D} := \{z \in \mathbb{C} : |z| < 1\}$ and let $H^2(\mathbb{D})$ denote the space of all holomorphic functions $f : \mathbb{D} \to \mathbb{C}$ such that

$$\|f\| := \left(\sup_{0<r<1} \frac{1}{2\pi} \int_0^{2\pi} |f(re^{i\theta})|^2 d\theta \right)^{1/2} < \infty.$$

This space is called the **Hardy space**.

(i) If $f \in H^2(\mathbb{D})$ is expressed as a Taylor series $f(z) = \sum_{n=0}^{\infty} a_n z^n$ (for $z \in \mathbb{D}$), then prove that

$$\|f\| = \left(\sum_{n=0}^{\infty} |a_n|^2 \right)^{1/2}.$$

(ii) If $f, g \in H^2(\mathbb{D})$ are expressed as $f(z) = \sum_{n=0}^{\infty} a_n z^n$ and $g(z) = \sum_{n=0}^{\infty} b_n z^n$ (for $z \in \mathbb{D}$), prove that the formula

$$\langle f, g \rangle := \sum_{n=0}^{\infty} a_n \overline{b_n}$$

is a well-defined inner product on $H^2(\mathbb{D})$.

(iii) Prove that $H^2(\mathbb{D})$ is isometrically isomorphic to ℓ^2 and hence a Hilbert space.

EXERCISE 3.45 Let $H^2(\mathbb{D})$ denote the Hardy space on the unit disc described in Exercise 3.44. For $\alpha \in \mathbb{D}$, let $\varphi_\alpha : H^2(\mathbb{D}) \to \mathbb{C}$ denote the linear functional

$$f \mapsto f(\alpha).$$

(i) Prove that φ_α is bounded and that $\|\varphi_\alpha\| \leq \frac{1}{\sqrt{1-|\alpha|^2}}$.

(ii) Determine a function $k_\alpha \in H^2(\mathbb{D})$ such that $f(\alpha) = \langle f, k_\alpha \rangle$ for all $f \in H^2(\mathbb{D})$. Use this function to prove that equality holds in part (i).

The phenomenon described above for $H^2(\mathbb{D})$ is a special case of the following notion.

3.6.3 DEFINITION Given a set X, we say that **H** is a *reproducing kernel Hilbert space* on X if it satisfies the following conditions:

(a) **H** is a vector space of $\mathbb{K}$-valued functions on X, with pointwise operations.

(b) **H** is endowed with an inner product $\langle \cdot, \cdot \rangle$ that makes it a Hilbert space.

(c) For each $y \in X$, the evaluation map $f \mapsto f(y)$ defines a bounded linear functional on **H**.

If this happens, then for each $y \in X$, there is a unique function $k_y \in \mathbf{H}$ such that $f(y) = \langle f, k_y \rangle$ for all $f \in \mathbf{H}$ (by the Riesz Representation Theorem). The function $K : X \times X \to \mathbb{K}$ given by

$$K(x, y) := k_y(x)$$

is called the *reproducing kernel* for **H**.

EXERCISE 3.46 Let **H** be a reproducing kernel Hilbert space over a set X, with reproducing kernel K.

(i) Prove that $K(y, x) = \overline{K(x, y)}$ for all $x, y \in X$.

(ii) For any $y_1, y_2, \ldots, y_n \in X$ and $\alpha_1, \alpha_2, \ldots, \alpha_n \in \mathbb{K}$, prove that $\sum_{i,j=1}^{n} K(y_i, y_j) \overline{\alpha_i} \alpha_j \geq 0$.

(iii) If $k_y \in \mathbf{H}$ is the function $k_y(x) := K(x, y)$ as above, then prove that span$\{k_y : y \in X\}$ is a dense subspace of **H**.

3.6.4 DEFINITION Let X be a set. A *kernel function* on X is a function $K : X \times X \to \mathbb{K}$ satisfying the following conditions.

(a) $K(y, x) = \overline{K(x, y)}$ for all $x, y \in X$ and

(b) For any $y_1, y_2, \ldots, y_n \in X$ and $\alpha_1, \alpha_2, \ldots, \alpha_n \in \mathbb{K}$, $\sum_{i,j=1}^{n} K(y_i, y_j)\overline{\alpha_i}\alpha_j \geq 0$.

EXERCISE 3.47 Let K be a kernel function on a set X. The purpose of this exercise is to construct a reproducing kernel Hilbert space having K as its reproducing kernel.

(i) Let **F** denote the vector space of all $\mathbb{K}$-valued functions on X, with pointwise operations. For each $y \in X$, let $k_y \in \mathbf{F}$ denote the function $k_y(x) := K(x, y)$. Set $\mathbf{W} := \mathrm{span}\{k_y : y \in X\}$. For two elements $f = \sum_{i=1}^{n} \alpha_i k_{y_i}$ and $g = \sum_{j=1}^{m} \beta_j k_{z_j}$ in **W**, define

$$(f, g) := \sum_{i=1}^{n} \sum_{j=1}^{n} \alpha_i \overline{\beta_j} K(z_j, y_i).$$

Prove that this is a well-defined semi-inner product on **W** (see Exercise 3.3).

Note: In general, f and g may be representable as such linear combinations in multiple ways. You need to prove that this inner product is independent of how f and g are represented.

(ii) For any $f \in \mathbf{W}$ and $x \in X$, prove that $|f(x)|^2 \leq K(x, x)(f, f)$. Conclude that $(\cdot, \cdot)$ is an inner product on **W**.

Hint: Use the Cauchy–Schwarz Inequality from Exercise 3.3.

(iii) Let **H** be the completion of **W** with respect to this inner product (as in Exercise 3.13). Prove that **H** may be realized as a subset of **F**.

(iv) Prove that **H** is a reproducing kernel Hilbert space over X, with reproducing kernel K.

The purpose of the next exercise is to show that a reproducing kernel Hilbert space is completely determined by its reproducing kernel. Together with Exercise 3.47, we conclude that to each kernel function K, there is one and only one reproducing kernel Hilbert space, which has K as its reproducing kernel.

EXERCISE 3.48 Let $\mathbf{H}_1$ and $\mathbf{H}_2$ be two reproducing kernel Hilbert spaces on a common set X, with the same reproducing kernel K. For $i = 1, 2$, set $\mathbf{W}_i := \mathrm{span}\{k_y \in \mathbf{H}_i : y \in X\}$ (with k_y defined as above).

(i) If $f \in \mathbf{W}_1 = \mathbf{W}_2$ and $x \in X$, prove that the value $f(x)$ is independent of whether we regard it as an element of $\mathbf{W}_1$ or $\mathbf{W}_2$. Furthermore, prove that $\|f\|_{\mathbf{H}_1} = \|f\|_{\mathbf{H}_2}$.

(ii) Prove that $\mathbf{H}_1 = \mathbf{H}_2$.

> *Hint:* Let $f \in \mathbf{H}_1$, and (f_n) be a sequence in $\mathbf{W}_1$ such that $f_n \to f$ in $\mathbf{H}_1$. Prove that there exists $g \in \mathbf{H}_2$ such that $\|f_n - g\|_{\mathbf{H}_2} \to 0$ and conclude that $f = g$.

(iii) Prove that $\|f\|_{\mathbf{H}_1} = \|f\|_{\mathbf{H}_2}$ for any $f \in \mathbf{H}_1$.

EXERCISE 3.49 Let $\mathbf{H}$ be a separable, infinite dimensional reproducing kernel Hilbert space over a set X and let $\{e_n : n \in \mathbb{N}\}$ be an orthonormal basis for $\mathbf{H}$. Show that the reproducing kernel of $\mathbf{H}$ is given by the formula

$$K(x,y) = \sum_{n=1}^{\infty} \overline{e_n(x)} e_n(y).$$

EXERCISE 3.50 Let $\mathbf{H}$ be the Hilbert space described in Exercise 3.22.

(i) For each $x \in [0,1]$, prove that $|f(x)| \leq \|f\|\sqrt{x}$ for all $f \in \mathbf{H}$. Conclude that $\mathbf{H}$ is a reproducing kernel Hilbert space on $[0,1]$.

> *Hint:* If $f \in \mathbf{H}$, then $f(x) = \int_0^x f'(t)dt$.

(ii) Show that the function $K : [0,1] \times [0,1] \to \mathbb{R}$ given by

$$K(x,y) := \begin{cases} (1-y)x & : \text{if } x \leq y, \\ (1-x)y & : \text{if } x \geq y. \end{cases}$$

is the reproducing kernel of $\mathbf{H}$ (and with this you can show that the estimate obtained in part (i) is *not* optimal).

Additional Reading

- Given the long and storied history of the Fourier series, there is no shortage of good books on the subject. However, my personal go-to book is this one.

  Stein, Elias M. and Shakarchi, Rami. (2003). *Fourier Analysis.* Vol. 1. Princeton Lectures in Analysis. An introduction. Princeton, NJ: Princeton University Press. ISBN: 0-691-11384-X.

- In the problems immediately after Exercise 3.45, we have developed (to an extent) some basic ideas surrounding reproducing kernel Hilbert spaces. Their theory is deep and touches a variety of allied fields such as complex analysis and partial differential equations. While many articles exist on the subject (and even a nice book by Paulsen and Raghupathi [45]), perhaps the best place to start remains a paper by Aronszajn, written after the Second World War.

Aronszajn, Nachman (1950). Theory of reproducing kernels. *Transactions of the American Mathematical Society*, 68, 337–404. ISSN: 0002-9947. DOI: 10.2307/1990404.

Chapter 4

Dual Spaces

4.1 The Duals of L^p Spaces

Given a normed linear space **E**, we have briefly encountered the dual space **E***, consisting of bounded linear functionals on **E**. We know that it is a Banach space (Theorem 2.3.13), but beyond that we do not know much. In the finite dimensional case, **E** and **E*** are isomorphic as vector spaces. However, in the infinite dimensional case, this need not be true. Moreover, even in the finite dimensional case, an arbitrary isomorphism may not be isometric!

Now, if **E** is a Hilbert space, then there is an isometric isomorphism **E** $\cong$ **E*** by the Riesz Representation Theorem. This is not true for an arbitrary Banach space. However, the core idea of that proof (specifically, the construction of the map Δ) is applicable when studying the dual space of ℓ^p or $L^p[a,b]$ for $1 \leq p \leq \infty$. This is the focus of this section, where we explicitly determine these dual spaces, giving new meaning to Hölder's Inequality and revisiting some beautiful measure theory along the way.

4.1.1 DEFINITION Throughout this section, fix $1 \leq p, q \leq \infty$, where q is the conjugate exponent of p (in other words, $\frac{1}{p} + \frac{1}{q} = 1$ if $1 < p < \infty$, $q = \infty$ when $p = 1$ and vice-versa).

(i) For each $y \in \ell^q$, define $\varphi_y : \ell^p \to \mathbb{K}$ by

$$\varphi_y((x_j)) := \sum_{n=1}^{\infty} x_n y_n.$$

111

For $1 < p < \infty$, φ_y is well-defined by Hölder's Inequality (Theorem 2.1.18) and

$$|\varphi_y(x)| \leq \|x\|_p \|y\|_q.$$

If $p = 1$ or $p = \infty$, the inequality is obvious. Hence $\varphi_y \in (\ell^p)^*$ and $\|\varphi_y\| \leq \|y\|_q$. Furthermore, for any $y, z \in \ell^q$, and $\alpha \in \mathbb{K}$, we have

$$\varphi_{y+z} = \varphi_y + \varphi_z, \text{ and } \varphi_{\alpha y} = \alpha \varphi_y.$$

Therefore we obtain a linear operator $\Delta : \ell^q \to (\ell^p)^*$ given by

$$y \mapsto \varphi_y,$$

satisfying $\|\Delta(y)\| \leq \|y\|_q$ for all $y \in \ell^q$.

(ii) Similarly, for each $g \in L^q[a,b]$, we define $\varphi_g : L^p[a,b] \to \mathbb{K}$ by

$$f \mapsto \int_a^b f(t)g(t)dt.$$

Once again φ_g is well-defined by Hölder's Inequality and we get a linear operator

$$\Delta : L^q[a,b] \to (L^p[a,b])^*,$$

satisfying $\|\Delta(g)\| \leq \|g\|_q$ for all $g \in L^q[a,b]$.

The goal of this section is to understand when this map Δ is an isomorphism.

Aside. While it is true that ℓ^p and $L^p[a,b]$ are both special case of abstract L^p spaces, we treat these two cases separately to reinforce the idea and also show the subtle differences between these two cases. It is true, though, that you are about to see two separate proofs of the same fact.

We first introduce some useful notation: For $x \in \mathbb{K}$, we write

$$\text{sgn}(x) := \begin{cases} \frac{|x|}{x} & : \text{if } x \neq 0, \\ 0 & : \text{if } x = 0. \end{cases}$$

so that $\text{sgn}(x)x = |x|$ for all $x \in \mathbb{K}$. For $j \in \mathbb{N}$, we will write $e_j = (0,0,\ldots,0,1,0,\ldots)$ (with 1 in the jth position) as in Example 1.1.7. Note that $e_j \in \ell^p$ for all $1 \leq p \leq \infty$.

4.1.2 PROPOSITION *For $1 \leq p \leq \infty$, the map $\Delta : \ell^q \to (\ell^p)^*$ is isometric.*

Proof. In all cases, we know that $\|\varphi_y\| \leq \|y\|_q$, so it suffices to prove the reverse inequality. We break the proof into three cases.

(i) If $p = 1$ and $q = \infty$: Let $y = (y_1, y_2, \ldots) \in \ell^\infty$. Then

$$|y_j| = |\varphi_y(e_j)| \leq \|\varphi_y\|\|e_j\|_1 = \|\varphi_y\|.$$

Hence, $\|y\|_\infty \leq \|\varphi_y\|$ as required.

(ii) If $p = \infty$ and $q = 1$: Let $y = (y_1, y_2, \ldots) \in \ell^1$. Then for each $n \in \mathbb{N}$, define $x^n = (x_j^n)$ by

$$x_j^n = \begin{cases} \operatorname{sgn}(y_j) & : \text{if } 1 \leq j \leq n, \\ 0 & : \text{otherwise.} \end{cases}$$

Then $x^n \in \ell^\infty$ and $\|x^n\|_\infty \leq 1$. Furthermore,

$$\sum_{j=1}^n |y_j| = \sum_{j=1}^n x_j^n y_j = \varphi_y(x^n) \leq \|\varphi_y\|\|x^n\|_\infty \leq \|\varphi_y\|.$$

Hence $\|y\|_1 \leq \|\varphi_y\|$.

(iii) If $1 < p < \infty$: Let $y = (y_1, y_2, \ldots) \in \ell^q$. Then for each $n \in \mathbb{N}$, define $x^n = (x_j^n)$ by

$$x_j^n = \begin{cases} \operatorname{sgn}(y_j)|y_j|^{q-1} & : \text{if } 1 \leq j \leq n, \\ 0 & : \text{otherwise.} \end{cases}$$

Then $x^n \in \ell^p$ and

$$\sum_{j=1}^n |y_j|^q = \sum_{j=1}^n x_j^n y_j = \sum_{j=1}^\infty x_j^n y_j = \varphi_y(x^n).$$

Also

$$|\varphi_y(x^n)| \leq \|\varphi_y\|\|x^n\|_p = \|\varphi_y\| \left(\sum_{j=1}^n |y_j|^{qp-p} \right)^{1/p} = \|\varphi_y\| \left(\sum_{j=1}^n |y_j|^q \right)^{1/p},$$

because $qp - p = p(q-1) = q$. Therefore we conclude that

$$\left(\sum_{j=1}^n |y_j|^q \right)^{1/q} \leq \|\varphi_y\|.$$

This is true for all $n \in \mathbb{N}$ and so $\|y\|_q \leq \|\varphi_y\|$. $\square$

We now show that this map Δ is surjective, provided $p \neq \infty$. This assumption is crucial and we will return to it later in the chapter. For now recall that c_{00} is dense in ℓ^p for $p \neq \infty$ (Proposition 2.3.4) and it is this fact that gets used in the proof.

4.1.3 THEOREM *If $1 \leq p < \infty$, then the map $\Delta : \ell^q \to (\ell^p)^*$ is an isometric isomorphism.*

Proof.   By Proposition 4.1.2, it suffices to show that Δ is surjective. So fix $\varphi \in (\ell^p)^*$ and we want to construct $y \in \ell^q$ such that $\varphi = \varphi_y$. If $e_j \in \ell^p$ are defined as above, we set

$$y := (\varphi(e_1), \varphi(e_2), \ldots).$$

(i) If $p = 1$, then $q = \infty$: For each $i \in \mathbb{N}$,

$$|y_i| = |\varphi(e_i)| \leq \|\varphi\| \|e_i\|_1 = \|\varphi\|.$$

Hence $y \in \ell^\infty$, and so $\varphi_y \in (\ell^1)^*$. Now if $x \in c_{00}$, write $x = (x_1, x_2, \ldots, x_n, 0, 0, \ldots)$, then

$$\varphi(x) = \sum_{i=1}^{n} x_i \varphi(e_i) = \sum_{i=1}^{n} x_i y_i = \varphi_y(x).$$

Now φ and φ_y are both continuous functions that agree on c_{00}, which is dense in ℓ^1 (Proposition 2.3.4). It follows that $\varphi = \varphi_y$.

(ii) If $1 < p < \infty$, then $1 < q < \infty$: For $n \in \mathbb{N}$, write

$$y^n := (y_1, y_2, \ldots, y_n, 0, 0, \ldots).$$

Then $y^n \in c_{00} \subset \ell^q$, so we may consider the corresponding linear functional $\varphi_{y^n} \in (\ell^p)^*$. For each $n \in \mathbb{N}$ and $x \in \ell^p$, let $x^n := (x_1, x_2, \ldots, x_n, 0, 0, \ldots)$. Then

$$\varphi_{y^n}(x) = \sum_{j=1}^{n} x_j y_j = \varphi \left(\sum_{j=1}^{n} x_j e_j \right) = \varphi(x^n).$$

Hence for any $x \in \ell^p$, we have

$$|\varphi_{y^n}(x)| \leq \|\varphi\| \|x^n\|_p \leq \|\varphi\| \|x\|_p$$

and therefore $\|\varphi_{y^n}\| \leq \|\varphi\|$ for all $n \in \mathbb{N}$. Since Δ is an isometry, it follows that

$$\left(\sum_{i=1}^{n} |y_i|^q\right)^{1/q} = \|y^n\|_q \leq \|\varphi\|$$

for all $n \in \mathbb{N}$. We conclude that $y \in \ell^q$ and that $\|y\|_q \leq \|\varphi\|$. Now that $y \in \ell^q$, we know that $\varphi_y \in (\ell^p)^*$. Furthermore, for any $x \in c_{00}$, write $x = (x_1, x_2, \ldots, x_n, 0, 0, \ldots)$. Then

$$\varphi(x) = \sum_{i=1}^{n} x_i \varphi(e_i) = \varphi_y(x)$$

and so $\varphi = \varphi_y$ on c_{00}. Once again, since c_{00} is dense in ℓ^p, it follows that $\varphi = \varphi_y$.

$\square$

Note that the same proofs work if we replace ℓ^p by $\mathbf{E} = (\mathbb{K}^n, \|\cdot\|_p)$ as well. In fact, for any $1 \leq p \leq \infty$, the map $\Delta : (\mathbb{K}^n, \|\cdot\|_q) \to \mathbf{E}^*$ is an injective linear map. Since both spaces have the same dimension, we conclude that Δ is an isomorphism (even if $p = \infty$!).

4.1.4 COROLLARY *Let $1 \leq p \leq \infty$ and let q be the conjugate exponent of p. If $\mathbf{E} = (\mathbb{K}^n, \|\cdot\|_p)$, then $\mathbf{E}^* \cong (\mathbb{K}^n, \|\cdot\|_q)$.*

Of course, we know from linear algebra that $\mathbf{E}^* \cong \mathbb{K}^n$ as a vector space. The point of this result is that we now know what the operator norm on $\mathbf{E}^*$ translates to under this isomorphism. We now turn to the function space analogues of the ℓ^p spaces and prove the same sort of result. If g is a function, we write $\mathrm{sgn}(g)$ for the function $t \mapsto \mathrm{sgn}(g(t))$.

4.1.5 PROPOSITION *For $1 \leq p \leq \infty$, the map $\Delta : L^q[a,b] \to (L^p[a,b])^*$ is isometric.*

Proof. Fix $g \in L^q[a,b]$. We know that $\|\varphi_g\| \leq \|g\|_q$, so it suffices to prove the reverse inequality. Once again we break it into three cases.

(i) If $p = 1$ and $q = \infty$: For $n \in \mathbb{N}$, define

$$E_n = \{t \in [a,b] : |g(t)| > \|\varphi_g\| + 1/n\}$$

and set $f_n = \mathrm{sgn}(g)\chi_{E_n}$. Then $\|f_n\|_1 = m(E_n) < \infty$ and

$$|\varphi_g(f_n)| = \int_{E_n} |g(t)|dt > (\|\varphi_g\| + 1/n)\, m(E_n).$$

However, $|\varphi_g(f_n)| \leq \|\varphi_g\| m(E_n)$ and so $m(E_n) = 0$ for each $n \in \mathbb{N}$. Therefore

$$m(\{t \in [a,b] : |g(t)| > \|\varphi_g\|\}) = m\left(\bigcup_{n=1}^{\infty} E_n\right) = 0.$$

Thus $\|g\|_\infty \leq \|\varphi_g\|$ as required.

(ii) If $p = \infty$ and $q = 1$: Let $f = \mathrm{sgn}(g)$. Then $\|f\|_\infty \leq 1$ and

$$\|g\|_1 = \int_a^b |g(t)| dt = \varphi_g(f) = |\varphi_g(f)| \leq \|\varphi_g\| \|f\|_\infty.$$

Therefore $\|g\|_1 \leq \|\varphi_g\|$.

(iii) If $1 < p < \infty$: If we set $f = \mathrm{sgn}(g)|g|^{q-1}$, then

$$\int_a^b |f(t)|^p dt = \int_a^b |g(t)|^{(q-1)p} dt = (\|g\|_q)^q = \int_a^b |g(t)|^q dt = \varphi_g(f).$$

In particular, $f \in L^p[a,b]$ and $\|f\|_p = \|g\|_q^{q/p}$. Also, $(\|g\|_q)^q \leq \|\varphi_g\| \|f\|_p$, which implies that

$$\|g\|_q = (\|g\|_q)^{q-q/p} \leq \|\varphi_g\|.$$

$\square$

In order to determine when the map Δ is surjective in this case, we need to revisit Lebesgue's Fundamental Theorem of Calculus. While we do not prove this result (as it would take us too far off course), we will discuss the philosophy behind it and point you to Royden [49] for further information on the subject. Consider the classical Fundamental Theorem of Calculus, which consists of two statements:

(i) Let $f \in C[a,b]$ and define

$$F(x) = \int_a^x f(t) dt.$$

Then F is differentiable on (a,b), and $F' = f$.

(ii) If $F : [a,b] \to \mathbb{K}$ is continuously differentiable on $[a,b]$, then F' is Riemann-integrable and for all $x \in [a,b]$,

$$F(x) = F(a) + \int_a^x F'(t) dt.$$

Lebesgue's Fundamental Theorem of Calculus is an answer to the corresponding questions for L^1 functions:

(i) Suppose $f \in L^1[a, b]$, then we may define

$$F(x) = \int_a^x f(t)dt.$$

Is F differentiable, and, if so, is it true that $F' = f$?

(ii) Suppose $F : [a, b] \to \mathbb{C}$ is a function, under what conditions is F differentiable and $F' \in L^1[a, b]$? Furthermore, if this is true, then does it follow that

$$F(x) = F(a) + \int_a^x F'(t)dt$$

holds for all $x \in [a, b]$?

The answer to the first of these questions is the Lebesgue Differentiation Theorem (see [49, Section 5.3] for the proof).

4.1.6 Theorem (Lebesgue's Differentiation Theorem) *Let $f \in L^1[a, b]$ and define*

$$F(x) = \int_a^x f(t)dt.$$

Then F is differentiable a.e. and $F' = f$ a.e.

The next, rather commonly used fact must be known to you. The definition that follows is the correct notion one needs to answer the second part of the question raised earlier.

4.1.7 Lemma *Let $f \in L^1[a, b]$. Then for any $\epsilon > 0$ there exists $\delta > 0$ such that for any measurable set $E \subset [a, b]$,*

$$m(E) < \delta \Rightarrow \int_E |f| < \epsilon.$$

4.1.8 Definition A function $F : [a, b] \to \mathbb{K}$ is said to be *absolutely continuous* if for any $\epsilon > 0$, there exists $\delta > 0$ such that for any finite collection $\{[a_i, b_i] : 1 \leq i \leq n\}$ of non-overlapping subintervals of $[a, b]$,

$$\sum_{i=1}^n (b_i - a_i) < \delta \Rightarrow \sum_{i=1}^n |F(b_i) - F(a_i)| < \epsilon.$$

It is easy to see that an absolutely continuous function is necessarily continuous and there are examples of continuous functions that are not absolutely continuous. The most famous of these is the Cantor–Lebesgue function (see Wheeden and Zygmund [61, Section 3.1]).

4.1.9 EXAMPLE Let $F : [a,b] \to \mathbb{K}$ be a function.

(i) If F is Lipschitz continuous, then F is absolutely continuous. This is because if $L > 0$ such that $|F(x) - F(y)| \leq L|x - y|$ for all $x, y \in [a,b]$, then for $\epsilon > 0$, we may choose $\delta := \epsilon/2L$.

(ii) If F is continuously differentiable on $[a,b]$, then F is absolutely continuous. This is because, by the Mean-Value Theorem, F is Lipschitz continuous with Lipschitz constant $L := \sup_{x \in [a,b]} |F'(x)|$.

(iii) If there exists $f \in L^1[a,b]$ such that

$$F(x) = \int_a^x f(t)dt,$$

then it follows from Lemma 4.1.7 that F is absolutely continuous.

Lebesgue's Fundamental Theorem of Calculus now completes this circle of ideas very succinctly (once again see Royden [49] for the details).

4.1.10 THEOREM (LEBESGUE'S FUNDAMENTAL THEOREM OF CALCULUS) *Let $F : [a,b] \to \mathbb{K}$ be an absolutely continuous function. Then*

(i) F is differentiable a.e. and $F' \in L^1[a,b]$.

(ii) Furthermore, for almost every $x \in [a,b]$,

$$F(x) = F(a) + \int_a^x F'(t)dt.$$

We are now in a position to use Lebesgue's theorem to determine the dual space of $L^p[a,b]$. To that end we need a lemma that will serve us well. The assumption that $p \neq \infty$ in the following lemma is the first hint that the dual of $L^\infty[a,b]$ will be more difficult to determine than that of the other L^p spaces (just as it was in the case of sequence spaces).

4.1.11 LEMMA *Let $1 \leq p < \infty$ and let $g \in L^1[a,b]$. Suppose that there exists $M > 0$ such that*

$$\left| \int_a^b f(t)g(t)dt \right| \leq M\|f\|_p$$

for all bounded functions $f \in L^p[a,b]$. Then $g \in L^q[a,b]$ and $\|g\|_q \leq M$.

Proof. Once again we break the proof into two cases.

(i) If $p = 1$: Let $n \in \mathbb{N}$ and consider

$$E_n = \{x \in [a,b] : |g(x)| > M + 1/n\}.$$

We wish to prove that $m(E_n) = 0$. If we set let $f_n = \operatorname{sgn}(g)\chi_{E_n}$, then $\|f_n\|_1 = m(E_n)$ and

$$\left| \int_a^b f_n(t)g(t)dt \right| = \int_{E_n} |g(t)|dt \geq (M + 1/n)m(E_n).$$

However,

$$\left| \int_a^b f_n(t)g(t)dt \right| \leq M\|f_n\|_1 = Mm(E_n)$$

and therefore $m(E_n) = 0$. This is true for all $n \in \mathbb{N}$, and hence

$$m(\{t \in [a,b] : |g(t)| > M\}) = m\left(\bigcup_{n=1}^{\infty} E_n \right) = 0.$$

Thus $g \in L^\infty[a,b]$ and $\|g\|_\infty \leq M$ as required.

(ii) If $1 < p < \infty$: Since $g \in L^1[a,b]$, it follows that $|g(x)| < \infty$ a.e. For $n \in \mathbb{N}$, define

$$g_n(x) = \begin{cases} g(x) & : \text{ if } |g(x)| \leq n, \\ 0 & : \text{ otherwise.} \end{cases}$$

Then each g_n is measurable, bounded and $g_n \to g$ pointwise a.e. If we set $f_n := \operatorname{sgn}(g_n)|g_n|^{q-1}$, then

$$\|f_n\|_p^p = \int_a^b |g_n(t)|^{(q-1)p}dt = \int_a^b |g_n(t)|^q dt = \|g_n\|_q^q.$$

In particular, $f_n \in L^p[a,b]$ and $\|f_n\|_p = \|g_n\|_q^{q/p}$. Also, $|g_n|^q = f_n g_n = f_n g$ and hence

$$\|g_n\|_q^q = \int_a^b f_n(t)g(t)dt \leq M\|f_n\|_p = M\|g_n\|_q^{q/p}.$$

Therefore $\|g_n\|_q \leq M$ for all $n \in \mathbb{N}$. Finally, by Fatou's Lemma,

$$\int_a^b |g(t)|^q dt \leq \liminf \int_a^b |g_n(t)|^q dt \leq M^q.$$

We conclude that $g \in L^q[a,b]$ and $\|g\|_q \leq M$. □

We are *almost* ready to describe the dual space of $L^p[a,b]$ for $1 \leq p < \infty$. The next lemma is the penultimate step in that process and a crucial one. It allows us to construct an absolutely continuous function from a bounded linear functional on $L^p[a,b]$.

4.1.12 LEMMA *Let $1 \leq p < \infty$ and $\varphi \in (L^p[a,b])^*$. Define $F : [a,b] \to \mathbb{K}$ by*

$$F(x) = \varphi(\chi_{[a,x]}).$$

Then F is absolutely continuous.

Proof.   We wish to show that for any $\epsilon > 0$ there exists $\delta > 0$ such that for any finite collection $\{[a_i, b_i] : 1 \leq i \leq n\}$ of non-overlapping subintervals of $[a,b]$,

$$\sum_{i=1}^{n}(b_i - a_i) < \delta \Rightarrow \sum_{i=1}^{n}|F(b_i) - F(a_i)| < \epsilon.$$

So assume φ is non-zero, fix $\epsilon > 0$ and let $\{[a_i, b_i] : 1 \leq i \leq n\}$ be a finite collection of non-overlapping intervals in $[a,b]$. Define

$$f = \sum_{i=1}^{n} \operatorname{sgn}(F(b_i) - F(a_i))\chi_{[a_i,b_i]}.$$

Then

$$\int_{a}^{b} |f(t)|^p dt = \sum_{i=1}^{n}|b_i - a_i|,$$

and

$$\varphi(f) = \sum_{i=1}^{n} \operatorname{sgn}(F(b_i) - F(a_i))\varphi(\chi_{[a_i,b_i]}) = \sum_{i=1}^{n}|F(b_i) - F(a_i)|$$

since $\varphi(\chi_{[b_i,a_i]}) = F(b_i) - F(a_i)$. Therefore

$$\sum_{i=1}^{n}|F(b_i) - F(a_i)| \leq \|\varphi\|\|f\|_p = \|\varphi\|(\sum_{i=1}^{n}|b_i - a_i|)^{1/p}.$$

We conclude that $\delta := \epsilon^p/2\|\varphi\|^p$ works. $\square$

Recall that a function $f : [a,b] \to \mathbb{K}$ is said to be a ***step function*** if it is a finite linear combination of characteristic functions of sub-intervals of $[a,b]$. In particular, a step function is a simple function. Let $\mathcal{S}$ denote the set of all step functions on

$[a, b]$. Then it is clear that $S \subset L^p[a, b]$ for any $1 \le p \le \infty$. The next lemma is a small piece of the puzzle we need to complete our proof and we leave it as an exercise for the reader.

4.1.13 LEMMA *Let S denote the set of all step functions on an interval $[a, b]$. If $1 \le p < \infty$, then S is dense in $L^p[a, b]$. Furthermore, if $f \in L^p[a, b]$ is a bounded function, then there is a sequence $(f_n) \subset S$ and $M > 0$ such that $f_n \to f$ in $L^p[a, b]$ and $|f_n| \le M$ for all $n \in \mathbb{N}$.*

We are finally in a position to prove the main result of the section.

4.1.14 THEOREM (RIESZ REPRESENTATION THEOREM (F. RIESZ, 1909)) *If $1 \le p < \infty$, then the map $\Delta : L^q[a, b] \to (L^p[a, b])^*$ is an isometric isomorphism.*

Proof. By Proposition 4.1.5, it suffices to prove that Δ is surjective. So fix $\varphi \in (L^p[a, b])^*$ and we want to construct $g \in L^q[a, b]$ such that $\varphi = \varphi_g$. As you might expect, we start with the function $F : [a, b] \to \mathbb{C}$ given by

$$F(x) := \varphi(\chi_{[a,x]}).$$

By Lemma 4.1.12, F is absolutely continuous. Hence by Lebesgue's Fundamental Theorem of Calculus there exists $g \in L^1[a, b]$ such that

$$\varphi(\chi_{[a,x]}) = F(x) = \int_a^x g(t)dt = \int_a^b \chi_{[a,x]}(t)g(t)dt.$$

Hence if $f \in L^p[a, b]$ is a step function, then by linearity it follows that

$$\varphi(f) = \int_a^b f(t)g(t)dt.$$

Now let $S \subset L^p[a, b]$ denote the subspace of step functions. If $f \in L^p[a, b]$ is a bounded function, then by Lemma 4.1.13 there is a sequence $(f_n) \in S$ and $M > 0$ such that $f_n \to f$ in $L^p[a, b]$ and $|f_n| \le M$ for all $n \in \mathbb{N}$. Replacing (f_n) by a subsequence if necessary, we may assume that $f_n \to f$ pointwise a.e. Hence $f_n g \to fg$ pointwise a.e. and $|f_n g| \le M|g| \in L^1[a, b]$ for all $n \in \mathbb{N}$. By the Dominated

Convergence Theorem,

$$\int_a^b f(t)g(t)dt = \lim_{n\to\infty} \int_a^b f_n(t)g(t)dt = \lim_{n\to\infty} \varphi(f_n).$$

Since $\|f_n\|_p \to \|f\|_p$ (by Remark 2.1.2), it follows that

$$\left| \int_a^b f(t)g(t)dt \right| = \lim_{n\to\infty} |\varphi(f_n)| \le \lim_{n\to\infty} \|\varphi\| \|f_n\|_p = \|\varphi\| \|f\|_p.$$

Therefore the inequality

$$\left| \int_a^b f(t)g(t)dt \right| \le \|\varphi\| \|f\|_p$$

holds for all bounded measurable functions $f \in L^p[a,b]$. By Lemma 4.1.11 we conclude that $g \in L^q[a,b]$ and $\|g\|_q \le \|\varphi\|$.

Now consider the bounded linear functional $\varphi_g \in (L^p[a,b])^*$ and observe that $\varphi(f) = \varphi_g(f)$ for all $f \in \mathcal{S}$. Since $\mathcal{S}$ is dense in $L^p[a,b]$, it follows that $\varphi = \varphi_g$. This concludes the proof.   □

The Riesz Representation Theorem explicitly identifies the dual space of $L^p[a,b]$ (and also ℓ^p) for $1 \le p < \infty$. However, in the course of the proof, we have conspicuously left out the case of $p = \infty$. In fact, the proof relied on various facts (such as the density of step functions, the Dominated Convergence Theorem and others) which only work for $p < \infty$. In order to resolve this question, we need to now explain why the map $\Delta : L^1[a,b] \to (L^\infty[a,b])^*$ is not surjective (it is, of course, isometric and therefore injective).

In order to do this, we need a recipe to construct bounded linear functionals with prescribed conditions. The most general such theorem is the Hahn–Banach Extension Theorem, which is what we turn to next.

4.2   The Hahn–Banach Extension Theorem

So now we know something about the duals of the ℓ^p and L^p spaces, but what about a general normed linear space? If you think about it, we do not even know if the dual space is non-zero! Of course, if $\mathbf{E}$ is finite dimensional, then $\mathbf{E}^*$ is non-zero. However, for infinite dimensional spaces, the obvious 'dual basis' to a given Hamel basis may not yield *bounded* linear functionals (indeed, they will not – see Exercise 5.33). Therefore we need a way to *construct* bounded linear functionals.

The Hahn–Banach Theorem is the most general way to do so. Its usefulness cannot be overstated and it is one of the most important results in the subject.

Suppose $\mathbf{E}$ is a normed linear space and $\mathbf{F} < \mathbf{E}$ a subspace. Given a bounded linear functional $\varphi : \mathbf{F} \to \mathbb{K}$, we would like to construct a bounded linear functional $\psi : \mathbf{E} \to \mathbb{K}$ such that

$$\psi|_{\mathbf{F}} = \varphi.$$

In other words, ψ would be a continuous extension of φ. Furthermore, we would like $\|\psi\| = \|\varphi\|$. Hence ψ would be a *norm-preserving extension* of φ. We have already seen two situations in which this is possible: If $\overline{\mathbf{F}} = \mathbf{E}$, then there is a unique continuous extension of φ by Proposition 3.2.4; and if $\mathbf{E}$ is a Hilbert space, then such is a continuous, norm-preserving extension exists by Corollary 3.2.6.

However, in the general case, we need a new idea to construct such an extension. If $\operatorname{codim}(\mathbf{F}) = 1$, then we can write

$$\mathbf{E} = \{x + \alpha e : x \in \mathbf{F}, \alpha \in \mathbb{K}\}$$

for some $e \in \mathbf{E} \setminus \mathbf{F}$. Hence we only need to define $\psi(e) \in \mathbb{K}$ in such a way that

$$|\psi(z)| \leq \|\varphi\| \|z\| \tag{4.1}$$

for all $z \in \mathbf{E}$ (this would automatically imply that $\|\psi\| = \|\varphi\|$). Note that the real issue here is not the existence of an extension. An extension may be constructed by defining $\psi(e)$ to be *anything* in $\mathbb{K}$. However, it is preserving the inequality of Equation 4.1 (and thereby ensuring that the extension is continuous) that is the difficult part.

In order to ensure that the result we obtain is widely applicable, it makes sense to prove it in a little more generality than what is needed in this specific instance. Recall from Definition 2.5.1 that a seminorm is a function $p : \mathbf{E} \to \mathbb{R}_+$ on a vector space that has all the requirements of a norm except that it need not be positive definite. The next lemma is the most basic step in the theorem and the one that needs the most attention. Since it needs the underlying field to have an order structure, the proof only applies to real-valued linear functionals.

4.2.1 LEMMA *Let $\mathbf{E}$ be a vector space over $\mathbb{R}$ and $p : \mathbf{E} \to \mathbb{R}$ be a seminorm on $\mathbf{E}$. Let $\mathbf{F} < \mathbf{E}$ be a subspace of $\mathbf{E}$ of codimension one and let $\varphi : \mathbf{F} \to \mathbb{R}$ be a linear functional on $\mathbf{F}$ satisfying*

$$\varphi(x) \leq p(x)$$

for all $x \in F$. Then there exists a linear functional $\psi : E \to \mathbb{R}$ such that $\psi|_F = \varphi$ and

$$\psi(x) \leq p(x)$$

for all $x \in E$.

Proof.   Since $\mathrm{codim}(F) = 1$, there exists $e \in E \setminus F$ such that every $z \in E$ can be expressed in the form $z = x + \alpha e$ for some (unique) $x \in F$ and $\alpha \in \mathbb{R}$. As mentioned above, we need to determine a real number

$$t := \psi(e)$$

so that $\psi(z) \leq p(z)$ may hold for all $z \in E$. In order to do so, we first assume that such an extension ψ exists. We use it to determine conditions that this real number 't' must satisfy. Having arrived at this *necessary* condition, we then proceed to show that such a real number exists and complete the proof by observing that this necessary condition is, in fact, sufficient as well.

(i) Suppose $\psi : E \to \mathbb{R}$ exists such that $\psi|_F = \varphi$ and $\psi(z) \leq p(z)$ for all $z \in E$. Then set $t := \psi(e)$ so that $\psi(x + \alpha e) = \varphi(x) + \alpha t \leq p(x + \alpha e)$. Consider the two cases:

(a) If $\alpha > 0$, then t must satisfy the inequality

$$t \leq \frac{p(x + \alpha e) - \varphi(x)}{\alpha}$$

for all $x \in F$. Hence

$$t \leq t_1 := \inf \left\{ \frac{p(x + \alpha e) - \varphi(x)}{\alpha} : x \in F, \alpha \in \mathbb{R}_{>0} \right\}.$$

(b) If $\alpha < 0$, we write $\beta = -\alpha > 0$. Then for any $y \in F$,

$$\psi(y - \beta e) = \varphi(y) - \beta t \leq p(y - \beta e).$$

Hence t must satisfy the inequality

$$t \geq \frac{\varphi(y) - p(y - \beta e)}{\beta}$$

for all $y \in \mathbf{F}$. Therefore

$$t \geq t_2 := \sup \left\{ \frac{\varphi(y) - p(y - \beta e)}{\beta} : y \in \mathbf{F}, \beta \in \mathbb{R}_{>0} \right\}.$$

However, this is only possible if $t_2 \leq t_1$ so let us verify this fact.

(ii) We want to show that for all $x, y \in \mathbf{F}$ and all $\alpha, \beta > 0$, the inequality

$$\frac{\varphi(y) - p(y - \beta e)}{\beta} \leq \frac{p(x + \alpha e) - \varphi(x)}{\alpha}$$

holds. Cross-multiplying, this amounts to proving that

$$\varphi(\alpha y) + \varphi(\beta x) \leq \beta p(x + \alpha e) + \alpha p(y - \beta e).$$

Since p is a seminorm, this reduces to proving that

$$\varphi(\alpha y + \beta x) \leq p(\beta x + \alpha \beta e) + p(\alpha y - \alpha \beta e).$$

Now this last inequality follows from a calculation:

$$\begin{aligned} \varphi(\alpha y + \beta x) &= \varphi(\alpha y - \alpha \beta e + \alpha \beta e + \beta x) \\ &= \varphi(\alpha y - \alpha \beta e) + \varphi(\beta x + \alpha \beta e) \\ &\leq p(\alpha y - \alpha \beta e) + p(\beta x + \alpha \beta e). \end{aligned}$$

Therefore we conclude that $t_2 \leq t_1$ as desired.

(iii) Having discussed the necessary condition in Step (i), we now construct the extension. Consider t_1 and t_2 as above so that $t_2 \leq t_1$. We choose $t \in \mathbb{R}$ such that $t_2 \leq t \leq t_1$ and *define* $\psi : \mathbf{E} \to \mathbb{R}$ by

$$\psi(x + \alpha e) := \varphi(x) + \alpha t.$$

Then this map ψ is linear and satisfies the condition that $\psi(z) \leq p(z)$ for all $z \in \mathbf{E}$. $\square$

Having completed the codimension one case, we need Zorn's Lemma to complete the argument.

4.2.2 THEOREM (HAHN–BANACH THEOREM: REAL CASE (HAHN, 1927, AND BANACH, 1929)) *Let* $\mathbf{E}$ *be a vector space over* $\mathbb{R}$ *and* $p : \mathbf{E} \to \mathbb{R}$ *be a seminorm on* $\mathbf{E}$. *Let* $\mathbf{F} < \mathbf{E}$ *be a subspace and* $\varphi : \mathbf{F} \to \mathbb{R}$ *a linear functional such that*

$$\varphi(x) \leq p(x)$$

for all $x \in \mathbf{F}$. *Then there exists a linear functional* $\psi : \mathbf{E} \to \mathbb{R}$ *such that*

$$\psi|_{\mathbf{F}} = \varphi$$

and $\psi(x) \leq p(x)$ *for all* $x \in \mathbf{E}$.

Proof.   If $n := \operatorname{codim}(\mathbf{F}) < \infty$, then we repeat Lemma 4.2.1 inductively.

If not, then we appeal to Zorn's Lemma. Define $\mathcal{F}$ to be the set of all pairs $(\mathbf{W}, \psi_{\mathbf{W}})$, where $\mathbf{W} < \mathbf{E}$ is a subspace containing $\mathbf{F}$ and $\psi_{\mathbf{W}} : \mathbf{W} \to \mathbb{R}$ is a linear functional on $\mathbf{W}$ such that

$$\psi_{\mathbf{W}}|_{\mathbf{F}} = \varphi$$

and $\psi_{\mathbf{W}}(x) \leq p(x)$ for all $x \in \mathbf{W}$. The set $\mathcal{F}$ is clearly non-empty since $(\mathbf{F}, \varphi) \in \mathcal{F}$. Define a partial order $\leq$ on $\mathcal{F}$ by setting $(\mathbf{W}_1, \psi_{\mathbf{W}_1}) \leq (\mathbf{W}_2, \psi_{\mathbf{W}_2})$ if and only if

$$\mathbf{W}_1 \subset \mathbf{W}_2 \text{ and } \psi_{\mathbf{W}_2}|_{\mathbf{W}_1} = \psi_{\mathbf{W}_1}.$$

It is easy to check that $\mathcal{F}$ now becomes a partially ordered set. We now verify that Zorn's Lemma is applicable: Let $\mathcal{C}$ be a totally ordered subset of $\mathcal{F}$. Define

$$\mathbf{W}_0 := \bigcup_{(\mathbf{W}, \psi_{\mathbf{W}}) \in \mathcal{C}} \mathbf{W}.$$

Then since $\mathcal{C}$ is totally ordered, $\mathbf{W}_0$ is a subspace. Define $\psi_0 : \mathbf{W}_0 \to \mathbb{R}$ by

$$\psi_0(x) := \psi_{\mathbf{W}}(x)$$

if $x \in \mathbf{W}$. Then one can check that ψ_0 is well-defined (again since $\mathcal{C}$ is totally ordered). Furthermore, $\psi_0(x) \leq p(x)$ holds for all $x \in \mathbf{W}_0$. Hence $(\mathbf{W}_0, \psi_0) \in \mathcal{F}$ and it is clearly an upper bound for $\mathcal{C}$.

Zorn's Lemma now tells us that $\mathcal{F}$ has a maximal element, which we denote by $(\mathbf{F}_0, \varphi_0)$. Now we claim that $\mathbf{F}_0 = \mathbf{E}$ and this is where we need the previous lemma. Suppose not, then there exists $e \in \mathbf{E} \setminus \mathbf{F}_0$. Define $\mathbf{F}_1 := \operatorname{span}(\mathbf{F}_0 \cup \{e\})$. Then by Lemma 4.2.1 we may extend φ_0 to a linear map $\varphi_1 : \mathbf{F}_1 \to \mathbb{R}$ satisfying

$$\varphi_1(x) \leq p(x)$$

for all $x \in \mathbf{F}_1$. This would mean that $(\mathbf{F}_1, \varphi_1) \in \mathcal{F}$, which would contradict the maximality of $(\mathbf{F}_0, \varphi_0)$. Hence $\mathbf{F}_0 = \mathbf{E}$, and $\psi = \varphi_0$ is the required linear functional. $\qquad\square$

We now wish to prove the analogous theorem for complex vector spaces. In order to do this, we need a natural way to pass from complex linear functionals to real linear functionals and that is what the next lemma does for us.

4.2.3 LEMMA *Let* **E** *be a complex vector space.*

(i) *If* $\varphi : \mathbf{E} \to \mathbb{R}$ *is an* $\mathbb{R}$*-linear functional, then*

$$\widehat{\varphi}(x) := \varphi(x) - i\varphi(ix)$$

is a $\mathbb{C}$*-linear functional*

(ii) *If* $\psi : \mathbf{E} \to \mathbb{C}$ *is a* $\mathbb{C}$*-linear functional, then* $\varphi := \mathrm{Re}(\psi)$ *is a* $\mathbb{R}$*-linear functional and* $\widehat{\varphi} = \psi$.

(iii) *If* p *is a seminorm and* φ *and* $\widehat{\varphi}$ *are as in part (i), then the inequality*

$$|\varphi(x)| \le p(x)$$

holds for all $x \in \mathbf{E}$ *if and only if*

$$|\widehat{\varphi}(x)| \le p(x)$$

holds for all $x \in \mathbf{E}$.

(iv) *If* **E** *is a normed linear space and* φ *and* $\widehat{\varphi}$ *are as above, then* $\|\varphi\| = \|\widehat{\varphi}\|$.

Proof.   The proofs of the first two statements are left as an exercise.

(iii) Suppose that the inequality

$$|\widehat{\varphi}(x)| \le p(x)$$

holds for all $x \in \mathbf{E}$. Then, clearly, the same inequality holds with φ. Conversely, suppose

$$|\varphi(x)| \le p(x)$$

holds for all $x \in \mathbf{E}$, then fix $x \in \mathbf{E}$ and choose $\lambda \in \mathbb{C}$ with $|\lambda| = 1$ such that $|\widehat{\varphi}(x)| = \lambda\widehat{\varphi}(x)$. Then

$$|\widehat{\varphi}(x)| = \widehat{\varphi}(\lambda x) = \mathrm{Re}(\widehat{\varphi}(\lambda x)) = \varphi(\lambda x) \le p(\lambda x) = |\lambda|p(x) = p(x).$$

Thus $|\widehat{\varphi}(x)| \le p(x)$ for all $x \in \mathbf{E}$.

(iv) Let $p(x) := \|\varphi\| \|x\|$. Then $|\varphi(x)| \leq p(x)$ for all $x \in \mathbf{E}$. By part (iii), $|\widehat{\varphi}(x)| \leq p(x)$ for all $x \in \mathbf{E}$ and so $\|\widehat{\varphi}\| \leq \|\varphi\|$. The reverse inequality is obvious since $|\varphi(x)| = |\operatorname{Re}(\widehat{\varphi}(x))| \leq |\widehat{\varphi}(x)|$ for all $x \in \mathbf{E}$. $\qquad\square$

The transition from the real case to the complex case is now easy.

Let $\mathbf{E}$ be a normed linear space and $p : \mathbf{E} \to \mathbb{R}_+$ be a seminorm on $\mathbf{E}$. Let $\mathbf{F} < \mathbf{E}$ be a subspace and $\varphi : \mathbf{F} \to \mathbb{K}$ a linear functional such that

$$|\varphi(x)| \leq p(x)$$

for all $x \in \mathbf{F}$. Then there exists a linear functional $\psi : \mathbf{E} \to \mathbb{K}$ such that $\psi|_\mathbf{F} = \varphi$ and

$$|\psi(x)| \leq p(x)$$

for all $x \in \mathbf{E}$.

Proof.

(i) Suppose $\mathbf{E}$ is an $\mathbb{R}$-vector space: By Theorem 4.2.2, there exists a linear functional $\psi : \mathbf{E} \to \mathbb{R}$ such that

$$\psi(x) \leq p(x)$$

for all $x \in \mathbf{E}$. Then

$$-\psi(x) = \psi(-x) \leq p(-x) = p(x)$$

holds for all $x \in \mathbf{E}$, and so $|\psi(x)| \leq p(x)$ holds.

(ii) Suppose $\mathbf{E}$ is a $\mathbb{C}$-vector space: Define $f : \mathbf{F} \to \mathbb{R}$ by $f = \operatorname{Re}(\varphi)$. Then f is linear and

$$|f(x)| \leq p(x)$$

for all $x \in \mathbf{F}$. By part (i) there is a linear functional $g : \mathbf{E} \to \mathbb{R}$ such that

$$g|_\mathbf{F} = f$$

and $|g(x)| \leq p(x)$ for all $x \in \mathbf{E}$. Define $\psi : \mathbf{E} \to \mathbb{C}$ by $\psi := \widehat{g}$ as in Lemma 4.2.3. Then ψ is $\mathbb{C}$-linear and satisfies

$$|\psi(x)| \leq p(x)$$

for all $x \in \mathbf{E}$. This completes the proof. $\qquad\square$

The first theorem extending linear functionals was proved by F. Riesz (1910-11), who wanted an analogue of the Riesz–Fischer Theorem (part (v) of Theorem 3.5.5) for L^p spaces when $p \neq 2$. The next such result was due to Helley (1912), who introduced the important step of enlarging the original subspace by one vector (Lemma 4.2.1). It was in the independent work of Hahn (1927) and Banach (1929) that we see the emergence of the complete proof of the theorem for real Banach spaces. Both of them used Helly's idea and added to it the use of transfinite induction (Zorn's Lemma would not exist until 1935).

The complex version of the theorem was proved by Bohnenblust and Sobczyk (1938). They were also the first to refer to it as the **Hahn–Banach Theorem**.

The next corollary is the way the Hahn–Banach Theorem gets used most of the time. In fact, each corollary we state here is often taken to be a part of a suite of results collectively termed 'the Hahn–Banach Theorem'. This slight abuse of language is indicative of how ubiquitous the theorem is in the literature.

4.2.5 COROLLARY *Let* $\mathbf{E}$ *be a normed linear space and* $\mathbf{F} < \mathbf{E}$ *a subspace. Then for any bounded linear functional* $\varphi \in \mathbf{F}^*$ *there exists* $\psi \in \mathbf{E}^*$ *such that*

$$\psi|_\mathbf{F} = \varphi$$

and $\|\psi\| = \|\varphi\|$.

Proof. Apply the Hahn–Banach Theorem with $p(x) := \|\varphi\|\|x\|$ □

4.2.6 COROLLARY *Let* $\mathbf{E}$ *be a normed linear space,* $\{e_1, e_2, \ldots, e_n\} \subset \mathbf{E}$ *be a finite, linearly independent set and* $\{\alpha_1, \alpha_2, \ldots, \alpha_n\} \subset \mathbb{K}$ *be arbitrary scalars. Then there exists a bounded linear functional* $\psi \in \mathbf{E}^*$ *such that*

$$\psi(e_i) = \alpha_i$$

for all $1 \leq i \leq n$.

Proof. Take $\mathbf{F} := \text{span}(\{e_1, e_2, \ldots, e_n\})$ and define $\varphi : \mathbf{F} \to \mathbb{K}$ by

$$\varphi(e_i) = \alpha_i,$$

extended linearly to all of $\mathbf{F}$. Since $\mathbf{F}$ is finite dimensional, φ is a bounded linear functional on $\mathbf{F}$ (by Corollary 2.4.9). We may now apply Corollary 4.2.5 to get a linear functional $\psi \in \mathbf{E}^*$ satisfying the required conditions. □

The next corollary is an interesting one. It states that for any vector x in a normed linear space $\mathbf{E}$, there is a bounded linear functional $\psi \in \mathbf{E}^*$ which satisfies $|\psi(x)| = \|x\|$. In other words, the norm of the vector can be realized by evaluation by bounded linear functionals. In particular, it resolves one important question we mentioned at the start of the section; we finally see that $\mathbf{E}^* \neq \{0\}$ whenever $\mathbf{E} \neq \{0\}$. That such a basic fact required a proof speaks volumes for the subtleties of the subject.

4.2.7 COROLLARY *Let $\mathbf{E}$ be a normed linear space and $x \in \mathbf{E}$. Then*

$$\|x\| = \sup\{|\psi(x)| : \psi \in \mathbf{E}^*, \|\psi\| \leq 1\}.$$

Furthermore, this supremum is attained.

Proof.   Set $\alpha := \sup\{|\psi(x)| : \psi \in \mathbf{E}^*, \|\psi\| \leq 1\}$. Then, it is easy to see that $\alpha \leq \|x\|$. Conversely, set $\mathbf{F} := \operatorname{span}(x)$. Define $\varphi : \mathbf{F} \to \mathbb{K}$ by

$$\beta x \mapsto \beta\|x\|.$$

Then $\varphi \in \mathbf{F}^*$; indeed, it is easy to check that $\|\varphi\| = 1$. By Corollary 4.2.5 there exists $\psi \in \mathbf{E}^*$ such that $\psi(x) = \|x\|$ and $\|\psi\| = 1$. Hence $\|x\| = \alpha$ as desired. $\square$

If $\mathbf{E}$ is a normed linear space and $\varphi \in \mathbf{E}^*$, then we know that

$$\|\varphi\| = \sup\{|\varphi(y)| : y \in \mathbf{E}, \|y\| \leq 1\}. \tag{4.2}$$

Corollary 4.2.7 may be thought of as a mirror image of this fact except that the norm in Equation 4.2 may not be attained.

The next corollary, and the last of this section, states that the dual space *separates points* of the underlying normed linear space. It is a straightforward consequence of Corollary 4.2.7.

4.2.8 COROLLARY *If $x, y \in \mathbf{E}$ are two vectors such that $\psi(x) = \psi(y)$ for all $\psi \in \mathbf{E}^*$, then $x = y$.*

We will revisit the Hahn–Banach Theorem in Chapter 6, where we will explore the deep relationship between a normed linear space and its dual space. We will establish a geometric version of the theorem, which exploits the notion of convexity. For now, though, let us give some useful consequences of the theorem.

4.3 Duals of Subspaces and Quotient Spaces

Let $\mathbf{E}$ be a normed linear space and $\mathbf{F} < \mathbf{E}$ be a subspace. We wish to understand the relationship between the dual spaces $\mathbf{E}^*$ and $\mathbf{F}^*$. For any bounded linear functional $\varphi \in \mathbf{E}^*$, it is clear that its restriction $\varphi|_{\mathbf{F}}$ is a bounded linear functional on $\mathbf{F}$ and that $\|\varphi|_{\mathbf{F}}\| \leq \|\varphi\|$. This gives us a bounded linear map $R : \mathbf{E}^* \to \mathbf{F}^*$ given by

$$\varphi \mapsto \varphi|_{\mathbf{F}}.$$

The Hahn–Banach Theorem merely says that this map is surjective.

4.3.1 DEFINITION The space $\ker(R) := \{\varphi \in \mathbf{E}^* : \varphi|_{\mathbf{F}} = 0\}$ is called the **annihilator** of $\mathbf{F}$ and is denoted by $\mathbf{F}^{\perp}$. Note that $\mathbf{F}^{\perp}$ is a subspace of $\mathbf{E}^*$ and that it is closed.

If $\mathbf{E}$ is a Hilbert space, then for any $\varphi \in \mathbf{E}^*$ there exists $y \in \mathbf{E}$ such that

$$\varphi(x) = \langle x, y \rangle$$

for all $x \in \mathbf{E}$ (by Theorem 3.2.2). Therefore $\varphi|_{\mathbf{F}} = 0$ if and only if $y \in \mathbf{F}^{\perp}$ (as defined in Definition 3.1.1). This justifies the abuse of notation, although one should keep in mind that when $\mathbf{E}$ is not a Hilbert space, $\mathbf{F}^{\perp}$ is a subset of $\mathbf{E}^*$ (and not of $\mathbf{E}$).

4.3.2 PROPOSITION *Let $\mathbf{F}$ be a subspace of a normed linear space $\mathbf{E}$. Then the map $R : \mathbf{E}^* \to \mathbf{F}^*$ given by*

$$\varphi \mapsto \varphi|_{\mathbf{F}},$$

induces an isometric isomorphism $\widehat{R} : \mathbf{E}^*/\mathbf{F}^{\perp} \to \mathbf{F}^*$ given by

$$\varphi + \mathbf{F}^{\perp} \mapsto \varphi|_{\mathbf{F}}.$$

Proof. As mentioned above, $\mathbf{F}^{\perp}$ is a closed subspace of $\mathbf{E}^*$. Hence $\mathbf{E}^*/\mathbf{F}^{\perp}$ is a normed linear space (by Proposition 2.5.2). That $\widehat{R}$ exists follows from the First Isomorphism Theorem. Furthermore, since R is surjective, $\widehat{R}$ is an isomorphism of vector spaces.

It remains to show that $\widehat{R}$ is isometric. So fix $\varphi \in \mathbf{E}^*$. Then for any $\psi \in \mathbf{F}^{\perp}$, we have $\varphi|_{\mathbf{F}} = (\varphi + \psi)|_{\mathbf{F}}$. Hence

$$\|R(\varphi)\| \leq \|\varphi + \psi\|.$$

This is true for all $\psi \in \mathbf{F}^\perp$ and therefore

$$\|\widehat{R}(\varphi + \mathbf{F}^\perp)\| = \|R(\varphi)\| \le \|\varphi + \mathbf{F}^\perp\|.$$

For the reverse inequality, set $\eta := \varphi|_{\mathbf{F}}$. Then $\eta \in \mathbf{F}^*$ so there exists $\psi \in \mathbf{E}^*$ such that $\psi|_{\mathbf{F}} = \eta$ and $\|\psi\| = \|\eta\|$ by the Hahn–Banach Theorem. Hence

$$\|\widehat{R}(\psi + \mathbf{F}^\perp)\| = \|\eta\| = \|\psi\| \ge \|\psi + \mathbf{F}^\perp\|.$$

But $\psi - \varphi \in \mathbf{F}^\perp$ so that $\psi + \mathbf{F}^\perp = \varphi + \mathbf{F}^\perp$. We conclude that $\|\widehat{R}(\varphi + \mathbf{F}^\perp)\| \ge \|\varphi + \mathbf{F}^\perp\|$, as desired. $\qquad\square$

The next theorem may be viewed as a strengthening of Corollary 4.2.8. There we were able to separate points in a normed linear space by a bounded linear functional. Here one separates a closed subspace from a point outside it by means of a bounded linear functional.

4.3.3 PROPOSITION *Let* $\mathbf{E}$ *be an normed linear space,* $\mathbf{F} < \mathbf{E}$ *a subspace,* $x_0 \in \mathbf{E} \setminus \mathbf{F}$ *and suppose that*

$$d := d(x_0, \mathbf{F}) > 0.$$

Then there exists $\varphi \in \mathbf{E}^*$ *such that*

  (a) $\varphi(x) = 0$ *for all* $x \in \mathbf{F}$,

  (b) $\varphi(x_0) = 1$ *and*

  (c) $\|\varphi\| = d^{-1}$.

Proof. Since $d(x_0, \mathbf{F}) = d(x_0, \overline{\mathbf{F}}) > 0$, we may assume without loss of generality that $\mathbf{F}$ is closed. If $\pi : \mathbf{E} \to \mathbf{E}/\mathbf{F}$ denotes the natural quotient map, then $\|\pi(x_0)\| = d$. Hence by Corollary 4.2.7 there exists $\psi \in (\mathbf{E}/\mathbf{F})^*$ such that $\|\psi\| = 1$ and

$$\psi(x_0 + \mathbf{F}) = \|x_0 + \mathbf{F}\| = d.$$

Define $\varphi : \mathbf{E} \to \mathbb{K}$ by

$$\varphi := d^{-1} \psi \circ \pi.$$

Then $\varphi \in \mathbf{E}^*$ satisfies conditions (a) and (b). To verify condition (c), observe that $\|\pi\| \le 1$ and therefore

$$|\varphi(y)| \le d^{-1} \|\psi\| \|\pi(y)\| \le d^{-1} \|y\|$$

for all $y \in \mathbf{E}$. Hence $\|\varphi\| \leq d^{-1}$. However, $\|\psi\| = 1$ and hence

$$\sup\{|\psi(z + \mathbf{F})| : \|z + \mathbf{F}\| = 1\} = 1.$$

Therefore there exists a sequence $(x_n + \mathbf{F}) \in \mathbf{E}/\mathbf{F}$ such that $\|x_n + \mathbf{F}\| = 1$ for all $n \in \mathbb{N}$ and

$$\lim_{n \to \infty} |\psi(x_n + \mathbf{F})| = 1.$$

Let $y_n \in \mathbf{F}$ such that $\|x_n + y_n\| < 1 + 1/n$ for each $n \in \mathbb{N}$ so that

$$d^{-1}|\psi(x_n + \mathbf{F})| = |\varphi(x_n + y_n)| \leq \|\varphi\|\|x_n + y_n\| \leq \|\varphi\|(1 + 1/n).$$

Letting $n \to \infty$, we conclude that $d^{-1} \leq \|\varphi\|$. $\qquad\square$

The next corollary is remarkably widely used in the subject and is another one of those results that is simply subsumed under the title of 'Hahn–Banach Theorem'.

4.3.4 COROLLARY *Let* $\mathbf{E}$ *be an normed linear space, and* $\mathbf{F} < \mathbf{E}$ *be a subspace. Then*

$$\overline{\mathbf{F}} = \bigcap_{\varphi \in \mathbf{F}^{\perp}} \ker(\varphi).$$

In other words, a vector $x \in \mathbf{E}$ *belongs to* $\overline{\mathbf{F}}$ *if and only if* $\varphi(x) = 0$ *for all* $\varphi \in \mathbf{F}^{\perp}$.

Proof. Clearly,

$$\mathbf{W} := \bigcap_{\varphi \in \mathbf{F}^{\perp}} \ker(\varphi)$$

is a closed subspace containing $\mathbf{F}$ and hence $\overline{\mathbf{F}} \subset \mathbf{W}$. Conversely, if $x_0 \notin \overline{\mathbf{F}}$, then $d(x_0, \mathbf{F}) > 0$. By Proposition 4.3.3 there exists $\varphi \in \mathbf{E}^*$ such that $\mathbf{F} \subset \ker(\varphi)$ and

$$\varphi(x_0) \neq 0.$$

Hence $x_0 \notin \mathbf{W}$. We conclude that $\mathbf{W} \subset \overline{\mathbf{F}}$ as well. $\qquad\square$

As discussed above, the notion of the annihilator $\mathbf{F}^{\perp}$ arises most naturally in the context of Hilbert spaces. Along those lines, the next result may be thought of as an analogue of Corollary 3.1.14. Its proof is an immediate consequence of Corollary 4.3.4.

4.3.5 COROLLARY *Let* $\mathbf{E}$ *be an normed linear space and* $\mathbf{F} < \mathbf{E}$ *a subspace. Then* $\mathbf{F}$ *is dense in* $\mathbf{E}$ *if and only if* $\mathbf{F}^{\perp} = \{0\}$.

Now let $\mathbf{F} < \mathbf{E}$ be a closed subspace so that $\mathbf{E}/\mathbf{F}$ is a normed linear space. We wish to identify $(\mathbf{E}/\mathbf{F})^*$ in terms of $\mathbf{F}$ and $\mathbf{E}$. Note that $\pi : \mathbf{E} \to \mathbf{E}/\mathbf{F}$ is continuous. Hence if ψ is a bounded linear functional on $\mathbf{E}/\mathbf{F}$, then $\psi \circ \pi : \mathbf{E} \to \mathbb{K}$ is bounded and linear. This gives us a map $Q : (\mathbf{E}/\mathbf{F})^* \to \mathbf{E}^*$, given by

$$\psi \mapsto \psi \circ \pi.$$

This map is clearly linear as well. The next theorem shows that Q is injective and identifies its range.

4.3.6 THEOREM *Let $\mathbf{F}$ be a closed subspace of a normed linear space $\mathbf{E}$. Then there is an isometric isomorphism $Q : (\mathbf{E}/\mathbf{F})^* \to \mathbf{F}^\perp$.*

Proof.   Note that for any $\psi \in (\mathbf{E}/\mathbf{F})^*$,

$$Q(\psi)|_{\mathbf{F}} = \psi \circ \pi|_{\mathbf{F}} = 0$$

since $\pi|_{\mathbf{F}} = 0$. Hence $Q(\psi) \in \mathbf{F}^\perp$. Since $\|\pi\| \leq 1$, we have

$$\|Q(\psi)\| \leq \|\psi\|. \tag{4.3}$$

Conversely, by Proposition 2.2.11 applied to $\mathbf{E}/\mathbf{F}$, we have

$$\|\psi\| = \sup\{|\psi(x + \mathbf{F})| : x + \mathbf{F} \in \mathbf{E}/\mathbf{F}, \|x + \mathbf{F}\| = 1\}.$$

Therefore there exists a sequence $(x_n) \subset \mathbf{E}$ such that $\|x_n + \mathbf{F}\| = 1$ for all $n \in \mathbb{N}$ and

$$\lim_{n \to \infty} |\psi(x_n + \mathbf{F})| = \|\psi\|.$$

For each $n \in \mathbb{N}$ there exists $y_n \in \mathbf{F}$ such that $\|x_n + y_n\| < 1 + 1/n$. Hence

$$|Q(\psi)(x_n + y_n)| = |\psi(x_n + y_n + \mathbf{F})| = |\psi(x_n + \mathbf{F})|.$$

This implies that

$$\|Q(\psi)\|(1 + 1/n) \geq |\psi(x_n + \mathbf{F})|.$$

Passing to the limit, we conclude that $\|Q(\psi)\| \geq \|\psi\|$. Together with Equation 4.3, it follows that $\|Q(\psi)\| = \|\psi\|$, proving that Q is isometric.

We now show that Q is surjective: If $\varphi \in \mathbf{F}^\perp$, then $\varphi \in \mathbf{E}^*$ and $\varphi|_{\mathbf{F}} = 0$. Hence φ induces a map $\psi : \mathbf{E}/\mathbf{F} \to \mathbb{K}$, given by

$$x + \mathbf{F} \mapsto \varphi(x).$$

This map is clearly well-defined and linear. Furthermore, since $\psi \circ \pi$ is bounded, it follows from Corollary 2.5.6 that $\psi \in (\mathbf{E}/\mathbf{F})^*$. By construction, $Q(\psi) = \varphi$ and hence Q is surjective. $\qquad\square$

4.4 Separability and Reflexivity

Let us now return to an unresolved question from the first section of this chapter. Recall that for each $y = (y_n) \in \ell^1$, the map $\varphi_y : \ell^\infty \to \mathbb{K}$, given by $(x_k) \mapsto \sum_{n=1}^\infty x_n y_n$ is a bounded linear functional. Furthermore, the map

$$\Delta : \ell^1 \to (\ell^\infty)^*$$

given by $\Delta(y) := \varphi_y$, is an isometry by Proposition 4.1.2. Similarly, we have an isometry

$$\Delta : L^1[a,b] \to (L^\infty[a,b])^*.$$

Earlier in the chapter, we had hinted that these maps may not be surjective. Armed with the Hahn–Banach Theorem, the proof of this fact is now at hand. To drive the point home, we give two proofs. We will first explain why these maps *ought not* to be surjective (for abstract reasons) and then give examples of bounded linear functionals that lie outside their ranges.

The first of these proofs stems from the following result. Recall that a topological space is said to be separable if it has a countable dense set.

4.4.1 PROPOSITION *Let* $\mathbf{E}$ *be an normed linear space. If* $\mathbf{E}^*$ *is separable, then* $\mathbf{E}$ *is separable.*

Proof. If $\mathbf{E}^*$ is separable, then it is easy to see that the unit sphere

$$S := \{\varphi \in \mathbf{E}^* : \|\varphi\| = 1\}$$

is separable. We choose a countable set $\{\varphi_1, \varphi_2, \ldots\} \subset S$ that is dense in S. For each $n \in \mathbb{N}$, $\|\varphi_n\| = 1$. Therefore there exist vectors $x_n \in \mathbf{E}$ such that $\|x_n\| = 1$ and

$$|\varphi_n(x_n)| > 1/2.$$

Let $\mathbf{F} = \mathrm{span}\{x_n : n \geq 1\}$. If we set $\mathbb{K}_0 = \mathbb{Q}$ or $\mathbb{Q} \times \mathbb{Q}$ according as $\mathbb{K} = \mathbb{R}$ or $\mathbb{C}$, then

$$D := \mathrm{span}_{\mathbb{K}_0}\{x_n : n \geq 1\}$$

is countable and dense in $\mathbf{F}$. Hence $\mathbf{F}$ is separable. We wish to prove that $\overline{\mathbf{F}} = \mathbf{E}$, and in order to do that, we would like to appeal to Corollary 4.3.5. Suppose $\varphi \in \mathbf{F}^\perp$

is a non-zero bounded linear operator, then we may assume that $\|\varphi\| = 1$. Since $\varphi(x_n) = 0$ for all $n \in \mathbb{N}$, it follows that

$$\|\varphi - \varphi_n\| \geq |\varphi(x_n) - \varphi_n(x_n)| > 1/2$$

for all $n \in \mathbb{N}$. This contradicts the fact that $\{\varphi_n : n \in \mathbb{N}\}$ is dense in S. Thus $\mathbf{F}^\perp = \{0\}$ and $\mathbf{F}$ is dense in $\mathbf{E}$ by Corollary 4.3.5. We conclude that $\mathbf{E}$ is separable. $\qquad\square$

4.4.2 COROLLARY *The maps*

$$\Delta : \ell^1 \to (\ell^\infty)^* \text{ and}$$

$$\Delta : L^1[a,b] \to (L^\infty[a,b])^*$$

are not surjective

Proof. If Δ were an isomorphism, then $(\ell^\infty)^*$ would be isomorphic to ℓ^1. However, ℓ^1 is separable so this would imply ℓ^∞ was separable, which contradicts Example 2.3.16. Similarly, $(L^\infty[a,b])^* \not\cong L^1[a,b]$. $\qquad\square$

We now give 'concrete' examples of bounded linear functionals that lie outside the range of Δ. That such functionals exist is, of course, a consequence of the Hahn–Banach Theorem. The first example we construct (of a bounded linear functional on ℓ^∞) is quite a bit more general than what we need for our purposes. However, this class of functionals is sufficiently important that they warrant such attention.

Let $S : \ell^\infty \to \ell^\infty$ denote the left-shift operator

$$S(x_1, x_2, x_3, \ldots) = (x_2, x_3, \ldots).$$

Let c denote the vector space of all convergent sequences in $\mathbb{K}$ (see Exercise 2.27), thought of as a closed subspace of ℓ^∞. Define $\varphi : c \to \mathbb{K}$ by

$$\varphi((x_j)) := \lim_{n \to \infty} x_n$$

This is well-defined, bounded and linear. Therefore by the Hahn–Banach Theorem, we may extend this to a bounded linear functional on ℓ^∞. We wish to construct a particular kind of extension so that we may have some additional properties. Such a linear functional is referred to as a *Banach limit*.

THEOREM (BANACH LIMITS) *There exists a bounded linear functional* ψ : $\ell^\infty \to \mathbb{K}$ *satisfying the following properties:*

(i) *If* $x = (x_j) \in c$, *then* $\psi(x) = \lim_{n \to \infty} x_n$.

(ii) *If* $S : \ell^\infty \to \ell^\infty$ *denotes the left-shift operator, then* $\psi(x) = \psi(S(x))$ *for all* $x \in \ell^\infty$.

(iii) $\|\psi\| = 1$.

Proof. Define

$$\mathbf{F} := \{x - S(x) : x \in \ell^\infty\}.$$

Then $\mathbf{F}$ is clearly a subspace of ℓ^∞. If $\mathbf{1} := (1, 1, 1, \ldots)$, then we claim that $d(\mathbf{1}, \mathbf{F}) = 1$. Firstly, observe that

$$d(\mathbf{1}, \mathbf{F}) \leq \|\mathbf{1}\| = 1.$$

Now, if $x = (x_n) \in \ell^\infty$ and $z = (z_n) := x - S(x) + \mathbf{1}$, then $z_n = x_n - x_{n+1} + 1$. Therefore for any $N \in \mathbb{N}$, we have

$$\|z\| \geq \frac{1}{N} \sum_{i=1}^{N} |z_i| \geq \frac{1}{N} \left| \sum_{i=1}^{N} z_i \right| = \left| \frac{1}{N}(x_{N+1} - x_1) + 1 \right| \geq 1 - \frac{|x_{N+1} - x_1|}{N}.$$

Since $|x_{N+1} - x_1| \leq 2\|x\|$, we may let $N \to \infty$ to conclude that $\|z\| \geq 1$. Hence $d(\mathbf{1}, \mathbf{F}) \geq 1$ as well and this proves the claim.

By Proposition 4.3.3 there is a bounded linear functional $\psi : \ell^\infty \to \mathbb{K}$ such that $\psi|_{\mathbf{F}} = 0$, $\psi(\mathbf{1}) = 1$ and $\|\psi\| = 1$. We verify that ψ has the remaining two properties as well.

(ii) For any $x \in \ell^\infty, x - S(x) \in \mathbf{F}$. Therefore $\psi(x) = \psi(S(x))$.

(i) Now if $x = (x_j) \in c$, then set

$$\alpha := \lim_{n \to \infty} x_n.$$

Then for any $\epsilon > 0$ there exists $N \in \mathbb{N}$ such that $|x_n - \alpha| < \epsilon$ for all $n \geq N$. Set

$$y = (x_N, x_{N+1}, x_{N+2}, \ldots) = S^N(x),$$

where S^N denotes the composition of S with itself N times. It follows from property (b) that

$$\psi(x) = \psi(y).$$

However, by construction, $\|y - \alpha \mathbf{1}\| \leq \epsilon$. Hence

$$|\psi(x) - \alpha| = |\psi(y) - \alpha| = |\psi(y - \alpha \mathbf{1})| \leq \|y - \alpha \mathbf{1}\| \leq \epsilon.$$

This is true for any $\epsilon > 0$ and therefore $\psi(x) = \alpha$. $\qquad\qquad$ □

We now return to our original problem of showing that the maps $\Delta : \ell^1 \to (\ell^\infty)^*$ and $\Delta : L^1[a, b] \to L^\infty[a, b]$ are not surjective.

4.4.4 EXAMPLE

(i) Let $\psi \in (\ell^\infty)^*$ be a Banach limit, as constructed in Theorem 4.4.3. We claim that $\psi \notin \Delta(\ell^1)$: Suppose there exists $y = (y_j)$ such that $\psi = \Delta(y) = \varphi_y$, then

$$\lim_{i \to \infty} x_i = \sum_{n=1}^{\infty} x_n y_n$$

for all $(x_j) \in c$. In particular, for $j \in \mathbb{N}$ fixed, consider $e_j \in \ell^\infty$ (as defined in Example 1.1.7). Then $e_j \in c_0 \subset c$ so

$$y_j = \psi(e_j) = 0.$$

Thus $y = 0$. But if this were true, it would imply that $\psi = 0$. However, $\psi \neq 0$ since $\psi(\mathbf{1}) = 1$. Therefore there is no $y \in \ell^1$ such that $\varphi = \Delta(y)$.

(ii) Consider $C[a, b] \subset L^\infty[a, b]$ and define $\varphi : C[a, b] \to \mathbb{K}$ by

$$\varphi(f) := f(b).$$

Then φ is a bounded linear functional so by the Hahn–Banach Theorem, there exists a bounded linear functional $\psi : L^\infty[a, b] \to \mathbb{K}$ such that

$$\psi(f) = f(b)$$

for all $f \in C[a, b]$. We claim that $\psi \notin \Delta(L^1[a, b])$. Suppose there exists $g \in L^1[a, b]$ such that $\psi = \varphi_g$, then g would satisfy

$$f(b) = \int_a^b f(t)g(t)dt$$

for all $f \in C[a,b]$. For $n \in \mathbb{N}$, let $f_n \in C[a,b]$ be given by

$$f_n(t) = \begin{cases} 0 & : \text{if } a \leq t \leq b - 1/n, \\ 1 & : \text{if } t = b, \\ \text{linear} & : \text{otherwise.} \end{cases}$$

Then for any $t \in [a,b)$,

$$\lim_{n \to \infty} f_n(t) = 0.$$

Hence $f_n \to 0$ pointwise a.e. Since $f_n \in L^\infty[a,b]$ and $g \in L^1[a,b]$, we may apply Dominated Convergence Theorem to conclude that

$$\lim_{n \to \infty} \psi(f_n) = \lim_{n \to \infty} \int_a^b f_n(t)g(t)dt = 0.$$

However, $\psi(f_n) = f_n(b) = 1$ for all $n \in \mathbb{N}$. This contradiction shows that $\psi \neq \varphi_g$ for any $g \in L^1[a,b]$.

Having sorted out this question, we turn to the other property mentioned in the title of this section – reflexivity. The idea is to turn the tables on the dual space and let the underlying normed linear space **E** *act* on **E***. This action leads to a natural inclusion of **E** into its *double dual*, the dual of **E***. We wish to understand this map and what it tells us about **E**.

4.4.5 DEFINITION Let **E** be an normed linear space.

(i) The *double dual* of **E** is the dual of **E*** and is denoted by $\mathbf{E}^{**} := (\mathbf{E}^*)^*$.

(ii) For each $x \in \mathbf{E}$, define $\hat{x} : \mathbf{E}^* \to \mathbb{K}$ by

$$\hat{x}(\varphi) := \varphi(x).$$

Note that $\hat{x}$ is a linear functional on **E*** and

$$\|\hat{x}\| = \sup\{|\varphi(x)| : \varphi \in \mathbf{E}^*, \|\varphi\| = 1\} = \|x\| \tag{4.4}$$

by Corollary 4.2.7. Therefore $\hat{x} \in \mathbf{E}^{**}$.

(iii) Define $J : \mathbf{E} \to \mathbf{E}^{**}$ by

$$J(x) := \widehat{x}.$$

Then J is a linear transformation, which is isometric by Equation 4.4.

(iv) $\mathbf{E}$ is said to be *reflexive* if J is an isomorphism from $\mathbf{E}$ to $\mathbf{E}^{**}$.

4.4.6 EXAMPLE

(i) If $\mathbf{E}$ is finite dimensional, then it is reflexive.

Proof. If $\mathbf{E}$ is finite dimensional, then

$$\dim(\mathbf{E}) = \dim(\mathbf{E}^*) = \dim(\mathbf{E}^{**}).$$

Since J is injective, it must be surjective.   □

(ii) Every Hilbert space is reflexive.

Proof. Given the Riesz Representation Theorem, this proof is almost tautological. Let $\mathbf{H}$ be a Hilbert space and define $\Delta : \mathbf{H} \to \mathbf{H}^*$ by $\Delta(y) := \varphi_y$ where

$$\varphi_y(x) := \langle x, y \rangle.$$

Then Δ is an conjugate–linear isomorphism of normed linear spaces. In particular, $\mathbf{H}^*$ is a Hilbert space under the inner product

$$(\varphi_y, \varphi_z) := \langle z, y \rangle.$$

Hence if $T \in \mathbf{H}^{**}$, then by the Riesz Representation Theorem applied to $\mathbf{H}^*$ there exists $\varphi \in \mathbf{H}^*$ such that

$$T(\psi) = (\psi, \varphi)$$

for all $\psi \in \mathbf{H}^*$. By the Riesz Representation Theorem there exists $y \in \mathbf{H}$ such that $\varphi = \varphi_y$. For any $z \in \mathbf{H}$, taking $\psi = \varphi_z$ in the above equation gives

$$T(\varphi_z) = (\varphi_z, \varphi_y) = \langle y, z \rangle.$$

Now consider $\widehat{y} \in \mathbf{H}^{**}$ and observe that

$$\widehat{y}(\varphi_z) = \varphi_z(y) = \langle y, z \rangle.$$

Therefore $T = \widehat{y}$ and so J is surjective.   □

(iii) For $1 < p < \infty$, ℓ^p is reflexive.

Proof. Let $q \in (1, \infty)$ be such that $1/p + 1/q = 1$ and let $\Delta_p : \ell^q \to (\ell^p)^*$ be the map from Definition 4.1.1. Similarly, let $\Delta_q : \ell^p \to (\ell^q)^*$ denote the corresponding map obtained by letting ℓ^p act on ℓ^q. Then by Theorem 4.1.3, both maps induce isomorphisms

$$\Delta_p : \ell^q \cong (\ell^p)^* \text{ and } \Delta_q : \ell^p \cong (\ell^q)^*.$$

Now if $T \in (\ell^p)^{**}$, then $T : (\ell^p)^* \to \mathbb{K}$ is bounded and linear. Hence $T \circ \Delta_p : \ell^q \to \mathbb{K}$ is bounded and linear. Since Δ_q is surjective there exists $x \in \ell^p$ such that

$$\Delta_q(x) = T \circ \Delta_p.$$

Hence for any $y \in \ell^q$, we have

$$\sum_{n=1}^{\infty} x_n y_n = \Delta_q(x)(y) = T(\Delta_p(y)).$$

However, we observe that

$$\widehat{x}(\Delta_p(y)) = \widehat{x}(\varphi_y) = \varphi_y(x) = \sum_{n=1}^{\infty} x_n y_n.$$

Since Δ_p is surjective, we conclude that $\widehat{x} = T$. Therefore $J : \ell^p \to (\ell^p)^{**}$ is surjective. $\square$

(iv) A similar argument shows that $L^p[a, b]$ is reflexive, provided $1 < p < \infty$.

Aside. When discussing reflexivity, one must be a little careful. There is an example (due to James [28]) of a Banach space $\mathbf{E}$ that is isomorphic to $\mathbf{E}^{**}$ but fails to be reflexive! The point is that reflexivity requires a *specific* map to be an isomorphism. The mere existence of an isomorphism does not guarantee reflexivity.

Once again we are left wondering about ℓ^1 and ℓ^∞ (and their continuous analogues). In order to resolve this, we need to know some properties that reflexive spaces enjoy.

4.4.7 PROPOSITION *Let $\mathbf{E}$ be a reflexive space. Then $\mathbf{E}$ is separable if and only if $\mathbf{E}^*$ is separable.*

Proof.   If $\mathbf{E}^*$ is separable, then $\mathbf{E}$ is separable by Proposition 4.4.1. Conversely, if $\mathbf{E}$ is separable and reflexive, then $\mathbf{E}^{**} = (\mathbf{E}^*)^*$ is separable. Hence $\mathbf{E}^*$ is separable by Proposition 4.4.1. $\square$

4.4.8 PROPOSITION   *Let $\mathbf{E}$ be a reflexive space and $\mathbf{F} < \mathbf{E}$ a closed subspace. Then $\mathbf{F}$ is reflexive.*

Proof.   We have a isometric maps $J_{\mathbf{F}} : \mathbf{F} \to \mathbf{F}^{**}$ and $J_{\mathbf{E}} : \mathbf{E} \to \mathbf{E}^{**}$. Assuming $J_{\mathbf{E}}$ is surjective, we want to show that $J_{\mathbf{F}}$ is surjective. So if $T \in \mathbf{F}^{**}$, then we wish to show that there exists $x \in \mathbf{F}$ such that $T = J_{\mathbf{F}}(x)$. Consider $T : \mathbf{F}^* \to \mathbb{K}$ as a bounded linear functional and define $S : \mathbf{E}^* \to \mathbb{K}$ by

$$S(\varphi) := T(\varphi|_{\mathbf{F}}).$$

Then S is clearly a linear functional and

$$|S(\varphi)| = |T(\varphi|_{\mathbf{F}})| \leq \|T\| \|\varphi|_{\mathbf{F}}\| \leq \|T\| \|\varphi\|.$$

Hence S is a bounded and therefore in $\mathbf{E}^{**}$. Since $\mathbf{E}$ is reflexive there exists $x \in \mathbf{E}$ such that

$$S = J_{\mathbf{E}}(x).$$

We claim that $x \in \mathbf{F}$ and that $T = J_{\mathbf{F}}(x)$. Suppose $x \notin \mathbf{F}$. Then by Proposition 4.3.3 there exists $\varphi \in \mathbf{E}^*$ such that $\varphi|_{\mathbf{F}} = 0$ and $\varphi(x) = 1$. However, this would imply that

$$1 = \varphi(x) = J_{\mathbf{E}}(x)(\varphi) = S(\varphi) = T(\varphi|_{\mathbf{F}}) = T(0) = 0.$$

This is a contradiction and therefore it must happen that $x \in \mathbf{F}$. Now for any $\varphi \in \mathbf{F}^*$, we choose a Hahn–Banach extension $\psi \in \mathbf{E}^*$ of φ. Then $\psi(x) = \varphi(x)$ since $x \in \mathbf{F}$. Therefore

$$T(\varphi) = T(\psi|_{\mathbf{F}}) = S(\psi) = J_{\mathbf{E}}(x)(\psi) = \psi(x) = \varphi(x) = J_{\mathbf{F}}(x)(\varphi).$$

We conclude that $J_{\mathbf{F}}$ is surjective. $\square$

   We are now in a position to prove the lack of reflexivity of the remaining familiar spaces.

4.4.9 EXAMPLE

(i) Any reflexive space is necessarily complete because the dual space of any normed linear space is complete. Therefore $(c_{00}, \|\cdot\|_p)$ is not reflexive for any $1 \le p \le \infty$ (see Proposition 2.3.4 and Exercise 2.26).

(ii) For the same reason, $(C[a,b], \|\cdot\|_p)$ is not reflexive, if $1 \le p < \infty$ (by Proposition 2.3.12).

(iii) ℓ^1 is separable, but its dual space ℓ^∞ is not separable (by Example 2.3.16). Therefore ℓ^1 is not reflexive. Similarly, $L^1[a,b]$ is not reflexive either.

(iv) $C[a,b]$ (equipped with the supremum norm) is not reflexive because $C[a,b]^*$ is not separable (Exercise 4.31), while $C[a,b]$ itself is separable.

(v) $L^\infty[a,b]$ is not reflexive since it has a closed subspace $C[a,b]$ that is not reflexive.

(vi) From Exercise 4.13, there is an isometric isomorphism $\ell^1 \to (c_0)^*$. Hence

$$(c_0)^{**} \cong \ell^\infty.$$

Now c_{00} is dense in c_0 (by Exercise 2.26), which tells us that c_0 is separable (by Remark 2.3.15). Since ℓ^∞ is not separable, it follows that c_0 is not reflexive.

(vii) Since c_0 is a closed subspace of ℓ^∞, it follows from Proposition 4.4.8 that ℓ^∞ is not reflexive.

We must now point out an important gap in our knowledge. While we know that $C[a,b]^*$ and $(\ell^\infty)^*$ are both non-separable, we do not as yet know what they are. We will study these spaces in detail in Chapter 9. For now it suffices to know that both these dual spaces are vector spaces of *measures* and each such measure induces a linear functional by integration.

With that mysterious comment, we move on to more tractable topics.

4.5 The Transpose of an Operator

Given a normed linear space **E**, we have now begun to understand the role of the dual space **E***. In many ways, it is the mirror image of **E**, and properties of **E** are reflected in those of **E*** and vice-versa. Now we will expand this relationship to include linear operators. Given an operator T between normed linear spaces, we will define a natural *dual* operator between the corresponding dual spaces. This operator mirrors the properties of T and allows us to prove interesting results about T simply knowing its relationship with its transpose.

This idea truly comes to fruition in the context of Hilbert spaces, where the transpose of an operator is replaced by its *adjoint*. The deep relationship between an operator and its adjoint leads to some of the most striking results in the subject and will be the focus of our attention later in the book. For now our goal is modest. We will define the transpose and give some useful examples.

4.5.1 DEFINITION Let $T : \mathbf{E} \to \mathbf{F}$ be a linear operator between two normed linear space. The *transpose* of T is the operator $T' : \mathbf{F}^* \to \mathbf{E}^*$ defined by the formula

$$T'(\varphi)(x) := \varphi(T(x)).$$

If $\varphi \in \mathbf{F}^*$, then $\varphi \circ T : \mathbf{E} \to \mathbb{K}$ is a continuous linear map since it is the composition of two such functions. Thus $T'(\varphi) \in \mathbf{E}^*$. Furthermore, if $\varphi_1, \varphi_2 \in \mathbf{F}^*$, then

$$T'(\varphi_1 + \varphi_2) = T'(\varphi_1) + T'(\varphi_2)$$

as can be seen by applying both sides to elements of $\mathbf{E}$. Similarly, T' respects scalar multiplication and we have confirmed that T' is a linear map. In other words, our definition makes sense and we may proceed without further ado.

4.5.2 EXAMPLE If $\mathbf{E}$ and $\mathbf{F}$ are both finite dimensional, we fix two ordered bases $\Lambda_{\mathbf{E}} = \{e_1, e_2, \ldots, e_n\}$ and $\Lambda_{\mathbf{F}} = \{f_1, f_2, \ldots, f_m\}$ of $\mathbf{E}$ and $\mathbf{F}$, respectively. Let $A = (a_{i,j})$ be the matrix associated to T in these bases. In other words, for each $1 \leq j \leq n$,

$$T(e_j) = \sum_{i=1}^{m} a_{i,j} f_j$$

Let $\Lambda_{\mathbf{E}^*} = \{\varphi_1, \varphi_2, \ldots, \varphi_n\}$ and $\Lambda_{\mathbf{F}^*} = \{\psi_1, \psi_2, \ldots, \psi_m\}$ be the ordered dual bases to $\Lambda_{\mathbf{E}}$ and $\Lambda_{\mathbf{F}}$, respectively, and let $B = (b_{i,j})$ be the associated matrix for the operator T' in these bases. In other words, for each $1 \leq k \leq m$,

$$T'(\psi_k) = \sum_{\ell=1}^{n} b_{\ell,k} \varphi_\ell.$$

Then

$$b_{j,i} = T'(\psi_i)(e_j) = \psi_i(T(e_j)) = \sum_{i=1}^{m} a_{i,j} \psi_i(f_j) = a_{i,j}.$$

Therefore the matrix B is the transpose of the matrix A in the sense that you must be familiar with from linear algebra.

The next example is a simple generalization of this idea to infinite dimensions.

4.5.3 EXAMPLE Let $1 < p < \infty$ and $1 < r < \infty$ and let $T \in \mathcal{B}(\ell^p, \ell^r)$ be a bounded, linear operator. Define an infinite matrix $M = (a_{i,j})$ by the formula

$$T(e_j) = (a_{1,j}, a_{2,j}, \ldots).$$

Under the identification $\Delta_p : \ell^q \xrightarrow{\cong} (\ell^p)^*$ and $\Delta_r : \ell^s \xrightarrow{\cong} (\ell^r)^*$ (where q and s are the conjugate exponents to p and r, respectively), we wish to determine the corresponding matrix for the operator T', thought of as a linear map from ℓ^s to ℓ^q. To be precise, we wish to understand the map

$$\Delta_p^{-1} \circ T' \circ \Delta_r : \ell^s \to \ell^q.$$

For clarity, we write $\Lambda_{\mathbf{E}} := \{e_1, e_2, \ldots\}$ for the standard basis in ℓ^p, and $\Lambda_{\mathbf{F}} := \{f_1, f_2, \ldots\}$ for the corresponding vectors in ℓ^r. We write $\Lambda_{\mathbf{E}^*} := \{\varphi_1, \varphi_2, \ldots\} \subset (\ell^p)^*$ where

$$\varphi_n((x_j)) = x_n.$$

Then $\Lambda_{\mathbf{E}^*}$ plays the role of the dual basis to $\Lambda_{\mathbf{E}}$. Similarly, we define $\Lambda_{\mathbf{F}^*} := \{\psi_1, \psi_2, \ldots\} \subset (\ell^r)^*$ where $\psi_m((y_k)) := y_m$. As before, we have

$$T'(\psi_i)(e_j) = \psi_i(T(e_j)) = \sum_{i=1}^{\infty} a_{i,j} \psi_i(f_j) = a_{i,j}.$$

Therefore $T'(\psi_i) = \sum_{j=1}^{\infty} a_{i,j} \varphi_j$. Composing with $\Delta_p^{-1} : (\ell^p)^* \to \ell^q$, we have

$$\Delta_p^{-1} \circ T'(\psi_i) = (a_{i,1}, a_{i,2}, \ldots).$$

If we write $\{g_1, g_2, \ldots\}$ for the standard basis of ℓ^s, then $\Delta_r(g_i) = \psi_i$ so that

$$\Delta_p^{-1} \circ T' \circ \Delta_r(g_i) = (a_{i,1}, a_{i,2}, \ldots).$$

In other words, $\Delta_p^{-1} \circ T' \circ \Delta_r$ is associated to the matrix $(a_{j,i})$, which is the transpose of the matrix M.

Let us now look at a continuous analogue of this example and attempt a coordinate-free approach to the question of computing the transpose.

4.5.4 EXAMPLE   For simplicity, we fix $1 < p < \infty$ and a measurable function $K : [0,1] \times [0,1] \to \mathbb{K}$ such that

$$\alpha := \sup_{t \in [0,1]} \int_0^1 |K(s,t)| ds < \infty, \text{ and } \beta := \sup_{s \in [0,1]} \int_0^1 |K(s,t)| dt < \infty.$$

Then we let q denote the conjugate exponent of p. For any $f \in L^p[0,1]$, we have

$$\int_0^1 |K(s,t)||f(t)|dt = \int_0^1 |K(s,t)|^{1/q} |K(s,t)|^{1/p} |f(t)| dt$$

$$\leq \beta^{1/q} \left(\int_0^1 |K(s,t)||f(t)|^p dt \right)^{1/p}$$

by Hölder's Inequality. Hence

$$\int_0^1 \left(\int_0^1 |K(s,t)||f(t)|dt \right)^p ds \leq \beta^{p/q} \int_0^1 \int_0^1 |K(s,t)||f(t)|^p dt ds$$

$$= \beta^{p/q} \int_0^1 \int_0^1 |K(s,t)||f(t)|^p ds dt$$

$$\leq \beta^{p/q} \alpha \|f\|_p^p.$$

Hence we may define an integral operator $T : L^p[0,1] \to L^p[0,1]$ by

$$T(f)(s) := \int_0^1 K(s,t) f(t) dt$$

and T is a bounded linear operator with $\|T\| \leq \beta^{1/q} \alpha^{1/p}$.

We now compute the transpose T', thought of as an operator on $L^q[0,1]$, via the identification $\Delta_p : L^q[0,1] \to (L^p[0,1])^*$. For any $g \in L^q[0,1]$ and $f \in L^p[0,1]$, we have

$$T'(\Delta_p(g))(f) = \Delta_p(g)(T(f)) = \int_0^1 T(f)(s) g(s) ds$$

$$= \int_0^1 \int_0^1 K(s,t) f(t) g(s) dt ds$$

$$= \int_0^1 \int_0^1 K(s,t) g(s) f(t) ds dt$$

by the Fubini–Tonelli Theorem (which is applicable because the integrand is in $L^1([0,1]^2)$ by Hölder's Inequality). Define $S : L^q[0,1] \to L^q[0,1]$ by

$$S(g)(t) := \int_0^1 K(s,t)g(s)ds.$$

Then, by the calculation above, S is a bounded linear operator, with $\|S\| \leq \alpha^{1/p}\beta^{1/q}$. Furthermore,

$$T'(\Delta_p(g))(f) = \int_0^1 S(g)(t)f(t)dt = \Delta_p(S(g))(f).$$

This is true for any $f \in L^p[0,1]$ and $g \in L^q[0,1]$, so $\Delta_p^{-1} \circ T' \circ \Delta_p = S$. Once again we see that the transpose of T is an integral operator whose kernel is the 'transpose' of the kernel K of T.

4.5.5 EXAMPLE Let **E** be a normed linear space and $\mathbf{F} < \mathbf{E}$ be a subspace of **E**. If $T : \mathbf{F} \to \mathbf{E}$ denotes the inclusion map, then $T' : \mathbf{E}^* \to \mathbf{F}^*$ is the restriction map R from Proposition 4.3.2.

Similarly, if **F** is closed and $\pi : \mathbf{E} \to \mathbf{E}/\mathbf{F}$ is the quotient map, then $\pi' : (\mathbf{E}/\mathbf{F})^* \to \mathbf{E}^*$ is the map Q from Theorem 4.3.6.

4.5.6 PROPOSITION *If $T \in \mathcal{B}(\mathbf{E}, \mathbf{F})$, then $T' \in \mathcal{B}(\mathbf{F}^*, \mathbf{E}^*)$ and $\|T'\| = \|T\|$.*

Proof. For any $\varphi \in \mathbf{F}^*$, $T'(\varphi)$ is a bounded linear functional on **E**. If $x \in \mathbf{E}$, then

$$|T'(\varphi)(x)| = |\varphi(T(x))| \leq \|\varphi\|\|T(x)\| \leq \|\varphi\|\|T\|\|x\|.$$

Thus $\|T'(\varphi)\| \leq \|T\|\|\varphi\|$. This is true for any $\varphi \in \mathbf{F}^*$ and thus T' is a bounded linear operator with $\|T'\| \leq \|T\|$. Conversely, if $x \in \mathbf{E}$, then by Corollary 4.2.7 there is a bounded linear functional $\varphi_0 \in \mathbf{F}^*$ such that $\|\varphi_0\| = 1$ and $|\varphi_0(T(x))| = \|T(x)\|$. Therefore

$$\|T(x)\| = |T'(\varphi_0)(x)| \leq \|T'(\varphi_0)\|\|x\| \leq \|T'\|\|\varphi_0\|\|x\| = \|T'\|\|x\|.$$

Thus $\|T\| \leq \|T'\|$ holds as well. $\qquad\square$

The proof of the next result is entirely straightforward and we leave it as an exercise for the reader.

4.5.7 PROPOSITION *Let* **E**, **F**, *and* **W** *be normed linear spaces.*

(i) *Let S and T* $\in \mathcal{B}(\mathbf{E}, \mathbf{F})$ *and* $\alpha \in \mathbb{K}$*. Then* $(T + S)' = T' + S'$ *and* $(\alpha S)' = \alpha S'$*.*

(ii) *If* $S \in \mathcal{B}(\mathbf{E}, \mathbf{F})$ *and* $T \in \mathcal{B}(\mathbf{F}, \mathbf{W})$*, then* $(TS)' = S'T'$*.*

Given a subspace **F** of a normed linear space **E**, we had defined its annihilator $\mathbf{F}^{\perp}$ in Definition 4.3.1 by $\mathbf{F}^{\perp} = \{\varphi \in \mathbf{E}^* : \varphi|_{\mathbf{F}} = 0\}$. The next definition is meant to be 'dual' to this.

4.5.8 DEFINITION Let **E** be a normed linear space and **W** be a subspace of $\mathbf{E}^*$. Define

$$\mathbf{W}^{\perp} := \{x \in \mathbf{E} : \varphi(x) = 0 \text{ for all } \varphi \in \mathbf{W}\}.$$

In contrast with Definition 4.3.1, $\mathbf{W}^{\perp}$ is a subspace of **E**, not of $\mathbf{E}^{**}$. At this point, we must apologize for the abuse of notation. However, this is standard practice and it is probably for the best if you simply confirm that this notation makes sense in the case when **E** is a Hilbert space.

We end this section with a simple fact relating an operator to its transpose. We will see many more results like this as we go along, but this should give you a sense for how these arguments are meant to go.

4.5.9 PROPOSITION *Let* $T : \mathbf{E} \to \mathbf{F}$ *be a bounded linear operator and let* $T' : \mathbf{F}^* \to \mathbf{E}^*$ *be its transpose. Then*

(i) $\ker(T') = \text{Range}(T)^{\perp}$*.*

(ii) $\ker(T) = \text{Range}(T')^{\perp}$*.*

Proof.   Part (i) is an easy exercise for you to try (Exercise 4.40). For Part (ii), we prove containment both ways. If $x \in \ker(T)$, then for any $\varphi \in \mathbf{F}^*, T'(\varphi)(x) = \varphi(T(x)) = 0$. Hence $x \in \text{Range}(T')^{\perp}$. Thus

$$\ker(T) \subset \text{Range}(T')^{\perp}.$$

Conversely, if $x \in \text{Range}(T')^{\perp}$, then for any $\varphi \in \mathbf{F}^*, T'(\varphi)(x) = \varphi(T(x)) = 0$. Since $\mathbf{F}^*$ separates points of **F** (by Corollary 4.2.8), this forces $T(x) = 0$. Hence $x \in \ker(T)$ and this proves that $\text{Range}(T')^{\perp} \subset \ker(T)$ as well.   □

This concludes our discussion on the dual space for the moment. We will return to it in Chapter 6, armed with a few more tools. Until then do try the problems given below and pay close attention to the variety of uses of the Hahn–Banach Theorem.

4.6 Exercises

The Duals of L^p Spaces

EXERCISE 4.1 Let $y = (y_1, y_2, \ldots)$ be a sequence in $\mathbb{K}$. For each $x = (x_n) \in \ell^\infty$, suppose that the series $\sum_{i=1}^\infty x_i y_i$ converges. Furthermore, suppose that the map $\varphi : \ell^\infty \to \mathbb{K}$ given by

$$\varphi((x_n)) := \sum_{i=1}^{\infty} x_i y_i$$

defines a bounded linear map on ℓ^∞. Prove that $y \in \ell^1$ and that $\|y\|_1 = \|\varphi\|$.

EXERCISE 4.2 (⚒) Prove Lemma 4.1.13.

EXERCISE 4.3 For $0 < p < 1$, define $L^p[0,1]$ exactly as in Definition 2.1.15. Define a metric $d_p : L^p[0,1] \times L^p[0,1] \to \mathbb{R}_+$ by

$$d_p(f,g) := \int_0^1 |f(x) - g(x)|^p dx.$$

The aim of this exercise is to show that the only linear functional on $(L^p[0,1], d_p)$ that is continuous is the zero map (succinctly, $L^p[0,1]^* = \{0\}$).

(i) Suppose $\varphi : L^p[0,1] \to \mathbb{K}$ is a non-zero, continuous linear functional. Let $f \in L^p[0,1]$ be such that $|\varphi(f)| \geq 1$. Prove that there exists $s \in [0,1]$ such that

$$\int_0^s |f(x)|^p dx = \frac{1}{2}\int_0^1 |f(x)|^p dx.$$

(ii) With $s \in [0,1]$ as above, set $g_1 := 2f\chi_{[0,s]}$ and $g_2 = 2f\chi_{(s,1]}$. Prove that $|\varphi(g_i)| \geq 1$ for some $i \in \{1,2\}$ and that $d_p(g_i, 0) = 2^{p-1} d_p(f,0)$.

(iii) Use this idea to construct a sequence (f_n) in $L^p[0,1]$ such that

$$\lim_{n\to\infty} d_p(f_n, 0) = 0$$

and $|\varphi(f_n)| \geq 1$ for all $n \in \mathbb{N}$.

(iv) Conclude that $L^p[0,1]^* = \{0\}$.

The Hahn–Banach Extension Theorem

EXERCISE 4.4 Prove parts (i) and (ii) of Lemma 4.2.3.

EXERCISE 4.5 Let $\mathbf{E} := \mathbb{R}^2$, equipped with the supremum norm.

(i) Let $\mathbf{F} := \{(x, y) \in \mathbf{E} : y = 0\}$ and let $\varphi : \mathbf{F} \to \mathbb{R}$ be given by $\varphi(x, y) = x$. Prove that φ has a unique norm-preserving extension to $\mathbf{E}$.

(ii) Let $\mathbf{F} := \{(x, y) \in \mathbf{E} : x = y\}$ and let $\varphi : \mathbf{F} \to \mathbb{R}$ be given by $\varphi(x, y) = x$. Determine all norm-preserving extensions of φ to $\mathbf{E}$. Observe that there are infinitely many such extensions.

EXERCISE 4.6 (✘) Let $\mathbf{E}$ be a normed linear space and $\{e_1, e_2, \ldots, e_n\}$ be a finite, linearly independent set. Prove that there are bounded linear functionals $\{\varphi_1, \varphi_2, \ldots, \varphi_n\} \subset \mathbf{E}^*$ such that

$$\varphi_i(e_j) = \delta_{i,j} = \begin{cases} 1 & : \text{ if } i = j, \\ 0 & : \text{ if } i \neq j, \end{cases}$$

for all $1 \leq i, j \leq n$.

Note: The analogous statement does not hold for an *infinite* linearly independent set – have a look at Exercise 5.33.

EXERCISE 4.7 Let $\mathbf{E}$ and $\mathbf{F}$ be normed linear spaces with $\mathbf{E}$ non-zero. If $\mathcal{B}(\mathbf{E}, \mathbf{F})$ is a Banach space, then prove that $\mathbf{F}$ is complete.

EXERCISE 4.8 (✘) Let $\mathbf{E}$ and $\mathbf{W}$ be normed linear spaces with $\mathbf{W}$ finite dimensional. If $\mathbf{F} < \mathbf{E}$ is a subspace and $T : \mathbf{F} \to \mathbf{W}$ is a bounded linear operator, then prove that there is a bounded linear map $S : \mathbf{E} \to \mathbf{W}$ such that

$$S|_{\mathbf{F}} = T.$$

Give an example to show that this need not happen if $\mathbf{W}$ is infinite dimensional.

EXERCISE 4.9 Let $\mathbf{E}$ be a normed linear space and $T : \mathbf{E} \to \ell^\infty$ be a bounded linear map.

(i) Prove that there is a sequence $(\varphi_n) \subset \mathbf{E}^*$ such that

$$T(x) = (\varphi_1(x), \varphi_2(x), \ldots).$$

Furthermore, prove that $\|T\| = \sup_{n \in \mathbb{N}} \|\varphi_n\|$.

(ii) If $\mathbf{F} < \mathbf{E}$ is a subspace and $T : \mathbf{F} \to \ell^\infty$ is a bounded linear map, then prove that there is an extension $S : \mathbf{E} \to \ell^\infty$ such that $S|_{\mathbf{F}} = T$ and $\|S\| = \|T\|$.

4.6.1 DEFINITION A normed linear space $\mathbf{E}$ is said to be *strictly convex* if, for any distinct $x, y \in \mathbf{E}$ with $\|x\| = \|y\| = 1$, we have

$$\left\| \frac{x+y}{2} \right\| < 1.$$

(To understand the terminology, interpret this as a statement about the unit ball.)

EXERCISE 4.10

(i) Prove that any Hilbert space is strictly convex.

(ii) If $1 < p < \infty$, then show that ℓ^p is strictly convex.

Hint: Read the proof of Hölder's Inequality carefully.

(iii) Prove that ℓ^1 and ℓ^∞ are not strictly convex (see Exercise 2.4).

Note: Parts (ii) and (iii) are true for the function spaces $L^p[a, b]$ as well.

EXERCISE 4.11 (TAYLOR, 1939) Let $\mathbf{E}$ be an normed linear space such that $\mathbf{E}^*$ is strictly convex. If $\mathbf{F} < \mathbf{E}$ and $\varphi \in \mathbf{F}^*$, prove that there exists a *unique* $\psi \in \mathbf{E}^*$ such that $\psi|_{\mathbf{F}} = \varphi$ and

$$\|\psi\| = \|\varphi\|.$$

In other words, φ has a unique norm-preserving extension.

EXERCISE 4.12 Let $(\mathbf{E}, \| \cdot \|_{\mathbf{E}})$ be a separable normed linear space.

(i) Prove that there is a countable collection $\{\varphi_1, \varphi_2, \dots\} \subset \mathbf{E}^*$ such that

$$\|x\|_{\mathbf{E}} = \sup_{n \in \mathbb{N}} |\varphi_n(x)|$$

for each $x \in \mathbf{E}$.

Hint: Take a countable dense subset in the unit sphere $S_{\mathbf{E}}$ and apply Corollary 4.2.7 to each element of it.

(ii) For $x \in \mathbf{E}$, define

$$\|x\|_0 := \left(\sum_{n=1}^{\infty} \frac{|\varphi_n(x)|^2}{2^n} \right)^{1/2}.$$

Prove that $\| \cdot \|_0$ is a norm on $\mathbf{E}$ and that it is strictly convex.

(iii) Prove that there is a strictly convex norm $\|\cdot\|_c$ on $\mathbf{E}$, which is equivalent to $\|\cdot\|_{\mathbf{E}}$.

Note: It is a theorem due to Bourgain [5] that ℓ^∞/c_0 does not admit an equivalent strictly convex norm. Therefore the conclusion of Exercise 4.12 does not hold for non-separable spaces (see Exercise 4.22).

EXERCISE 4.13 (⚒) Let $\mathbf{E} = (c_0, \|\cdot\|_\infty)$ and $\mathbf{F} = \ell^1$. For each $y = (y_n) \in \mathbf{F}$, define $\varphi_y : \mathbf{E} \to \mathbb{K}$ by

$$\varphi_y((x_j)) := \sum_{n=1}^{\infty} x_n y_n.$$

From Exercise 2.14, we know that $\varphi_y \in \mathbf{E}^*$ and that $\|\varphi_y\| = \|y\|_1$. Hence we get a map $\Delta : \mathbf{F} \to \mathbf{E}^*$ given by

$$y \mapsto \varphi_y,$$

which is linear and isometric. Prove that this map is an isomorphism.

EXERCISE 4.14 In Exercise 4.13, we have constructed an isometric isomorphism $\Delta_0 : \ell^1 \to (c_0)^*$. By Theorem 4.1.3, we have an isometric isomorphism $\Delta_1 : \ell^\infty \to (\ell^1)^*$. Therefore we have an isometric isomorphism

$$\Gamma : c_0^{**} \to \ell^\infty.$$

If $J : c_0 \to c_0^{**}$ is the natural map from Definition 4.4.5, then show that the composition $\Gamma \circ J : c_0 \to \ell^\infty$ is simply the natural inclusion map $c_0 \hookrightarrow \ell^\infty$.

EXERCISE 4.15 Let $\mathbf{E}$ be a normed linear space, $\mathbf{F} < \mathbf{E}$ be a subspace and $\varphi \in \mathbf{F}^*$ be a bounded linear functional on $\mathbf{F}$. Let

$$S_\varphi := \{\psi \in \mathbf{E}^* : \psi|_{\mathbf{F}} = \varphi \text{ and } \|\psi\| = \|\varphi\|\}.$$

In other words, S_φ is the collection of all norm-preserving extensions of φ. Prove that S_φ is a non-empty, convex, closed and bounded subset of $\mathbf{E}^*$. Give an example to show that S_φ may not be compact.

The Dual of Subspaces and Quotient Spaces

EXERCISE 4.16 Let $\mathbf{E}$ be a normed linear space such that for any $\varphi \in \mathbf{E}^*$ there exists $x \in \mathbf{E}$ such that $\|x\| = 1$ and $|\varphi(x)| = \|\varphi\|$. Let $\mathbf{F} < \mathbf{E}$ be a proper closed subspace of $\mathbf{E}$. Prove that there exists a vector $x_0 \in \mathbf{E}$ such that $\|x_0\| = 1$ and $d(x_0, \mathbf{F}) = 1$.

In other words, if every element of $\mathbf{E}^*$ attains its norm on the unit sphere of $\mathbf{E}$, then the conclusion of Riesz' Lemma holds for $t = 1$. In particular, this applies to reflexive spaces by Exercise 4.28.

EXERCISE 4.17 Let $\mathbf{E}$ be a normed linear space and $\mathbf{F}$ be a subspace of $\mathbf{E}$. For any $x \in \mathbf{E}$, prove that

$$d(x, \mathbf{F}) = \sup\{|\psi(x)| : \psi \in \mathbf{F}^{\perp}, \|\psi\| \leq 1\}.$$

Separability and Reflexivity

EXERCISE 4.18 Let $\mathbf{E}$ be a normed linear space and S be any subset of $\mathbf{E}$. If $\mathbf{E}$ is separable, then prove that S is separable.

Note: The analogous statement is true for metric spaces, but it is not true for arbitrary topological spaces, so be careful!

EXERCISE 4.19 (⚒)

(i) Follow the line of reasoning of Exercise 4.13 to show that there is an isometric isomorphism $\Delta : \ell^1 \xrightarrow{\cong} c^*$ (compare this result with Exercise 2.38).

(ii) Prove that c is not reflexive.

EXERCISE 4.20 Let $\mathbf{E}$ be a normed linear space and $\mathbf{F} < \mathbf{E}$ be a closed subspace of $\mathbf{E}$.

(i) If $\mathbf{E}$ is separable, then prove that $\mathbf{E}/\mathbf{F}$ is separable.

(ii) Give a (non-trivial) example where $\mathbf{E}/\mathbf{F}$ is separable, but $\mathbf{E}$ is not.

EXERCISE 4.21 (⚒) Let $\mathbf{E}$ be a normed linear space and $\mathbf{F} < \mathbf{E}$ be a closed subspace of $\mathbf{E}$. If $\mathbf{F}$ and $\mathbf{E}/\mathbf{F}$ are both separable, then prove that $\mathbf{E}$ is separable.

EXERCISE 4.22 Consider c_0 as a closed subspace of ℓ^∞.

(i) Let $S := \{(x_n) \in \ell^\infty : x_k \in \{0, 1\} \text{ for all } k \in \mathbb{N}\}$. If $x = (x_n), y = (y_n) \in S$ are such that $x_k \neq y_k$ for infinitely many values of $k \in \mathbb{N}$, then prove that $\|x - y + c_0\| = 1$ (see Exercise 2.42).

(ii) Prove that ℓ^∞ / c_0 is not separable (this also follows from Exercise 4.21).

EXERCISE 4.23 If $1 \leq p < \infty$, prove that $\mathcal{B}(\ell^p) = \mathcal{B}(\ell^p, \ell^p)$ is not separable.

Hint: Construct an isometric linear map $T : \ell^\infty \to \mathcal{B}(\ell^p)$.

EXERCISE 4.24 For any normed linear space $\mathbf{E}$, show that $(\mathbf{E}^*)^{**} = (\mathbf{E}^{**})^*$.

EXERCISE 4.25 (⚒) If $\mathbf{E}$ is a reflexive normed linear space, then prove that $\mathbf{E}^*$ is reflexive.

EXERCISE 4.26 (⚒) If $\mathbf{E}$ is a Banach space such that $\mathbf{E}^*$ is reflexive, then prove that $\mathbf{E}$ is reflexive.

Hint: Use Exercise 4.25 and the fact that the natural map $J : \mathbf{E} \to \mathbf{E}^{**}$ is isometric.

EXERCISE 4.27 If $\mathbf{E}$ is reflexive and $\mathbf{F} < \mathbf{E}$ is a closed subspace, then prove that $\mathbf{E}/\mathbf{F}$ is reflexive.

Hint: Use Theorem 4.3.6 and Exercise 4.26.

EXERCISE 4.28 (⚒) Let $\mathbf{E}$ be a reflexive Banach space. For every $\varphi \in \mathbf{E}^*$, prove that there exists $x \in \mathbf{E}$ such that $\|x\| = 1$ and $\varphi(x) = \|\varphi\|$. Furthermore, if $\mathbf{E}$ is assumed to be strictly convex (see Definition 4.6.1), then prove that this point x must be unique.

EXERCISE 4.29 Give an example of a bounded linear functional $\varphi \in (\ell^1)^*$ such that $|\varphi(x)| < \|\varphi\|\|x\|$ for all $x \in \ell^1$.

EXERCISE 4.30 Use Exercise 4.13 to determine the form of every bounded linear functional $\varphi \in (c_0)^*$ which attains its norm on the closed unit ball.

EXERCISE 4.31 (⚒) Let $\mathbf{E} = (C[a,b], \|\cdot\|_\infty)$ and fix $t \in [a,b]$. Define $\varphi_t : \mathbf{E} \to \mathbb{K}$ by $\varphi_t(f) := f(t)$. Prove that $\varphi_t \in \mathbf{E}^*$ and that $\|\varphi_t\| = 1$. Use the collection $\{\varphi_t : t \in [a,b]\}$ to show that $\mathbf{E}^*$ is not separable.

EXERCISE 4.32 Let $\mathbf{E}$ be a reflexive normed linear space and let $f : [0,1] \to \mathbf{E}$ be a continuous function. For each $\varphi \in \mathbf{E}^*$, the function $\varphi \circ f : [0,1] \to \mathbb{K}$ is continuous. Therefore we may define $T : \mathbf{E}^* \to \mathbb{K}$ by

$$T(\varphi) := \int_0^1 \varphi(f(t))dt.$$

(i) Prove that T is linear and bounded. Conclude that there exists a unique vector $x \in \mathbf{E}$ such that

$$\varphi(x) = \int_0^1 \varphi(f(t))dt$$

for all $\varphi \in \mathbf{E}^*$. This vector is denoted by

$$\int_0^1 f(t)dt.$$

(ii) Prove that

$$\left\| \int_0^1 f(t)dt \right\| \leq \int_0^1 \|f(t)\|dt.$$

Note: These results are true even if $\mathbf{E}$ is not reflexive. However, the proof is much longer (see [34, Section 3.3]).

The Transpose of an Operator

EXERCISE 4.33 (✖) Prove Proposition 4.5.7.

EXERCISE 4.34 Let $T : \ell^1 \to c_0$ be the map $T((x_j)) := (y_1, y_2, \ldots)$, where

$$y_n := \sum_{m=n}^{\infty} x_m.$$

Observe that T is a bounded, linear operator. Under the identifications of Exercise 4.13 and Theorem 4.1.3, determine the transpose map $T' : \ell^1 \to \ell^\infty$.

EXERCISE 4.35 Let $\mathbf{E}$ be a normed linear space and $\varphi \in \mathbf{E}^*$. Compute φ' as an element in $\mathcal{B}(\mathbb{K}, \mathbf{E}^*)$.

EXERCISE 4.36 Let $T : \ell^1 \to \ell^1$ be the right-shift operator given by

$$T((x_k)) := (0, x_1, x_2, \ldots).$$

Under the identification $\ell^\infty \cong (\ell^1)^*$, determine the transpose map $T' : \ell^\infty \to \ell^\infty$.

EXERCISE 4.37 Let $\mathbf{E}$ be a normed linear space and $\mathbf{W}$ be a subspace of $\mathbf{E}^*$.

(i) Prove that $\mathbf{W}^\perp$ is a closed subspace of $\mathbf{E}$.

(ii) If $\mathbf{F}$ is subspace of $\mathbf{E}$, then prove that $(\mathbf{F}^\perp)^\perp = \overline{\mathbf{F}}$.

(iii) If $\mathbf{W}$ is a subspace of $\mathbf{E}^*$, then show that $(\mathbf{W}^\perp)^\perp \supset \overline{\mathbf{W}}$.

EXERCISE 4.38

(i) If $\mathbf{E}$ is reflexive and $\mathbf{W}$ is subspace of $\mathbf{E}^*$, then prove that $(\mathbf{W}^\perp)^\perp = \overline{\mathbf{W}}$.

(ii) Let $\mathbf{E} = \ell^1$ and let $\mathbf{W} = c_0$, thought of as a subspace of $(\ell^1)^* \cong \ell^\infty$. Prove that

$$\mathbf{W} \neq (\mathbf{W}^\perp)^\perp.$$

EXERCISE 4.39 Let $\Delta : L^1[a,b] \to (L^\infty[a,b])^*$ be the map given in Definition 4.1.1. Treating $\mathbf{F} := C[a,b]$ as a closed subspace of $L^\infty[a,b]$, prove that

$$\Delta(L^1[a,b]) \cap \mathbf{F}^\perp = \{0\}.$$

This is another way of seeing that Δ is not surjective.

Hint: See the proof of Proposition 2.3.12.

EXERCISE 4.40 (⚒) Prove Part (i) of Proposition 4.5.9.

EXERCISE 4.41 If $T : \mathbf{E} \to \mathbf{F}$ is a topological isomorphism, then prove that $T' : \mathbf{F}^* \to \mathbf{E}^*$ is also a topological isomorphism and that

$$(T^{-1})' = (T')^{-1}.$$

EXERCISE 4.42 (⚒) Let $T : \mathbf{E} \to \mathbf{F}$ be a linear operator and $T'' := (T')' : \mathbf{E}^{**} \to \mathbf{F}^{**}$ be the transpose of T'. If $J_{\mathbf{E}} : \mathbf{E} \to \mathbf{E}^{**}$ and $J_{\mathbf{F}} : \mathbf{F} \to \mathbf{F}^{**}$ denote the canonical inclusions, then prove that

$$T'' \circ J_{\mathbf{E}} = J_{\mathbf{F}} \circ T.$$

Banach Limits and Amenable Groups

4.6.2 DEFINITION A Banach limit is a bounded linear functional $\psi : \ell^\infty \to \mathbb{K}$ satisfying the properties listed in Theorem 4.4.3). A sequence $x = (x_n) \in \ell^\infty$ is said to be ***almost convergent*** if there exists $L \in \mathbb{K}$ such that $\psi(x) = L$ for any Banach limit ψ. If that happens, we write

$$\mathrm{Lim}(x) = L.$$

EXERCISE 4.43

(i) Show that the sequence $x = (1, 0, 1, 0, \ldots)$ is almost convergent.

Hint: Consider $u := x + S(x)$, where $S : \ell^\infty \to \ell^\infty$ denotes the left-shift operator.

Let us now generalize this: A sequence $x = (x_n) \in \ell^\infty$ is said to be **periodic** if there exist natural numbers N and p such that $x_{n+p} = x_n$ for all $n \geq N$.

(ii) If $x \in \ell^\infty$ is periodic, then prove that x is almost convergent. In fact, if N and p are as above, then prove that

$$\text{Lim}(x) = \frac{1}{p}(x_N + x_{N+1} + \ldots + x_{N+p-1}).$$

EXERCISE 4.44 Consider ℓ^∞ as the real vector space of real-valued sequences and let $\psi : \ell^\infty \to \mathbb{R}$ be a Banach limit.

(i) If $x = (x_n) \in \ell^\infty$ is such that $x_n \geq 0$ for all $n \in \mathbb{N}$, then prove that $\psi(x) \geq 0$.

Hint: If $0 \leq x_n \leq 1$ for all $n \in \mathbb{N}$, then $\|\mathbf{1} - x\| \leq 1$.

(ii) If $x = (x_n) \in \ell^\infty$ is a real-valued sequence, then prove that

$$\liminf_{n \to \infty} x_n \leq \psi(x) \leq \limsup_{n \to \infty} x_n.$$

Hint: For the first inequality, let $\alpha := \liminf_{n \to \infty} x_n$. Then for any $\epsilon > 0$ there exists $N \in \mathbb{N}$ such that $x_n > \alpha - \epsilon$ for all $n \geq N$.

4.6.3 REMARK The proof of Theorem 4.4.3 also applies to $\ell^\infty(\mathbb{Z})$ rather than $\ell^\infty = \ell^\infty(\mathbb{N})$. In other words, there is a bounded linear functional $\psi : \ell^\infty(\mathbb{Z}) \to \mathbb{K}$ satisfying all the properties given in Theorem 4.4.3 (note that the shift operator $S : \ell^\infty(\mathbb{Z}) \to \ell^\infty(\mathbb{Z})$ is defined as $S((x_j))_n := x_{n+1}$).

EXERCISE 4.45 Consider $\ell^\infty(\mathbb{Z})$ as a real vector space of real-valued sequences and let $\psi : \ell^\infty(\mathbb{Z}) \to \mathbb{R}$ be a Banach limit as described above. For $A \subset \mathbb{Z}$, let $\chi_A \in \ell^\infty(\mathbb{Z})$ denote the sequence

$$(\chi_A)_n := \begin{cases} 1 & : \text{if } n \in A, \\ 0 & : \text{otherwise.} \end{cases}$$

Define $\mu : 2^{\mathbb{Z}} \to \mathbb{R}$ by

$$\mu(A) := \psi(\chi_A).$$

Prove the following properties of μ:

(i) μ is *positive*: $\mu(A) \geq 0$ for all $A \subset \mathbb{Z}$.

(ii) μ is *finitely additive*: If $A, B \subset \mathbb{Z}$ are disjoint, then $\mu(A \sqcup B) = \mu(A) + \mu(B)$.

(iii) μ is *normalized*: $\mu(\mathbb{Z}) = 1$.

(iv) μ is *translation invariant*: For each $A \subset \mathbb{Z}$ and $m \in \mathbb{Z}$, $\mu(A + m) = \mu(A)$.

(v) μ is not countably additive.

4.6.4 DEFINITION A countable group G is said to be **amenable** if there is a set function $\mu : 2^G \to \mathbb{R}_+$ such that

(i) μ is positive, finitely additive and normalized (in the sense of Exercise 4.45).

(ii) μ is left-translation invariant: If $A \subset G$ and $g \in G$, then $\mu(gA) = \mu(A)$, where $gA := \{gx : x \in A\}$.

With a slight abuse of terminology, such set functions are also called *finitely additive measures*.

EXERCISE 4.46 Prove that a finite group is amenable.

EXERCISE 4.47 Let $\mathbb{F}_2$ denote the free group on two generators $\{a, b\}$. Prove that $\mathbb{F}_2$ is not amenable.

Hint: For $x \in \{a, b, a^{-1}, b^{-1}\}$, let W_x be the set of words in $\mathbb{F}_2$ starting with x.

EXERCISE 4.48 Prove that a subgroup of an amenable group is amenable.

Hint: Let μ be a measure on the group G that witnesses amenability. If H is a subgroup of G, let M be a set consisting of exactly one representative from each right coset of H in G. Now define $\nu : 2^H \to \mathbb{R}$ by

$$\nu(A) := \mu\left(\bigcup_{m \in M} Am\right).$$

Additional Reading

- The result of Exercise 4.11 is a theorem due to Taylor from 1939.

  Taylor, Angus Ellis (1939). The extension of linear functionals. *Duke Mathematical Journal*, 5, 538–547. ISSN: 0012-7094.

  The converse is also true and was proved in a one-page paper!

  Foguel, Shaul R. (1958). On a theorem by A. E. Taylor. *Proceedings of the American Mathematical Society*, 325. ISSN: 0002-9939. DOI: 10.2307/2033162.

  The Taylor–Foguel theorem was further explored by Phelps, who proved that a subspace has the property that every bounded linear functional on it has a unique extension if and only if its annihilator has a certain 'best approximation property' (á la Theorem 3.1.6).

  Phelps, Robert Ralph (1960). Uniqueness of Hahn–Banach extensions and unique best approximation. *Transactions of the American Mathematical Society*, 95, 238–255. ISSN: 0002-9947. DOI: 10.2307/1993289.

- The space of almost convergent sequences (see Exercise 4.43) was studied by Lorentz.

  Lorentz, George Gunter (1948). A contribution to the theory of divergent sequences. *Acta Mathematica*, 80, 167–190. ISSN: 0001-5962. DOI: 10.1007/BF02393648.

  In this paper, he characterizes such sequences and uses that characterization to show that many interesting classes of sequences are almost convergent – so much so that the set of almost convergent sequences is not separable!

- In Exercise 4.45, we have proved that the integers $(\mathbb{Z}, +)$ form an amenable group. These groups have been studied extensively because amenability is reflected in a variety of ways in the operator algebras associated to the group (we will see operator algebras later in the book). A good place to start learning about these objects is this book by Runde.

  Runde, Volker (2002). *Lectures on Amenability*. Vol. 1774. Lecture Notes in Mathematics. Berlin: Springer–Verlag. ISBN: 3-540-42852-6. DOI: 10.1007/b82937.

Chapter 5

Operators on Banach Spaces

5.1 Baire Category Theorem

In this chapter, we explore the theory of bounded linear operators. As we will see, the completeness of the underlying spaces plays an important role. We have already seen one example of this in Theorem 2.3.13, where we showed that $\mathcal{B}(\mathbf{E}, \mathbf{F})$ is a Banach space whenever $\mathbf{F}$ is complete. Here, that simple minded use of completeness gives way to a deeper understanding with the introduction of the Baire Category Theorem. This innocuous-looking result allows us to prove two results that are fundamental to the subject: the Principle of Uniform Boundedness (Theorem 5.2.1) and the Open Mapping Theorem (Theorem 5.3.4). These two theorems and their many applications are the main focus of this chapter.

5.1.1 THEOREM (BAIRE CATEGORY THEOREM (BAIRE, 1899)) *Let (X, d) be a complete metric space and $\{V_n\}_{n=1}^{\infty}$ be a countable collection of open, dense subsets of X. Then*

$$G := \bigcap_{n=1}^{\infty} V_n$$

is dense in X.

Proof. As is usual, we write $B(x, r) := \{y \in X : d(y, x) < r\}$ and $B[x, r] := \{y \in X : d(y, x) \leq r\}$.

Let U be a non-empty open set. We wish to prove that $G \cap U \neq \emptyset$. We do this by inductively constructing a sequence whose limit point lies in this intersection.

161

To begin with, $U \cap V_1$ is non-empty and open. Hence there exists $x_1 \in U \cap V_1$ and $r_1 > 0$ such that $B(x_1, r_1) \subset U \cap V_1$. By shrinking r_1 if need be, we may assume that $r_1 < 1$ and

$$\overline{B(x_1, r_1)} = B[x_1, r_1] \subset U \cap V_1.$$

Since V_2 is dense in X, $B(x_1, r_1) \cap V_2 \neq \emptyset$. Once again there exists $x_2 \in X$ and $r_2 > 0$ such that $r_2 < \frac{1}{2}$ and

$$B[x_2, r_2] \subset B(x_1, r_1) \cap V_2.$$

Observe that $B[x_2, r_2] \subset U \cap (V_1 \cap V_2)$. Thus proceeding, we obtain a sequence $(x_n) \subset X$ and $(r_n) \subset \mathbb{R}_+$ such that

  (i) $B[x_n, r_n] \subset B(x_{n-1}, r_{n-1})$,

  (ii) $r_n < 1/n$ and

  (iii) $B[x_n, r_n] \subset U \cap \left[\bigcap_{i=1}^{n-1} V_i \right]$.

For $m \in \mathbb{N}$ fixed and all $n > m$, we have $d(x_n, x_m) < r_m < 1/m$ and hence (x_n) is Cauchy. Since X is complete, there exists $x_0 \in X$ such that $x_n \to x_0$. We claim that $x_0 \in U \cap G$. To see this, note that for all $n \geq m$,

$$x_n \in B[x_m, r_m].$$

Since this set is closed, it follows that

$$x_0 \in B[x_m, r_m] \subset U \bigcap \left[\bigcap_{i=1}^{m-1} V_i \right].$$

This is true for all $m \in \mathbb{N}$ and hence $x_0 \in U \cap G$. $\qquad\square$

Let (X, d) be a metric space and $A \subset X$. We say that A is ***nowhere dense*** if $\overline{A}$ has empty interior (equivalently, if $X \setminus \overline{A}$ is a dense open set). Note that A is nowhere dense if and only if $\overline{A}$ is nowhere dense. The Baire Category Theorem may now be restated as

5.1.2 COROLLARY *A complete metric space cannot be written as a countable union of nowhere dense sets.*

Aside. A space that is a countable union of nowhere dense set is said to be of the *first category*; while a space that cannot be expressed as such a union is said to be of the *second category*. Thus Baire's theorem states that a complete metric space is of the second category and hence its name. This use of the word *category* is historical and has slowly fallen out of use.

Let us now see how Baire's theorem applies to the study of normed linear spaces. The next lemma is a simple use of the scaling trick and you might have already proved it in Exercise 2.13.

5.1.3 LEMMA *Every closed, proper subspace of a normed linear space is nowhere dense.*

We are now in a position to prove a result we have hinted at a few times now. What is remarkable about this fact is that, under a topological assumption on the space, we are able to conclude a purely algebraic fact.

5.1.4 COROLLARY *If* **E** *is an infinite dimensional Banach space, then* **E** *cannot have a countable Hamel basis.*

Proof. Suppose $A := \{e_n : n \in \mathbb{N}\}$ is a countable subset of **E**, then define $\mathbf{F}_n := \operatorname{span}\{e_1, e_2, \ldots, e_n\}$. Since $\mathbf{F}_n$ is finite dimensional, $\mathbf{F}_n$ is both closed and a proper subspace of **E**. Therefore it is nowhere dense by Lemma 5.1.3. By Corollary 5.1.2,

$$\mathbf{E} \neq \bigcup_{n=1}^{\infty} \mathbf{F}_n.$$

In particular, **E** contains at least one element that cannot be expressed as a finite linear combination of elements in A. □

5.1.5 EXAMPLE

(i) Let **H** be an infinite dimensional separable Hilbert space and Λ an orthonormal basis for **H**, then Λ cannot be a Hamel basis for **H** (see Exercise 3.25).

(ii) There is no norm on c_{00} that makes it a Banach space.

5.2 The Principle of Uniform Boundedness

The next theorem is one of those rare results in mathematics where pointwise boundedness of a collection of functions implies uniform boundedness. It is worth pointing out that, in the hypotheses of this result, the domain space needs to be complete, while the codomain need not.

5.2.1 THEOREM (PRINCIPLE OF UNIFORM BOUNDEDNESS (HAHN, 1922, AND BANACH AND STEINHAUS, 1927)) *Let* $\mathbf{E}$ *be a Banach space and* $\mathbf{F}$ *be any normed linear space. Let* $\mathcal{G} \subset \mathcal{B}(\mathbf{E}, \mathbf{F})$ *be a collection of bounded linear operators such that*

$$\sup_{T \in \mathcal{G}} \|T(x)\| < \infty$$

for each $x \in \mathbf{E}$. *Then* $\sup_{T \in \mathcal{G}} \|T\| < \infty$.

Proof.   For $n \in \mathbb{N}$, define

$$B_n := \bigcap_{T \in \mathcal{G}} \{x \in \mathbf{E} : \|T(x)\| \leq n\}.$$

Then each B_n is closed because it is the intersection of a family of closed sets. By hypothesis,

$$\mathbf{E} = \bigcup_{n=1}^{\infty} B_n.$$

By the Baire Category Theorem, there exists $N \in \mathbb{N}$ such that B_N has non-empty interior. Therefore there exists $x_0 \in \mathbf{E}$ and $r > 0$ such that $B(x_0, r_0) \subset B_N$. Now, we apply the scaling trick: For any non-zero $x \in \mathbf{E}$, set

$$z := \frac{r_0 x}{2\|x\|} + x_0 \in B(x_0, r_0).$$

For any $T \in \mathcal{G}$, $\|T(z)\| \leq N$. Unwrapping this, we get

$$\|T(x)\| = \frac{2\|x\|}{r_0} \|T(z) - T(x_0)\| \leq \frac{2\|x\|}{r_0} (N + \|T(x_0)\|).$$

If $M := \frac{2}{r_0}(N + \|T(x_0)\|)$, then for any $x \in \mathbf{E}$ and any $T \in \mathcal{G}$, we have $\|T(x)\| \leq M\|x\|$. Thus $\sup\{\|T\| : T \in \mathcal{G}\} \leq M < \infty$. $\qquad\square$

Simple examples (see Exercise 5.2) show that completeness of the domain space is essential for the conclusion of Theorem 5.2.1 to hold. The next result is often also referred to as the principle of uniform boundedness.

5.2.2 THEOREM (BANACH–STEINHAUS THEOREM, 1927) *Let* $\mathbf{E}$ *be a Banach space and* $\mathbf{F}$ *be any normed linear space. Suppose* $(T_n)_{n=1}^{\infty} \subset \mathcal{B}(\mathbf{E}, \mathbf{F})$ *is a sequence of bounded operators such that, for each* $x \in \mathbf{E}$, *the sequence* $(T_n(x))_{n=1}^{\infty}$ *is convergent in* $\mathbf{F}$. *Then the map* $T : \mathbf{E} \to \mathbf{F}$ *defined by*

$$T(x) := \lim_{n \to \infty} T_n(x)$$

is a bounded linear map and $\|T\| \leq \liminf_{n \to \infty} \|T_n\|$.

Proof. It is easy to see that the map T defined above is linear; we need only prove that it is bounded. For each $x \in \mathbf{E}, (T_n(x))_{n=1}^{\infty}$ is convergent and hence bounded. By the Principle of Uniform Boundedness, there exists $M > 0$ such that $\|T_n\| \leq M$ for all $n \in \mathbb{N}$. For each $x \in \mathbf{E}$,

$$\|T(x)\| = \lim_{n \to \infty} \|T_n(x)\| \leq M\|x\|.$$

Hence $T \in \mathcal{B}(\mathbf{E},\mathbf{F})$ and $\|T\| \leq M$. The norm inequality is a refinement of this argument and is left as an exercise for the reader (Exercise 5.3). $\square$

One important caveat though: The operators (T_n) and T defined in Theorem 5.2.2 have but a tenuous relationship. The convergence is pointwise and not necessarily in the operator norm, as the following example shows. Let $T_n : \ell^2 \to \ell^2$ be given by

$$T_n((x_m)) := (x_1, x_2, \ldots, x_n, 0, 0, \ldots).$$

For any $x \in \ell^2, T_n(x) \to x$. However, if $\{e_j : j \in \mathbb{N}\}$ denotes the standard orthonormal basis of ℓ^2, then

$$\|T_n - I\| \geq \|T_n(e_{n+1}) - e_{n+1}\| = 1$$

and so $T_n \not\to I$ in the operator norm of $\mathcal{B}(\ell^2)$.

5.2.3 REMARK We will now use the Principle of Uniform Boundedness to study the Fourier series. Specifically, we would like to know if the Fourier series of a continuous function converges pointwise to the original function. We know from Theorem 3.5.5 that the series converges in the L^2 norm and therefore a *subsequence* converges pointwise. Furthermore, we had seen in Exercise 3.42 that the Fourier series of a continuously differentiable function must converge uniformly to the original function. However, the question of pointwise convergence for an arbitrary continuous function still remains.

To study this, we define $\mathbf{E} := \{f \in C[-\pi, \pi] : f(-\pi) = f(\pi)\}$ and note that $\mathbf{E}$ is a Banach space when equipped with the supremum norm (since it is closed in $C[-\pi, \pi]$). For each $f \in \mathbf{E}$, the nth Fourier coefficient of f is given by

$$\widehat{f}(n) = \frac{1}{2\pi} \int_{-\pi}^{\pi} f(t)e^{-int}dt.$$

We define the partial sums of the Fourier series by

$$s_m(f)(x) := \sum_{n=-m}^{m} \widehat{f}(n)e^{inx}.$$

We know that $\lim_{m \to \infty} \|s_m(f) - f\|_2 = 0$. The question is, is it true that $s_m(f) \to f$ pointwise? To understand this, we consider $\varphi_m : \mathbf{E} \to \mathbb{K}$ by

$$f \mapsto s_m(f)(0).$$

Now define $\varphi : \mathbf{E} \to \mathbb{K}$ by $\varphi(f) := f(0)$ and we ask if it is true that $\varphi_m(f) \to \varphi(f)$. To answer this question, we introduce an auxiliary function that will allow us to study the operator s_m as an integral operator. Note that

$$s_m(f)(x) = \frac{1}{2\pi} \int_{-\pi}^{\pi} f(y) \left(\sum_{k=-m}^{m} e^{ik(x-y)} \right) dy = \frac{1}{2\pi} \int_{-\pi}^{\pi} f(y) D_m(x-y) dy,$$

where $D_m : \mathbb{R} \to \mathbb{K}$ is the function

$$D_m(t) = \sum_{n=-m}^{m} e^{int}.$$

Then $D_m(-t) = D_m(t)$ and considering $D_m(t)e^{it/2} - D_m(t)e^{-it/2}$, we see that

$$D_m(t) = \begin{cases} \frac{\sin(m+1/2)t}{\sin(t/2)} & : \text{if } t \neq 2k\pi, \\ 2m+1 & : \text{if } t = 2k\pi. \end{cases}$$

The function D_m is called the mth **Dirichlet kernel** and will play an important role for us.

5.2.4 LEMMA

$$\lim_{m \to \infty} \int_{-\pi}^{\pi} |D_m(t)| dt = \infty.$$

Proof. The proof follows from a series of inequalities. For any $t \in \mathbb{R}$, $|\sin(t)| \leq |t|$. Therefore

$$\int_{-\pi}^{\pi} |D_m(t)| dt = 2 \int_{0}^{\pi} |D_m(t)| dt$$

$$\geq 4 \int_{0}^{\pi} \frac{|\sin((m+1/2)t)|}{t} dt$$

$$= 4 \int_{0}^{(m+1/2)\pi} \frac{|\sin(t)|}{t} dt$$

$$\geq 4 \int_0^{m\pi} \frac{|\sin(t)|}{t} dt$$

$$= 4 \sum_{k=1}^{m} \int_{(k-1)\pi}^{k\pi} \frac{|\sin(t)|}{t} dt$$

$$\geq 4 \sum_{k=1}^{m} \int_{(k-1)\pi}^{k\pi} \frac{|\sin(t)|}{k\pi} dt$$

$$= \frac{8}{\pi} \sum_{k=1}^{m} \frac{1}{k} \to \infty.$$

$\square$

5.2.5 LEMMA *Let $\varphi_m : \mathbf{E} \to \mathbf{C}$ be given by $\varphi_m(f) := s_m(f)(0)$. Then*

$$\|\varphi_m\| = \frac{1}{2\pi} \int_{-\pi}^{\pi} |D_m(t)| dt.$$

Proof. By Remark 5.2.3, since $D_m(-s) = D_m(s)$, we have

$$\varphi_m(f) = \frac{1}{2\pi} \int_{-\pi}^{\pi} f(s) D_m(s) ds$$

and hence it follows that

$$\|\varphi_m\| \leq \frac{1}{2\pi} \int_{-\pi}^{\pi} |D_m(t)| dt.$$

For the reverse inequality, let $E_m := \{t \in [-\pi, \pi] : D_m(t) \geq 0\}$. Then since $D_m(-t) = D_m(t)$, it follows that E_m is symmetric about 0. For $k \in \mathbb{N}$, define

$$f_k(t) = \frac{1 - kd(t, E_m)}{1 + kd(t, E_m)}.$$

Then $f_k \in C[-\pi, \pi]$ and $f_k(\pi) = f_k(-\pi)$. Hence $f_k \in \mathbf{E}$ and

$$\lim_{k \to \infty} f_k(t) = \begin{cases} 1 & : \text{ if } t \in E_m, \\ -1 & : \text{ if } t \notin E_m. \end{cases}$$

Since $\|f_k\|_\infty \leq 1$, it follows by the Dominated Convergence Theorem that

$$\lim_{k \to \infty} |\varphi_m(f_k)| = \frac{1}{2\pi} \int_{-\pi}^{\pi} |D_m(t)| dt.$$

This proves the reverse inequality that we needed. $\square$

5.2.6 THEOREM *There exists a function $f \in \mathbf{E}$ whose Fourier series does not converge pointwise to f.*

Proof. Consider $\varphi_m : \mathbf{E} \to \mathbb{K}$ as above. If it were true that $\lim_{n \to \infty} \varphi_m(f) = f(0)$ for each $f \in \mathbf{E}$, then it must happen that

$$\sup_{m \in \mathbb{N}} \|\varphi_m(f)\| < \infty$$

for all $f \in \mathbf{E}$. By the Principle of Uniform Boundedness, this would imply that

$$\sup_{m \in \mathbb{N}} \|\varphi_m\| < \infty.$$

This contradicts Lemma 5.2.4 and Lemma 5.2.5. Therefore there must exist at least one function $f \in \mathbf{E}$ such that $\varphi_m(f) \not\to f(0)$. $\square$

In fact, the conclusion of Theorem 5.2.6 can be strengthened. One can show that the set

$$\{f \in \mathbf{E} : s_m(f)(0) \not\to f(0)\}$$

is dense in $\mathbf{E}$. In fact, the set $\{f \in \mathbf{E} : s_m(f)(0) \to f(0)\}$ is also dense in $\mathbf{E}$! (see Exercise 5.5 and Exercise 3.42).

In the year 1807, **Joseph Fourier** (1768–1830) submitted a memoir to the Paris Academy of Sciences, in which he laid out a method to study heat propagation. The memoir was reviewed by Laplace, Lagrange, Poisson and Monge and was found to be lacking in rigour. Specifically, the idea that an infinite sum of functions can add up to an arbitrary function was rejected. The fault, though, was not entirely Fourier's. There was, at the time, simply no good definition of a function, or of a convergent series! However, the originality and wide applicability of Fourier's methods exerted an important influence on the mathematical community.

Several unsuccessful attempts were made to resolve this question of convergence: by Poisson (1820), Cauchy (1823) and Fourier (in his treatise of 1822). However, it was Lejeune Dirchlet (1829) who first proved that the Fourier series of a 2π-periodic, piecewise smooth function converged to the original function pointwise at all points of continuity.

The development of the Fourier series is intimately tied to the historical roots of mathematical analysis in the 19th century. It forced mathematicians to rigourously define the notion of a function, convergence of a series, integration and many other ideas that are now integral to the subject.

5.3   The Open Mapping and Closed Graph Theorems

Given a bijective linear map between vector spaces, it is easy to see that the inverse is also linear. The analogous statement is true for group and ring homomorphisms as well. Unfortunately, this breaks down in topology: The inverse of a continuous bijection need not be continuous.

In our situation, we have already seen an example (Exercise 2.24) of a bijective, bounded linear map whose inverse is not continuous. One of the main results of this section offers us a simple and effective condition under which such a map has a continuous inverse: one simply needs both the domain and range to be complete. This result (the Bounded Inverse Theorem) is itself a consequence of the more general Open Mapping Theorem. The latter is our first stop in this section.

For a normed linear space $\mathbf{E}$, subsets $A, B \subset \mathbf{E}$ and a scalar $\lambda \in \mathbb{K}$, we define $A + B := \{a + b : a \in A, b \in B\}$ and $\lambda A := \{\lambda a : a \in A\}$. We begin with a small observation.

5.3.1 LEMMA. *If A is convex, then $A + A = 2A$.*

Proof.   It is clear that $2A \subset A + A$. For the reverse inclusion, let $x, y \in A$. Then since A is convex, $\frac{x+y}{2} \in A$. Therefore

$$x + y = 2\frac{x + y}{2} \in 2A.$$

This proves the reverse inclusion as well.   □

The proof of the Open Mapping Theorem is long, so we break it into a few steps. This also allows us to highlight how completeness is used in the proof. Before we begin, recall that $B_{\mathbf{E}}(x, r)$ denotes the open ball of radius r in $\mathbf{E}$ centered at x.

5.3.2 LEMMA *Let $\mathbf{E}$ be an normed linear space, $\mathbf{F}$ be a Banach space and $T \in \mathcal{B}(\mathbf{E}, \mathbf{F})$ be a surjective, bounded linear map. Then for every $r > 0$ there exists $s > 0$ such that*

$$B_{\mathbf{F}}(0, s) \subset \overline{T(B_{\mathbf{E}}(0, r))}.$$

Proof. Fix $r > 0$. For $n \in \mathbb{N}$, define

$$B_n := n \overline{T(B_{\mathbf{E}}(0, r))}.$$

Then B_n is closed and

$$\mathbf{F} = T(\mathbf{E}) = \bigcup_{n=1}^{\infty} B_n.$$

Since $\mathbf{F}$ is complete, by the Baire Category Theorem (Corollary 5.1.2), there exists $N \in \mathbb{N}$ such that B_N has non-empty interior. Since the map $y \mapsto Ny$ is a homeomorphism of $\mathbf{E}$, B_1 must contain an open set. Thus there exists $y_0 \in \mathbf{F}$ and $s_0 > 0$ such that

$$B_{\mathbf{F}}(y_0, s_0) \subset B_1.$$

In particular, $y_0 \in B_1$ and so $-y_0 \in B_1$ as well. Now, for any $y \in B_{\mathbf{F}}(0, s_0)$, we have

$$y = (y + y_0) + (-y_0) \in B_{\mathbf{F}}(y_0, s_0) + B_1 \subset B_1 + B_1.$$

Since B_1 is convex, Lemma 5.3.1 implies that $B_{\mathbf{F}}(0, s_0) \subset 2B_1$. Therefore $s = s_0/2$ works. □

In the next lemma, we need the domain to be complete, which allows us to strengthen the conclusion of Lemma 5.3.2.

5.3.3 LEMMA *Let $\mathbf{E}$ and $\mathbf{F}$ be Banach spaces and $T \in \mathcal{B}(\mathbf{E}, \mathbf{F})$ be a surjective, bounded linear map. Then for every $r > 0$ there exists $s > 0$ such that*

$$B_{\mathbf{F}}(0, s) \subset T(B_{\mathbf{E}}(0, r)).$$

Proof. For each natural number $n \in \mathbb{N}$, set $r_n := r/2^{n+1}$. By Lemma 5.3.2, there exists $s_n > 0$ such that

$$B_{\mathbf{F}}(0, s_n) \subset \overline{T(B_{\mathbf{E}}(0, r_n))}. \tag{5.1}$$

Furthermore, we may choose s_n inductively in such a way that $\lim_{n \to \infty} s_n = 0$. We claim that

$$B_{\mathbf{F}}(0, s_1) \subset T(B_{\mathbf{E}}(0, r)).$$

Fix $y \in B_{\mathbf{F}}(0, s_1)$, then $y_1 \in \overline{T(B_{\mathbf{E}}(0, r_1))}$ by Equation 5.1. Therefore there exists $x_1 \in B_{\mathbf{E}}(0, r_1)$ such that

$$\|T(x_1) - y\| < s_2.$$

Now, $T(x_1) - y \in B_{\mathbf{F}}(0, s_2)$ and $B_{\mathbf{F}}(0, s_2) \subset \overline{T(B_{\mathbf{E}}(0, r_2))}$. Therefore there exists $x_2 \in B_{\mathbf{E}}(0, r_2)$ such that

$$\|T(x_2) + T(x_1) - y\| < s_3.$$

Thus proceeding, we obtain a sequence $(x_n) \subset \mathbf{E}$ such that, for each $n \in \mathbb{N}$, $x_n \in B_{\mathbf{E}}(0, r_n)$ and

$$\|T(x_n) + T(x_{n-1}) + \ldots + T(x_1) - y\| < s_{n+1}. \tag{5.2}$$

Now, note that

$$\sum_{n=1}^{\infty} \|x_n\| \leq \sum_{n=1}^{\infty} \frac{r}{2^{n+1}} = \frac{r}{2} < \infty.$$

Since $\mathbf{E}$ is a Banach space (by Proposition 2.3.8), there exists $z \in \mathbf{E}$ such that

$$z = \sum_{n=1}^{\infty} x_n.$$

By Equation 5.2, we have that

$$\left\| T\left(\sum_{i=1}^{n} x_i \right) - y \right\| < s_{n+1}.$$

By assumption, $\lim_{n \to \infty} s_n = 0$. Since T is continuous, it follows that $T(z) = y$. Note that $\|z\| \leq \frac{r}{2} < r$, so $y \in T(B_{\mathbf{E}}(0, r))$. This proves that $B_{\mathbf{F}}(0, s_1) \subset T(B_{\mathbf{E}}(0, r))$. $\qquad \square$

The proof of the main theorem is now straight-forward. Recall that an ***open map*** is one that maps open sets to open sets.

5.3.4 THEOREM (OPEN MAPPING THEOREM (BANACH, 1932)) *Let* **E** *and* **F** *be Banach spaces and* $T \in \mathcal{B}(\mathbf{E}, \mathbf{F})$ *be a surjective, bounded linear map. Then* T *is an open map.*

Proof. Let $V \subset \mathbf{E}$ be an open set. We wish to prove that $T(V)$ is open. So let $y \in T(V)$ and write $y = T(x)$ for some $x \in V$. Since V is open, there exists $r > 0$ such that $B_{\mathbf{E}}(x, r) \subset V$. By Lemma 5.3.3, there exists $s > 0$ such that

$$B_{\mathbf{F}}(0, s) \subset T(B_{\mathbf{E}}(0, r)).$$

Hence

$$B_{\mathbf{F}}(y, s) = y + B_{\mathbf{F}}(0, s) \subset y + T(B_{\mathbf{E}}(0, r)) = T(x + B_{\mathbf{E}}(0, r)) = T(B_{\mathbf{E}}(x, r)) \subset T(V).$$

Thus for each $y \in T(V)$, there exists $s > 0$ such that $B_{\mathbf{F}}(y, s) \subset T(V)$. In other words $T(V)$ is an open set.   □

5.3.5 THEOREM (BOUNDED INVERSE THEOREM) *Let* **E** *and* **F** *be Banach spaces. If* $T \in \mathcal{B}(\mathbf{E}, \mathbf{F})$ *is bijective, then* T^{-1} *is continuous.*

Proof. Since T is bijective, T^{-1} is a well-defined linear operator. That T^{-1} is continuous follows from the Open Mapping Theorem.   □

The First Isomorphism Theorem now finally takes the form that one would expect and the proof follows immediately from Theorem 2.5.9 and the Bounded Inverse Theorem.

5.3.6 COROLLARY (FIRST ISOMORPHISM THEOREM) *Let* **E** *and* **F** *be Banach spaces and* $T \in \mathcal{B}(\mathbf{E}, \mathbf{F})$ *be a surjective, bounded linear operator. Then the map* $\widehat{T} : \mathbf{E}/\ker(T) \to$ **F** *given by*

$$x + \ker(T) \mapsto T(x)$$

is a topological isomorphism.

5.3.7 EXAMPLE The examples given below show that the completeness assumptions of the Open Mapping Theorem cannot be avoided:

   (i) In Exercise 2.24, we constructed an example of an injective map $T : \ell^2 \to \ell^2$, whose inverse (defined on Range(T)) is not continuous. The reason for this, of course, is that Range(T) is not complete.

(ii) Let $\mathbf{E} := (C[0,1], \|\cdot\|_\infty)$ and $\mathbf{F}$ be the subspace

$$\mathbf{F} := \{f \in C^1[0,1] : f(0) = 0\},$$

equipped with the supremum norm. Note that $\mathbf{F}$ is not closed (and hence not complete). Define $T : \mathbf{E} \to \mathbf{F}$ by

$$T(f)(x) := \int_0^x f(t)dt$$

Then T is well-defined, bounded and bijective. However, $T^{-1} : \mathbf{F} \to \mathbf{E}$ is the map

$$f \mapsto f',$$

which is not bounded (Exercise 5.18).

(iii) If $\iota : (c_{00}, \|\cdot\|_1) \to (c_{00}, \|\cdot\|_\infty)$ to be the identity map, then ι is clearly bijective and bounded. However, the two norms are not equivalent (see Example 2.4.5). Therefore the inverse map is not continuous.

Before we move on, we give one more interesting application of the Open Mapping Theorem. To begin with, let us extract the following result buried in the proof of that theorem.

5.3.8 LEMMA *Let $\mathbf{E}$ be a Banach space, $\mathbf{F}$ be any normed linear space and $T \in \mathcal{B}(\mathbf{E}, \mathbf{F})$ be a bounded linear operator. If*

$$B_{\mathbf{F}}[0,1] \subset \overline{T(B_{\mathbf{E}}[0,1])},$$

then T is surjective.

Proof. By linearity, it suffices to show that the range of T contains the unit sphere in $\mathbf{F}$. So fix $y \in \mathbf{F}$ such that $\|y\| = 1$, then by hypothesis there exists $x_1 \in \mathbf{E}$ such that $\|x_1\| \le 1$ and

$$\|y - T(x_1)\| < \frac{1}{2}.$$

Now, $2(y - T(x_1)) \in B_{\mathbf{F}}[0,1]$, so there exists $x_2 \in \mathbf{E}$ with $\|x_2\| \le 1$ such that

$$\|2(y - T(x_1)) - T(x_2)\| < \frac{1}{4}.$$

Hence

$$\left\| y - T\left(x_1 + \frac{x_2}{2}\right) \right\| < \frac{1}{8}.$$

Thus proceeding, we obtain a sequence $(x_n) \subset B_{\mathbf{E}}[0,1]$ such that, for each $n \in \mathbb{N}$,

$$\left\| y - T\left(\sum_{j=1}^{n} \frac{x_j}{2^{j-1}}\right) \right\| < \frac{1}{2^{n+1}}.$$

As before, Proposition 2.3.8 ensures that the series

$$z := \sum_{j=1}^{\infty} \frac{x_j}{2^{j-1}}$$

converges in $\mathbf{E}$. By continuity of T, we see that $T(z) = y$. $\qquad\square$

We had proved in Theorem 3.4.6 that every Hilbert space $\mathbf{H}$ may be realised as $\ell^2(A)$ for some set A. The next result may be thought of as the analogue for Banach spaces.

5.3.9 THEOREM *Let $\mathbf{E}$ be a Banach space and A be a dense subset of $B_{\mathbf{E}}[0,1]$. Then there is a closed subspace $\mathbf{F}$ of $\ell^1(A)$ and an isometric isomorphism*

$$\ell^1(A)/\mathbf{F} \cong \mathbf{E}.$$

In particular, every separable Banach space is isometrically isomorphic to a quotient of ℓ^1.

Proof. Define $T : \ell^1(A) \to \mathbf{E}$ by

$$T(f) := \sum_{e \in A} f(e)e,$$

where the sum is defined as in Remark 3.3.13. Then it is easy to see that T is linear, and

$$\|T(f)\| \le \sum_{e \in A} |f(e)| \|e\| \le \sum_{e \in A} |f(e)| = \|f\|_1. \tag{5.3}$$

Furthermore, for any $e \in A$, the function $\delta_e \in \ell^1(A)$ clearly satisfies $T(\delta_e) = e$. Since A is dense in $B_{\mathbf{E}}[0,1]$, we conclude that

$$B_{\mathbf{E}}[0,1] \subset \overline{T(B_{\ell^1(A)}[0,1])}.$$

Therefore T is surjective (and an open map) by Lemma 5.3.8. If we set $\mathbf{F} := \ker(T)$, then the First Isomorphism Theorem gives us a topological isomorphism

$$\widehat{T} : \ell^1(A)/\mathbf{F} \to \mathbf{E}.$$

Now, choose $f \in \ell^1(A)$ such that $f \notin \mathbf{F}$. For $g \in \mathbf{F}$, $\|T(f)\| = \|T(f+g)\| \leq \|f+g\|_1$ by Equation 5.3. Hence

$$\|\widehat{T}(f + \mathbf{F})\| = \|T(f)\| \leq \|f + \mathbf{F}\|.$$

Now suppose $x := \widehat{T}(f + \mathbf{F})$, then x is non-zero since $\widehat{T}$ is injective. We may thus define $y := x/\|x\|$, so that $\|y\| = 1$. By the proof of Lemma 5.3.8, there exists $h \in \ell^1(A)$ such that $\|h\|_1 \leq 1$ and $T(h) = y$. Therefore if $f_0 := \|x\|h$, then $T(f_0) = x$ and

$$\|f + \mathbf{F}\| = \|f_0 + \mathbf{F}\| \leq \|f_0\|_1 \leq \|x\| = \|\widehat{T}(f + \mathbf{F})\|.$$

We conclude that $\widehat{T}$ is an isometric isomorphism.

Finally, observe that if $\mathbf{E}$ is separable, then we may take A to be a countable set, in which case there is a natural isomorphism $\ell^1(A) \cong \ell^1$. $\qquad\square$

We now turn to the Closed Graph Theorem, which gives us an easy way to *check* if a linear map is continuous or not. To put it in perspective, if we wish to show that a function $f : X \to Y$ (between metric spaces) is continuous, then we could check sequential continuity. In other words, for a sequence $(x_n) \subset X$, if $x_n \to x$, we would check that $f(x_n) \to f(x)$. A priori, though, we do not know if the sequence $(f(x_n))$ is convergent at all! This complicates matters, because one has to contend with the possibility that $(f(x_n))$ does not converge anywhere (let alone to $f(x)$).

The Closed Graph Theorem allows us to circumvent this issue (in the context of linear maps between Banach spaces). Let us see how this works.

5.3.10 DEFINITION Let X and Y be topological spaces and $f : X \to Y$ be a function. The *graph of f* is the set

$$G(f) := \{(x, f(x)) : x \in X\} \subset X \times Y.$$

5.3.11 LEMMA *Let X and Y be two metric spaces and $f : X \to Y$ be a continuous map. Then $G(f)$ is closed in $X \times Y$ (where $X \times Y$ is equipped with a product metric).*

Proof. Choose a sequence $(x_n, f(x_n)) \in G(f)$ such that $(x_n, f(x_n)) \to (x, y)$ in $X \times Y$. Then by definition of the product topology, $x_n \to x$ in X and $f(x_n) \to y$ in

Y. Since f is continuous, $f(x_n) \to f(x)$ as well and since Y is Hausdorff, it follows that

$$y = f(x).$$

Therefore $(x, y) \in G(f)$.   □

It is quite easy to construct a discontinuous (non-linear) function $f : \mathbb{R} \to \mathbb{R}$ whose graph is closed. However, the Closed Graph Theorem does provide a converse of Lemma 5.3.11 in the context of linear maps.

5.3.12 DEFINITION Let $\mathbf{E}$ and $\mathbf{F}$ be normed linear spaces and $T : \mathbf{E} \to \mathbf{F}$ be a linear operator (not necessarily bounded). The function $\|\cdot\|_G : \mathbf{E} \to \mathbb{R}_+$

$$\|x\|_G := \|x\|_{\mathbf{E}} + \|T(x)\|_{\mathbf{F}}$$

defines a norm on $\mathbf{E}$ and is called the **graph norm** on $\mathbf{E}$ with respect to T. Note that

(i) $\|x\|_{\mathbf{E}} \leq \|x\|_G$ for all $x \in \mathbf{E}$ and

(ii) $\|\cdot\|_{\mathbf{E}} \sim \|\cdot\|_G$ if and only if T is bounded.

The next lemma is a simple consequence of the Bounded Inverse Theorem and its proof is relegated to the exercises.

5.3.13 LEMMA *Let $\mathbf{E}$ be a vector space that is a Banach space with respect to two norms $\|\cdot\|_1$ and $\|\cdot\|_2$. Suppose that there exists $c > 0$ such that*

$$\|x\|_1 \leq c\|x\|_2$$

for all $x \in \mathbf{E}$. Then $\|\cdot\|_1 \sim \|\cdot\|_2$.

There is an important caveat to Lemma 5.3.13: The space $\mathbf{E}$ must be complete with respect to each norm independently. For instance, if we take $\mathbf{E} := C[0,1]$ and equip it with $\|\cdot\|_\infty$ and $\|\cdot\|_1$, then $(C[0,1], \|\cdot\|_\infty)$ is complete and $\|f\|_1 \leq \|f\|_\infty$ for all $f \in \mathbf{E}$. However, the two norms are not equivalent.

5.3.14 THEOREM (CLOSED GRAPH THEOREM (BANACH, 1932)) *Let $\mathbf{E}$ and $\mathbf{F}$ be Banach spaces and $T : \mathbf{E} \to \mathbf{F}$ be a linear operator. If $G(T)$ is closed in $\mathbf{E} \times \mathbf{F}$, then T is continuous.*

Proof. Consider the Graph norm $\|\cdot\|_G$ on **E** with respect to T, as above. By Lemma 5.3.13, it suffices to show that $(\mathbf{E}, \|\cdot\|_G)$ is a Banach space. So, suppose $(x_n) \subset \mathbf{E}$ is Cauchy with respect to $\|\cdot\|_G$. Then $\|x_n - x_m\|_{\mathbf{E}} \leq \|x_n - x_m\|_G$ and

$$\|T(x_n) - T(x_m)\|_{\mathbf{F}} \leq \|x_n - x_m\|_G.$$

Hence (x_n) is Cauchy in **E** and $(T(x_n))$ is Cauchy in **F**. Since both spaces are complete, there exist $x \in \mathbf{E}$ and $y \in \mathbf{F}$ such that $\lim_{n \to \infty} x_n = x$ and

$$\lim_{n \to \infty} T(x_n) = y.$$

Since $G(T)$ is closed, this implies that $T(x) = y$. We conclude that $\|x_n - x\|_G \to 0$. Hence $(\mathbf{E}, \|\cdot\|_G)$ is a Banach space and Lemma 5.3.13 implies that

$$\|\cdot\|_G \sim \|\cdot\|_{\mathbf{E}}.$$

Therefore T is bounded. □

As mentioned before, the Closed Graph Theorem gives us an easy way to determine if a linear operator $T : \mathbf{E} \to \mathbf{F}$ between Banach spaces is continuous. Let us now give an example of how this gets used.

Let **F** be a normed linear space. A subset $\mathcal{G} \subset \mathbf{F}^*$ of bounded linear functionals on **F** is said to **separate points** of **F** if, for any distinct pair of elements $x, y \in \mathbf{F}$, there exists $\varphi \in \mathcal{G}$ such that $\varphi(x) \neq \varphi(y)$.

5.3.15 LEMMA *Let* **E** *and* **F** *be Banach spaces and* $\mathcal{G} \subset \mathbf{F}^*$ *a collection of bounded linear functionals on* **F** *that separates points of* **F**. *If* $T : \mathbf{E} \to \mathbf{F}$ *is a linear operator such that* $\varphi \circ T : \mathbf{E} \to \mathbb{K}$ *is continuous for each* $\varphi \in \mathcal{G}$, *then* T *is continuous.*

Proof. This is a simple application of the Closed Graph Theorem. Suppose $(x_n) \subset \mathbf{E}$ is a sequence such that $x_n \to x$ in **E** and $T(x_n) \to y$ in **F**. We wish to prove that $T(x) = y$. So fix $\varphi \in \mathcal{G}$. Since $\varphi \circ T$ is continuous,

$$\lim_{n \to \infty} \varphi(T(x_n)) = \varphi(T(x)).$$

Since φ is continuous on **F**, it follows that

$$\lim_{n \to \infty} \varphi(T(x_n)) = \varphi(y).$$

Therefore $\varphi(T(x)) = \varphi(y)$ and this happens for each $\varphi \in \mathcal{G}$. Since $\mathcal{G}$ separates points of **F**, $T(x) = y$. Hence $G(T)$ is closed and T is continuous. □

5.3.16 EXAMPLE Let $M := (a_{i,j})_{i,j \in \mathbb{N}}$ be an infinite matrix with entries in $\mathbb{K}$ and let $1 \le p, r \le \infty$ be real numbers. For each $x = (x_1, x_2, \ldots) \in \ell^p$, let $T(x) = (y_1, y_2, \ldots)$ be the sequence given by

$$y_i := \sum_{j=1}^{\infty} a_{i,j} x_j.$$

Suppose that $T(x) \in \ell^r$ for each $x \in \ell^p$. Then we claim that T defines a bounded linear operator from ℓ^p to ℓ^r. Linearity is obvious, so we prove boundedness.

(i) For each $i \in \mathbb{N}$, the series $\sum_{j=1}^{\infty} a_{i,j} x_j$ is convergent in $\mathbb{K}$ for each $x = (x_n) \in \ell^p$. Taking finite sums, we may define linear functionals $\varphi_n^i : \ell^p \to \mathbb{K}$ given by

$$\varphi_n^i((x_j)) := \sum_{j=1}^{n} a_{i,j} x_j.$$

Each φ_n^i is then a bounded linear functional. Define $\psi_i : \ell^p \to \mathbb{K}$ by

$$\psi_i((x_j)) := \sum_{j=1}^{\infty} a_{i,j} x_j = \lim_{n \to \infty} \varphi_n^i((x_j)).$$

By the Banach–Steinhaus Theorem, each ψ_i is a bounded linear functional. Let q denote the conjugate exponent of p. If $1 \le p < \infty$, there is a natural isomorphism $\ell^q \cong (\ell^p)^*$ (Theorem 4.1.3) which ensures that $(a_{i,1}, a_{i,2}, \ldots) \in \ell^q$. If $p = \infty$, Exercise 4.1 ensures that $(a_{i,1}, a_{i,2}, \ldots) \in \ell^1$. Therefore in either case, $(a_{i,1}, a_{i,2}, \ldots) \in \ell^q$ for each $i \in \mathbb{N}$.

(ii) Now let s be the conjugate exponent of r and let $\Delta : \ell^s \to (\ell^r)^*$ be the injective map from Proposition 4.1.2. For each $i \in \mathbb{N}$, let $e_i \in \ell^s$ denote the sequence $(0, 0, \ldots, 0, 1, 0, \ldots)$ as before. Then

$$\mathcal{G} := \{\Delta(e_i) : i \in \mathbb{N}\}$$

is a family of bounded linear functionals that separates points of ℓ^r (members of $\mathcal{G}$ are nothing but the coordinate projections on ℓ^r). Now, the conclusion of part (i) simply says that

$$\psi_i = \Delta(e_i) \circ T$$

is a bounded linear functional for each $i \in \mathbb{N}$. By Lemma 5.3.15, we conclude that $T : \ell^p \to \ell^r$ is bounded.

The continuous analog of this example is given in Exercise 5.25.

5.4 Fourier Series of L^1 Functions

In Chapter 3, we had constructed an isometric isomorphism $U : L^2[-\pi, \pi] \to \ell^2(\mathbb{Z})$ given by $f \mapsto \widehat{f}$, where

$$\widehat{f}(n) := \frac{1}{2\pi} \int_{-\pi}^{\pi} f(t) e^{-int} dt.$$

Now,

$$|\widehat{f}(n)| \leq \frac{1}{2\pi} \|f\|_1. \tag{5.4}$$

for each $n \in \mathbb{N}$. Therefore the Fourier coefficients are well-defined even if $f \in L^1[-\pi, \pi]$. However, if we are to extend the domain of U to include $L^1[-\pi, \pi]$, we need to redefine our codomain as well. Equation 5.4 tells us to use a Banach space of sequences with the *supremum* norm. With a little hindsight, we set

$$c_0(\mathbb{Z}) := \{(x_n)_{n \in \mathbb{Z}} : \lim_{n \to \pm\infty} x_n = 0\}.$$

Note that $(c_0(\mathbb{Z}), \| \cdot \|_\infty)$ is a Banach space (Exercise 2.26).

5.4.1 PROPOSITION *If $f \in L^1[-\pi, \pi]$, then $\widehat{f} \in c_0(\mathbb{Z})$. Furthermore, the map $f \mapsto \widehat{f}$ defines a bounded linear operator*

$$T : L^1[-\pi, \pi] \to c_0(\mathbb{Z}).$$

Proof. Consider the restriction of the Fourier transform

$$T_0 : C[-\pi, \pi] \to \ell^2(\mathbb{Z}).$$

The inequality in Equation 5.4 shows that

$$\|\widehat{f}\|_\infty \leq \frac{1}{2\pi} \|f\|_1$$

for each $f \in C[-\pi, \pi]$. Therefore we may think of T_0 as a bounded linear transformation from $(C[-\pi, \pi], \| \cdot \|_1)$ to $(c_0(\mathbb{Z}), \| \cdot \|_\infty)$. Since $(C[-\pi, \pi], \| \cdot \|_1)$ is dense in $L^1[-\pi, \pi]$ and $c_0(\mathbb{Z})$ is a Banach space, it follows from Proposition 3.2.4 that T_0 extends uniquely to a linear transformation

$$T : L^1[-\pi, \pi] \to c_0(\mathbb{Z}).$$

This is the desired map. $\square$

5.4.2 REMARK   If $f \in L^1[-\pi, \pi]$, $\hat{f} \in c_0(\mathbb{Z})$. Therefore

$$\lim_{n \to \pm\infty} \int_{-\pi}^{\pi} f(t)e^{-int}dt = \lim_{n \to \pm\infty} \int_{-\pi}^{\pi} f(t)\cos(nt)dt = \lim_{n \to \pm\infty} \int_{-\pi}^{\pi} f(t)\sin(nt)dt = 0.$$

This is also called the Riemann–Lebesgue Lemma. It was originally proved by Riemann in the context of bounded Riemann-integrable functions in 1852 (published posthumously in 1867) and by Lebesgue in the context of L^1 functions in 1903.

The next lemma is what we need to show that the map T is injective.

5.4.3 LEMMA   *If $f \in L^1[-\pi, \pi]$ is such that*

$$\int_E f(t)dt = 0$$

for every measurable set $E \subset [-\pi, \pi]$, then $f = 0$ a.e.

Proof.   By definition,

$$\int_E f(t)dt = \int_E \mathrm{Re}(f)(t)dt + i \int_E \mathrm{Im}(f)(t)dt.$$

Therefore to prove the result, we may as well assume that f is real-valued. Now set

$$E_n := \{x \in [-\pi, \pi] : f(x) > 1/n\}.$$

Then E_n is measurable and

$$0 = \int_{E_n} f(t)dt > \frac{m(E_n)}{n}.$$

Therefore $m(E_n) = 0$ for each $n \in \mathbb{N}$, so that $E := \{x \in [-\pi, \pi] : f(x) > 0\} = \bigcup_{n=1}^{\infty} E_n$ has measure zero. In other words, $f(x) \leq 0$ almost everywhere. Similarly, $f(x) \geq 0$ almost everywhere as well and thus $f = 0$ a.e. $\square$

5.4.4 PROPOSITION   *The Fourier transform $T : L^1[-\pi, \pi] \to c_0(\mathbb{Z})$ is injective.*

Proof. For $n \in \mathbb{Z}$, define $e_n(t) := e^{int}$. Set $\mathcal{A} := \text{span}\{e_n : n \in \mathbb{Z}\}$, $\mathbf{E} := \{f \in C[-\pi, \pi] : f(-\pi) = f(\pi)\}$ and let $f \in L^1[-\pi, \pi]$ be such that

$$\int_{-\pi}^{\pi} f(t)e^{-int}dt = 0$$

for all $n \in \mathbb{Z}$. We wish to verify that the hypothesis of Lemma 5.4.3 holds and then conclude that $f = 0$ in $L^1[-\pi, \pi]$. To begin with, observe that

$$\int_{-\pi}^{\pi} f(t)g(t)dt = 0$$

for all $g \in \mathcal{A}$. If $g \in \mathbf{E}$, then by Proposition 3.5.2, there is a sequence $(g_n) \subset \mathcal{A}$ such that $\|g_n - g\|_\infty \to 0$. By the Dominated Convergence Theorem,

$$\int_{-\pi}^{\pi} f(t)g(t)dt = \lim_{n \to \infty} \int_{-\pi}^{\pi} f(t)g_n(t)dt = 0.$$

Now, if $F \subset [-\pi, \pi]$ is a measurable set, then there is a sequence $(g_n) \subset \mathbf{E}$ such that $g_n \to \chi_F$ with respect to $\|\cdot\|_1$. Furthermore, we may arrange it so that $\|g_n\|_\infty \leq 1$ (see the proof of Proposition 2.3.12). Hence there is a subsequence (g_{n_k}) such that $g_{n_k} \to \chi_F$ pointwise a.e. and $\|g_{n_k}\|_\infty \leq 1$ for all $k \in \mathbb{N}$. Once again, by the Dominated Convergence Theorem,

$$\int_F f(t)dt = \int_{-\pi}^{\pi} f(t)\chi_F(t)dt = \lim_{k \to \infty} \int_{-\pi}^{\pi} f(t)g_{n_k}(t)dt = 0.$$

By Lemma 5.4.3, $f = 0$ in $L^1[-\pi, \pi]$ and thus T is injective. $\qquad \square$

In the case of L^2 functions, we had a clear cut isomorphism from $L^2[-\pi, \pi]$ to $\ell^2(\mathbb{Z})$. However, in the case of L^1 functions, the range of the Fourier transform is little more mysterious. This is because the Bounded Inverse Theorem throws a spanner in the works.

5.4.5 THEOREM *The Fourier transform* $T : L^1[-\pi, \pi] \to c_0(\mathbb{Z})$ *is not surjective.*

Proof.   Suppose T were surjective, then by the Bounded Inverse Theorem, there would exist $c > 0$ such that

$$\|T(f)\|_\infty \geq c\|f\|_1 \tag{5.5}$$

for all $f \in L^1[-\pi, \pi]$. However, if $n \in \mathbb{N}$, consider the nth Dirichlet kernel

$$D_n(t) = \sum_{k=-n}^{n} e^{ikt}.$$

Then

$$T(D_n)_j = \begin{cases} \frac{1}{2\pi} & : \text{if } -n \leq j \leq n, \\ 0 & : \text{otherwise.} \end{cases}$$

so that $\|T(D_n)\|_\infty = 1/2\pi$ for all $n \in \mathbb{N}$. However, by Lemma 5.2.4, $\|D_n\|_1 \to \infty$ as $n \to \infty$. Hence Equation 5.5 cannot be satisfied. $\square$

Before we conclude this section, let us give another interesting application of the Bounded Inverse Theorem to the theory of Fourier series. This time it is a *characterization* of the norm on $L^1[-\pi, \pi]$.

5.4.6 EXAMPLE   Consider two properties that the norm $\|\cdot\|_1$ satisfies:

(i) $(L^1[-\pi, \pi], \|\cdot\|_1)$ is a Banach space and

(ii) for each $n \in \mathbb{Z}$, the nth Fourier coefficient $f \mapsto \widehat{f}(n)$ defines a bounded linear functional on $L^1[-\pi, \pi]$ (by Equation 5.4).

We claim that $\|\cdot\|_1$ is (upto equivalence), the *only* norm with these two properties. So suppose $\|\cdot\|_D$ is a norm on $L^1[-\pi, \pi]$ such that both these properties are satisfied. Then consider the identity map

$$\iota : (L^1[-\pi, \pi], \|\cdot\|_1) \to (L^1[-\pi, \pi], \|\cdot\|_D).$$

Suppose $(f_k) \subset L^1[-\pi, \pi]$ is a sequence such that $\|f_k - f\|_1 \to 0$ and $\|\iota(f_k) - g\|_D \to 0$. Then by the continuity of the Fourier coefficients (in both norms), we conclude that

$$\widehat{f}(n) = \lim_{n \to \infty} \widehat{f_k}(n) = \widehat{g}(n).$$

Since the Fourier transform is injective, $g = f = \iota(f)$. By the Closed Graph Theorem, ι is continuous (remember, $\|\cdot\|_D$ is also a complete norm). Thus there exists $\beta > 0$ such that

$$\|f\|_D \leq \beta \|f\|_1$$

for all $f \in L^1[-\pi, \pi]$. By the Bounded Inverse Theorem (specifically, Lemma 5.3.13) it follows that the two norms are equivalent. Thus the two properties listed above characterize the norm $\|\cdot\|_1$.

5.5 Schauder Bases for Banach Spaces

As we have seen before, a Hamel basis is somewhat unwieldy for infinite dimensional vector spaces. For Hilbert spaces, we found a satisfactory replacement in the notion of an orthonormal basis. For arbitrary (separable) Banach spaces, Schauder bases play an analogous role.

5.5.1 DEFINITION (SCHAUDER, 1927) Let **E** be a normed linear space. A countable set $\{e_1, e_2, \ldots\}$ is said to be a ***Schauder basis*** for **E** if for each $x \in \mathbf{E}$, there exist unique scalars $\{a_1, a_2, \ldots\}$ such that

$$x = \sum_{n=1}^{\infty} a_n e_n. \tag{5.6}$$

The emphasis in Definition 5.5.1 is on the *uniqueness* of this expression, for that allows us to make the following definition.

5.5.2 DEFINITION For each $n \in \mathbb{N}$, let $\varphi_n : \mathbf{E} \to \mathbb{K}$ be the unique function such that, for each $x \in \mathbf{E}$,

$$x = \sum_{n=1}^{\infty} \varphi_n(x) e_n.$$

In other words, $\varphi_n(x) = a_n$ if Equation 5.6 holds. The maps $\{\varphi_n\}_{n=1}^{\infty}$ are called the ***coordinate functionals*** of **E** with respect to the Schauder basis $\{e_n\}_{n=1}^{\infty}$.

This terminology is justified because these are linear (Exercise 5.36). That they are *bounded* (when **E** is a Banach space) is a consequence of the Bounded Inverse Theorem and we will prove it now. To put this result in perspective, let us mention one fact: If **E** is an infinite dimensional Banach space and $\{e_\gamma : \gamma \in \Gamma\}$ is a Hamel basis, then at least one of the corresponding coordinate functionals must be discontinuous (Exercise 5.33).

5.5.3 THEOREM *Let **E** be a Banach space and $\{e_1, e_2, \ldots\}$ be a Schauder basis for **E**. Then each coordinate functional φ_n is bounded.*

Proof.   The goal of the proof, broadly speaking, is to construct a (vector space) copy of $\mathbf{E}$ and equip it with a norm in which each φ_n is bounded. That 'copy' is the space $\mathbf{F}$ of all scalar sequences (a_i) such that the series

$$\sum_{i=1}^{\infty} a_i e_i$$

converges in $\mathbf{E}$. For simplicity of notation, we define the partial sums of this series to be $Q_n : \mathbf{F} \to \mathbf{E}$, given by

$$Q_n((a_i)) := \sum_{i=1}^{n} a_i e_i.$$

For a sequence $a = (a_i) \in \mathbf{F}$, define

$$\|a\|_{\mathbf{F}} := \sup_{n \in \mathbb{N}} \left\| \sum_{i=1}^{n} a_i e_i \right\| = \sup_{n \in \mathbb{N}} \|Q_n(a)\|.$$

This is well-defined because $(Q_n(a))_{n=1}^{\infty}$ is a convergent sequence (and hence bounded). Furthermore, the uniqueness condition of Definition 5.5.1 ensures that $a_n = 0$ for all $n \in \mathbb{N}$ whenever $\|a\|_{\mathbf{F}} = 0$. The triangle inequality is easy to check, so $\|\cdot\|_{\mathbf{F}}$ defines a norm on $\mathbf{F}$.

We claim that $(\mathbf{F}, \|\cdot\|_{\mathbf{F}})$ is a Banach space. To prove this, we follow the logic of Proposition 2.3.3 and break it into steps for convenience.

(i)   Let (x^k) be a Cauchy sequence in $\mathbf{F}$ and write $x^k = (x_1^k, x_2^k, \ldots)$. Then, for each $n \in \mathbb{N}$,

$$|x_n^k - x_n^m| \|e_n\| = \|(Q_n(x^k) - Q_{n-1}(x^k)) - (Q_n(x^m) - Q_{n-1}(x^m))\|$$

$$\leq 2 \sup_{j \in \mathbb{N}} \|Q_j(x^k) - Q_j(x^m)\| \tag{5.7}$$

$$= 2\|x^k - x^m\|_{\mathbf{F}}.$$

Therefore (x_n^k) is Cauchy in $\mathbb{K}$ and there exists $y_n \in \mathbb{K}$ such that

$$y_n = \lim_{k \to \infty} x_n^k.$$

(ii) To prove that $y = (y_n) \in \mathbf{F}$, it suffices to show that (z_n) is Cauchy in $(\mathbf{E}, \|\cdot\|)$, where

$$z_n := \sum_{i=1}^{n} y_i e_i.$$

To see this, fix $\epsilon > 0$, then there exists $N \in \mathbb{N}$ such that $\|x^k - x^m\|_{\mathbf{F}} < \epsilon$ for all $k, m \geq N$. Since $x^N \in \mathbf{F}$, there exists $M \in \mathbb{N}$ such that, for all $n \geq \ell \geq M$,

$\|\sum_{i=\ell}^n x_i^N e_i\| < \epsilon$. Following the logic of Equation 5.7, we see that

$$\left\| \sum_{i=\ell}^n x_i^k e_i - \sum_{i=\ell}^n x_i^N e_i \right\| < 2\epsilon$$

holds for all $k \geq N$ and all $n \geq \ell \geq M$. Thus

$$\left\| \sum_{i=\ell}^n x_i^k e_i \right\| < 3\epsilon$$

holds for all $k \geq N$ and all $n \geq \ell \geq M$. Letting $k \to \infty$, we conclude that

$$\left\| \sum_{i=\ell}^n y_i e_i \right\| \leq 3\epsilon$$

holds for all $n \geq \ell \geq M$. Thus the sequence (z_n) defined above is Cauchy and thus convergent.

(iii) It remains to show that (x^k) converges to y in **F**. For $n \in \mathbb{N}$ fixed, Equation 5.7 implies that

$$\left\| \sum_{i=1}^n x_i^k e_i - \sum_{i=1}^n x_i^m e_i \right\| < \epsilon$$

if $k, m \geq N$. Letting $k \to \infty$, we conclude that

$$\left\| \sum_{i=1}^n y_i e_i - \sum_{i=1}^n x_i^m e_i \right\| \leq \epsilon$$

for all $m \geq N$. Hence $\|y - x^m\|_{\mathbf{F}} \leq \epsilon$ for all $m \geq N$. Thus (x^k) converges to y in **F** and we have shown that $(\mathbf{F}, \|\cdot\|_{\mathbf{F}})$ is a Banach space.

Now we define $T : \mathbf{F} \to \mathbf{E}$ by

$$T((a_i)) := \sum_{n=1}^{\infty} a_n e_n.$$

Then, T is well-defined (by definition of **F**), linear and bijective (because $\{e_n\}$ is a Schauder basis of **E**). Finally, observe that

$$\|T((a_i))\| = \left\| \sum_{n=1}^{\infty} a_n e_n \right\| = \lim_{k \to \infty} \left\| \sum_{n=1}^k a_n e_n \right\| \leq \sup_{k \in \mathbb{N}} \|Q_k((a_i))\| = \|(a_i)\|_{\mathbf{F}}.$$

Hence T is a bounded operator as well. By the Bounded Inverse Theorem, T^{-1} is bounded as well. Now for each $n \in \mathbb{N}$, the linear functional φ_n satisfies the following inequality (as in Equation 5.7)

$$|\varphi_n(x)|\|e_n\| \leq 2\|T^{-1}(x)\| \leq 2\|T^{-1}\|\|x\|.$$

Hence φ_n is a bounded linear functional. $\square$

5.5.4 EXAMPLE

 (i) If **E** is a finite dimensional normed linear space, then any Hamel basis of **E** is a Schauder basis.

 (ii) If **E** is a separable Hilbert space, then any orthonormal basis of **E** is a Schauder basis.

 (iii) In fact, if **E** is a normed linear space that admits a Schauder basis, then **E** must be separable. The proof of this is similar to Theorem 3.4.8 (where we proved the analogous statement for Hilbert spaces) and is left as an exercise for the reader (Exercise 5.37).

 (iv) If **E** $= \ell^p$ for $1 \leq p < \infty$, or c_0, then the standard basis $\{e_1, e_2, \ldots\}$ is a Schauder basis.

 (v) For $n \geq 0$, let $\zeta_n \in C[0,1]$ be the function $\zeta_n(t) := t^n$. Then the set $\{\zeta_0, \zeta_1, \ldots\}$ *does not* form a Schauder basis for $C[0,1]$. This is because any function that can be expressed as a series as in Equation 5.6 must be differentiable (because the convergence of the power series is uniform). However, not every element in $C[0,1]$ is differentiable.

We now state a result that should remind you of the Fourier expansion of an element of a Hilbert space with respect to an orthonormal basis (specifically, Exercise 3.36).

5.5.5 THEOREM   *Let* **E** *be a Banach space with a Schauder basis* $\{e_n\}_{n=1}^{\infty}$ *and let* $\varphi_n :$ **E** $\to \mathbb{K}$ *be the coordinate linear functional associated to* e_n. *For* $n \in \mathbb{N}$, *define* $P_n :$ **E** $\to$ **E** *by*

$$P_n(x) = \sum_{i=1}^{n} \varphi_i(x)e_i.$$

Then

 (i) *Each* P_n *is a bounded linear operator.*

 (ii) *For any* $x \in$ **E**, $\lim_{n \to \infty} P_n(x) = x$.

 (iii) $\sup_{n \in \mathbb{N}} \|P_n\| < \infty$.

The proof is a direct consequence of Theorem 5.5.3 and the Principle of Uniform Boundedness. Therefore we leave its proof as an exercise (Exercise 5.38). What is interesting about this theorem though, is that it has a converse, which is a useful way to *recognize* when a set is a Schauder basis.

5.5.6 PROPOSITION *Let* $\mathbf{E}$ *be a Banach space and* $\{e_1, e_2, \ldots\}$ *be a countable set of non-zero vectors in* $\mathbf{E}$. *Then* $\{e_n\}$ *is a Schauder basis for* $\mathbf{E}$ *if and only if both the following conditions are met.*

(a) *There exists a constant* $M > 0$ *such that for any* $n < m$ *and any scalars* $\alpha_1, \alpha_2, \ldots, \alpha_m$, *one has*

$$\left\| \sum_{i=1}^{n} \alpha_i e_i \right\| \leq M \left\| \sum_{i=1}^{m} \alpha_i e_i \right\|. \tag{5.8}$$

(b) $\mathrm{span}(\{e_n\})$ *is dense in* $\mathbf{E}$.

The least number M satisfying the inequality in Equation 5.8 is called the **basis constant** of $\{e_n\}$.

Proof. Suppose that $\{e_n\}_{n=1}^{\infty}$ is a Schauder basis, then the set is linearly independent and therefore every element must be non-zero. Condition (b) holds by definition and condition (a) holds by Theorem 5.5.5 by taking $M := \sup_{n \in \mathbb{N}} \|P_n\|$, with P_n as defined in that theorem.

For the converse, suppose that $\{e_n\}$ is a countable set of non-zero vectors such that (a) and (b) hold. We wish to prove that $\{e_n\}$ forms a Schauder basis for $\mathbf{E}$.

(i) To begin with, we claim that the set $\{e_n\}$ is linearly independent. Suppose $\alpha_1, \alpha_2, \ldots, \alpha_n$ are scalars such that $\sum_{i=1}^{n} \alpha_i e_i = 0$, then by Equation 5.8,

$$\sum_{i=1}^{n-1} \alpha_i e_i = 0.$$

Subtracting, we see that $\alpha_n e_n = 0$, which implies that $\alpha_n = 0$ since $e_n \neq 0$. Proceeding in this way, we conclude that $\alpha_j = 0$ for all $1 \leq i \leq n$.

(ii) Now set $\mathbf{F} := \mathrm{span}(\{e_n\})$. Then $\mathbf{F}$ is dense in $\mathbf{E}$ by condition (b). For each $n \in \mathbb{N}$, define $T_n : \mathbf{F} \to \mathbf{E}$ by

$$T_n \left(\sum_{i=1}^{N} \alpha_i e_i \right) := \sum_{j=1}^{n} \alpha_j e_j,$$

where $N \geq n$. Note that this definition makes sense because $\{e_n\}$ is linearly independent. Each T_n is clearly linear and $\|T_n\| \leq M$ by Equation 5.8.

(iii) Since **E** is a Banach space, T_n extends to a bounded linear operator $S_n : \mathbf{E} \to$ **E** such that $\|S_n\| \leq M$ (by Proposition 3.2.4). Now define $\varphi_n : \mathbf{E} \to \mathbb{K}$ by the formula

$$\varphi_n(x)e_n = S_n(x) - S_{n-1}(x)$$

(where S_0 is taken to be the zero operator). Observe that $|\varphi_n(x)| \leq \frac{2M}{\|e_n\|}\|x\|$ for each $x \in \mathbf{E}$, so φ_n is a bounded linear functional. Also,

$$S_n(x) = \sum_{i=1}^{n} \varphi_i(x)e_i.$$

(iv) If $x \in \mathbf{E}$, we claim that $x = \sum_{i=1}^{\infty} \varphi_i(x)e_i$. To see this, choose a sequence (x_k) in **F** such that $x_k \to x$ and fix $\epsilon > 0$. Now choose $K \in \mathbb{N}$ such that $\|x - x_K\| < \epsilon$ and choose $N \in \mathbb{N}$ large enough so that

$$S_N(x_K) = x_K.$$

We can do this because $x_K \in \mathbf{F}$. Then $S_n(x_K) = x_K$ for all $n \geq N$ and

$$\left\| \sum_{i=1}^{n} \varphi_i(x)e_i - x \right\| = \|S_n(x) - x\|$$

$$\leq \|S_n(x) - S_n(x_K)\| + \|S_n(x_K) - x_K\| + \|x_K - x\|$$

$$\leq M\|x - x_K\| + \|x_K - x\| \leq (M+1)\epsilon.$$

This is true for any $\epsilon > 0$ and all $n \geq N$, so we conclude that

$$x = \sum_{i=1}^{\infty} \varphi_i(x)e_i. \tag{5.9}$$

(v) Finally, we wish to prove that such an expression is unique. So suppose $x = \sum_{i=1}^{\infty} \alpha_i e_i$ for some scalars $\alpha_i \in \mathbb{K}$. Then fix $n \in \mathbb{N}$ and observe that $\varphi_n(e_j) = 0$ if $n \neq j$. Using the continuity of φ_n, we see that

$$\alpha_n = \sum_{i=1}^{\infty} \alpha_i \varphi_n(e_i) = \varphi_n \left(\sum_{i=1}^{\infty} \alpha_i e_i \right) = \varphi_n(x).$$

Therefore the expression in Equation 5.9 is unique.

Thus we have proved that $\{e_n\}$ is a Schauder basis for **E**. $\square$

We are now in a position to construct a Schauder basis for $L^p[0,1]$ when $1 \leq p < \infty$ (Note that $L^\infty[0,1]$ cannot have a Schauder basis because it is not separable). Before we begin, we need a couple of useful notions.

Recall that a *dyadic rational* is a rational number which can be expressed in the form $\frac{n}{2^m}$ for integers n and m. Such rational numbers are dense in $\mathbb{R}$. Therefore if we choose the characteristic function χ_E of an interval $E \subset [0,1]$, then χ_E is a limit (in $L^p[0,1]$) of a sequence χ_{E_n}, where E_n is an interval whose end-points are dyadic rationals. Therefore we arrive at the following fact, which is a slight strengthening of Lemma 4.1.13.

5.5.7 LEMMA *Let S_d denote the linear span of functions of the form χ_E, where E is a half-open interval of $[0,1]$ whose end-points are dyadic rationals. If $1 \leq p < \infty$, then S_d is dense in $L^p[0,1]$.*

5.5.8 DEFINITION Let A be a convex subset of a real vector space and let $f : A \to [-\infty, +\infty]$ be a function. We say that f is *convex* if, whenever $x, y \in A$ and $0 \leq t \leq 1$, then

$$f(tx + (1-t)y) \leq tf(x) + (1-t)f(y).$$

Notice that the function $t \mapsto |t|^p$ is convex on $\mathbb{R}$ if $1 \leq p < \infty$.

5.5.9 EXAMPLE (HAAR, 1909) Fix $1 \leq p < \infty$ and let $\{h_1, h_2, \ldots\} \subset L^p[0,1]$ be functions defined as $h_1 \equiv 1$ and

$$h_{2^n + \ell}(t) = \begin{cases} 1 & : \text{if } \frac{2\ell - 2}{2^{n+1}} < t < \frac{2\ell - 1}{2^{n+1}}, \\ -1 & : \text{if } \frac{2\ell - 1}{2^{n+1}} < t < \frac{2\ell}{2^{n+1}}, \\ 0 & : \text{otherwise.} \end{cases}$$

for $n \geq 0$ and $\ell \in \{1, 2, \ldots, 2^n\}$. We now show that the set $\{h_i\}_{i=1}^\infty$ forms a Schauder basis for $L^p[0,1]$ by verifying the hypotheses of Proposition 5.5.6.

(a) Observe that

$$\chi_{[0,1/2)} = \frac{h_1 + h_2}{2}, \chi_{(1/2,1]} = \frac{h_1 - h_2}{2}, \chi_{[0,1/4)} = \frac{\chi_{[0,1/2)} + h_3}{2},$$

and so on. Therefore span$(\{h_i\})$ contains the set S_d described in Lemma 5.5.7 and is thus dense in $L^p[0,1]$.

(b) Since the function $t \mapsto |t|^p$ is convex on $\mathbb{R}$,

$$|b+a|^p + |b-a|^p \geq 2|b|^p \tag{5.10}$$

for any $a, b \in \mathbb{R}$. Now if $n \in \mathbb{N}$ is fixed and $\alpha_1, \alpha_2, \ldots, \alpha_{n+1}$ are scalars, then consider the functions

$$f = \sum_{i=1}^{n} \alpha_i h_i \text{ and } g = \sum_{j=1}^{n+1} \alpha_j h_j.$$

These two functions differ only on the support of h_{n+1}, which is an interval I. On this interval, f has a constant value, say b. Furthermore, g has value $(b + \alpha_{n+1})$ on the first half of I and $(b - \alpha_{n+1})$ on the second half of I. We write I^+ and I^- for the first and second half of I, respectively, and observe that $m(I^+) = m(I^-) =: \beta$. Therefore

$$\|g\|_p^p - \|f\|_p^p = \int_I (|g|^p - |f|^p)$$
$$= \int_{I^+} (|b + \alpha_{n+1}|^p - |b|^p) + \int_{I^-} (|b - \alpha_{n+1}|^p - |b|^p)$$
$$= (|b + \alpha_{n+1}|^p + |b - \alpha_{n+1}|^p - 2|b|^p) \beta$$
$$\geq 0$$

where the last inequality holds by Equation 5.10. Therefore

$$\left\| \sum_{i=1}^{n} \alpha_i h_i \right\| \leq \left\| \sum_{j=1}^{n+1} \alpha_j h_j \right\|.$$

Condition (b) of Proposition 5.5.6 now follows by induction.

Thus $\{h_i\}_{i=1}^{\infty}$ is a Schauder basis for $L^p[0,1]$, called the **Haar system**.

One may use similar ideas to construct a Schauder basis for $C[0,1]$, which we describe in Exercise 5.39.

5.6 Compact Operators

From linear algebra, we know that every linear operator between finite dimensional spaces may be associated to a matrix. This association allows us to analyze such operators with computational tools, such as determinants and polynomials. We now introduce a class of operators that are closely related to matrices. As a result, they serve as a useful bridge between matrix theory and the study of operators on infinite dimensional spaces.

Henceforth we write $B_{\mathbf{E}} := \{x \in \mathbf{E} : \|x\| \leq 1\}$ for the closed unit ball in a normed linear space $\mathbf{E}$.

5.6.1 DEFINITION An operator $T : \mathbf{E} \to \mathbf{F}$ between two normed linear spaces is said to be *compact* if $T(B_{\mathbf{E}})$ has compact closure.

5.6.2 REMARK

(i) There is nothing sacrosanct about the unit ball in this definition. An operator is compact if and only if $T(A)$ has compact closure for every norm-bounded subset A of $\mathbf{E}$.

(ii) Also, if $T(B_{\mathbf{E}})$ has compact closure, then it is bounded. Therefore every compact operator is a bounded operator.

(iii) If $T : \mathbf{E} \to \mathbf{F}$ is a compact operator, then $T(B_{\mathbf{E}})$ is totally bounded. Conversely, if $\mathbf{F}$ is complete and $T(B_{\mathbf{E}})$ is totally bounded, then T is compact.

(iv) If $\mathbf{E}$ and $\mathbf{F}$ are both finite dimensional, then *every* linear operator $T : \mathbf{E} \to \mathbf{F}$ is compact, because $T(B_{\mathbf{E}})$ is necessarily a bounded subset of $\mathbf{F}$ and thus has compact closure by Theorem 2.4.16.

(v) At the other extreme, if $\mathbf{E}$ is infinite dimensional, then the identity operator on $\mathbf{E}$ is not compact.

The next lemma gives us an ample supply of compact operators. Recall that an operator $T : \mathbf{E} \to \mathbf{F}$ is said to be have *finite rank* if $\text{Range}(T)$ is finite dimensional.

5.6.3 LEMMA *If $T : \mathbf{E} \to \mathbf{F}$ is a bounded operator of finite rank, then it is compact.*

Proof. Since T is bounded, $T(B_{\mathbf{E}})$ is a bounded subset of $\text{Range}(T)$. Furthermore, $\text{Range}(T)$ is a closed subspace of $\mathbf{F}$, since it is finite dimensional. Therefore $\overline{T(B_{\mathbf{E}})}$ is a closed and bounded subset of $\text{Range}(T)$. By Theorem 2.4.16, it must be compact in $\text{Range}(T)$ and therefore in $\mathbf{F}$. $\qquad\square$

If $\mathbf{E}$ and $\mathbf{F}$ are normed linear spaces, we will henceforth write $\mathcal{F}(\mathbf{E}, \mathbf{F})$ for the set of all bounded, finite rank operators from $\mathbf{E}$ to $\mathbf{F}$ and we write $\mathcal{K}(\mathbf{E}, \mathbf{F})$ for the space of all compact operators from **E** to **F**. By Lemma 5.6.3,

$$\mathcal{F}(\mathbf{E}, \mathbf{F}) \subset \mathcal{K}(\mathbf{E}, \mathbf{F}) \subset \mathcal{B}(\mathbf{E}, \mathbf{F}).$$

If $\mathbf{E} = \mathbf{F}$, we write $\mathcal{F}(\mathbf{E}) := \mathcal{F}(\mathbf{E}, \mathbf{E})$ and $\mathcal{K}(\mathbf{E}) := \mathcal{K}(\mathbf{E}, \mathbf{E})$.

5.6.4 PROPOSITION *Let* **E** *and* **F** *be normed linear spaces. Then*

(i) $\mathcal{K}(\mathbf{E}, \mathbf{F})$ *is a subspace of* $\mathcal{B}(\mathbf{E}, \mathbf{F})$.

(ii) *If* **F** *is a Banach space, then* $\mathcal{K}(\mathbf{E}, \mathbf{F})$ *is closed in* $\mathcal{B}(\mathbf{E}, \mathbf{F})$.

Proof.

(i) It is clear that a scalar multiple of a compact operator is compact. Therefore we let $S, T \in \mathcal{K}(\mathbf{E}, \mathbf{F})$ and prove that $S + T \in \mathcal{K}(\mathbf{E}, \mathbf{F})$. We wish to prove that $Y := (S + T)(B_\mathbf{E})$ has compact closure. Since we are in a metric space, it suffices to show that any sequence in Y has a subsequence that converges in **F**. To do this, fix a sequence $(y_n) \subset Y$ and write

$$y_n := (S + T)(x_n)$$

for some sequence $(x_n) \subset B_\mathbf{E}$. Since S is compact, there is a subsequence $(S(x_{n_k}))_{k=1}^\infty$ that converges to a point $y_1 \in \mathbf{F}$. Since T is compact, $(T(x_{n_k}))_{k=1}^\infty$ has a subsequence, say $(T(x_{n_{k_j}}))_{j=1}^\infty$ that converges to a point $y_2 \in \mathbf{F}$. However, $(S(x_{n_{k_j}}))_{j=1}^\infty$ must converges to y_1, since it is a subsequence of $(S(x_{n_k}))_{k=1}^\infty$. Therefore

$$y_{n_{k_j}} = S(x_{n_{k_j}}) + T(x_{n_{k_j}}) \to y_1 + y_2.$$

This is true for any sequence (y_n) of Y and thus Y has compact closure.

(ii) Suppose $(T_n) \subset \mathcal{K}(\mathbf{E}, \mathbf{F})$ is a sequence such that $\lim_{n\to\infty} \|T_n - T\| = 0$. We wish to prove that $T(B_\mathbf{E})$ is totally bounded. To that end, fix $\epsilon > 0$ and choose $T_N \in \mathcal{K}(\mathbf{E}, \mathbf{F})$ such that $\|T_N - T\| < \frac{\epsilon}{3}$. Since $T_N(B_\mathbf{E})$ is totally bounded, there are finitely many points $\{x_1, x_2, \ldots, x_k\} \subset B_\mathbf{E}$ such that

$$T_N(B_\mathbf{E}) \subset \bigcup_{i=1}^{k} B\left(T_N(x_i), \frac{\epsilon}{3}\right).$$

If $y \in T(B_\mathbf{E})$, then we write $y = T(x)$ for some $x \in B_\mathbf{E}$ and choose $1 \leq i \leq k$ such that $T(x) \in B\left(T_N(x_i), \frac{\epsilon}{3}\right)$. Now observe that

$$\|T(x) - T(x_i)\| \leq \|T(x) - T_N(x)\| + \|T_N(x) - T_N(x_i)\| + \|T_N(x_i) - T(x_i)\| < \epsilon.$$

This implies that

$$T(B_\mathbf{E}) \subset \bigcup_{i=1}^{k} B\left(T(x_i), \epsilon\right).$$

This is true for any $\epsilon > 0$, proving that $T(B_\mathbf{E})$ is totally bounded. Now is where we use our hypothesis: Since $\mathbf{F}$ is complete, $T(B_\mathbf{E})$ has compact closure.

□

In the absence of completeness in Part (ii) of Proposition 5.6.4, one can only conclude that $T(B_\mathbf{E})$ is totally bounded. Exercise 5.42 gives an example where $T(B_\mathbf{E})$ need not have compact closure despite this. Let us now give some interesting examples of compact operators.

5.6.5 EXAMPLE Let $\mathbf{E}$ and $\mathbf{F}$ be two normed linear spaces, $y \in \mathbf{F}$ and $\varphi \in \mathbf{E}^*$. Define $T : \mathbf{E} \to \mathbf{F}$ by

$$T(x) = \varphi(x)y.$$

Then T is a rank one, bounded operator and is thus compact. More generally, given a finite set $\{y_1, y_2, \ldots, y_n\} \subset \mathbf{F}$ and $\{\varphi_1, \varphi_2, \ldots, \varphi_n\} \subset \mathbf{E}^*$, we may define $T : \mathbf{E} \to \mathbf{F}$ by

$$T(x) = \sum_{i=1}^{n} \varphi_i(x)y_i.$$

Then T is a finite rank, bounded operator.

5.6.6 EXAMPLE Let $\alpha := (\alpha_n) \in \ell^\infty$ be a bounded sequence and let $T : \ell^2 \to \ell^2$ be the operator given by

$$T((x_j)) := (\alpha_1 x_1, \alpha_2 x_2, \ldots).$$

We had seen in Exercise 2.15 that T is a bounded operator and that $\|T\| = \|\alpha\|_\infty$. We claim that T is compact if $\alpha \in c_0$.

To see this, fix $n \in \mathbb{N}$ and define $T_n : \ell^2 \to \ell^2$ by

$$T_n((x_j)) := (\alpha_1 x_1, \alpha_2 x_2, \ldots, \alpha_n x_n, 0, 0, \ldots).$$

Then T_n is a bounded operator and has finite rank. Therefore T_n is compact. Now if $x \in \ell^2$, then

$$\|(T - T_n)(x)\|_2^2 = \sum_{i=n+1}^{\infty} |\alpha_i|^2 |x_i|^2 \leq \left(\sup_{j \geq n+1} |\alpha_j|^2 \right) \|x\|_2^2.$$

Hence $\|T - T_n\| \leq \sup_{j \geq n+1} |\alpha_j|$. Since $\lim_{j \to \infty} \alpha_j = 0$, T is a limit of compact operators and is thus compact by Proposition 5.6.4.

The next example is one that we had encountered in Example 2.2.16 and later in Example 5.3.16. There we had used matrices to describe bounded operators on the ℓ^p spaces. Now we try to determine when such an operator may be compact.

5.6.7 EXAMPLE Let $M := (a_{i,j})_{i,j \in \mathbb{N}}$ be an infinite matrix with entries in $\mathbb{K}$. For each $j \in \mathbb{N}$, define

$$\gamma_j := \sum_{i=1}^{\infty} |a_{i,j}|.$$

Assume that $\alpha := \sup\{\gamma_j : j \in \mathbb{N}\} < \infty$. Let $T : \ell^1 \to \ell^1$ be the induced operator as described in Example 2.2.16. In other words, if $x = (x_j) \in \ell^1$, then $T(x) = (y_1, y_2, \ldots)$, where

$$y_i := \sum_{j=1}^{\infty} a_{i,j} x_j.$$

We know that $T : \ell^1 \to \ell^1$ is a well-defined and bounded operator. We claim that T is compact if $\lim_{j \to \infty} \gamma_j = 0$. To see this, fix $n \in \mathbb{N}$ and define $T_n : \ell^1 \to \ell^1$ by

$$T_n(x)_i := \sum_{j=1}^{n} a_{i,j} x_j.$$

Observe that $T_n(x) = \sum_{i=1}^{n} x_i T(e_i)$ and thus T_n is a bounded operator of finite rank. Now for any $x \in \ell^1$, we have

$$\|(T - T_n)(x)\|_1 = \sum_{i=1}^{\infty} \left| \sum_{j=n+1}^{\infty} a_{i,j} x_j \right| \leq \sum_{i=1}^{\infty} \sum_{j=n+1}^{\infty} |a_{i,j}| |x_j|$$

$$= \sum_{j=n+1}^{\infty} \sum_{i=1}^{\infty} |a_{i,j}| |x_j| \leq \left(\sup_{j \geq n+1} |\gamma_j| \right) \|x\|_1$$

(notice that we may interchange summation signs by Proposition 1.2.13). Therefore $\|T - T_n\| \leq \sup_{j \geq n+1} |\gamma_j|$. Since $\lim_{j \to \infty} \gamma_j = 0$, T is a limit of finite rank, bounded operators. Since ℓ^1 is a Banach space, T is compact by Proposition 5.6.4.

We now discuss the integral operators on L^2 that we had seen in Example 2.2.17. What is remarkable is that any such operator is compact!

5.6.8 EXAMPLE Let $\mathbf{H} = L^2[0,1]$ and $K \in L^2([0,1]^2)$. Consider the associated integral operator $T : \mathbf{H} \to \mathbf{H}$ given by

$$T(f)(x) = \int_0^1 K(x,y)f(y)dy.$$

In Example 2.2.17, we had seen that T is a bounded operator on $\mathbf{H}$ and that

$$\|T\| \leq \|K\|_2.$$

We will use these facts to prove that T is, in fact, a compact operator. We will do so by building T up from simpler operators of the same type.

(i) Suppose that there exist finitely many functions $h_1, h_2, \ldots, h_n$ and $g_1, g_2, \ldots, g_n \in C[0,1]$ such that

$$K(x,y) = \sum_{i=1}^n h_i(x)g_i(y).$$

Then $T(f) = \sum_{i=1}^n \langle f, \overline{g_i} \rangle h_i$. Thus T has finite rank and is bounded and is thus compact.

(ii) Now we consider

$$\mathcal{A} = \left\{ (x,y) \mapsto \sum_{i=1}^n h_i(x)g_i(y) : h_i, g_j \in C[0,1] \right\} \subset C([0,1]^2).$$

This is an algebra in $C([0,1]^2)$ which contains the constant function $\mathbf{1}$, separates points of $[0,1]^2$ and is closed under complex conjugation. Therefore by the Stone–Weierstrass Theorem (Theorem A.1.2), $\mathcal{A}$ is dense in $C([0,1]^2)$ with respect to the supremum norm. Hence if $K \in C([0,1]^2)$, there is a sequence (K_n) in $\mathcal{A}$ such that

$$\|K_n - K\|_\infty \to 0.$$

Therefore, $\|K_n - K\|_2 \to 0$ and we may define $T_n \in \mathcal{B}(\mathbf{H})$ by

$$T_n(f)(x) = \int_0^1 K_n(x,y)f(y)dy.$$

Then by Example 2.2.17,

$$\|T_n - T\| \leq \|K_n - K\|_2 \to 0.$$

Since $T_n \in \mathcal{K}(\mathbf{H})$ for all $n \in \mathbb{N}$ (by part (i)), it follows that $T \in \mathcal{K}(\mathbf{H})$ by Proposition 5.6.4.

(iii) Finally, if $K \in L^2([0,1]^2)$, then by Proposition 2.3.12, there is a sequence $(K_n) \subset C([0,1]^2)$ such that $\|K_n - K\|_2 \to 0$. Approximating T as we did above, we see that T is a limit of compact operators and is thus compact.

Now that we have produced all these examples, we realize that there is a common theme here. All compact operators we produce are limits of finite rank operators. This begs the question, are *all* compact operators of this form? To an extent, it is true.

5.6.9 THEOREM *Let $\mathbf{E}$ be a normed linear space and $\mathbf{F}$ be a Banach space that admits a Schauder basis. Then a bounded operator $T : \mathbf{E} \to \mathbf{F}$ is compact if and only if it is a limit of bounded, finite rank operators.*

Proof. By Lemma 5.6.3 and Proposition 5.6.4, we know that any operator that is a limit of finite rank operators must be compact. Conversely, suppose $\mathbf{F}$ admits a Schauder basis $\{e_n\}$ and let $T : \mathbf{E} \to \mathbf{F}$ be a compact operator. By Theorem 5.5.5, there are bounded linear maps $P_n : \mathbf{F} \to \mathbf{F}$ such that

$$\lim_{n \to \infty} P_n(y) = y$$

for any $y \in \mathbf{F}$ and $M := \sup_{n \in \mathbb{N}} \|P_n\| < \infty$. Furthermore, each P_n has finite rank by construction. Therefore if we define $T_n : \mathbf{E} \to \mathbf{F}$ by

$$T_n := P_n \circ T,$$

then each T_n is a bounded operator of finite rank. We claim that $\lim_{n \to \infty} \|T_n - T\| = 0$. First, observe that if $x \in \mathbf{E}$, then

$$\lim_{n \to \infty} T_n(x) = \lim_{n \to \infty} P_n(T(x)) = T(x).$$

Now, fix $\epsilon > 0$. Consider the totally bounded set $T(B_{\mathbf{E}})$ and choose a finite set $\{x_1, x_2, \ldots, x_k\}$ such that

$$T(B_{\mathbf{E}}) \subset \bigcup_{i=1}^{k} B\left(T(x_i), \epsilon\right).$$

For each $1 \le i \le k$, choose $N_i \in \mathbb{N}$ such that $\|T_n(x_i) - T(x_i)\| < \epsilon$ for all $n \ge N_i$. For $x \in B_{\mathbf{E}}$, choose $1 \le i \le k$ such that $T(x) \in B\left(T(x_i), \epsilon\right)$. Then for $n \ge N :=$

$\max\{N_1, N_2, \ldots, N_k\}$, we have

$$\|T(x) - T_n(x)\| \leq \|T(x) - T(x_i)\| + \|T(x_i) - T_n(x_i)\| + \|T_n(x_i) - T_n(x)\|$$
$$\leq \epsilon + \epsilon + \|P_n(T(x_i) - T(x))\|$$
$$\leq 2\epsilon + M\|T(x_i) - T(x)\|$$
$$\leq (2 + M)\epsilon.$$

This is true for any $x \in B_E$ and thus $\|T_n - T\| \leq (2 + M)\epsilon$ for all $n \geq N$. We conclude that $\lim_{n \to \infty} \|T_n - T\| = 0$. $\qquad\square$

Our next theorem is one that we will use a great deal later in the book. It is essentially a direct corollary of Theorem 5.6.9, but it needs one more ingredient, the proof of which we will relegate to the exercises.

5.6.10 LEMMA *Let* $T : \mathbf{E} \to \mathbf{F}$ *be a compact operator. Then* $\overline{\text{Range}(T)}$ *is a separable subspace of* $\mathbf{F}$.

5.6.11 COROLLARY *Let* $\mathbf{E}$ *be a normed linear space and* $\mathbf{F}$ *be a Hilbert space. Then a bounded operator* $T : \mathbf{E} \to \mathbf{F}$ *is compact if and only if it is a limit of bounded, finite rank operators.*

Proof. As before, a limit of bounded, finite rank operators must be compact. Therefore we start with a compact operator $T : \mathbf{E} \to \mathbf{F}$ and prove that it is a limit of finite rank operators. By Lemma 5.6.10, $\overline{\text{Range}(T)}$ is a separable Hilbert space. Hence it contains a countable orthonormal basis, which is a Schauder basis. Theorem 5.6.9 now completes the proof. $\qquad\square$

For the final result of this chapter, we need one additional ingredient. It is one of the most useful compactness theorems in Real Analysis and you will find a compact proof of it in Rudin [50].

5.6.12 DEFINITION Let (X, d) be a compact metric space and $\mathcal{G} \subset C(X)$ be a family of continuous functions on X. We say that $\mathcal{G}$ is *equicontinuous* if for every $\epsilon > 0$, there is a $\delta > 0$ such that for any pair of points $x, y \in X$ and any $f \in \mathcal{G}$,

$$d(x, y) < \delta \Rightarrow |f(x) - f(y)| < \epsilon.$$

The Bolzano–Weierstrass Theorem states that a bounded sequence of real numbers has a convergent subsequence. The theorem we now state may be thought of as an analogue of that result for the space of continuous functions.

5.6.13 THEOREM (ASCOLI, 1883, AND ARZELÀ, 1889) *Let* X *be a compact metric space and let* $\mathbf{E} := C(X)$, *equipped with the supremum norm. If* $(f_n) \subset \mathbf{E}$ *is a bounded sequence that is equicontinuous, then* (f_n) *contains a convergent subsequence.*

The next theorem will be particularly useful to us when we study operators on Hilbert spaces in Chapter 8. For the moment, though, it serves as a neat application of the Arzelà–Ascoli Theorem.

5.6.14 THEOREM (SCHAUDER, 1930) *Let* $T : \mathbf{E} \to \mathbf{F}$ *be a compact operator between two normed linear spaces. Then the transpose* $T' : \mathbf{F}^* \to \mathbf{E}^*$ *is a compact operator.*

Proof.   Let $B_{\mathbf{F}^*}$ denote the closed unit ball in $\mathbf{F}^*$. We wish to prove that $T'(B_{\mathbf{F}^*})$ has compact closure in $\mathbf{E}^*$. Since $\mathbf{E}^*$ is a metric space, it suffices to show that every sequence in $T'(B_{\mathbf{F}^*})$ has a subsequence that converges in $\mathbf{E}^*$. To that end, let $(\varphi_n) \subset T'(B_{\mathbf{F}^*})$ be a sequence and for each $n \in \mathbb{N}$, write

$$\varphi_n = T'(\psi_n) = \psi_n \circ T$$

for some $\psi_n \in B_{\mathbf{F}^*}$. Let $X := \overline{T(B_{\mathbf{E}})}$, then X is a compact metric space by hypothesis. We will write $\| \cdot \|_\infty$ to denote the supremum norm on $C(X)$. We write $f_n := \psi_n|_X$, then (f_n) is a sequence in $C(X)$. We verify that (f_n) satisfies the hypotheses of the Arzelà–Ascoli Theorem.

(i) For any $x, y \in X$, we have

$$|f_n(x) - f_n(y)| = |\psi_n(x) - \psi_n(y)| \le \|\psi_n\| \|x - y\| \le \|x - y\|.$$

Therefore (f_n) is equicontinuous.

(ii) Since T is a bounded operator, $T(B_{\mathbf{E}})$ is a bounded subset of $\mathbf{F}$. Therefore there exists $M \ge 0$ such that $\|y\| \le M$ for all $y \in X$. Hence for each $n \in \mathbb{N}$,

$$\|f_n\|_\infty = \sup_{y \in X} |\psi_n(y)| \le \|\psi_n\| \sup_{y \in X} \|y\| \le M.$$

Thus (f_n) is a norm-bounded subset of $C(X)$ as well.

By the Arzelà–Ascoli Theorem, there is a subsequence $(f_{n_k})_{k=1}^{\infty}$ that is Cauchy in $C(X)$. For any $k, \ell \in \mathbb{N}$, we have

$$
\begin{aligned}
\|\varphi_{n_k} - \varphi_{n_\ell}\| &= \|\psi_{n_k} \circ T - \psi_{n_\ell} \circ T\| \\
&= \sup\{|\psi_{n_k}(T(x)) - \psi_{n_\ell}(T(x))| : x \in B_{\mathbf{E}}\} \\
&= \sup\{|\psi_{n_k}(y) - \psi_{n_\ell}(y)| : y \in T(B_{\mathbf{E}})\} \\
&\leq \|f_{n_k} - f_{n_\ell}\|_{\infty}.
\end{aligned}
$$

Thus $(\varphi_{n_k})_{k=1}^{\infty}$ is Cauchy in $\mathbf{E}^*$. Since $\mathbf{E}^*$ is complete, $(\varphi_{n_k})_{k=1}^{\infty}$ converges in $\mathbf{E}^*$. $\qquad\square$

5.7 Exercises

Baire Category Theorem

EXERCISE 5.1 Recall that $\ell^p \subset c_0$ for all $1 \leq p < \infty$. Prove that

$$
c_0 \neq \bigcup_{1 \leq p < \infty} \ell^p.
$$

Principle of Uniform Boundedness

EXERCISE 5.2 (✼) Let $\mathbf{E} := (c_{00}, \|\cdot\|_{\infty})$ and for each $n \in \mathbb{N}$, let $\varphi_n : \mathbf{E} \to \mathbb{K}$ be the linear functional

$$
\varphi_n((x_j)) := \sum_{k=1}^{n} x_k.
$$

(i) Prove that, for each $x \in \mathbf{E}$, $\sup_{n \in \mathbb{N}} |\varphi_n(x)| < \infty$.

(ii) Prove that $\sup_{n \in \mathbb{N}} \|\varphi_n\| = \infty$.

EXERCISE 5.3 (✼)

(i) Complete the proof of Theorem 5.2.2: Prove that the operator $T \in \mathcal{B}(\mathbf{E}, \mathbf{F})$ defined in that theorem satisfies $\|T\| \leq \liminf_{n \to \infty} \|T_n\|$.

(ii) Give an example to show that strict inequality can hold in Part (i).

EXERCISE 5.4 Let **E** be a Banach space and **F** be a normed linear space. Let $\mathcal{G} \subset \mathcal{B}(\mathbf{E}, \mathbf{F})$ be a family of bounded linear operators such that

$$\sup_{T \in \mathcal{G}} \|T\| = \infty.$$

Then prove that there exists a dense G_δ-set $D \subset \mathbf{E}$ such that

$$\sup_{T \in \mathcal{G}} \|T(x)\| = \infty.$$

for any $x \in D$.

EXERCISE 5.5 Let $\mathbf{E} = \{f \in C[-\pi, \pi] : f(\pi) = f(-\pi)\}$ and for $f \in \mathbf{E}$, let $s_m(f)$ denote the mth partial sum of the Fourier series of f as in Remark 5.2.3. Prove that

$$G = \{f \in \mathbf{E} : s_m(f)(0) \nrightarrow f(0)\}$$

is a dense subset of **E**.

5.7.1 DEFINITION A subset D of a normed linear space **E** is said to be **weakly bounded** if, for each bounded linear functional $\varphi \in \mathbf{E}^*$, the set

$$\varphi(D) := \{\varphi(x) : x \in D\}$$

is a bounded subset of $\mathbb{K}$.

EXERCISE 5.6 Prove that a weakly bounded set in a normed linear space is bounded.

Hint: Consider the set $J(D)$, where $J : \mathbf{E} \to \mathbf{E}^{**}$ is the natural inclusion.

EXERCISE 5.7 Let $(a_n) \subset \mathbb{K}$ be a sequence of scalars such that for any given $(x_n) \in c_0$, the series

$$\sum_{n=1}^{\infty} a_n x_n$$

converges in $\mathbb{K}$. Prove that $(a_n) \in \ell^1$.

Open Mapping and Bounded Inverse Theorems

EXERCISE 5.8 Let $T : \mathbf{E} \to \mathbf{F}$ be a linear map between two normed linear spaces. If T is open, then prove that T is surjective.

EXERCISE 5.9 Let $T : \mathbf{E} \to \mathbf{F}$ be a linear map between two normed linear spaces such that $\ker(T)$ is closed in $\mathbf{E}$.

(i) Let $\widehat{T} : \mathbf{E}/\ker(T) \to \mathbf{F}$ denote the map induced by T as in the First Isomorphism Theorem. Prove that T is an open map if and only if $\widehat{T}$ is open.

(ii) Conclude that the Bounded Inverse Theorem implies the Open Mapping Theorem.

EXERCISE 5.10 Prove that the Closed Graph Theorem implies the Bounded Inverse Theorem.

EXERCISE 5.11 Let $\varphi : \mathbf{E} \to \mathbb{K}$ be a linear functional on a normed linear space $\mathbf{E}$ (not necessarily continuous). If φ is non-zero, then prove that φ is an open map.

EXERCISE 5.12 Let $\mathbf{E}$ and $\mathbf{F}$ be normed linear spaces with $\mathbf{F}$ finite dimensional. If $T : \mathbf{E} \to \mathbf{F}$ is a surjective, linear map, then prove that T is an open map.

Hint: By Theorem 2.4.8, you may assume that $\mathbf{F} = (\mathbb{K}^n, \|\cdot\|_\infty)$.

EXERCISE 5.13 (⚒) Prove Lemma 5.3.13.

EXERCISE 5.14 (⚒) Let $\mathbf{E}$ and $\mathbf{F}$ be Banach spaces and $T : \mathbf{E} \to \mathbf{F}$ be a bounded linear operator. Let $\pi : \mathbf{E} \to \mathbf{E}/\ker(T)$ be the quotient map and let $\widehat{T} : \mathbf{E}/\ker(T) \to \mathbf{F}$ be the map given by

$$x + \ker(T) \mapsto T(x).$$

Prove that the following statements are equivalent:

(i) $\widehat{T} : \mathbf{E}/\ker(T) \to \text{Range}(T)$ is a topological isomorphism.

(ii) $\text{Range}(T)$ is closed in $\mathbf{F}$.

(iii) There exists a constant $\beta > 0$ such that $\|\pi(x)\| \leq \beta\|T(x)\|$ for all $x \in \mathbf{E}$.

EXERCISE 5.15 Let $\mathbf{E}$ and $\mathbf{F}$ be Banach spaces and $T \in \mathcal{B}(\mathbf{E}, \mathbf{F})$ be a bounded, linear operator such that $\text{Range}(T)$ is closed in $\mathbf{F}$. Let $T' \in \mathcal{B}(\mathbf{F}^*, \mathbf{E}^*)$ be the transpose of T. Show that $\text{Range}(T') = \ker(T)^\perp$.

Hint: If $\varphi \in \ker(T)^{\perp}$, the equation $\psi(T(x)) = \varphi(x)$ determines a linear functional on Range(T). Use Exercise 5.14 to show that it is bounded.

EXERCISE 5.16 (⚒) Let **E** and **F** be Banach spaces and $T \in \mathcal{B}(\mathbf{E}, \mathbf{F})$ be a surjective, bounded, linear operator. If $T' \in \mathcal{B}(\mathbf{F}^*, \mathbf{E}^*)$ denotes the transpose of T, then prove that T' is *bounded below*. In other words, prove that there is a constant $c > 0$ such that $\|T'(\varphi)\| \geq c\|\varphi\|$ for all $\varphi \in \mathbf{F}^*$.

Hint: Choose $c > 0$ such that $B_{\mathbf{F}}[0, c] \subset T(B_{\mathbf{E}}[0, 1])$ (which exists by the Open Mapping Theorem).

EXERCISE 5.17 Show that ℓ^1 is not topologically isomorphic to a quotient of ℓ^p for any $1 < p < \infty$ (in contrast with Theorem 5.3.9).

Closed Graph Theorem

EXERCISE 5.18 (⚒) Let $\mathbf{E} := (C[0,1], \|\cdot\|_\infty)$ and $\mathbf{F} := \{f \in C^1[0,1] : f(0) = 0\}$, also equipped with the supremum norm. Define $T : \mathbf{E} \to \mathbf{F}$ by

$$T(f)(x) := \int_0^x f(t)dt.$$

Then T is well-defined (by the Fundamental Theorem of Calculus), bounded and injective. Prove that T^{-1} has a closed graph, but is not a bounded operator.

EXERCISE 5.19 Let $T : \mathbf{E} \to \mathbf{F}$ be a bounded linear map between two Banach spaces. If Range(T) has finite codimension, then prove that Range(T) is closed.

Hint: If $n := \text{codim}(\text{Range}(T))$, then extend T to a bijective map $S : \mathbf{E} \oplus \mathbb{K}^n \to \mathbf{F}$.

EXERCISE 5.20 Show that for every infinite dimensional normed linear space, there is a linear subspace of finite codimension that is not closed.

EXERCISE 5.21 Give an example of a discontinuous linear operator between two normed linear spaces whose graph is closed.

EXERCISE 5.22

(i) Let $(\mathbf{F}, \|\cdot\|_{\mathbf{F}})$ be a Banach space and I be any set. Define $\ell^\infty(I, \mathbf{F})$ to be the space of all bounded functions $f : I \to \mathbf{F}$, equipped with the supremum norm

$$\|f\|_\infty := \sup_{i \in I} \|f(i)\|_{\mathbf{F}}.$$

Prove that $\ell^\infty(I, \mathbf{F})$ is a Banach space.

(ii) Use the space in part (i) to prove that the Closed Graph Theorem implies the Principle of Uniform Boundedness.

EXERCISE 5.23 Let E be a Banach space and $S, T \in \mathcal{B}(E)$ with S injective. If $T(E) \subset S(E)$, then prove that there exists a bounded operator $R \in \mathcal{B}(E)$ such that $T = SR$.

5.24 EXERCISE (HELLINGER AND TOEPLITZ, 1910) *Let $\mathbf{H}$ be a Hilbert space and $T : \mathbf{H} \to \mathbf{H}$ be a linear map such that $\langle Tx, y \rangle = \langle x, Ty \rangle$ holds for all $x, y \in \mathbf{H}$. Prove that T is bounded.*

EXERCISE 5.25 Let $1 \le p < \infty$ and q be the conjugate exponent of p. let $K : [a,b] \times [a,b] \to \mathbb{K}$ be a measurable function such that $K(s, \cdot) \in L^q[a,b]$ for almost every $s \in [a,b]$. For each $f \in L^p[a,b]$, the expression

$$T(f)(s) := \int_a^b K(s,t)f(t)dt$$

makes sense almost everywhere. Suppose that $T(f) \in L^p[a,b]$ for each $f \in L^p[a,b]$, then prove that T defines a bounded linear operator from $L^p[a,b]$ to $L^p[a,b]$.

Fourier Series

5.7.2 DEFINITION

(i) For $n \ge 0$, let D_n denote the nth Dirichlet kernel. The nth **Fejér kernel** is given by

$$F_n(x) := \frac{D_0(x) + D_1(x) + \ldots + D_{n-1}(x)}{n}$$

(ii) Let $f \in L^1[-\pi, \pi]$. For $n \ge 0$, let $s_n(f)$ denote the nth partial sum of the Fourier series of f.

$$s_n(f)(x) := \sum_{k=-n}^{n} \widehat{f}(k)e^{ikx}.$$

The nth **Cesáro mean** of the Fourier series is given by

$$\sigma_n(f)(x) = \frac{s_0(f)(x) + s_1(f)(x) + \ldots + s_{n-1}(f)(x)}{n}.$$

EXERCISE 5.26

(i) For any $f \in C[-\pi, \pi]$, prove that

$$\sigma_n(f)(x) = \frac{1}{2\pi} \int_{-\pi}^{\pi} f(x-y) F_n(y) dy.$$

(ii) Prove that $\sigma_n : C[-\pi, \pi] \to C[-\pi, \pi]$ is a bounded linear operator, when $C[-\pi, \pi]$ is equipped with the supremum norm.

EXERCISE 5.27 Prove that

$$F_n(x) = \frac{1}{n} \frac{\sin^2(nx/2)}{\sin^2(x/2)}.$$

Conclude that F_n is a non-negative function and that $\frac{1}{2\pi} \int_{-\pi}^{\pi} F_n(x) dx = 1$.

EXERCISE 5.28 For any $\pi \geq \delta > 0$, prove that

$$\lim_{n \to \infty} \sup_{\delta \leq |x| \leq \pi} F_n(x) = 0.$$

5.29 EXERCISE (FEJÉR, 1904) Let $\mathbf{E} := \{f \in C[-\pi, \pi] : f(-\pi) = f(\pi)\}$, equipped with the supremum norm. Prove that

$$\lim_{n \to \infty} \sigma_n(f) = f$$

for any $f \in \mathbf{E}$ (in other words, the Fourier series of f is *uniformly Cesáro summable*).

Hint: For $f \in \mathbf{E}$ and $\epsilon > 0$, choose $\delta > 0$ such that $|y| < \delta$ implies $|f(x-y) - f(x)| < \epsilon$ for all $x \in [-\pi, \pi]$. Now use Exercise 5.28.

EXERCISE 5.30 Let $f : [-\pi, \pi] \to \mathbb{K}$ is an integrable function and $x \in [-\pi, \pi]$ be a fixed point. If $\lim_{n \to \infty} s_n(f)(x) = f(x)$, then prove that

$$\lim_{n \to \infty} \sigma_n(f)(x) = f(x).$$

EXERCISE 5.31 Let $\| \cdot \|_D$ be a norm on $C[a,b]$ such that

(i) $(C[a,b], \| \cdot \|_D)$ is a Banach space and

(ii) for each $t \in [a,b]$, the evaluation map $f \mapsto f(t)$ is a continuous linear functional on $(C[a,b], \| \cdot \|_D)$.

Then prove that $\| \cdot \|_D$ is equivalent to $\| \cdot \|_\infty$.

EXERCISE 5.32 Let $1 \leq p \leq \infty$ and $\|\cdot\|_D$ be a norm on ℓ^p such that

(i) $(\ell^p, \|\cdot\|_D)$ is a Banach space and

(ii) for each $j \in \mathbb{N}$, the evaluation map $(x_n) \mapsto x_j$ is a continuous linear functional on $(\ell^p, \|\cdot\|_D)$.

Then prove that $\|\cdot\|_D$ is equivalent to $\|\cdot\|_p$.

Schauder Basis

EXERCISE 5.33 (⚒) Let **E** be an infinite dimensional Banach space and $\{e_\alpha : \alpha \in J\}$ be a (necessarily uncountable) Hamel basis. To each $\alpha \in J$, we may associate a coordinate linear functional φ_α exactly as in Definition 5.5.2. Prove that there exists at least one $\alpha \in J$ such that φ_α is discontinuous.

Hint: Choose a sequence (e_{α_n}) of mutually distinct basis vectors and take

$$x := \sum_{n=1}^{\infty} 2^{-n} \frac{e_{\alpha_n}}{\|e_{\alpha_n}\|}.$$

EXERCISE 5.34 Consider c, the space of convergent sequences. For $n \in \mathbb{N}$, let $e_n \in c$ be the sequence $(0,0,\ldots,1,0,\ldots)$ as before. Prove that $\{e_n\}_{n=1}^{\infty} \cup \{\mathbf{1}\}$ is a Schauder basis for c.

EXERCISE 5.35 Let α, β be two non-zero scalars such that $0 < |\beta| < |\alpha|$. Let $\{e_1, e_2, \ldots\} \subset \ell^2$ be defined by

$$\begin{aligned}
e_1 &:= (\alpha, \beta, 0, 0, \ldots), \\
e_2 &:= (0, \alpha, \beta, 0, 0, \ldots), \\
e_3 &:= (0, 0, \alpha, \beta, 0, 0, \ldots), \\
&\vdots
\end{aligned}$$

Prove that $\{e_n\}_{n=1}^{\infty}$ is a Schauder basis for ℓ^2.

EXERCISE 5.36 (⚒) Let **E** be a normed linear space and $\{e_n\}_{n=1}^{\infty}$ be a Schauder basis for **E**. Let $\{\varphi_n\}_{n=1}^{\infty}$ be the coordinate functionals of **E** with respect to this Schauder basis.

(i) Prove that the set $\{e_n\}_{n=1}^{\infty}$ is linearly independent.

(ii) Prove that each φ_n is linear.

EXERCISE 5.37 (⚒) Prove that any normed linear space that has a Schauder basis is separable.

EXERCISE 5.38 (⚒) Prove Theorem 5.5.5.

EXERCISE 5.39 (Faber, 1910) The aim of this exercise is to construct a Schauder basis for $(C[0,1], \|\cdot\|_\infty)$. Enumerate the dyadic rationals in $[0,1]$ as

$$0, 1, \frac{1}{2}, \frac{1}{4}, \frac{3}{4}, \frac{1}{8}, \frac{3}{8}, \frac{5}{8}, \frac{7}{8}, \frac{1}{16}, \ldots,$$

written as $(r_i)_{i=1}^\infty$. Define a sequence $(f_n)_{n=0}^\infty \subset C[0,1]$ as follows: Set $f_0 \equiv 1$, $f_1(t) = t$ and for $n \geq 2$, let $f_n(r_n) = 1$, $f_n(r_j) = 0$ if $j \neq n$ and f_n be linear between any two neighbours among the dyadic rationals. Explicitly,

$$f_{2^n+\ell}(t) = \begin{cases} 1 & : \text{ if } t = \frac{2\ell-1}{2^{n+1}}, \\ 0 & : \text{ if } t \notin \left(\frac{2\ell-2}{2^{n+1}}, \frac{2l}{2^{n+1}}\right), \\ \text{linear} & : \text{ otherwise.} \end{cases}$$

for $n \geq 0$ and $l \in \{1, 2, \ldots, 2^n\}$.

(i) Prove that $\text{span}(\{f_n : n \geq 0\})$ is dense in $C[0,1]$.

Hint: Functions in $C[0,1]$ are uniformly continuous and $\{r_i\}_{i=1}^\infty$ is dense in $[0,1]$.

(ii) If $p = \sum_{k=0}^n \alpha_k f_k \in \text{span}(\{f_n : n \geq 0\})$, then prove that

$$\|p\|_\infty = \max\{|p(r_i)| : 1 \leq i \leq n\}.$$

Conclude that, if we fix $n < m$ and scalars $\alpha_1, \alpha_2, \ldots, \alpha_m$, then

$$\left\| \sum_{i=1}^n \alpha_i f_i \right\| \leq \left\| \sum_{i=1}^m \alpha_i f_i \right\|.$$

(iii) Conclude that $\{f_n : n \geq 0\}$ is a Schauder basis for $C[0,1]$. This is called the ***Faber–Schauder basis*** for $C[0,1]$.

Compact Operators

EXERCISE 5.40 Let **E** and **F** be normed linear spaces and let $\mathcal{F}(\mathbf{E}, \mathbf{F})$ denote the set of all bounded, finite rank operators from **E** to **F**. Prove that $\mathcal{F}(\mathbf{E}, \mathbf{F})$ is a vector subspace of $\mathcal{B}(\mathbf{E}, \mathbf{F})$.

EXERCISE 5.41 Let $\mathbf{E}$ be a normed linear space and $U, V \subset \mathbf{E}$.

(i) If $\overline{U}$ and $\overline{V}$ are both compact, then prove that $\overline{U+V}$ is compact.

(ii) Use this to give an alternate proof of Proposition 5.6.4.

EXERCISE 5.42 (⚒) Let $\mathbf{E} = \mathbf{F} = (c_{00}, \|\cdot\|_2)$. Define $T : \mathbf{E} \to \mathbf{F}$ by

$$T((x_j)) := \left(\frac{x_1}{1}, \frac{x_2}{2}, \frac{x_3}{3}, \ldots\right).$$

(i) Prove that T is a bounded linear operator that is not compact.

(ii) Express T as a limit of compact (finite rank) operators.

Hence the completeness assumption in Part (ii) of Proposition 5.6.4 is necessary.

EXERCISE 5.43 (⚒) Prove Lemma 5.6.10.

EXERCISE 5.44 Let $M := (a_{i,j})_{i,j\in\mathbb{N}}$ be an infinite matrix whose entries are in $\mathbb{K}$. For $i \in \mathbb{N}$, set

$$\delta_i := \sum_{j=1}^{\infty} |a_{i,j}|.$$

Assume that $\beta := \sup_{i\in\mathbb{N}} \delta_i < \infty$. For $x = (x_1, x_2, \ldots) \in \ell^\infty$, define a sequence $T(x) = (y_1, y_2, \ldots)$ by

$$y_i = \sum_{j=1}^{\infty} a_{i,j} x_j.$$

By Exercise 2.22, T defines a bounded linear operator $T : \ell^\infty \to \ell^\infty$. Prove that T is compact if $\lim_{i\to\infty} \delta_i = 0$.

EXERCISE 5.45 (⚒) Let $M := (a_{i,j})_{i,j\in\mathbb{N}}$ be an infinite matrix whose entries are in $\mathbb{K}$. For $i, j \in \mathbb{N}$, define

$$\delta_i := \sum_{j=1}^{\infty} |a_{i,j}| \text{ and } \gamma_j := \sum_{i=1}^{\infty} |a_{i,j}|.$$

Assume that $\alpha := \sup_{j\in\mathbb{N}} \gamma_j < \infty$ and $\beta := \sup_{i\in\mathbb{N}} \delta_i < \infty$. Fix $1 < p < \infty$. For $x = (x_1, x_2, \ldots) \in \ell^p$, define $T(x) = (y_1, y_2, \ldots)$ by

$$y_i = \sum_{j=1}^{\infty} a_{i,j} x_j.$$

(i) Prove that T defines a bounded linear operator $T : \ell^p \to \ell^p$ and that $\|T\| \leq \alpha^{1/p}\beta^{1/q}$, where q is the conjugate exponent of p.

(ii) If either (γ_j) or (δ_i) tends to zero, then prove that T is a compact operator.

EXERCISE 5.46 (⚒) Let $S : \mathbf{E} \to \mathbf{F}$ and $T : \mathbf{F} \to \mathbf{W}$ be two linear operators between Banach spaces. If either S or T is compact, then prove that the composition $T \circ S : \mathbf{E} \to \mathbf{W}$ is compact.

EXERCISE 5.47 Prove the following converse of Theorem 5.6.14: Let $T : \mathbf{E} \to \mathbf{F}$ be a bounded operator such that $T' : \mathbf{F}^* \to \mathbf{E}^*$ is compact. If $\mathbf{F}$ is a Banach space, then prove that T is a compact operator.

Hint: Use Theorem 5.6.14 together with Exercise 4.42 to show that $T(B_{\mathbf{E}})$ is totally bounded.

Complemented Subspaces

EXERCISE 5.48 (⚒) Let $\mathbf{E}$ be a vector space and $\mathbf{F} < \mathbf{E}$ be a subspace. Prove that there is a subspace $\mathbf{F}'$ of $\mathbf{E}$ such that $\mathbf{E} = \mathbf{F} + \mathbf{F}'$ and $\mathbf{F} \cap \mathbf{F}' = \{0\}$. When this happens, we say that $\mathbf{F}'$ is an ***algebraic complement*** of $\mathbf{F}$.

5.7.3 DEFINITION Let $\mathbf{E}$ be a normed linear space and $\mathbf{F} < \mathbf{E}$ a closed subspace. Let $\mathbf{F}' < \mathbf{E}$ be an algebraic complement of $\mathbf{F}$ and suppose further that $\mathbf{F}'$ is a closed subspace of $\mathbf{E}$. Then there is a natural map $T : \mathbf{F} \oplus_1 \mathbf{F}' \to \mathbf{E}$ given by

$$T(x, y) := x + y,$$

where $\mathbf{F} \oplus_1 \mathbf{F}'$ is the normed linear space described in Exercise 2.5. We say that $\mathbf{F}'$ is a ***topological complement*** of $\mathbf{F}$ if T is a topological isomorphism. If such a closed subspace $\mathbf{F}'$ exists, then we say that $\mathbf{F}$ is ***complemented*** in $\mathbf{E}$.

EXERCISE 5.49 Let $\mathbf{E}$ be a normed linear space and $\mathbf{F} < \mathbf{E}$ be a finite dimensional subspace. Prove that $\mathbf{F}$ is complemented in $\mathbf{E}$.

Hint: See Exercise 4.8.

EXERCISE 5.50 (⚒) Let $\mathbf{H}$ be a Hilbert space and $\mathbf{M} < \mathbf{H}$ be a closed subspace. Prove that $\mathbf{M}$ is complemented in $\mathbf{H}$.

EXERCISE 5.51 Let $\mathbf{E}$ be a Banach space and suppose $\mathbf{F} < \mathbf{E}$ is a closed subspace. If $\mathbf{F}'$ is an algebraic complement of $\mathbf{F}$ that is also closed, then prove that $\mathbf{F}'$ is a topological complement of $\mathbf{F}$ in $\mathbf{E}$.

EXERCISE 5.52 This exercise is another way to view Exercise 5.51, as a statement about idempotent linear operators: Let $\mathbf{E}$ be a Banach space and $\mathbf{F}_1$ and $\mathbf{F}_2$ be two closed subspaces of $\mathbf{E}$ which are algebraic complements of each other (see Exercise 5.48). In other words, every $x \in \mathbf{E}$ can be expressed uniquely in the form $x = x_1 + x_2$, with $x_1 \in \mathbf{F}_1$ and $x_2 \in \mathbf{F}_2$. For $i = 1, 2$, define $P_i : \mathbf{E} \to \mathbf{E}$ by $P_i(x) = x_i$.

(i) Prove that P_1 and P_2 are both linear maps.

(ii) Prove that $P_1^2 = P_1, P_2^2 = P_2$ and $P_1 P_2 = P_2 P_1 = 0$.

(iii) Prove that P_1 and P_2 are continuous.

(1) Conclude that $\mathbf{F}_1$ and $\mathbf{F}_2$ are topological complements of each other.

Additional Reading

- For a deep dive into many topics related to Schauder bases, a good book is that of Lindenstrauss and Tzafriri [39]. However, perhaps the best place to start is this article by R.C. James.

  James, Robert C. (1982). Bases in Banach spaces. *The American Mathematical Monthly*, 89.9, 625–640.

- In Theorem 5.6.9, we showed that any compact operator on a Banach space is a limit of finite rank operators, provided the Banach space has a Schauder basis. In 1932, Banach asked if this result was true under the weaker assumption that the Banach space was separable. In 1973, Enflo constructed a separable, reflexive Banach space $\mathbf{E}$ with the property that $\mathcal{F}(\mathbf{E})$ is not dense in $\mathcal{K}(\mathbf{E})$. In the process, he also proved the existence of a separable Banach space without a Schauder basis.

  Enflo, Per (1973). A counterexample to the approximation problem in Banach spaces. *Acta Mathematica*, 130, 309–317. ISSN: 0001-5962. DOI: 10.1007/BF02392270.

- In Exercise 5.50, we have proved that every closed subspace of a Hilbert space is complemented. We will prove (see Exercise 6.7) that there is a closed subspace of ℓ^1 which is not complemented. Therefore this property does not extend to arbitrary Banach spaces. It is an interesting fact, due to Lindenstrauss and Tzafriri, that this property *characterizes* Hilbert spaces. If a

Banach space has the property that every closed subspace is complemented, then it must be a Hilbert space!

Lindenstrauss, Joram, and Tzafriri, Lior (1971). On the complemented subspaces problem. *Israel Journal of Mathematics*, 9, 263–269. ISSN: 0021-2172. DOI: 10.1007/BF02771592.

Chapter 6

Weak Topologies

6.1 Weak Convergence

In Chapter 4, we saw a number of interesting ways in which a normed linear space interacts with its dual space. Notably, the Hahn–Banach Theorem gave us a powerful way to construct bounded linear functionals. In this chapter, we explore this relationship further and once again the Hahn–Banach Theorem plays a central role.

We will begin by exploring the notion of weak convergence. The ideas developed in this section will, together with a geometric version of the Hahn–Banach Theorem, help us define new topologies on a normed linear space and its dual space. These topologies are weaker (and more forgiving) than the norm topology, which allows us to prove some powerful compactness theorems.

6.1.1 DEFINITION Let $\mathbf{E}$ be an normed linear space. A sequence $(x_n) \subset \mathbf{E}$ is said to *converge weakly* to $x \in \mathbf{E}$ if

$$\varphi(x_n) \to \varphi(x)$$

for all $\varphi \in \mathbf{E}^*$. If this happens, we write $x_n \xrightarrow{w} x$.

In this chapter, if $x_n \to x$ in the norm, then we say that $x_n \to x$ *strongly* and we write $x_n \xrightarrow{s} x$.

6.1.2 REMARK Some immediate observations are in order. Let $(x_n) \subset \mathbf{E}$ be a sequence. If $x_n \xrightarrow{w} x$ and $x_n \xrightarrow{w} y$, then $x = y$. This is because $\mathbf{E}^*$ separates points of $\mathbf{E}$ (Corollary 4.2.8). Furthermore, if $x_n \xrightarrow{w} x$, then any subsequence (x_{n_k}) of (x_n) also converges weakly to x.

6.1.3 EXAMPLE

(i) If $x_n \xrightarrow{s} x$, then $x \xrightarrow{w} x$. This is because every element of $\mathbf{E}^*$ is continuous with respect to the norm topology.

(ii) If $\mathbf{E}$ is a finite dimensional normed linear space and $x_n \xrightarrow{w} x$, then $x_n \xrightarrow{s} x$.

Proof. By Corollary 2.4.11, we may assume without loss of generality that $\mathbf{E} = (\mathbb{K}^m, \|\cdot\|_1)$ for some $m \in \mathbb{N}$. For each $1 \leq i \leq m$, the projection maps $\pi_i : \mathbf{E} \to \mathbb{K}$ are bounded linear functionals. Hence $\pi_i(x_n) \to \pi_i(x)$. Then

$$\lim_{n\to\infty} \|x_n - x\|_1 = \lim_{n\to\infty} \sum_{i=1}^{m} |\pi_i(x_n) - \pi_i(x)| = 0.$$

Hence $x_n \xrightarrow{s} x$. ☐

(iii) If $\mathbf{H}$ is an infinite dimensional Hilbert space and $(e_n) \subset \mathbf{H}$ an orthonormal sequence, then for any $x \in \mathbf{H}$,

$$\lim_{n\to\infty} \langle e_n, x \rangle = 0$$

by the Riemann–Lebesgue Lemma (Corollary 3.3.11). By the Riesz Representation Theorem, $e_n \xrightarrow{w} 0$.

However, we claim that (e_n) does not converge strongly to *any* point in $\mathbf{H}$. If $x \in \mathbf{H}$ were such that $e_n \xrightarrow{s} x$, then $e_n \xrightarrow{w} x$, whence $x = 0$. But $\|e_n\| = 1$ for all $n \in \mathbb{N}$, so the continuity of the norm (Remark 2.1.2) implies that $\|x\| = 1$. This contradiction shows that (e_n) is not strongly convergent.

As shown in this last example, if (x_n) converges weakly to x, then $(\|x_n\|)$ may not converge to $\|x\|$. This is in sharp contrast with strong convergence. However, the next result does give us some control on the norms of the elements.

6.1.4 LEMMA

Let $\mathbf{E}$ be an normed linear space and $x_n \xrightarrow{w} x$, then (x_n) is bounded and

$$\|x\| \leq \liminf_{n\to\infty} \|x_n\|.$$

Proof. Consider the map $J : \mathbf{E} \to \mathbf{E}^{**}$ given by $J(x) := \widehat{x}$, where $\widehat{x} : \mathbf{E}^* \to \mathbb{K}$ is given by

$$\widehat{x}(\varphi) := \varphi(x).$$

By hypothesis, $\widehat{x_n}(\varphi) \to \widehat{x}(\varphi)$ for all $\varphi \in \mathbf{E}^*$. By the Banach–Steinhaus Theorem applied to the Banach space $\mathbf{E}^*$, $(\|\widehat{x_n}\|)$ is a bounded sequence and

$$\|\widehat{x}\| \leq \liminf \|\widehat{x_n}\|.$$

Now the result follows from the fact that J is an isometry. ☐

As we have seen above, weak convergence does not imply strong convergence (outside the finite dimensional case). The next result is a useful sufficient condition that works in the context of Hilbert spaces.

6.1.5 PROPOSITION *Let* $\mathbf{E}$ *be an inner product space and* $(x_n) \subset \mathbf{E}$ *be a sequence such that* $x_n \xrightarrow{w} x$. *If* $\limsup_{n\to\infty} \|x_n\| \leq \|x\|$, *then* $x_n \xrightarrow{s} x$.

Proof. By Lemma 6.1.4, it follows that $\|x\| = \lim_{n\to\infty} \|x_n\|$. Also, since $y \mapsto \langle y, x \rangle$ is a bounded linear functional,

$$\lim_{n\to\infty} \langle x_n, x \rangle = \langle x, x \rangle = \|x\|^2.$$

Hence

$$\lim_{n\to\infty} \|x - x_n\|^2 = \|x\|^2 - 2 \lim_{n\to\infty} \mathrm{Re}\langle x, x_n \rangle + \lim_{n\to\infty} \|x_n\|^2 = 0.$$
$\square$

6.1.6 PROPOSITION *Let* $\mathbf{E}$ *be an normed linear space,* $(x_n) \subset \mathbf{E}$ *be a bounded sequence and* $\mathcal{G} \subset \mathbf{E}^*$ *be such that* $\mathrm{span}(\mathcal{G})$ *is a norm dense subset of* $\mathbf{E}^*$. *Suppose* $x \in \mathbf{E}$ *is a vector such that*

$$\varphi(x_n) \to \varphi(x)$$

for all $\varphi \in \mathcal{G}$. *Then* $x_n \xrightarrow{w} x$

Proof. By assumption, $\varphi(x_n) \to \varphi(x)$ for all $\varphi \in \mathbf{F} := \mathrm{span}(\mathcal{G})$. Now if $\psi \in \mathbf{E}^*$ and $\epsilon > 0$, then there exists $\varphi \in \mathbf{F}$ such that $\|\psi - \varphi\| < \epsilon$. Then we look to exploit the inequality

$$|\psi(x_n) - \psi(x)| \leq |\psi(x_n) - \varphi(x_n)| + |\varphi(x_n) - \varphi(x)| + |\varphi(x) - \psi(x)|. \tag{6.1}$$

By hypothesis, there exists $M > 0$ such that $\|x_n\| \leq M$ for all $n \in \mathbb{N}$. Also, there exists $N \in \mathbb{N}$ such that $|\varphi(x_n) - \varphi(x)| < \epsilon$ for all $n \geq N$. Plugging all this back in Equation 6.1, we get

$$|\psi(x_n) - \psi(x)| \leq M\epsilon + \epsilon + \epsilon\|x\|,$$

which holds for all $n \geq N$. Since this is true for any $\epsilon > 0$, $\psi(x_n) \to \psi(x)$. $\square$

6.1.7 COROLLARY *Let* $\mathbf{E} = c_0$ *or* ℓ^p *with* $1 < p < \infty$ *and let* $(x^n) \subset \mathbf{E}$ *be a bounded sequence such that*

$$x_j^n \to x_j$$

for each $j \in \mathbb{N}$. *Then* $x^n \xrightarrow{w} x$.

Proof.   We first assume $\mathbf{E} = \ell^p$ for $1 < p < \infty$, since the other case is similar. For each $j \in \mathbb{N}$, the evaluation map $\varphi_j : \mathbf{E} \to \mathbb{K}$ is given by

$$\varphi_j((y_n)) := y_j$$

and let $\mathcal{G} := \{\varphi_j : j \in \mathbb{N}\}$. By assumption, $\varphi(x^n) \to \varphi(x)$ for each $\varphi \in \mathcal{G}$, so we look to apply Proposition 6.1.6. Recall that we have an isomorphism $\Delta : \ell^q \to (\ell^p)^*$, where $1/p + 1/q = 1$. Under this isomorphism,

$$e_j \mapsto \varphi_j.$$

Since $1 < q < \infty$, c_{00} is dense in ℓ^q. Therefore $\mathrm{span}(\mathcal{G}) = \Delta(c_{00})$ is dense in $\mathbf{E}^*$. The conclusion now follows from Proposition 6.1.6.

For the case of $\mathbf{E} = c_0$, the argument is identical, except we use the isomorphism $\Delta : \ell^1 \to (c_0)^*$ proved in Exercise 4.13.   □

6.1.8 EXAMPLE   Let $\mathbf{E} = \ell^\infty$ and

$$x^k := (\underbrace{1, 1, 1, \ldots, 1}_{k \text{ times}}, 0, 0, \ldots).$$

For $j \in \mathbb{N}$, consider the evaluation linear functional $\varphi_j \in \mathbf{E}^*$ given by $\varphi_j((y_n)) := y_j$. Then $\lim_{k \to \infty} \varphi_j(x^k) = 1$. Therefore if $x^k \xrightarrow{w} x$, then it follows that

$$x = (1, 1, 1, \ldots) = \mathbf{1}.$$

However, let $\psi \in (\ell^\infty)^*$ be a Banach limit (Example 4.4.4) so that

$$\psi((x_j)) = \lim_{n \to \infty} x_n$$

for all $(x_j) \in c$. Then in particular, $\psi(x^k) = 0$ for all $n \in \mathbb{N}$. Since $\psi(\mathbf{1}) = 1$, it follows that (x^k) is not weakly convergent. However, (x^k) is bounded in ℓ^∞. Thus the conclusion of Corollary 6.1.7 does not hold for ℓ^∞. Once again, this failure is down to the fact that c_{00} is not dense in $(\ell^\infty)^*$, since the latter is not separable.

Recall that the standard orthonormal basis $(e_n) \subset \ell^2$ is an example of a weakly convergent sequence that is not strongly convergent. It is a remarkable fact that this phenomenon cannot occur in ℓ^1 and is responsible for many interesting counterexamples in the subject (for instance, see Exercise 6.7).

6.1.9 THEOREM (SCHUR, 1921) *A weakly convergent sequence in ℓ^1 is strongly convergent.*

Proof. Let $(x^n) \subset \ell^1$ be a sequence such that $x^n \xrightarrow{w} x$. Then we wish to prove that $x^n \xrightarrow{s} x$. By translation, we may assume that $x = 0$. We prove the result by contradiction, in the following steps.

(i) Suppose (x^n) *does not* converge strongly to 0, then there would be a subsequence (x^{n_j}) of (x^n) and an $\epsilon > 0$ such that $\|x^{n_j}\| \geq \epsilon$ for all $j \in \mathbb{N}$. Replacing the original sequence by the subsequence and by re-scaling the sequence, we may assume that $\|x^n\| \geq 1$ for all $n \in \mathbb{N}$.

(ii) Since each evaluation map is a bounded linear functional, $\lim_{n \to \infty} x_j^n = 0$ for all $j \in \mathbb{N}$. Since $x^1 \in \ell^1$, there exists $K_1 \in \mathbb{N}$ such that $\sum_{i=K_1+1}^{\infty} |x_i^1| < \frac{1}{5}$. Furthermore,

$$\lim_{n \to \infty} \sum_{i=1}^{K_1} |x_i^n| = 0.$$

(since this is a finite sum). Therefore there exists $n_2 \in \mathbb{N}$ such that $\sum_{i=1}^{K_1} |x_i^{n_2}| < \frac{1}{5}$.

(iii) Now since $x^{n_2} \in \ell^1$, there exists $K_2 \in \mathbb{N}$ such that $K_2 > K_1$ and $\sum_{i=K_2+1}^{\infty} |x_i^{n_2}| < \frac{1}{5}$. By the argument of part (ii), there exists $n_3 \in \mathbb{N}$ such that $\sum_{i=1}^{K_2} |x_i^{n_3}| < \frac{1}{5}$. Continuing in this way, we inductively construct a subsequence $\{x^{n_1} = x^1, x^{n_2}, x^{n_3}, \ldots\}$ of (x^n) and an increasing sequence of integers $\{K_0 = 0, K_1, K_2, \ldots\}$ such that

$$\sum_{i=1}^{K_{j-1}} |x_j^n| < \frac{1}{5} \text{ and } \sum_{i=K_j+1}^{\infty} |x_i^{n_j}| < \frac{1}{5}$$

for all $j \in \mathbb{N}$.

(iv) Now define $y = (y_k)$ by

$$y_k := \operatorname{sgn}(x_k^{n_j}), \qquad (K_{j-1} < k \leq K_j).$$

(Recall the definition of the $\operatorname{sgn}(\cdot)$ function from the discussion preceding Proposition 4.1.2). Then $y \in \ell^\infty$ since $|y_k| \leq 1$ for all $k \in \mathbb{N}$. We consider the corresponding bounded linear functional $\varphi_y \in (\ell^1)^*$ given by

$$\varphi_y((z_n)) = \sum_{n=1}^{\infty} z_n y_n.$$

We claim that $(\varphi_y(x^{n_j}))$ does not converge to 0. To see this, consider the following inequalities

$$\sum_{k=1}^{\infty} x_k^{n_j} y_k = \sum_{k=K_{j-1}+1}^{K_j} x_k^{n_j} y_k + \sum_{k=1}^{K_{j-1}} x_k^{n_j} y_k + \sum_{k=K_j+1}^{\infty} x_k^{n_j} y_k$$

$$= \sum_{k=K_{j-1}+1}^{K_j} |x_k^{n_j}| + \sum_{k=1}^{K_{j-1}} x_k^{n_j} y_k + \sum_{k=K_j+1}^{\infty} x_k^{n_j} y_k$$

$$\geq \sum_{k=K_{j-1}+1}^{K_j} |x_k^{n_j}| - \sum_{k=1}^{K_{j-1}} |x_k^{n_j}| - \sum_{k=K_j+1}^{\infty} |x_k^{n_j}|$$

$$\geq \sum_{k=1}^{\infty} |x_k^{n_j}| - 2\sum_{k=1}^{K_{j-1}} |x_k^{n_j}| - 2\sum_{k=K_j+1}^{\infty} |x_k^{n_j}|$$

$$\geq \|x^{n_j}\|_1 - \frac{2}{5} - \frac{2}{5} \geq \frac{1}{5}.$$

We conclude that $\varphi_y(x^{n_j}) \nrightarrow 0$. Therefore (x^n) cannot converge weakly to 0, which is the contradiction we were seeking. $\square$

Schur's Theorem is a statement about weakly convergent sequences. As we will see later, this notion of convergence arises within the context of the weak topology on ℓ^1, which is necessarily weaker than the norm topology. Therefore there are sets whose weak closure and strong closure do not coincide (see Theorem 6.3.13). In particular, Schur's Theorem does not hold if we replace sequences by *nets*.

Aside. The argument given in Theorem 6.1.9 warrants some attention. One begins with a sequence $(x^n) \subset \ell^1$ which does not converge to 0 in the norm. This allows us to construct a linear functional φ_y on ℓ^1, which concentrates the 'mass' of the sequence within intervals $[K_{j-1}, K_j]$ for each $j \in \mathbb{N}$. One may visualize this element $y \in \ell^\infty$ as having a 'hump' in each of these intervals. For this reason, this is called the *gliding (or sliding) hump argument*. This method was (perhaps) first used by Lebesgue, when he constructed a continuous periodic function whose Fourier series diverges at a point. It was also used by Hahn to prove the Principle of Uniform Boundedness (see [59] for more on this).

Let us now use Proposition 6.1.6 to examine weak convergence for L^p spaces. The proof of the next lemma is a straightforward consequence of that result and is left as an exercise for the reader.

6.1.10 LEMMA *Let $1 < p < \infty$ and $(f_n) \subset L^p[a,b]$ be a bounded sequence. For any $f \in L^p[a,b]$, $f_n \xrightarrow{w} f$ if and only if*

$$\lim_{n \to \infty} \int_a^x f_n(t)\,dt = \int_a^x f(t)\,dt$$

for each $x \in [a,b]$.

Oddly enough, Lemma 6.1.10 does not hold for $p = 1$ (see Exercise 6.3). However, if $1 < p < \infty$, then it lends itself to some very useful applications. We explore one such, which relates weak convergence to pointwise convergence. By its very nature, we need access to Egoroff's Theorem. Once again we will state the theorem here and point you to Royden [49, Section 3.6] for its proof.

6.1.11 THEOREM (EGOROFF'S THEOREM (SEVERINI, 1910, AND EGOROFF, 1911)) *Let (f_n) be a sequence of measurable $\mathbb{K}$-valued functions on $[a,b]$ such that $\lim_{n \to \infty} f_n(x) = f(x)$ a.e. Then for any $\epsilon > 0$, there is a set $A \subset [a,b]$ such that $m([a,b] \setminus A) < \epsilon$ and $f_n \to f$ uniformly on A.*

6.1.12 PROPOSITION *Let $1 < p < \infty$ and $(f_n) \subset L^p[a,b]$ be a sequence such that $(\|f_n\|_p)$ is uniformly bounded. If $f : [a,b] \to \mathbb{K}$ is a function such that $\lim_{n \to \infty} f_n(x) = f(x)$ a.e., then $f \in L^p[a,b]$ and $f_n \xrightarrow{w} f$ in $L^p[a,b]$.*

Proof. Firstly, choose $M > 0$ such that $\|f_n\|_p \leq M$ for all $n \in \mathbb{N}$. Then by Fatou's Lemma applied to the sequence $(|f_n|^p)$, we see that $f \in L^p[a,b]$ and that $\|f\|_p \leq M$.

Now to prove weak convergence, we wish to verify the hypothesis of Lemma 6.1.10. So, fix $x \in [a,b]$ and $\epsilon > 0$. Let $\delta > 0$ be a positive number to be chosen later. Then, by Egoroff's Theorem, there is a measurable set $A \subset [a,b]$ such that $m([a,b] \setminus A) < \delta$ and $f_n \to f$ uniformly on A. Hence there is $N_0 \in \mathbb{N}$ such that

$$\sup_{x \in A} |f_n(x) - f(x)| \leq \frac{\epsilon}{2m(A)}$$

for all $n \geq N_0$. Therefore if $n \geq N_0$, we have

$$\left| \int_a^x f_n(t)dt - \int_a^x f(t)dt \right| \leq \int_a^x |f_n - f|$$

$$\leq \int_a^b |f_n - f|$$

$$= \int_A |f_n - f| + \int_{[a,b]\setminus A} |f_n - f|$$

$$\leq \frac{\epsilon}{2m(A)} m(A) + \|f_n - f\|_p m([a,b] \setminus A)^{1/q},$$

where the second term arises from Hölder's Inequality. Now if we choose $\delta > 0$ such that $0 < \delta < \left(\frac{\epsilon}{4M}\right)^q$, then we see that

$$\left| \int_a^x f_n(t)dt - \int_a^x f(t)dt \right| \leq \frac{\epsilon}{2} + 2M\delta^{1/q} < \epsilon$$

for all $n \geq N_0$. Thus we have verified Lemma 6.1.10 and completed the proof. □

Once again, the conclusion of Proposition 6.1.12 does not hold if $p = 1$ and we will leave it to you to find a counterexample (Exercise 6.5).

6.2   The Hahn–Banach Separation Theorem

In Chapter 4, we proved the Hahn–Banach Extension Theorem. The version for real vector spaces relied on Lemma 4.2.1, where we extended the linear functional from a subspace of codimension one to the whole space. Let us now revisit this argument, this time paying attention to the seminorm $p : E \to \mathbb{R}$. A closer look at the proof tells us that we needed two properties of p:

  (a) $p(\alpha x) = \alpha p(x)$ for all $x \in E$ and $\alpha \geq 0$ and

  (b) $p(x + y) \leq p(x) + p(y)$ for all $x, y \in E$.

We never used the fact that $p(\alpha x) = |\alpha| p(x)$ for all $\alpha \in \mathbb{R}$ (The analogous property *was* needed, however, for complex vector spaces).

6.2.1 DEFINITION   Let E be a vector space. A function $p : E \to \mathbb{R}$ is said to be a *sublinear functional* if it satisfies the properties (a) and (b) given above.

We conclude that, in the statement of the Hahn–Banach Theorem for real vector spaces (Theorem 4.2.2), we may assume that p is just a sublinear functional and not necessarily a seminorm. The next result tells us that sublinear functionals are plentiful and is the first hint that convexity might play an important role going forward.

6.2.2 THEOREM *Let C be a non-empty, convex, open subset of a normed linear space* **E** *such that* $0 \in C$. *For* $x \in$ **E**, *define*

$$p(x) := \inf\{t > 0 : t^{-1}x \in C\}.$$

Then

    (i) *There exists* $M > 0$ *such that* $0 \le p(x) \le M\|x\|$ *for all* $x \in$ **E**. *In particular,* $p(x) < \infty$.

    (ii) p *is a sublinear functional.*

    (iii) *For any* $x \in$ **E**, $p(x) < 1$ *if and only if* $x \in C$.

This function p *is called the* **Minkowski functional** *(or* **gauge***) of* C.

Proof.

    (i) Since $0 \in C$ and C is open, there exists $r > 0$ such that $B(0, r) \subset C$. Thus for any $x \in$ **E**,

$$\frac{r}{2\|x\|}x \in C.$$

Hence $p(x) \le \frac{2\|x\|}{r}$, so $M := 2/r$ works.

    (ii) If $x \in$ **E** and $\alpha > 0$, then $t^{-1}x \in C$ if and only if $(t\alpha)^{-1}\alpha x \in C$. From this, it follows that

$$p(\alpha x) = \alpha p(x).$$

Now if $x, y \in$ **E**, then we wish to prove that $p(x+y) \le p(x) + p(y)$. Fix $\epsilon > 0$ and choose $s, t > 0$ such that $s^{-1}x \in C$, $t^{-1}y \in C$, $s < p(x) + \epsilon$ and $t < p(y) + \epsilon$. Define

$$r := \frac{s}{s+t}.$$

Then $0 < r < 1$. Since C is convex,

$$(s+t)^{-1}(x+y) = r(s^{-1}x) + (1-r)(t^{-1}y) \in C.$$

Hence $p(x+y) \le s + t < p(x) + p(y) + 2\epsilon$. This is true for all $\epsilon > 0$ and thus $p(x+y) \le p(x) + p(y)$.

(iii) If $x \in \mathbf{E}$ such that $p(x) < 1$, then there exists $0 < t < 1$ such that $t^{-1}x \in C$. Since C is convex and $0 \in C$,

$$x = (1 - t)0 + t(t^{-1}x) \in C.$$

Conversely, if $x \in C$, then since C is open, there exists $r > 0$ such that $B(x, r) \subset C$. In particular,

$$x + \frac{r}{2\|x\|}x \in C.$$

Therefore if we set $t := 1 + \frac{r}{2\|x\|}$, then $t > 1$ and $p(x) \leq t^{-1}$. In particular, $p(x) < 1$. This completes the proof. $\qquad\square$

6.2.3 Proposition *Let $\mathbf{E}$ be a normed linear space over $\mathbb{R}$ and let $C \subset \mathbf{E}$ be a non-empty, convex, open set. If $x_0 \notin C$, then there exists $\psi \in \mathbf{E}^*$ such that*

$$\psi(x) < \psi(x_0)$$

for all $x \in C$.

Proof. We first assume that $0 \in C$. Let $\mathbf{F} := \text{span}(x_0)$ and let p denote the Minkowski functional of C. Define $\varphi : \mathbf{F} \to \mathbb{R}$ by $\varphi(\alpha x_0) = \alpha$. Since $x_0 \notin C$, $\alpha^{-1}(\alpha x_0) \notin C$ for any $\alpha > 0$ and hence

$$p(\alpha x_0) > \alpha = \varphi(\alpha x_0).$$

If $\alpha < 0$, then this equation holds trivially since $p(\alpha x_0) \geq 0$. Thus by the Hahn–Banach Theorem, there exists $\psi : \mathbf{E} \to \mathbb{R}$ such that $\psi|_\mathbf{F} = \varphi$ and

$$\psi(x) \leq p(x)$$

for all $x \in \mathbf{E}$. By Theorem 6.2.2, there exists $M > 0$ such that $p(x) \leq M\|x\|$ for all $x \in \mathbf{E}$. Hence

$$\psi(x) \leq M\|x\|$$

for all $x \in \mathbf{E}$. Replacing x by $-x$, the same inequality holds, so we conclude that ψ is a bounded linear functional. Now if $x \in C$, then

$$\psi(x) \leq p(x) < 1 = \psi(x_0).$$

Thus ψ is the desired linear functional.

Now we consider the case when $0 \notin C$. Fix $x_1 \in C$ and consider $D := C - x_1$. Then D is also open and convex, $0 \in D$ and $x_0 - x_1 \notin D$. By the first part of the theorem, there exists $\psi \in \mathbf{E}^*$ such that

$$\psi(y) < \psi(x_0 - x_1)$$

for all $y \in D$. Since ψ is linear, we conclude that $\psi(x) < \psi(x_0)$ for all $x \in C$. □

It is evident that convexity is necessary for Proposition 6.2.3 to work (simply visualize a counterexample in $\mathbb{R}^2$). The next example shows that openness is also unavoidable (unless $\mathbf{E}$ is finite dimensional – see Exercise 6.13).

6.2.4 EXAMPLE Let $\mathbf{E} := c_0$ and define

$$C = \{(x_n) \in c_0 : \text{there exists } N \in \mathbb{N} \text{ such that } x_N > 0 \text{ and } x_n = 0 \text{ for all } n > N\}.$$

Observe that C is convex and $0 \notin C$. Now suppose that there exists $\psi \in \mathbf{E}^*$ such that $\psi(x) < \psi(0) = 0$ for all $x \in C$. By Exercise 4.13, there exists $y = (y_n) \in \ell^1$ such that

$$\psi((x_n)) = \sum_{n=1}^{\infty} x_n y_n.$$

for all $(x_n) \in c_0$. Therefore for each $j \in \mathbb{N}$, $y_j = \psi(e_j) < 0$. However, if $x := (y_2, -y_1, 0, 0, \ldots) \in C$, then $\psi(x) = y_1 y_2 - y_2 y_1 = 0$. This contradicts our choice of ψ.

6.2.5 DEFINITION Let $\mathbf{E}$ be a vector space over $\mathbb{R}$, $\varphi : \mathbf{E} \to \mathbb{R}$ a non-zero linear functional and $\alpha \in \mathbb{R}$. The set

$$[\varphi = \alpha] := \{x \in \mathbf{E} : \varphi(x) = \alpha\}$$

is called an *affine hyperplane* of $\mathbf{E}$.

The geometric Hahn–Banach Theorems, which we now prove, may be thought of as linear analogues of Urysohn's Lemma (Remark 2.3.10). There, one is given two disjoint subsets of a topological space and one would like to separate them by means of a continuous function. Here, we are given disjoint subsets of a vector space and we would like to separate them by means of a linear functional. This is an analogy, of course, and should not be taken literally, but it should give you a sense for what the theorem is trying to achieve.

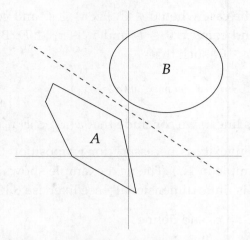

FIGURE 6.1   Separation of convex sets by a hyperplane

6.2.6 THEOREM (HAHN–BANACH SEPARATION THEOREM – I (HAHN, 1927, AND BANACH, 1929)) *Let* **E** *be a normed linear space over* $\mathbb{R}$ *and let* A *and* B *be two non-empty, disjoint, convex subsets of* **E**. *If* A *is open, then there exists* $\psi \in \mathbf{E}^*$ *and* $\alpha \in \mathbb{R}$ *such that*

$$\psi(a) \leq \alpha \leq \psi(b)$$

for all $a \in A$ *and* $b \in B$. *In other words, the closed hyperplane* $[\psi = \alpha]$ *separates* A *from* B.

Proof. Define $C := A - B$, then it is easy to see that C is convex because A and B are. Furthermore, C is open because

$$C = \bigcup_{b \in B} (A - b),$$

which is a union of sets each of which is homeomorphic to A. Finally, since A and B are disjoint, $0 \notin C$. By Proposition 6.2.3, there exists $\psi \in \mathbf{E}^*$ such that $\psi(x) < \psi(0) = 0$ for all $x \in C$. This implies that $\psi(a) < \psi(b)$ for all $a \in A$ and $b \in B$. Therefore any $\alpha \in \mathbb{R}$ satisfying

$$\sup_{a \in A} \psi(a) \leq \alpha \leq \inf_{b \in B} \psi(b)$$

will do the job. $\qquad\square$

6.2.7 THEOREM (HAHN–BANACH SEPARATION THEOREM – II) *Let* **E** *be a normed linear space over* $\mathbb{R}$ *and let* A *and* B *be two non-empty, disjoint, convex subsets on* **E**. *If* A *is closed and* B *is compact, then there exists* $\psi \in \mathbf{E}^*, \alpha \in \mathbb{R}$ *and* $\epsilon > 0$ *such that*

$$\psi(a) \leq \alpha - \epsilon < \alpha + \epsilon \leq \psi(b)$$

for all $a \in A$ *and* $b \in B$. *In other words, the closed hyperplane* $[\psi = \alpha]$ *strictly separates* A *from* B.

Proof. To begin with, we claim that there is a number $r > 0$ such that

$$[A + B(0, r)] \cap [B + B(0, r)] = \emptyset.$$

Geometrically, one may think of these sets as *thickening* both A and B, while still keeping them disjoint. Suppose not, then for all $n \in \mathbb{N}$, there exists $u_n \in [A + B(0, 1/n)] \cap [B + B(0, 1/n)]$. Write

$$u_n = a_n + x_n \text{ and } b_n + y_n,$$

where $a_n \in A, b_n \in B, \|x_n\| < 1/n$ and $\|y_n\| < 1/n$. Hence $\|a_n - b_n\| \leq \frac{2}{n}$. Since B is compact, there is a subsequence (b_{n_k}) of (b_n) and a point $b \in B$ such that $\lim_{k \to \infty} b_{n_k} = b$. Hence $\lim_{k \to \infty} a_{n_k} = b$ as well. Since A is closed,

$$b \in A \cap B.$$

This contradicts the fact that A and B are disjoint, thus proving the claim.

Now choose $r > 0$ such that $\tilde{A} := A + B(0, r)$ and $\tilde{B} := B + B(0, r)$ are disjoint. Note that both $\tilde{A}$ and $\tilde{B}$ are convex and open (as in the proof of the previous theorem). Hence by Theorem 6.2.6, there exists $\psi \in \mathbf{E}^*$ and $\alpha \in \mathbb{R}$, such that

$$\psi(u) \leq \alpha \leq \psi(v)$$

for all $u \in \tilde{A}$ and $v \in \tilde{B}$. Now for $a \in A$ and $z \in B[0, 1]$, $a + \frac{r}{2}z \in \tilde{A}$. Therefore

$$\psi(a) + \frac{r}{2}\psi(z) \leq \alpha.$$

This is true for every $z \in B[0, 1]$, so we conclude that

$$\psi(a) + \frac{r}{2}\|\psi\| \leq \alpha.$$

Similarly, for any $b \in B$, $\alpha \leq \psi(b) - \frac{r}{2}\|\psi\|$. Therefore $\epsilon := \frac{r}{2}\|\psi\|$ works. $\qquad \square$

We will end this section with the corresponding statements for complex normed linear spaces, which we will magnanimously leave for you to prove (Exercise 6.14).

6.2.8 THEOREM (HAHN–BANACH SEPARATION THEOREM – COMPLEX CASE)

Let **E** *be a normed linear space over* **C** *and let* A *and* B *be two non-empty, disjoint, convex subsets of* **E**.

(i) *If* A *is open, then there exists* $\psi \in \mathbf{E}^*$ *and* $\alpha \in \mathbb{R}$ *such that*

$$\mathrm{Re}(\psi)(a) \le \alpha \le \mathrm{Re}(\psi)(b)$$

for all $a \in A$ *and* $b \in B$.

(ii) *If* A *is closed and* B *is compact, then there exists* $\psi \in \mathbf{E}^*, \alpha \in \mathbb{R}$ *and* $\epsilon > 0$ *such that*

$$\mathrm{Re}(\psi)(a) \le \alpha - \epsilon < \alpha + \epsilon \le \mathrm{Re}(\psi)(b)$$

for all $a \in A$ *and* $b \in B$.

In this section, we have proved the Hahn–Banach Separation Theorem by appealing to the Extension Theorem (Theorem 4.2.2). However, these two results are, in fact, equivalent to each other. Some authors choose to prove the Separation Theorem first and use it to prove the Extension Theorem. The interested reader will find this approach laid out in Limaye [37].

6.3   The Weak Topology

As promised at the start of the chapter, we will now define a new topology on a normed linear space **E**. This topology is weaker (or *coarser*) than the norm topology, and plays well with the vector space operations. It is also the natural home for the notion of weak convergence.

We will begin with a notion you would have seen in point–set topology (perhaps in the context of the product topology). It is that of a weak topology defined by a class of functions.

6.3.1 LEMMA   *Let* X *be a set and* $\mathcal{B} \subset 2^X$ *be a collection of subsets of* X *satisfying two conditions:*

- $\bigcup_{B \in \mathcal{B}} B = X$.

- *For every* $B_1, B_2 \in \mathcal{B}$ *and* $x \in B_1 \cap B_2$, *there exists* $B_3 \in \mathcal{B}$ *such that* $x \in B_3$ *and* $B_3 \subset B_1 \cap B_2$.

Then there is a unique topology τ_B on X such that

(i) $B \subset \tau_B$ *and*

(ii) *if σ is any other topology such that $B \subset \sigma$, then $\tau_B \subset \sigma$.*

Furthermore, B is a basis for the topology τ_B.

Proof. Let $\mathcal{U}$ be the collection of all topologies on X which contain B. Then $2^X \in \mathcal{U}$, so $\mathcal{U} \neq \emptyset$. Therefore we may define

$$\tau_B = \bigcap_{\sigma \in \mathcal{U}} \sigma.$$

Then it is easy to see that τ_B is a topology and it satisfies (i) and (ii). That B is a basis for τ_B follows from the assumptions made on it. $\qquad\square$

6.3.2 PROPOSITION *Let X be any set, (Y, τ_Y) be a topological space and let $\mathcal{G}$ denote a collection of functions from X to Y. Then there is a unique topology $\tau_{\mathcal{G}}$ on X such that*

(i) *Each $f \in \mathcal{G}$ is continuous with respect to $\tau_{\mathcal{G}}$ and*

(ii) *If σ is any other topology on X such that $f : (X, \sigma) \to (Y, \tau_Y)$ is continuous for all $f \in \mathcal{G}$, then $\tau_{\mathcal{G}} \subset \sigma$.*

*In other words, $\tau_{\mathcal{G}}$ is the smallest topology that makes every member of $\mathcal{G}$ continuous. This topology is called the **weak topology** on X defined by $\mathcal{G}$.*

Proof. Define $B \subset 2^X$ to be the collection of all finite intersections of sets of the form $f^{-1}(U)$, for some $U \in \tau_Y$ and $f \in \mathcal{G}$. In other words, $B \in \mathcal{B}$ if and only if there exist finitely many functions $f_1, f_2, \ldots, f_n \in \mathcal{G}$ and open sets $U_1, U_2, \ldots, U_n \in \tau_Y$ such that

$$B = \bigcap_{i=1}^{n} f_i^{-1}(U_i).$$

Note that $f^{-1}(Y) = X$ for any $f \in \mathcal{G}$. Also, if $B_1, B_2 \in \mathcal{B}$, then $B_1 \cap B_2 \in \mathcal{B}$. Therefore Lemma 6.3.1 applies and we take $\tau_{\mathcal{G}} := \tau_B$.

It is now clear that each $f \in \mathcal{G}$ is continuous with respect to $\tau_{\mathcal{G}}$ (by construction). Furthermore, if σ is any other topology on X such that $f : (X, \sigma) \to (Y, \tau_Y)$ is continuous for each $f \in \mathcal{G}$, then σ will necessarily contain every member of $\mathcal{B}$. But $\mathcal{B}$ is a basis for $\tau_{\mathcal{G}}$, so $\tau_{\mathcal{G}} \subset \sigma$ must hold. $\qquad\square$

Before we begin our journey, let us pause to prove a very useful result in this context; one that will be used repeatedly throughout the book.

6.3.3 PROPOSITION *Let X be a set, (Y, τ_Y) be a topological space and let $\mathcal{G}$ be a collection of functions from X to Y. Let $\tau_{\mathcal{G}}$ be the weak topology on X defined by $\mathcal{G}$. If (Z, τ_Z) is any topological space, then a function*

$$g : (Z, \tau_Z) \to (X, \tau_{\mathcal{G}})$$

is continuous if and only if $f \circ g : (Z, \tau_Z) \to (Y, \tau_Y)$ is continuous for each $f \in \mathcal{G}$.

Proof. If $g : (Z, \tau_Z) \to (X, \tau_{\mathcal{G}})$ is continuous, then clearly $f \circ g : (Z, \tau_Z) \to (Y, \tau_Y)$ is continuous because $f : (X, \tau_{\mathcal{G}}) \to (Y, \tau_Y)$ is continuous by construction.

Now suppose $g : Z \to X$ is function such that $f \circ g : (Z, \tau_Z) \to (Y, \tau_Y)$ is continuous for each $f \in \mathcal{G}$. To prove that g is continuous, we choose an open set $U \in \tau_{\mathcal{G}}$ and prove that $g^{-1}(U) \in \tau_Z$. We may assume that U is a basic open set, so there would exist $\{f_1, f_2, \ldots, f_n\} \subset \mathcal{G}$ and open sets $\{U_1, U_2, \ldots, U_n\} \subset \tau_Y$ such that

$$U = \bigcap_{i=1}^{n} f_i^{-1}(U_i).$$

Then

$$g^{-1}(U) = \bigcap_{i=1}^{n} g^{-1}(f_i^{-1}(U_i)) = \bigcap_{i=1}^{n} (f_i \circ g)^{-1}(U_i)$$

and this set belongs to τ_Z by hypothesis. Therefore g is continuous. $\square$

6.3.4 DEFINITION Let $\mathbf{E}$ be a normed linear space and $Y = \mathbb{K}$ equipped with the usual topology. Let $\mathcal{G} := \mathbf{E}^*$, the set of all bounded linear functionals on $\mathbf{E}$. The *weak topology* on $\mathbf{E}$ is the topology defined by $\mathcal{G}$ by means of Proposition 6.3.2. This is denoted by

$$\sigma(\mathbf{E}, \mathbf{E}^*)$$

to indicate that it is the topology on $\mathbf{E}$ inherited from $\mathbf{E}^*$. Elements of $\sigma(\mathbf{E}, \mathbf{E}^*)$ are called *weakly open* sets and their complements are called *weakly closed* sets. A subset of $\mathbf{E}$ is said to be *weakly compact* if it is compact with respect to $\sigma(\mathbf{E}, \mathbf{E}^*)$.

For clarity, we will henceforth refer to the norm topology on $\mathbf{E}$ by $\sigma(\mathbf{E}, \|\cdot\|)$, members of which will be referred to as *norm-open* (or *strongly open*) sets. The terms *norm-closed* and *norm-compact* are defined analogously.

6.3.5 REMARK Some remarks are in order before we proceed:

(i) Since each $\varphi \in \mathbf{E}^*$ is continuous with respect to the norm, it follows that

$$\sigma(\mathbf{E}, \mathbf{E}^*) \subset \sigma(\mathbf{E}, \|\cdot\|),$$

since $\sigma(\mathbf{E}, \mathbf{E}^*)$ is the smallest topology with this property. In other words, we have the following implications:

$$\text{weakly open} \Rightarrow \text{norm-open,}$$
$$\text{weakly closed} \Rightarrow \text{norm-closed and}$$
$$\text{weakly compact} \Leftarrow \text{norm-compact}$$

(ii) For a set $A \subset \mathbf{E}$, we write

$$\overline{A}^w$$

to denote the intersection of all weakly closed set containing A. It is called the **weak closure** of A. We denote the norm-closure of A by $\overline{A}^{\|\cdot\|}$ if the context demands it. Observe that $\overline{A}^{\|\cdot\|} \subset \overline{A}^w$, since $\overline{A}^w$ is a norm-closed set that contains A.

(iii) We now describe basic open sets in $\sigma(\mathbf{E}, \mathbf{E}^*)$.

(a) If $\varphi \in \mathbf{E}^*$ and $\epsilon > 0$, then

$$\{x \in \mathbf{E} : |\varphi(x)| < \epsilon\}$$

is a (sub-)basic open neighbourhood of 0, since it is inverse image under φ of an open set in $\mathbb{K}$.

(b) In general, if $\varphi_1, \varphi_2, \ldots, \varphi_n \in \mathbf{E}^*$ and $\epsilon_i > 0$, then

$$\{x \in \mathbf{E} : |\varphi_i(x)| < \epsilon_i \text{ for all } 1 \le i \le n\}$$

is a basic open neighbourhood of 0. If $\epsilon = \min\{\epsilon_i : 1 \le i \le n\}$, then this neighbourhood contains

$$\{x \in \mathbf{E} : |\varphi_i(x)| < \epsilon \text{ for all } 1 \le i \le n\}.$$

(c) If $x_0 \in \mathbf{E}, \varphi \in \mathbf{E}^* \epsilon > 0$, then

$$\{x \in \mathbf{E} : |\varphi(x) - \varphi(x_0)| < \epsilon\}$$

is a (sub-)basic open neighbourhood of x_0. As argued above, every basic open neighbourhood of x_0 will contain one of the form

$$\{x \in \mathbf{E} : |\varphi_i(x) - \varphi_i(x_0)| < \epsilon \text{ for all } 1 \le i \le n\},$$

for some $\varphi_1, \varphi_2, \ldots, \varphi_n \in \mathbf{E}^*$ and $\epsilon > 0$ fixed.

(iv) Finally, one may use this description of open sets to prove that, for a sequence (x_n) in $\mathbf{E}$, $x_n \xrightarrow{w} x$ if and only if $x_n \to x$ with respect to $\sigma(\mathbf{E}, \mathbf{E}^*)$ (Exercise 6.15).

Let us begin with the simplest fact that one would want to prove about any new topology one encounters. In what follows, we write $B_{\mathbb{K}}(\alpha, r) := \{z \in \mathbb{K} : |z - \alpha| < r\}$ for the ball in $\mathbb{K}$ of radius r centered at α.

6.3.6 PROPOSITION *The weak topology $\sigma(\mathbf{E}, \mathbf{E}^*)$ is Hausdorff.*

Proof. If $x, y \in \mathbf{E}$ are distinct, then by Corollary 4.2.8, there exists $\varphi \in \mathbf{E}^*$ such that $\varphi(x) \ne \varphi(y)$. Define

$$\epsilon := \frac{|\varphi(x) - \varphi(y)|}{3}$$

and set $U := B_{\mathbb{K}}(\varphi(x), \epsilon)$ and $V := B_{\mathbb{K}}(\varphi(y), \epsilon)$. Then $\varphi^{-1}(U)$ and $\varphi^{-1}(V)$ are disjoint, weakly open neighbourhoods of x and y, respectively. □

We now introduce a notion that is closely related to that of a normed linear space, but one that affords a little more flexibility.

6.3.7 DEFINITION A *topological vector space* is a vector space $\mathbf{E}$ equipped with a Hausdorff topology, such that the addition map $a : \mathbf{E} \times \mathbf{E} \to \mathbf{E}$ and the scalar multiplication map $s : \mathbb{K} \times \mathbf{E} \to \mathbf{E}$ are both continuous (here, $\mathbb{K}$ is equipped with the usual norm topology and the product spaces are equipped with the product topologies).

It is obvious from Remark 2.1.2 that any normed linear space is a topological vector space, but there are topological vector spaces which are not normable (see Exercise 6.44). We will now show that the weak topology gives a normed linear space the structure of a topological vector space and that it is not even metrizable if the space is infinite dimensional!

6.3.8 PROPOSITION *The space $(\mathbf{E}, \sigma(\mathbf{E}, \mathbf{E}^*))$ is a topological vector space.*

Proof.   We only prove that a is continuous, as the proof for s is similar. Let $W \in \sigma(\mathbf{E}, \mathbf{E}^*)$ denote a weakly open set and $(x, y) \in a^{-1}(W)$ so that $z = a(x, y) = x + y \in W$. Then there exist $\varphi_1, \varphi_2, \ldots, \varphi_n \in \mathbf{E}^*$ and $\epsilon > 0$ such that

$$W' := \{w \in \mathbf{E} : |\varphi_i(w) - \varphi_i(z)| < \epsilon \text{ for all } 1 \le i \le n\} \subset W.$$

Define weakly open neighbourhoods U and V of x and y, respectively by

$$U := \{u \in \mathbf{E} : |\varphi_i(u) - \varphi_i(x)| < \epsilon/2 \text{ for all } 1 \le i \le n\} \text{ and}$$
$$V := \{v \in \mathbf{E} : |\varphi_i(v) - \varphi_i(y)| < \epsilon/2 \text{ for all } 1 \le i \le n\}.$$

If $(u, v) \in U \times V$, then $|\varphi_i(u + v) - \varphi_i(z)| < \epsilon$ for all $1 \le i \le n$. Hence $a(u, v) = u + v \in W'$. Thus $U \times V \subset a^{-1}(W)$ and $(x, y) \in U \times V$. This is true for any $(x, y) \in a^{-1}(W)$, proving that $a^{-1}(W)$ is open. $\square$

If $\mathbf{E}$ is a finite dimensional normed linear space, we know that the topology induced by the norm is simple enough; it is merely the product topology when $\mathbf{E}$ is identified with $\mathbb{K}^n$. Therefore the next result should not be very surprising.

6.3.9 THEOREM   *If $\mathbf{E}$ is finite dimensional, then the weak and norm topologies coincide.*

Proof.   Let $\sigma(\mathbf{E}, \mathbf{E}^*)$ and $\sigma(\mathbf{E}, \| \cdot \|)$ denote the weak and norm topologies, respectively. By definition, $\sigma(\mathbf{E}, \mathbf{E}^*) \subset \sigma(\mathbf{E}, \| \cdot \|)$. If $\mathbf{E}$ is finite dimensional, we wish to prove that $\sigma(\mathbf{E}, \| \cdot \|) \subset \sigma(\mathbf{E}, \mathbf{E}^*)$.

We may assume without loss of generality that $\mathbf{E} = (\mathbb{K}^n, \| \cdot \|_\infty)$. For $1 \le i \le n$, let $\pi_i : \mathbf{E} \to \mathbb{K}$ denote the coordinate projections. Then for any $x \in \mathbf{E}$ and $r > 0$,

$$B_{\mathbf{E}}(x, r) = \{y \in \mathbf{E} : |\pi_i(y) - \pi_i(x)| < r \text{ for all } 1 \le i \le n\}.$$

Thus $B_{\mathbf{E}}(x, r) \in \sigma(\mathbf{E}, \mathbf{E}^*)$. This is true for any basic open set $B_{\mathbf{E}}(x, r) \in \sigma(\mathbf{E}, \| \cdot \|)$, and hence $\sigma(\mathbf{E}, \| \cdot \|) \subset \sigma(\mathbf{E}, \mathbf{E}^*)$. $\square$

Let us now turn to the most fundamental way in which the Hahn–Banach Separation Theorem informs the weak topology.

6.3.10 THEOREM   *Let $\mathbf{E}$ be an normed linear space and $C \subset \mathbf{E}$ be a convex set. Then C is weakly closed if and only if it is norm-closed.*

Proof.   If C is weakly closed, then it is norm-closed by Remark 6.3.5. Conversely, if C is convex and norm-closed, then we wish to prove that C is weakly closed.

Suppose $x \notin C$, then by the Hahn–Banach Separation Theorem there exists $\psi \in \mathbf{E}^*, \alpha \in \mathbb{R}$ and $\epsilon > 0$ such that $\mathrm{Re}(\psi)(x) < \alpha - \epsilon$ and

$$\mathrm{Re}(\psi)(y) > \alpha + \epsilon$$

for all $y \in C$. Note that $\psi : (\mathbf{E}, \sigma(\mathbf{E}, \mathbf{E}^*)) \to \mathbb{K}$ is continuous by the very definition of the weak topology and $\mathrm{Re} : \mathbb{K} \to \mathbb{R}$ is continuous (it is the identity map if $\mathbb{K} = \mathbb{R}$). Hence

$$U = \{u \in \mathbf{E} : \mathrm{Re}(\psi(u)) < \alpha - \epsilon\}$$

is weakly open, $x \in U$ and $U \cap C = \emptyset$. This is true for any $x \notin C$, which proves that $\mathbf{E} \setminus C$ is weakly open, which is what we wanted to prove. $\square$

We will give two applications of Theorem 6.3.10. The first one, due to Mazur, deals with weak convergence. Before we state it, observe that if $\{A_\alpha\}_{\alpha \in J}$ is a family of convex subsets of a vector space $\mathbf{E}$, then

$$\bigcap_{\alpha \in J} A_\alpha$$

is also a convex set. This allows us to the make the following definition.

6.3.11 DEFINITION Let $\mathbf{E}$ be a real vector space and K be a subset of $\mathbf{E}$. The *convex hull* of K is the smallest convex set in $\mathbf{E}$ containing K (equivalently, it is the intersection of all convex sets containing K). It is denoted by $\mathrm{conv}(K)$.

6.3.12 COROLLARY (MAZUR, 1933) *Let $\mathbf{E}$ be a normed linear space and (x_n) be a sequence in $\mathbf{E}$ such that $x_n \xrightarrow{w} x$. Then there is a sequence $(y_n) \subset \mathrm{conv}(\{x_i : i \in \mathbb{N}\})$ such that $y_n \xrightarrow{s} x$.*

Proof. If $C := \mathrm{conv}(\{x_i\}_{i=1}^\infty)$, then $x \in \overline{C}^w$, the weak closure of C. However, the norm-closure of C is also a convex set (Exercise 6.10) and is thus weakly closed by Theorem 6.3.10. Therefore $\overline{C}^w \subset \overline{C}^{\|\cdot\|}$ and $x \in \overline{C}^{\|\cdot\|}$, which is what we needed to prove. $\square$

Our next application of Theorem 6.3.10 allows us to prove that the weak topology is not metrizable in general. In order to do that, we first construct an example of a set that is norm-closed, but not weakly closed.

6.3.13 THEOREM *Let $\mathbf{E}$ be an infinite dimensional normed linear space. If $S_\mathbf{E}$ and $B_\mathbf{E}$ denote the unit sphere and closed unit ball, respectively, then*

$$\overline{S_\mathbf{E}}^w = B_\mathbf{E}.$$

In particular, $S_\mathbf{E}$ is not weakly closed.

Proof. Since $B_{\mathbf{E}}$ is convex and norm-closed, $B_{\mathbf{E}}$ is weakly closed by Theorem 6.3.10. Since $S_{\mathbf{E}} \subset B_{\mathbf{E}}$, it follows that

$$\overline{S_{\mathbf{E}}}^w \subset B_{\mathbf{E}}.$$

To prove the converse, it suffices to choose $x_0 \in \mathbf{E}$ such that $\|x_0\| < 1$ and prove that $x_0 \in \overline{S_{\mathbf{E}}}^w$. So, let $U \in \sigma(\mathbf{E}, \mathbf{E}^*)$ be any weakly open neighbourhood of x_0. We wish to prove that $U \cap S_{\mathbf{E}} \neq \emptyset$.

We may assume without loss of generality that

$$U = \{x \in \mathbf{E} : |\varphi_i(x) - \varphi_i(x_0)| < \epsilon \text{ for all } 1 \leq i \leq n\}$$

for some $\varphi_1, \varphi_2, \ldots, \varphi_n \in \mathbf{E}^*$ and $\epsilon > 0$. Now define $T : \mathbf{E} \to \mathbb{K}^n$ by

$$T(x) := (\varphi_1(x), \varphi_2(x), \ldots, \varphi_n(x)).$$

Then T cannot be injective since $\dim(\mathbf{E}) = \infty$. Hence there exists a non-zero vector $y_0 \in \mathbf{E}$ such that $\varphi_i(y_0) = 0$ for all $1 \leq i \leq n$. This implies that $x_0 + t y_0 \in U$ for all $t \in \mathbb{R}$. Now consider $g : \mathbb{R}_+ \to \mathbb{R}_+$ given by

$$g(t) := \|x_0 + t y_0\|.$$

Then g is norm continuous (by Remark 2.1.2) and $g(0) = \|x_0\| < 1$. Since

$$t\|y_0\| \leq \|x_0\| + \|x_0 + t y_0\|,$$

it follows that $\lim_{t \to \infty} g(t) = \infty$. By the Intermediate Value Theorem, there exists $t_0 \in \mathbb{R}$ such that $\|x_0 + t_0 y_0\| = 1$. Thus $x_0 + t_0 y_0 \in S_{\mathbf{E}} \cap U$ and $S_{\mathbf{E}} \cap U \neq \emptyset$. $\square$

The previous proof has an interesting nugget of information: If $\mathbf{E}$ is infinite dimensional and U is a weakly open neighbourhood of $0 \in \mathbf{E}$, then U contains an entire one-dimensional subspace! In particular, no such open set can be contained in a norm bounded set. This is in remarkable contrast to the way we perceive open sets in a metric topology. Try Exercise 6.17 to understand this better.

6.3.14 COROLLARY *If $\mathbf{E}$ is an infinite dimensional normed linear space, then the weak topology on $\mathbf{E}$ is not metrizable. In particular,*

$$\sigma(\mathbf{E}, \mathbf{E}^*) \neq \sigma(\mathbf{E}, \|\cdot\|).$$

Proof.   Let us assume that $\sigma(\mathbf{E}, \mathbf{E}^*)$ was metrizable by a metric d. For $n \in \mathbb{N}$, let $S_n := \{x \in \mathbf{E} : \|x\| = n\}$ and $B_n := \{x \in \mathbf{E} : \|x\| \le n\}$. Since the map $x \mapsto nx$ is a weak homeomorphism (by Proposition 6.3.8), it follows from Theorem 6.3.13 that

$$\overline{S_n}^w = B_n$$

for all $n \in \mathbb{N}$. In particular, $0 \in \overline{S_n}^w$. Therefore there exists $x_n \in S_n$ such that

$$d(x_n, 0) < 1/n.$$

Then it follows that $x_n \xrightarrow{w} 0$. However, $\|x_n\| = n$ for each $n \in \mathbb{N}$ and any weakly convergent sequence must be norm bounded (Lemma 6.1.4). This contradiction proves that no such metric can exist. $\square$

6.4   Weak Sequential Compactness

One of the chief reasons to study the weak topology (and the weak-$*$ topology we will introduce later in the chapter) is that compact sets are hard to come by in the norm topology. One of the most used results in real analysis is that a bounded sequence in $\mathbb{R}$ has a convergent subsequence (the Bolzano–Weierstrass Theorem). Not having access to such a theorem for infinite dimensional Banach spaces robs us of a very useful line of reasoning. Now we seek to remedy this difficulty.

6.4.1 DEFINITION   Let $\mathbf{E}$ be an normed linear space. A subset K of $\mathbf{E}$ is said to be **weakly sequentially compact** if every sequence in K has a subsequence that converges weakly to a point in K.

Note: Since $\sigma(\mathbf{E}, \mathbf{E}^*)$ is not metrizable, this is *not* the same as saying that K is weakly compact.

6.4.2 LEMMA   *Let $\mathbf{E}$ be an normed linear space, $\mathbf{F}$ a Banach space and let $D \subset \mathbf{E}$ be a dense set. Suppose $(T_n) \subset \mathcal{B}(\mathbf{E}, \mathbf{F})$ is a sequence of bounded linear operators such that*

  (i) *there exists $M > 0$ such that $\|T_n\| \le M$ for all $n \in \mathbb{N}$ (in other words, (T_n) is norm-bounded),*

  (ii) *$(T_n(y))$ converges in $\mathbf{F}$ for each $y \in D$.*

Then $(T_n(x))$ converges in $\mathbf{F}$ for all $x \in \mathbf{E}$.

Proof. Fix $x \in \mathbf{E}$. Since $\mathbf{F}$ is Banach, it suffices to show that $(T_n(x))$ is Cauchy in $\mathbf{F}$. So fix $\epsilon > 0$, then there exists $y \in D$ such that $\|x - y\| < \epsilon$. Hence

$$\|T_n(x) - T_m(x)\| \le \|T_n(x) - T_n(y)\| + \|T_n(y) - T_m(y)\| + \|T_m(y) - T_m(x)\|$$
$$\le M\|x - y\| + \|T_n(y) - T_m(y)\| + M\|x - y\|.$$

Since $(T_n(y))$ is convergent (and hence Cauchy) in $\mathbf{F}$, it follows that $(T_n(x))$ is Cauchy. $\qquad\square$

6.4.3 THEOREM (HELLEY'S THEOREM (HELLEY, 1912, AND BANACH, 1932))

Let $\mathbf{E}$ be a separable Banach space and $(\varphi_n) \subset \mathbf{E}^*$ be a norm-bounded sequence. Then there exists a subsequence (φ_{n_k}) and $\varphi \in \mathbf{E}^*$ such that

$$\varphi_{n_k}(x) \to \varphi(x)$$

for all $x \in \mathbf{E}$. Furthermore, $\|\varphi\| \le \liminf_{n \to \infty} \|\varphi_n\|$.

Proof. Our goal is to use Cantor's diagonalization argument to produce a subsequence of (φ_n) that satisfies the hypotheses of Lemma 6.4.2. Let $M > 0$ such that $\|\varphi_n\| \le M$ for all $n \in \mathbb{N}$ and let $\{x_n\}$ be a countable dense subset of $\mathbf{E}$. We inductively construct subsequences as follows.

(i) Fix $x_1 \in \mathbf{E}$ and consider the sequence $(\varphi_n(x_1))_{n=1}^{\infty} \subset \mathbb{K}$. This is a bounded sequence and thus has a convergent subsequence. Hence there is a strictly increasing sequence of integers $(s(1,n))$ and a point $a_1 \in \mathbb{K}$ such that

$$\lim_{n \to \infty} \varphi_{s(1,n)}(x_1) = a_1.$$

(ii) Now fix $x_2 \in \mathbf{E}$ and consider the sequence $(\varphi_{s(1,n)}(x_2))_{n=1}^{\infty} \subset \mathbb{K}$. As before, there exists a subsequence $(s(2,n))$ of $(s(1,n))$, and $a_2 \in \mathbb{K}$ such that

$$\lim_{n \to \infty} \varphi_{s(2,n)}(x_2) = a_2.$$

Proceeding inductively, we construct strictly increasing sequences $(s(j,n)) \subset \mathbb{N}$ such that

- For each $j \ge 1$, $(s(j+1,n))$ is a subsequence of $(s(j,n))$
- For each $j \ge 1$, $\lim_{n \to \infty} \varphi_{s(j,n)}(x_j) = a_j$.

(iii) Now define $n_k := s(k,k)$, then $(n_k)_{k \geq j}$ is a subsequence of each $(s(j,n))_{n=1}^{\infty}$. Therefore

$$\lim_{k \to \infty} \varphi_{n_k}(x_j) = a_j$$

exists in $\mathbb{K}$. By Lemma 6.4.2, $(\varphi_{n_k}(x))$ converges in $\mathbb{K}$ for all $x \in \mathbf{E}$. The result now follows from the Banach–Steinhaus Theorem (Theorem 5.2.2). □

The next theorem is the main result of this section. Although it only applies to reflexive spaces, its usefulness comes from the fact that many stock-in-trade Banach spaces (particularly $L^p[a,b]$ for $1 < p < \infty$) are reflexive.

6.4.4 THEOREM (EBERLEIN, 1947) *If* $\mathbf{E}$ *is reflexive, then the closed unit ball*

$$B_{\mathbf{E}} := \{x \in E : \|x\| \leq 1\}$$

is weakly sequentially compact.

Proof. Choose a sequence $(x_n) \subset B_{\mathbf{E}}$. We wish construct a convergent subsequence of (x_n). To that end, set

$$\mathbf{F} := \overline{\text{span}\{x_n\}}.$$

Since $\mathbf{E}$ is reflexive, so is $\mathbf{F}$ (by Proposition 4.4.8). Furthermore, $\mathbf{F}$ is separable (Why?). Now if $\mathbf{W} := \mathbf{F}^*$, then $\mathbf{W}$ is separable (by Proposition 4.4.7). For each $n \in \mathbb{N}$, consider $\widehat{x}_n \in \mathbf{W}$ and note that

$$\|\widehat{x}_n\| = \|x_n\|.$$

Therefore it follows that $(\widehat{x}_n) \subset \mathbf{W}^*$ is a norm-bounded set. By Helley's Theorem, there is a subsequence (x_{n_k}) of (x_n) and a linear functional $T \in \mathbf{W}^*$ such that

$$\widehat{x_{n_k}}(\varphi) \to T(\varphi)$$

for all $\varphi \in \mathbf{W}$. Since $\mathbf{F}$ is reflexive, there exists $x \in \mathbf{F}$ such that $T = \widehat{x}$. We conclude that

$$\lim_{k \to \infty} \varphi(x_{n_k}) = \varphi(x)$$

for all $\varphi \in \mathbf{W} = \mathbf{F}^*$. However, if $\psi \in \mathbf{E}^*$, then we may set $\varphi := \psi|_{\mathbf{F}} \in \mathbf{F}^*$ and conclude that

$$\lim_{k \to \infty} \psi(x_{n_k}) = \psi(x)$$

for all $\psi \in \mathbf{E}^*$. In other words, (x_{n_k}) converges weakly to x. Finally, $x \in B_\mathbf{E}$ because $\|x\| \leq \liminf \|x_{n_k}\| \leq 1$ by Lemma 6.1.4. Therefore $B_\mathbf{E}$ is weakly sequentially compact. $\qquad\square$

6.4.5 COROLLARY *If $1 < p < \infty$, then any norm-bounded sequence in $L^p[a, b]$ has a weakly convergent subsequence.*

6.4.6 EXAMPLE Some examples are now necessary to understand the hypotheses of these two theorems.

(i) Separability is necessary for Helley's Theorem: Let $\mathbf{E} = \ell^\infty$ and $\varphi_n \in \mathbf{E}^*$ be given by

$$\varphi_n((x_k)) = x_n.$$

Then $\|\varphi_n\| = 1$ for all $n \in \mathbb{N}$, so $(\varphi_n) \subset \mathbf{E}^*$ is bounded. Let e_n denote the nth standard basis vector, then

$$|\varphi_n(e_n) - \varphi_m(e_n)| = 1$$

if $n \neq m$. Therefore (φ_n) cannot have a convergent subsequence.

(ii) Reflexivity is necessary for Theorem 6.4.4: If $\mathbf{E} = \ell^1$, then by Schur's Theorem (Theorem 6.1.9), every weakly convergent sequence is strongly convergent. Since the unit ball is not sequentially compact in the norm, it follows that the unit ball cannot be weakly sequentially compact either!

(iii) For another example, consider $\mathbf{E} = L^1[0, 1]$ and let $f_n \in \mathbf{E}$ be given by

$$f_n = n\chi_{[0,1/n]}.$$

Then $\|f_n\|_1 = 1$ for all $n \in \mathbb{N}$. We claim that (f_n) does not have a weakly convergent subsequence. Suppose $f \in L^1[0, 1]$ is such that $f_{n_k} \overset{w}{\to} f$ for some subsequence (f_{n_k}) of (f_n), then we wish to arrive at a contradiction.

(a) Let $K \subset (0, 1)$ be a compact set and $g = \chi_K \in L^\infty[0, 1]$. Then there exists $N \in \mathbb{N}$ such that $[0, 1/N] \cap K = \emptyset$. Therefore $f_n g = 0$ for all $n \geq N$. Hence

$$\int_K f = \int_0^1 fg = \lim_{k \to \infty} \int_0^1 f_{n_k} g = 0.$$

(b) Let $E \subset (0,1)$ be a measurable set and let $\epsilon > 0$. Then by Lemma 4.1.7, there exists $\delta > 0$ such that for any measurable $A \subset [0,1]$,

$$m(A) < \delta \Rightarrow \int_A |f| < \epsilon.$$

By the regularity of the Lebesgue measure, there is a compact set $K \subset E$ such that $m(E \setminus K) < \delta$. Hence

$$\left| \int_E f \right| = \left| \int_E f - \int_K f \right| \le \int_{E \setminus K} |f| < \epsilon.$$

This is true for all $\epsilon > 0$ and therefore

$$\int_E f = 0$$

for any measurable set $E \subset (0,1)$ and hence $f = 0$ a.e. by Lemma 5.4.3.

(c) However, for all $n \in \mathbb{N}$,

$$\int_0^1 f_n(t)dt = 1.$$

Since $\chi_{[0,1]} \in L^\infty[0,1]$ defines a bounded linear functional on $L^1[0,1]$ and $f_{n_k} \xrightarrow{w} f$, it must happen that

$$\int_0^1 f(t)dt = 1.$$

This contradiction proves that no such subsequence can exist.

6.5   The Weak-∗ Topology

In the previous section, we proved that the unit ball of a reflexive Banach space is weakly sequentially compact. A closer examination of the proof shows that it is really a statement about the dual space. Indeed, Helley's Theorem states that a norm-bounded sequence in the dual space (of a separable normed linear space) has a subsequence that converges *pointwise*. We now study this notion of convergence, which will lead us to the topology lurking behind the scenes.

6.5.1 DEFINITION   Let $\mathbf{E}$ be a normed linear space and $(\varphi_n) \subset \mathbf{E}^*$ be a sequence of bounded linear functionals on $\mathbf{E}$. We say that (φ_n) is *weak-∗ convergent* to a point

$\varphi \in \mathbf{E}^*$ if

$$\lim_{n \to \infty} \varphi_n(x) = \varphi(x)$$

for each $x \in \mathbf{E}$. If this happens, we write $\varphi_n \xrightarrow{w^*} \varphi$.

Let us process this definition with a few examples and remarks before we proceed. Firstly, $\mathbf{E}^*$ is a normed linear space when equipped with the operator norm. Therefore it now carries *three* different notions of convergence. Given a sequence $(\varphi_n) \subset \mathbf{E}^*$, we may speak of

- $\varphi_n \xrightarrow{s} \varphi$ if $\lim_{n \to \infty} \|\varphi_n - \varphi\| = 0$,
- $\varphi_n \xrightarrow{w} \varphi$ if $T(\varphi_n) \to T(\varphi)$ for each $T \in \mathbf{E}^{**}$ and
- $\varphi_n \xrightarrow{w^*} \varphi$ if $\varphi_n(x) \to \varphi(x)$ for each $x \in \mathbf{E}$.

We know that norm convergence implies weak convergence. Now if $x \in \mathbf{E}$, then $\widehat{x} \in \mathbf{E}^{**}$. Therefore

$$\text{weak convergence} \Rightarrow \text{weak-}* \text{ convergence.}$$

Moreover, if $\mathbf{E}$ is reflexive, then

$$\text{weak convergence} \Leftrightarrow \text{weak-}* \text{ convergence.}$$

6.5.2 EXAMPLE

(i) As an example, consider $\mathbf{E} = L^p[a,b]$ with $1 < p < \infty$ and $(\varphi_n) \in \mathbf{E}^*$. Then write $\varphi_n = \varphi_{g_n}$ for some $g_n \in L^q[a,b]$ (here, q is the conjugate exponent of p). Then (φ_n) is weak-$*$ convergent if and only if there exists $g \in L^q[a,b]$ such that

$$\int_a^b f g_n \to \int_a^b f g$$

for all $f \in L^p[a,b]$. Since $(L^q[a,b])^* \cong L^p[a,b]$, this is equivalent to requiring that $g_n \xrightarrow{w} g$ in the weak topology of $L^q[a,b]$.

(ii) Let us see an example where these two notions are distinct. Let $\mathbf{E} = \ell^1$ and let $(\varphi_n) \subset \mathbf{E}^*$ be given by

$$\varphi_n((x_j)) := \sum_{i=1}^{n} x_i.$$

If we set $\varphi((x_j)) = \sum_{i=1}^{\infty} x_i$, then $\varphi_n(x) \to \varphi(x)$ for all $x \in \mathbf{E}$. Hence $\varphi_n \xrightarrow{w^*} \varphi$. Under the isomorphism $\Delta : \ell^{\infty} \to (\ell^1)^*$, we know that $\varphi_n = \Delta(x^n)$, where

$$x^n = (\underbrace{1,1,1,\ldots,1}_{n \text{ times}},0,0,\ldots).$$

However, we had seen in Example 6.1.8 that (x^n) does not converge weakly in ℓ^{∞}. Hence (φ_n) cannot converge weakly in $\mathbf{E}^*$ either.

Let us now define the topology on $\mathbf{E}^*$ which provides a home for weak-$*$ convergence.

6.5.3 DEFINITION Let $\mathbf{E}$ be an normed linear space. For each $x \in \mathbf{E}$, consider $\widehat{x} : \mathbf{E}^* \to \mathbb{K}$ given by

$$\widehat{x}(\varphi) := \varphi(x).$$

Let $\mathcal{G} = \{\widehat{x} : x \in \mathbf{E}\} \subset \mathbf{E}^{**}$ and $Y = \mathbb{K}$ equipped with the usual topology. The weak topology on $\mathbf{E}^*$ defined by $\mathcal{G}$ is called the *weak-$*$ topology* on $\mathbf{E}^*$. We denote this topology by $\sigma(\mathbf{E}^*, \mathbf{E})$.

6.5.4 REMARK Let $\mathbf{E}$ be a normed linear space.

(i) $\mathbf{E}^*$ now has three topologies on it.

(a) The norm topology $\sigma(\mathbf{E}^*, \|\cdot\|)$, which is a metric topology whose metric is given by

$$d(\varphi, \psi) = \sup\{|\varphi(x) - \psi(x)| : x \in \mathbf{E}, \|x\| \leq 1\}$$

(b) The weak topology $\sigma(\mathbf{E}^*, \mathbf{E}^{**})$, obtained by the action of $\mathbf{E}^{**}$ on $\mathbf{E}^*$. Here, a basic open set containing φ_0 is of the form

$$\{\varphi \in \mathbf{E}^* : |T_i(\varphi) - T_i(\varphi_0)| < \epsilon \text{ for all } 1 \leq i \leq n\}$$

for some finite set $\{T_1, T_2, \ldots, T_n\} \subset \mathbf{E}^{**}$ and $\epsilon > 0$.

(c) The weak-$*$ topology $\sigma(\mathbf{E}^*, \mathbf{E})$, obtained by the action of $\mathbf{E}$ on $\mathbf{E}^*$. As argued before, a basic open set containing φ_0 is of the form

$$\{\varphi \in \mathbf{E}^* : |\varphi(x_i) - \varphi_0(x_i)| < \epsilon \text{ for all } 1 \leq i \leq n\}$$

for some finite set $\{x_1, x_2, \ldots, x_n\} \subset \mathbf{E}$ and $\epsilon > 0$.

(ii) By definition, $\sigma(\mathbf{E}^*, \mathbf{E})$ is the smallest topology that makes each element of $\{\widehat{x} : x \in \mathbf{E}\}$ continuous. Therefore we have the following containments.

$$\sigma(\mathbf{E}^*, \mathbf{E}) \subset \sigma(\mathbf{E}^*, \mathbf{E}^{**}) \subset \sigma(\mathbf{E}^*, \|\cdot\|).$$

In other words,

$$\text{weak-}* \text{ open} \Rightarrow \text{weakly open} \Rightarrow \text{norm-open,}$$
$$\text{weak-}* \text{ closed} \Rightarrow \text{weakly closed} \Rightarrow \text{norm-closed,}$$
$$\text{weak-}* \text{ compact} \Leftarrow \text{weakly compact} \Leftarrow \text{norm-compact.}$$

(iii) Furthermore, if $\mathbf{E}$ is reflexive, then $\sigma(\mathbf{E}^*, \mathbf{E}^{**}) = \sigma(\mathbf{E}^*, \mathbf{E})$.

The proof of the next result is similar to Exercise 6.15 and we will omit it.

6.5.5 PROPOSITION *For a sequence $(\varphi_n) \subset \mathbf{E}^*$, $\varphi_n \xrightarrow{w^*} \varphi$ if and only if $\varphi_n \to \varphi$ with respect to $\sigma(\mathbf{E}^*, \mathbf{E})$.*

In Proposition 6.3.8, we had proved that $(\mathbf{E}, \sigma(\mathbf{E}, \mathbf{E}^*))$ is a topological vector space. That the analogous statement is true for the weak-$*$ topology should come as no surprise.

6.5.6 PROPOSITION *The space $(\mathbf{E}^*, \sigma(\mathbf{E}^*, \mathbf{E}))$ is a topological vector space.*

Proof. We need to show that $\sigma(\mathbf{E}^*, \mathbf{E})$ is Hausdorff. To that end, fix $\varphi, \psi \in \mathbf{E}^*$ such that $\varphi \neq \psi$. Then there exists $x \in \mathbf{E}$ such that $\varphi(x) \neq \psi(x)$. If we set $\epsilon := \frac{|\varphi(x) - \psi(x)|}{3} > 0$, then the sets

$$U = \{\eta \in \mathbf{E}^* : |\eta(x) - \varphi(x)| < \epsilon\} \text{ and } V = \{\eta \in \mathbf{E}^* : |\eta(x) - \psi(x)| < \epsilon\}$$

are both weak-$*$ open sets. Furthermore, they are disjoint, U contains φ, V contains ψ and $U \cap V = \varnothing$. This proves that $\sigma(\mathbf{E}^*, \mathbf{E})$ is Hausdorff.

We now need to show that the maps $a : \mathbf{E}^* \times \mathbf{E}^* \to \mathbf{E}^*$ and $s : \mathbb{K} \times \mathbf{E}^* \to \mathbf{E}^*$ are both continuous with respect to $\sigma(\mathbf{E}^*, \mathbf{E})$. The proofs are similar to that of Proposition 6.3.8 and we encourage you to try it (Exercise 6.34). □

Let us now compare the three topologies on $\mathbf{E}^*$ and understand their relationship.

6.5.7 THEOREM *If* $\mathbf{E}$ *is finite dimensional, then* $\sigma(\mathbf{E}^*, \mathbf{E}) = \sigma(\mathbf{E}^*, \mathbf{E}^{**}) = \sigma(\mathbf{E}^*, \|\cdot\|)$.

Proof. Since $\mathbf{E}$ is finite dimensional, so is $\mathbf{E}^*$. Therefore $\sigma(\mathbf{E}^*, \mathbf{E}^{**}) = \sigma(\mathbf{E}^*, \|\cdot\|)$ by Theorem 6.3.9. However, $\mathbf{E}$ is also reflexive (Example 4.4.6), which implies that $\sigma(\mathbf{E}^*, \mathbf{E}) = \sigma(\mathbf{E}^*, \mathbf{E}^{**})$. $\qquad\square$

6.5.8 PROPOSITION *If* $\mathbf{E}$ *is infinite dimensional, then* $\sigma(\mathbf{E}^*, \mathbf{E})$ *is not metrizable. In particular,* $\sigma(\mathbf{E}^*, \mathbf{E}) \neq \sigma(\mathbf{E}^*, \|\cdot\|)$.

Proof. Suppose $\sigma(\mathbf{E}^*, \mathbf{E})$ were induced by a metric $d : \mathbf{E}^* \times \mathbf{E}^* \to [0, \infty)$. Then for each $n \in \mathbb{N}$, consider the set $V_n = \{\varphi \in \mathbf{E}^* : d(\varphi, 0) < \frac{1}{n}\}$, where 0 denotes the zero linear functional in $\mathbf{E}^*$. Since this set is weak-$*$ open, there exist vectors $x_1, x_2, \ldots, x_n \in \mathbf{E}$ and $\epsilon > 0$ such that

$$U_n := \{\varphi \in \mathbf{E}^* : |\varphi(x_i)| < \epsilon \text{ for all } 1 \le i \le n\} \subset V_n.$$

Let $\mathbf{F}_n := \operatorname{span}\{x_1, x_2, \ldots, x_n\}$, then $\mathbf{F}_n$ is a finite dimensional subspace of $\mathbf{E}$. In particular, $\mathbf{F}_n \neq \mathbf{E}$ and is closed. Hence by Proposition 4.3.3, there is a non-zero linear functional $\psi_n \in \mathbf{E}^*$ such that

$$\psi_n|_{\mathbf{F}_n} = 0.$$

Define $\varphi_n := n \frac{\psi_n}{\|\psi_n\|}$, so that $\varphi_n \in U_n$ since $\varphi_n(x_i) = 0$ for all $1 \le i \le n$. However, $U_n \subset V_n$, which implies that $d(\varphi_n, 0) < \frac{1}{n}$. Therefore $\varphi_n \xrightarrow{w^*} 0$. But $\|\varphi_n\| = n$, for each $n \in \mathbb{N}$, which contradicts the Banach–Steinhaus Theorem (Theorem 5.2.2). This proves that no such metric can exist. $\qquad\square$

Now we wish to compare the weak and weak-$*$ topologies on $\mathbf{E}^*$ and we begin with a small lemma from linear algebra that you may already be familiar with.

6.5.9 LEMMA *Let* $\mathbf{E}$ *be a vector space and* $\{\varphi, \varphi_1, \varphi_2, \ldots, \varphi_n\}$ *be a finite collection of linear functionals on* $\mathbf{E}$. *Suppose that*

$$\bigcap_{i=1}^{n} \ker(\varphi_i) \subset \ker(\varphi),$$

then there exist scalars $\alpha_1, \alpha_2, \ldots, \alpha_n$ *such that* $\varphi = \sum_{i=1}^{n} \alpha_i \varphi_i$.

Proof. Let $\mathbf{F} := \mathbb{K}^{n+1}$, equipped with the Euclidean inner product. Define $T : \mathbf{E} \to \mathbf{F}$ be the linear map given by

$$T(x) := (\varphi(x), \varphi_1(x), \varphi_2(x), \ldots \varphi_n(x)).$$

By hypothesis, the vector $e_1 = (1, 0, 0, \ldots, 0)$ does not belong to $\mathrm{Range}(T)$. Now $\mathrm{Range}(T)$ is a closed subspace of $\mathbf{F}$, so there is a bounded linear functional $\psi \in \mathbf{F}^*$ such that $\psi(e_1) = 1$ and

$$\psi(T(x)) = 0$$

for all $x \in \mathbf{E}$ (by Proposition 4.3.3). By the Riesz Representation Theorem, there is a tuple $z = (\alpha_0, \alpha_1, \ldots, \alpha_n) \in \mathbf{F}$ such that $\psi(y) = \langle y, z \rangle$ for all $y \in \mathbf{F}$. In particular,

$$\alpha_0 = \psi(e_1) = 1.$$

Now writing out the equation $\psi(T(x)) = 0$, we see that

$$\varphi(x) = \sum_{i=1}^{n} \alpha_i \varphi_i(x)$$

for all $x \in \mathbf{E}$. $\qquad\square$

6.5.10 PROPOSITION *If* $T : \mathbf{E}^* \to \mathbb{K}$ *is a linear functional, then the following are equivalent:*

(i) *T is continuous with respect to $\sigma(\mathbf{E}^*, \mathbf{E})$.*

(ii) *The set $\{\varphi \in \mathbf{E}^* : |T(\varphi)| < 1\}$ is weak-$*$ open.*

(iii) *There exists $x \in \mathbf{E}$ such that $T = \hat{x}$.*

Proof. Both the implications $(i) \Rightarrow (ii)$ and $(iii) \Rightarrow (i)$ are true by definition. Therefore it suffices to prove that $(ii) \Rightarrow (iii)$.

So, suppose that the set $U := \{\varphi \in \mathbf{E}^* : |T(\varphi)| < 1\}$ is weak-$*$ open. Since it contains $0 \in \mathbf{E}^*$, it must contain a weak-$*$ open set of the form

$$V = \{\psi \in \mathbf{E}^* : |\psi(x_i)| < \epsilon \text{ for all } 1 \le i \le n\}$$

for some finite set $\{x_1, x_2, \ldots, x_n\} \subset \mathbf{E}$ and some $\epsilon > 0$. Now consider the linear functionals $\hat{x}_i : \mathbf{E}^* \to \mathbb{K}, (1 \le i \le n)$. With a view to applying Lemma 6.5.9, we wish to prove that

$$\bigcap_{i=1}^{n} \ker(\widehat{x}_i) \subset \ker(T). \tag{6.2}$$

So fix $\psi \in \bigcap_{i=1}^{n} \ker(\widehat{x}_i)$, then for any $k \in \mathbb{N}$,

$$k\psi(x_i) = k\widehat{x}_i(\psi) = 0$$

for all $1 \le i \le n$. This implies that $k\psi \in V \subset U$ and thus $|T(k\psi)| < 1$. Hence

$$|T(\psi)| < \frac{1}{k}.$$

This is true for all $k \in \mathbb{N}$ and therefore $T(\psi) = 0$. Thus we have verified Equation 6.2. By Lemma 6.5.9, there are scalars $\alpha_1, \alpha_2, \dots, \alpha_n \in \mathbb{K}$ such that

$$T = \sum_{i=1}^{n} \alpha_i \widehat{x}_i.$$

Therefore if $x = \sum_{i=1}^{n} \alpha_i x_i$, then $T = \widehat{x}$. $\square$

We are now in a position to complete our comparison of the three topologies.

6.5.11 Theorem *A normed linear space* $\mathbf{E}$ *is reflexive if and only if* $\sigma(\mathbf{E}^*, \mathbf{E}) = \sigma(\mathbf{E}^*, \mathbf{E}^{**})$.

Proof. If $\mathbf{E}$ is reflexive, then the two topologies coincide by definition. Therefore, we assume that $\sigma(\mathbf{E}^*, \mathbf{E}) = \sigma(\mathbf{E}^*, \mathbf{E}^{**})$ and we prove that $\mathbf{E}$ is reflexive. To that end, choose $T \in \mathbf{E}^{**}$ and observe that

$$U := \{\varphi \in \mathbf{E}^* : |T(\varphi)| < 1\}$$

is an element of $\sigma(\mathbf{E}^*, \mathbf{E}^{**})$, since it is the inverse image under T of an open set in $\mathbb{K}$. By hypothesis, this means that $U \in \sigma(\mathbf{E}^*, \mathbf{E})$. By Proposition 6.5.10, $T = \widehat{x}$ for some $x \in \mathbf{E}$. This is true for any $T \in \mathbf{E}^{**}$ and thus $\mathbf{E}$ is reflexive. $\square$

6.6   Weak-∗ Compactness

At long last, we arrive at the conclusion of this saga. The main result of this section (the Banach–Alaoglu Theorem) is a powerful compactness result and the most important reason to discuss the weak-∗ topology. After proving this result, we will also take a look at reflexive spaces to see if we can gain some new insight into Helley's Theorem.

Before we discuss the theorem, we need one important result from point–set topology that you might be familiar with (see Munkres [42]).

Let $\{X_\alpha : \alpha \in J\}$ *be a family of compact topological spaces. Then the product space*

$$X = \prod_{\alpha \in J} X_\alpha$$

is compact when equipped with the product topology.

Now if $\mathbf{E}$ be an normed linear space, then for any $x \in \mathbf{E}$, we have a natural map $\widehat{x} : \mathbf{E}^* \to \mathbb{K}$ given by $\widehat{x}(\varphi) = \varphi(x)$. Define

$$A := \prod_{x \in \mathbf{E}} \mathbb{K}$$

and equip it with the product topology (denoted by τ). For each $x \in \mathbf{E}$, we write $\pi_x : A \to \mathbb{K}$ for the projection onto that corresponding coordinate. If $\alpha \in \mathbb{K}$ and $\epsilon > 0$, we write $B_{\mathbb{K}}(\alpha, \epsilon) = \{z \in \mathbb{K} : |z - \alpha| < \epsilon\}$. Then any basic open set in τ is of the form

$$\bigcap_{i=1}^{n} \pi_{x_i}^{-1}(B_{\mathbb{K}}(\alpha_i, \epsilon_i)),$$

for some $\alpha_i \in \mathbb{K}$ and $\epsilon_i > 0$ $(1 \leq i \leq n)$. Now we get a map $\Theta : \mathbf{E}^* \to A$ given by

$$\Theta(\varphi) = (\varphi(x))_{x \in \mathbf{E}}.$$

Notice that we do not care about the linear structure (or lack thereof) of A. Therefore we need only think of Θ as a function.

6.6.2 LEMMA *The map*

$$\Theta : (\mathbf{E}^*, \sigma(\mathbf{E}^*, \mathbf{E})) \to (A, \tau)$$

induces a homeomorphism $\mathbf{E}^* \to \Theta(\mathbf{E}^*)$.

Proof. It is clear that Θ is injective and therefore induces a bijection from $\mathbf{E}^*$ to $\Theta(\mathbf{E}^*)$. Thus it remains to prove that Θ is both continuous and an open map.

(i) For $x \in \mathbf{E}$, observe that

$$\pi_x \circ \Theta = \widehat{x}$$

is weak-$*$ continuous by definition. However, the product topology on A is the weak topology generated by the functions $\{\pi_x : x \in \mathbf{E}\}$. Therefore Θ is continuous by Proposition 6.3.3.

(ii) Let $V \in \sigma(\mathbf{E}^*, \mathbf{E})$ be a basic open set. We wish to show that $\Theta(V)$ is open in $\Theta(\mathbf{E}^*)$. By definition, there exists $\varphi \in \mathbf{E}^*$, finitely many vectors $x_1, x_2, \ldots, x_n \in \mathbf{E}$ and $\epsilon > 0$ such that

$$V = \{\psi \in \mathbf{E}^* : |\psi(x_i) - \varphi(x_i)| < \epsilon \text{ for all } 1 \leq i \leq n\}.$$

If $V_i := B_{\mathbb{K}}(\varphi(x_i), \epsilon)$, then $\psi \in V$ if and only if $\psi(x_i) \in V_i$ for all $1 \leq i \leq n$. Hence

$$\Theta(V) = \left[\bigcap_{i=1}^{n} \pi_{x_i}^{-1}(V_i) \right] \cap \Theta(\mathbf{E}^*),$$

which is open in $\Theta(\mathbf{E}^*)$. This proves that Θ is an open map. $\qquad\square$

The real insight that leads to the Banach–Alaoglu Theorem is to restrict both the domain and the codomain of this map Θ in such a way that Θ becomes a homeomorphism between two compact sets. To do this, we first note that if $x \in \mathbf{E}$, then the ball

$$B_{\mathbb{K}}[0, \|x\|] = \{z \in \mathbb{K} : |z| \leq \|x\|\}$$

is a compact subset of $\mathbb{K}$. Therefore if we define

$$B := \prod_{x \in \mathbf{E}} B_{\mathbb{K}}[0, \|x\|],$$

then B is a subset of A and it is compact by Tychonoff's Theorem (we denote the product topology on B by τ_B). Now if $\varphi \in \mathbf{E}^*$ is a linear functional such that $\|\varphi\| \leq 1$, then for any $x \in \mathbf{E}$,

$$|\varphi(x)| \leq \|x\|,$$

so that $\varphi(x) \in B_{\mathbb{K}}[0, \|x\|]$. Write $B_{\mathbf{E}^*}$ for the closed unit ball in $\mathbf{E}^*$. Then the restriction of Θ gives us a well-defined map $\Theta : B_{\mathbf{E}^*} \to B$ given by

$$\Theta(\varphi) = (\varphi(x))_{x \in \mathbf{E}}.$$

The point of the Banach–Alaoglu Theorem then is to use Θ to pass compactness from B to $B_{\mathbf{E}^*}$.

6.6.3 THEOREM (BANACH–ALAOGLU THEOREM) *If* **E** *is a normed linear space, then* $B_{\mathbf{E}^*}$ *is compact in the weak-$*$ topology.*

Proof. By Lemma 6.6.2, Θ induces a homeomorphism

$$B_{\mathbf{E}^*} \to \Theta(B_{\mathbf{E}^*}).$$

Thus it suffices to show that $\Theta(B_{\mathbf{E}^*})$ is compact in (A, τ). Since $\Theta(B_{\mathbf{E}^*}) \subset B$, it suffices to show that $\Theta(B_{\mathbf{E}^*})$ is closed in B. To that end, fix $z := (\alpha_x)_{x \in \mathbf{E}} \in \overline{\Theta(B_{\mathbf{E}^*})}$ and define $\psi : \mathbf{E} \to \mathbb{K}$ by

$$\psi(x) := \alpha_x.$$

(i) We wish to prove that ψ is linear. We show that

$$\psi(x + y) = \psi(x) + \psi(y)$$

for all $x, y \in \mathbf{E}$ (scalar multiplication is similarly checked). Fix $x, y \in \mathbf{E}$ and $\epsilon > 0$ and consider the set

$$U := \pi_x^{-1}(B_{\mathbb{K}}(\alpha_x, \epsilon)) \cap \pi_y^{-1}(B_{\mathbb{K}}(\alpha_y, \epsilon)) \cap \pi_{x+y}^{-1}(B_{\mathbb{K}}(\alpha_{x+y}, \epsilon)).$$

Then U is open in the product topology and $z \in U$ by definition. Since $z \in \overline{\Theta(B_{\mathbf{E}^*})}$,

$$U \cap \Theta(B_{\mathbf{E}^*}) \neq \emptyset.$$

Therefore we may choose $\varphi \in B_{\mathbf{E}^*}$ such that $\Theta(\varphi) \in U$. This implies that

$$|\varphi(x) - \alpha_x| < \epsilon, |\varphi(y) - \alpha_y| < \epsilon \text{ and } |\varphi(x+y) - \alpha_{x+y}| < \epsilon.$$

Since φ is linear, it follows that

$$|\psi(x+y) - \psi(x) - \psi(y)| = |\alpha_{x+y} - \alpha_x - \alpha_y| < 3\epsilon.$$

This is true for all $\epsilon > 0$ and hence $\psi(x+y) = \psi(x) + \psi(y)$ holds.

(ii) Since $\psi(x) = \alpha_x \in B_{\mathbb{K}}[0, \|x\|]$ for all $x \in \mathbf{E}$, it follows that $\psi \in B_{\mathbf{E}^*}$.

Thus $z = (\psi(x))_{x \in \mathbf{E}} \in \Theta(B_{\mathbf{E}^*})$ and $\Theta(B_{\mathbf{E}^*})$ is closed. Hence $B_{\mathbf{E}^*}$ is compact. $\square$

We now apply the Banach–Alaoglu Theorem to reflexive spaces, with a view to revisit Helley's Theorem.

6.6.4 LEMMA *If* **E** *is a reflexive normed linear space, then so is* **E***.

Proof. Let $J : \mathbf{E} \to \mathbf{E}^{**}$ and $\widetilde{J} : \mathbf{E}^* \to \mathbf{E}^{***}$ be the natural inclusions (see Definition 4.4.5). Assuming J is surjective, we wish to prove that $\widetilde{J}$ is surjective. To that end, consider $T \in (\mathbf{E}^*)^{**} = (\mathbf{E}^{**})^*$. Then

$$\varphi := T \circ J$$

is a bounded linear functional on **E**. For any vector $x \in \mathbf{E}$, we have

$$T(J(x)) = \varphi(x) = J(x)(\varphi) = \widetilde{J}(\varphi)(J(x)).$$

Since the collection $\{J(x) : x \in \mathbf{E}\}$ exhausts all of $\mathbf{E}^{**}$, it follows that $T = \widetilde{J}(\varphi)$. Thus $\widetilde{J}$ is surjective. $\qquad\square$

6.6.5 THEOREM *If* **E** *is either reflexive or separable, then* $B_{\mathbf{E}^*}$ *is weak-$*$ sequentially compact.*

Proof. If **E** is separable, then this follows directly from Helley's Theorem (Theorem 6.4.3).

If **E** is reflexive, then so is $\mathbf{E}^*$ by Lemma 6.6.4. Therefore $B_{\mathbf{E}^*}$ is sequentially compact in the weak topology $\sigma(\mathbf{E}^*, \mathbf{E}^{**})$ by Theorem 6.4.4. Since **E** is reflexive,

$$\sigma(\mathbf{E}^*, \mathbf{E}^{**}) = \sigma(\mathbf{E}^*, \mathbf{E}).$$

Thus $B_{\mathbf{E}^*}$ is weak-$*$ sequentially compact. $\qquad\square$

6.6.6 LEMMA *Let* **E** *be a normed linear space and* $J : \mathbf{E} \to \mathbf{E}^{**}$ *be the natural inclusion.*

(i) *J induces a continuous map*

$$J : (\mathbf{E}, \sigma(\mathbf{E}, \mathbf{E}^*)) \to (\mathbf{E}^{**}, \sigma(\mathbf{E}^{**}, \mathbf{E}^*))$$

between the weak topology on **E** *and the weak-$*$ topology on* **E****.

(ii) *If* **E** *is reflexive, then this map is a homeomorphism.*

Proof.

(i) Let $\widetilde{J} : \mathbf{E}^* \to (\mathbf{E}^*)^{**}$ be the natural inclusion of $\mathbf{E}^*$ into its double dual. For $\varphi \in \mathbf{E}^*$, $\widetilde{J}(\varphi)$ may be thought of as a bounded linear functional $\widetilde{J}(\varphi) : \mathbf{E}^{**} \to \mathbb{K}$ (as in Lemma 6.6.4). Therefore

$$\widetilde{J}(\varphi) \circ J : \mathbf{E} \to \mathbb{K}.$$

is a bounded linear functional on $\mathbf{E}$. Thus $\widetilde{J}(\varphi) \circ J$ is continuous with with respect to $\sigma(\mathbf{E}, \mathbf{E}^*)$. This is true for every $\varphi \in \mathbf{E}^*$. However, the weak-$*$ topology $\sigma(\mathbf{E}^{**}, \mathbf{E}^*)$ is the weak topology generated by the family $\{\widetilde{J}(\varphi) : \varphi \in \mathbf{E}^*\}$. Thus J must be continuous by Proposition 6.3.3.

(ii) Now assume $\mathbf{E}$ is reflexive, so that J is a bijection. Then it suffices to prove that J is an open map. If $V \in \sigma(\mathbf{E}, \mathbf{E}^*)$ is a basic open set, we wish to prove that $J(V) \in \sigma(\mathbf{E}^{**}, \mathbf{E}^*)$. By definition, there exists $x_0 \in \mathbf{E}$, a finite collection $\varphi_1, \varphi_2, \ldots, \varphi_n \in \mathbf{E}^*$ and $\epsilon > 0$ such that

$$V = \{x \in \mathbf{E} : |\varphi_i(x) - \varphi_i(x_0)| < \epsilon \text{ for all } 1 \leq i \leq n\}.$$

Then

$$J(V) = \{\widehat{x} \in \mathbf{E}^{**} : |\widehat{x}(\varphi_i) - \widehat{x_0}(\varphi_i)| < \epsilon \text{ for all } 1 \leq i \leq n\}.$$

Write $T_0 = \widehat{x_0}$ and use the fact that $\mathbf{E}$ is reflexive to conclude that

$$J(V) = \{T \in \mathbf{E}^{**} : |T(\varphi_i) - T_0(\varphi_i)| < \epsilon \text{ for all } 1 \leq i \leq n\}.$$

This is precisely the form of a basic open set in $\sigma(\mathbf{E}^{**}, \mathbf{E}^*)$. Thus J is an open map.

$\square$

The next corollary is best understood in conjunction with Theorem 6.4.4, where we proved that $B_\mathbf{E}$ is weakly sequentially compact provided $\mathbf{E}$ is reflexive.

6.6.7 COROLLARY *If $\mathbf{E}$ is reflexive, then $B_\mathbf{E}$ is weakly compact.*

Proof. Let $J : \mathbf{E} \to \mathbf{E}^{**}$ be the natural isomorphism. By Lemma 6.6.6, J induces a homeomorphism

$$J : (\mathbf{E}, \sigma(\mathbf{E}, \mathbf{E}^*)) \to (\mathbf{E}^{**}, \sigma(\mathbf{E}^{**}, \mathbf{E}^*)).$$

Now $B_{\mathbf{E}^{**}}$ is compact with respect to $\sigma(\mathbf{E}^{**}, \mathbf{E}^*)$ by the Banach–Alaoglu Theorem (applied to $\mathbf{E}^*$). However, since $\mathbf{E}$ is reflexive,

$$B_{\mathbf{E}^{**}} = \{T \in (\mathbf{E}^*)^* : \|T\| \leq 1\} = \{J(x) \in \mathbf{E}^{**} : \|x\| \leq 1\} = J(B_\mathbf{E}).$$

Thus $B_\mathbf{E}$ is compact with respect to $\sigma(\mathbf{E}, \mathbf{E}^*)$.

$\square$

The next example shows that reflexivity is necessary for Corollary 6.6.7 to work.

6.6.8 EXAMPLE   Let $\mathbf{E} = c_0$. Let

$$U_j := \{(x_n) \in B_{\mathbf{E}} : |x_j| < 1\},$$

which is weakly open in $B_{\mathbf{E}}$ (with the subspace topology). We will leave it as an exercise to prove that $\{U_j : j \in \mathbb{N}\}$ is an open cover for $B_{\mathbf{E}}$ which does not have a finite subcover (Exercise 6.38). Therefore $B_{\mathbf{E}}$ is not weakly compact.

If $\mathbf{E}$ is an infinite dimensional normed linear space, we had seen that $\sigma(\mathbf{E}^*, \mathbf{E})$ is not metrizable. Therefore it may seem that Helley's Theorem and the Banach–Alaoglu Theorem are unrelated. However, it is a fact that Helley's Theorem may be obtained as a consequence of the Banach–Alaoglu Theorem and the reason is the next lemma.

6.6.9 LEMMA   *Let $\mathbf{E}$ be a separable normed linear space. Then $B_{\mathbf{E}^*}$ is metrizable in the weak-$*$ topology.*

Proof.   Since $\mathbf{E}$ is separable, so is $B_{\mathbf{E}} := \{x \in \mathbf{E} : \|x\| \leq 1\}$. Let $\{x_n\}$ be a countable dense subset of $B_{\mathbf{E}}$. Define $d : B_{\mathbf{E}^*} \times B_{\mathbf{E}^*} \to [0, \infty)$ by

$$d(\varphi, \psi) := \sum_{n=1}^{\infty} \frac{|\varphi(x_n) - \psi(x_n)|}{2^n}.$$

For any $\varphi, \psi \in B_{\mathbf{E}^*}$, $|\varphi(x_n) - \psi(x_n)| \leq 2$ for any $n \in \mathbb{N}$. Therefore the series converges and the function d is well-defined. That d is a metric is easy to see (Exercise 6.39). We wish to prove that the metric topology coincides with the weak-$*$ topology on $B_{\mathbf{E}^*}$.

(i) Let $\varphi_0 \in B_{\mathbf{E}^*}$ be a fixed point and $V \in \sigma(\mathbf{E}^*, \mathbf{E})$ be a weak-$*$ open set containing φ_0. We wish to find $r > 0$ such that

$$U_d(\varphi_0, r) := \{\psi \in B_{\mathbf{E}^*} : d(\psi, \varphi_0) < r\} \subset V.$$

We may assume that V is a basic open set of the form

$$V = \{\psi \in B_{\mathbf{E}^*} : |\psi(y_i) - \varphi_0(y_i)| < \epsilon \text{ for all } 1 \leq i \leq n\}$$

for some finite set of vectors $\{y_1, y_2, \ldots, y_n\} \subset \mathbf{E}$ and some $\epsilon > 0$. By scaling the vectors, we may also assume that $\|y_i\| \leq 1$ for all $1 \leq i \leq n$. For each

$1 \leq i \leq n$, choose $n_i \in \mathbb{N}$ such that $\|y_i - x_{n_i}\| < \frac{\epsilon}{4}$ and choose $r > 0$ so that

$$2^{n_i} r < \frac{\epsilon}{2}$$

for all $1 \leq i \leq n$. Then if $\psi \in U_d(\varphi_0, r)$ and $1 \leq i \leq n$,

$$
\begin{aligned}
|\psi(y_i) - \varphi_0(y_i)| &\leq |\psi(y_i) - \psi(x_{n_i})| + |\psi(x_{n_i}) - \varphi_0(x_{n_i})| + |\varphi_0(x_{n_i}) - \varphi_0(y_i)| \\
&\leq \|\psi\| \|y_i - x_{n_i}\| + 2^{n_i} d(\psi, \varphi_0) + \|\varphi_0\| \|y_i - x_{n_i}\| \\
&\leq \|y_i - x_{n_i}\| + \frac{\epsilon}{2} + \|y_i - x_{n_i}\| \\
&< \epsilon.
\end{aligned}
$$

Thus $\psi \in V$. This is true for any $\psi \in U_d(\varphi_0, r)$ and thus $U_d(\varphi_0, r) \subset V$.

(ii) Now suppose $\varphi_0 \in B_{\mathbf{E}^*}$ and $r > 0$, then we wish to find a weak-$*$ open set U containing φ_0 such that

$$U \subset U_d(\varphi_0, r) = \{\psi \in B_{\mathbf{E}^*} : d(\psi, \varphi_0) < r\}.$$

To that end, choose $\epsilon := \frac{r}{2}$ and choose $N \in \mathbb{N}$ such that

$$\frac{1}{2^{N-1}} < \frac{r}{2}.$$

Then define

$$V := \{\psi \in B_{\mathbf{E}^*} : |\psi(x_i) - \varphi_0(x_i)| < \epsilon \text{ for all } 1 \leq i \leq N\}.$$

Then V is weak-$*$ open and V contains φ_0. Furthermore, if $\psi \in V$, then

$$
\begin{aligned}
d(\psi, \varphi_0) &= \sum_{i=1}^{N} \frac{|\psi(x_i) - \varphi_0(x_i)|}{2^i} + \sum_{j=N+1}^{\infty} \frac{|\psi(x_j) - \varphi_0(x_j)|}{2^j} \\
&\leq \epsilon \left(\sum_{i=1}^{N} \frac{1}{2^i} \right) + \sum_{j=N+1}^{\infty} \frac{\|\psi - \varphi_0\| \|x_j\|}{2^j} \\
&< \epsilon + 2 \left(\sum_{j=N+1}^{\infty} \frac{1}{2^j} \right) \\
&= \epsilon + \frac{1}{2^{N-1}} \\
&< r.
\end{aligned}
$$

This proves that $V \subset U_d(\varphi_0, r)$.

Thus the two topologies (when restricted to $B_{\mathbf{E}^*}$) coincide. □

It is now immediate that Helley's Theorem (Theorem 6.4.3) follows from the Banach–Alaoglu Theorem.

6.6.10 COROLLARY (HELLEY'S THEOREM) *If* $\mathbf{E}$ *is a separable normed linear space, then* $B_{\mathbf{E}^*}$ *is weak-∗ sequentially compact.*

In Theorem 6.4.4, we had proved that the unit ball $B_{\mathbf{E}}$ of a normed linear space $\mathbf{E}$ is weakly sequentially compact, provided $\mathbf{E}$ is reflexive. Can we deduce this from the Banach–Alaoglu Theorem? Since the weak topology is not metrizable, the answer should be 'No'. However, we have

6.6.11 THEOREM (ŠMULIAN, 1940, AND EBERLEIN, 1947) *A subset of a Banach space is weakly compact if and only if it is weakly sequentially compact.*

We do not prove this result here. The interested reader may look up Diestel [12] for a proof of this result (in fact, Diestel gives two!). We end this section with a summary of all the compactness results we have proved. For convenience, we label these properties below:

- (W) : $B_{\mathbf{E}}$ is weakly compact (with respect to $\sigma(\mathbf{E}, \mathbf{E}^*)$).

- (WS) : $B_{\mathbf{E}}$ is weakly sequentially compact.

- (W*) : $B_{\mathbf{E}^*}$ is weak-∗ compact (with respect to $\sigma(\mathbf{E}^*, \mathbf{E})$).

- (W*S) : $B_{\mathbf{E}^*}$ is weak-∗ sequentially compact.

Then we have the following relations:

	(W)	(WS)	(W*)	(W*S)
Any normed linear space	N (6.6.8)	N (6.4.6)	Y (6.6.3)	N (6.4.6)
Separable space	N (6.6.8)	N (6.4.6)	Y (6.6.3)	Y (6.4.3)
Reflexive space	Y (6.6.7)	Y (6.4.4)	Y (6.6.3)	Y (6.6.5)

The **Banach–Alaoglu Theorem** has an interesting history. The first formulation of a weak-$*$ compactness theorem was by Helley in 1912, who proved that $B_{\mathbf{E}^*}$ is weak-$*$ sequentially compact if $\mathbf{E} = C[a, b]$. This was based on earlier work of Hilbert (in ℓ^2) and F. Riesz (in ℓ^p for $1 < p < \infty$) and used a diagonal argument similar to the one described in Theorem 6.4.3. The theorem was extended to arbitrary separable spaces by Banach in his book of 1932 [3].

The proof of the weak-$*$ compactness theorem in full generality (Theorem 6.6.3) would have to wait till 1938, when it was proved almost simultaneously by Alaoglu and the Bourbaki group. However, it is likely that this version of the theorem was 'in the air' by then. In fact, Pietsch [47] states that as many as twelve people can legitimately claim priority to it!

6.7   Locally Convex Spaces

We have seen in Section 6.2 that the Hahn–Banach Separation Theorem is deeply related to the notion of convexity. We now revisit the theorem to have a closer look at what makes it tick. In the process, we will prove the theorem in a broader context; one that encompasses the weak and weak-$*$ topologies as well. As you might imagine, the notion of a topological vector space from Definition 6.3.7 now takes center stage. We begin by revisiting the Minkoski functional in this context.

6.7.1 THEOREM *Let C be a non-empty, convex, open subset of a topological vector space* $\mathbf{E}$ *such that* $0 \in C$. *For* $x \in \mathbf{E}$, *define*

$$p(x) := \inf\{t > 0 : t^{-1}x \in C\}.$$

Then p is a well-defined, continuous sublinear functional and $C = \{x \in \mathbf{E} : p(x) < 1\}$.

As before, this function p is called the Minkowski functional of C.

Proof.

(i) Fix $x \in \mathbf{E}$. The sequence (n^{-1}) converges to 0 in $\mathbb{K}$. Since scalar multiplication is continuous, $n^{-1}x \to 0$ in $\mathbf{E}$. Since C is open, there exists $N \in \mathbb{N}$ such that $N^{-1}x \in C$. Thus the set over which we are taking an infimum is non-empty and $p(x)$ is well-defined.

(ii) The proof that p is sublinear is identical to Theorem 6.2.2.

(iii) The proof that $C = \{x \in \mathbf{E} : p(x) < 1\}$ is also similar to that of Theorem 6.2.2, although part (i) is needed as well.

(iv) Finally, we wish to prove that p is continuous. So let $U \subset \mathbb{R}_+$ be open and $x \in p^{-1}(U)$. Choose $\epsilon > 0$ such that $V = \{z \in \mathbb{R}_+ : |p(x) - z| < \epsilon\} \subset U$. Now set

$$W := x + \epsilon C.$$

Since both addition and scalar multiplication are continuous, W is an open set. Finally, if $y \in W$, then $\epsilon^{-1}(y - x) \in C$. By part (iii) and sublinearity, it follows that $p(y - x) < \epsilon$. Since p is sublinear, we may conclude that

$$|p(y) - p(x)| \leq p(y - x) < \epsilon.$$

Therefore, $W \subset p^{-1}(U)$. Hence x is an interior point of $p^{-1}(U)$ and $p^{-1}(U)$ is open, proving that p is continuous. $\qquad\square$

Of course, we wish to use this theorem to construct linear functionals. If $\mathbf{E}$ is a topological vector space, we once again write $\mathbf{E}^*$ for the set of all *continuous* linear functionals on $\mathbf{E}$, where $\mathbb{K}$ is equipped with the usual topology.

6.7.2 PROPOSITION *Let $\mathbf{E}$ be a topological vector space over $\mathbb{R}$ and let $C \subset \mathbf{E}$ be a non-empty convex, open set. If $x_0 \notin C$, then there exists $\psi \in \mathbf{E}^*$ such that $\psi(x) < \psi(x_0)$ for all $x \in C$.*

Proof.   As in Proposition 6.2.3, we may assume without loss of generality that $0 \in C$. Now let p be the Minkowski functional of C. Following the proof of Proposition 6.2.3, we construct a linear functional $\psi : \mathbf{E} \to \mathbb{R}$ such that $\psi(x) < \psi(x_0)$ for all $x \in C$ and

$$\psi(x) \leq p(x)$$

for all $x \in \mathbf{E}$. It now remains to prove that ψ is continuous. Since ψ is linear, we simply prove continuity at 0. So let $U \subset \mathbb{R}$ be an open set containing 0 and let $\epsilon > 0$ such that $V = \{z \in \mathbb{R} : |z| < \epsilon\} \subset U$. Now set

$$W := (\epsilon C) \cap (-\epsilon C).$$

Once again, W is an open set. If $y \in W$, then $\epsilon^{-1}y \in C$ and therefore $p(y) < \epsilon$. Hence $\psi(y) < \epsilon$. However, $\epsilon^{-1}(-y) \in C$ as well and therefore $-\psi(y) < \epsilon$, proving that

$$|\psi(y)| < \epsilon.$$

This is true for all $y \in W$ and we have proved that $W \subset \psi^{-1}(U)$. This proves that $\psi^{-1}(U)$ is open and ψ is continuous. □

We thus arrive at the first Hahn–Banach Separation Theorem, whose proof goes through verbatim from Theorem 6.2.6.

6.7.3 THEOREM (HAHN–BANACH SEPARATION THEOREM – I)

Let **E** *be a topological vector space over* $\mathbb{R}$, *and let* A *and* B *be two non-empty, disjoint, convex subsets of* **E**. *If* A *is open, then there exists* $\psi \in \mathbf{E}^*$ *and* $\alpha \in \mathbb{R}$ *such that*

$$\psi(a) \leq \alpha \leq \psi(b)$$

for all $a \in A$ *and* $b \in B$.

Now you would expect the second Separation Theorem (Theorem 6.2.7) to also go through, but there is a catch! If you look through the proof carefully, we used the fact that 0 has a collection of basic open neighbourhoods $\{B(0, 1/n) : n \in \mathbb{N}\}$ with the property that each $B(0, 1/n)$ is *convex*. This collection of sets does not have a direct analogue in the context of topological vector spaces unless one places a further assumption. Recall (Definition 2.5.1) that a seminorm on a vector space **E** is a function $p : \mathbf{E} \to \mathbb{R}_+$ such that $p(\alpha x) = |\alpha| p(x)$ and $p(x + y) \leq p(x) + p(y)$ for all $x, y \in \mathbf{E}$ and all $\alpha \in \mathbb{K}$.

6.7.4 DEFINITION A topological vector space **E** is said to be *locally convex* if there is a family Γ of seminorms on **E** such that sets of the form

$$U_p^n := \left\{ x \in \mathbf{E} : p(x) < \frac{1}{n} \right\} \qquad (p \in \Gamma, n \in \mathbb{N})$$

form a basis for the topology at 0 (in other words, every such set is open and every open set containing 0 must contain one such set).

Let us take a moment to put this definition in perspective.

6.7.5 REMARK

(i) Let $\mathbf{E}$ is a locally convex space and Γ be a family of seminorms as above. For any $p \in \Gamma, n \in \mathbb{N}$ and $y \in \mathbf{E}$, the set

$$U_{p,y}^n = \left\{ x \in \mathbf{E} : p(x - y) < \frac{1}{n} \right\} = y + U_p^n$$

is open and the collection $\{ U_{p,y}^n : p \in \Gamma \}$ forms a basis for the topology at y. Notice that each such set is convex: If $x_1, x_2 \in U_{p,y}^n$ and $0 \le t \le 1$, then

$$p(tx_1 + (1 - t)x_2 - y) \le tp(x_1 - y) + (1 - t)p(x_2 - y) < \frac{1}{n}.$$

Therefore the topology on $\mathbf{E}$ has a basis consisting of convex sets.

(ii) In fact, more is true. If $\mathbf{E}$ is a topological vector space which has a basis consisting of convex sets, then one can construct a family Γ of seminorms as above. This is why they are termed *locally convex* spaces. The proof of this latter fact is omitted here. The interested reader will find it in Vol. 1 of Kadison and Ringrose [31, Theorem 1.2.6].

(iii) Notice that a locally convex space is, at heart, a topological vector space and is therefore Hausdorff. So if Γ is a family of seminorms as above, then it is easy to verify that

$$\bigcap_{p \in \Gamma} p^{-1}(\{0\}) = \{0\}.$$

The next few examples should tell you why we are interested in locally convex spaces.

6.7.6 EXAMPLE

(i) Let $\mathbf{E}$ be a normed linear space. Define $p : \mathbf{E} \to \mathbb{R}_+$ by

$$p(x) := \|x\|.$$

Then $\Gamma := \{p\}$ satisfies the conditions of Definition 6.7.4 and thus every normed linear space is a locally convex space.

(ii) Let $\mathbf{E}$ be a normed linear space and $\sigma(\mathbf{E}, \mathbf{E}^*)$ denote the weak topology. For each finite collection $F := \{\varphi_1, \varphi_2, \dots, \varphi_k\}$ of bounded linear functionals, define $p_F : \mathbf{E} \to \mathbb{R}_+$ by

$$x \mapsto \max\{|\varphi_1(x)|, |\varphi_2(x)|, \ldots, |\varphi_k(x)|\}.$$

By the description of the basic open sets in the weak topology in Remark 6.3.5, the collection $\Gamma := \{p_F : F \subset \mathbf{E}^* \text{ finite}\}$ forms a family of seminorms satisfying the requirements of Definition 6.7.4. Therefore $(\mathbf{E}, \sigma(\mathbf{E}, \mathbf{E}^*))$ is a locally convex space.

(iii) Similarly, if $\mathbf{E}$ is a normed linear space, then $(\mathbf{E}^*, \sigma(\mathbf{E}^*, \mathbf{E}))$ is a locally convex space (Exercise 6.47).

(iv) If $0 < p < 1$, then we may define ℓ^p as

$$\ell^p = \{(x_n) : x_i \in \mathbb{K} \text{ for all } i \in \mathbb{N} \text{ and } \sum_{i=1}^{\infty} |x_i|^p < \infty\}.$$

We may define $d_p : \ell^p \times \ell^p \to \mathbb{R}$ by

$$d_p((x_n), (y_n)) := \sum_{i=1}^{\infty} |x_i - y_i|^p.$$

In Exercise 2.11, you would have proved that this defines a metric on ℓ^p. In fact, it is a topological vector space, but it is not locally convex (try Exercise 6.44 and Exercise 6.45).

In order to prove the second Separation Theorem in this context, we need a couple of lemmas. The first of these is a direct consequence of the definition of a topological vector space and is left as an exercise (Exercise 6.48).

6.7.7 LEMMA *Let $\mathbf{E}$ be a topological vector space and let U be an open set containing 0. Then there exists an open set V containing 0 such that $V - V \subset U$.*

The next lemma is something you might have proved before (Exercise 2.6) in the context of normed linear spaces. Without a norm, the proof is much more complicated.

6.7.8 LEMMA *Let $\mathbf{E}$ be a topological vector space and C and D be two subsets of $\mathbf{E}$. If C is compact and D is closed, then $C + D$ is closed.*

Proof. Fix $p \in \overline{C + D}$ and we wish to prove that $p \in C + D$. For convenience, we write $\mathcal{N}(0)$ to be the collection of all open sets containing 0.

(i) Let $U \in \mathcal{N}(0)$. Then $p + U$ is open and contains p and thus $(p + U) \cap (C + D) \neq \varnothing$. Hence

$$A_U := (p + U - D) \cap C$$

is non-empty. Since $\mathcal{N}(0)$ is closed under finite intersections, it follows that the collection $\{\overline{A_U} : U \in \mathcal{N}(0)\}$ consists of closed subsets of C such that any finite subcollection has non-empty intersection (do verify this). Since C is compact,

$$\bigcap_{U \in \mathcal{N}(0)} \overline{A_U} \neq \varnothing.$$

(ii) Let q be a point in this intersection. If $V \in \mathcal{N}(0)$, $q + V$ is an open set containing q. Since $q \in \overline{A_V}$,

$$(q + V) \cap A_V \neq 0.$$

This implies that $(q + V) \cap (p + V - D) \neq \varnothing$. Hence

$$(q + V - V) \cap (p - D) \neq \varnothing.$$

(iii) If $U \in \mathcal{N}(0)$, choose $V \in \mathcal{N}(0)$ such that $V - V \subset U$ by Lemma 6.7.7. Then $(q + V - V) \subset (q + U)$, which implies that

$$(q + U) \cap (p - D) \neq \varnothing.$$

Rewriting this, we see that

$$(p - q + U) \cap D \neq \varnothing.$$

The collection $\{p - q + U : U \in \mathcal{N}(0)\}$ forms a basis of open sets at $(p - q)$. Therefore we must conclude that $(p - q) \in \overline{D}$.

(iv) Since D is closed, $(p - q) \in D$. However, by construction, $q \in C$. Therefore $p = q + (p - q) \in C + D$.

$\square$

This last lemma gives us the tools we need to prove the Separation Theorem. It would be instructive to compare this proof with that of Theorem 6.2.7. Also, do pay attention to how local convexity gets used.

6.7.9 THEOREM (HAHN–BANACH SEPARATION THEOREM – II) *Let* **E** *be a locally convex topological vector space over* $\mathbb{R}$ *and let* A *and* B *be two non-empty, disjoint, convex subsets of* **E**. *If* A *is closed and* B *is compact, then there exists* $\psi \in \mathbf{E}^*, \alpha \in \mathbb{R}$ *and* $\epsilon > 0$ *such that*

$$\psi(a) \leq \alpha - \epsilon \leq \alpha + \epsilon \leq \psi(b)$$

for all $a \in A$ *and* $b \in B$.

Proof.

(i) Given A and B as above, $(A - B)$ is closed by Lemma 6.7.8. Since $A \cap B = \varnothing, 0 \notin (A - B)$ as well. Therefore there is an open set U containing 0 such that $U \cap (A - B) = \varnothing$. Now choose V open such that $0 \in V$ and $V - V \subset U$ by Lemma 6.7.7. A short argument now shows that

$$(A + V) \cap (B + V) = \varnothing. \tag{6.3}$$

(ii) Since **E** is locally convex, there is a family Γ of seminorms on **E** satisfying the conditions of Definition 6.7.4. Therefore we may assume that V is of the form

$$V = \left\{ x \in \mathbf{E} : p(x) < \frac{1}{n} \right\}$$

for some $p \in \Gamma$ and $n \in \mathbb{N}$. Since p is a seminorm, V is convex. Therefore $(A + V)$ and $(B + V)$ are both open, disjoint and convex. By Theorem 6.7.3, there exists $\psi \in \mathbf{E}^*$ and $\alpha \in \mathbb{R}$ such that

$$\psi(u) \leq \alpha \leq \psi(v)$$

for all $u \in A + V$ and $v \in B + V$.

(iii) Choose $a_0 \in A$. Then for any $x \in V$, $\psi(a_0) + \psi(x) \leq \alpha$. Therefore

$$\epsilon := \sup_{x \in V} \psi(x) \leq \alpha - \psi(a_0)$$

is well-defined. Since p is a seminorm, $x \in V$ if and only if $-x \in V$. Therefore $\epsilon > 0$ as well. If $a \in A$, we have

$$\psi(a) + \psi(x) \leq \alpha$$

for all $x \in V$. Hence $\psi(a) + \epsilon \leq \alpha$.

(iv) Furthermore,

$$-\epsilon = -\sup_{x \in V} \psi(x) = \inf_{x \in V} \psi(-x) = \inf_{y \in V} \psi(y).$$

If $b \in B$ and $y \in V$, $\alpha \le \psi(b) + \psi(y)$. Therefore $\alpha \le \psi(b) - \epsilon$ as well. □

Finally, both the theorems have a complex version. The proof of this version is, once again, left as an exercise.

6.7.10 THEOREM (HAHN–BANACH SEPARATION THEOREM – COMPLEX CASE)

*Let **E** be a topological vector space over $\mathbb{C}$ and let A and B be two disjoint, non-empty, convex subsets of **E**.*

(i) If A is open, then there exists $\psi \in \mathbf{E}^*$ and $\alpha \in \mathbb{R}$ such that

$$\mathrm{Re}(\psi)(a) \le \alpha \le \mathrm{Re}(\psi)(b)$$

for all $a \in A$ and $b \in B$.

*(ii) If A is closed, B is compact and **E** is locally convex, then there exists $\psi \in \mathbf{E}^*$, $\alpha \in \mathbb{R}$ and $\epsilon > 0$ such that*

$$\mathrm{Re}(\psi)(a) \le \alpha - \epsilon \le \alpha + \epsilon \le \mathrm{Re}(\psi)(b)$$

for all $a \in A$ and $b \in B$.

Notice that the Separation Theorem has some unintended consequences. Singleton sets are both closed and compact in a Hausdorff space, so distinct points may be separated by continuous linear functionals.

6.7.11 COROLLARY *Let **E** be a locally convex topological vector space. Then **E*** separates points of **E**. In particular, **E*** $\ne \{0\}$.*

In Exercise 4.3, you had encountered a topological vector space whose dual space is trivial. In a rather roundabout way, we have now proved that that space ($L^p[0, 1]$ for $0 < p < 1$) is *not* locally convex.

6.8   The Krein–Milman Theorem

Thanks to the Hahn–Banach Separation Theorems, convex sets are a natural tool to analyze and understand weak topologies (through the lens of linear functionals). However, we do not have many tools to construct convex sets though. Recall

(Definition 6.3.11) that the convex hull of a set K is the smallest convex set containing K. It is denoted by conv(K). By Exercise 6.11,

$$\text{conv}(K) = \bigcup_{n=1}^{\infty} \left\{ \sum_{i=1}^{n} t_i x_i : x_i \in K, 0 \le t_i \le 1, \sum_{i=1}^{n} t_i = 1 \right\}.$$

This is the most natural way to construct convex sets and we often define convex sets as convex hulls.

Of course, it may well happen that conv$(K) = $ conv(K') even if $K \ne K'$. Therefore if we are given a convex set A, we would like to find a *minimal* generating set $K \subset A$ such that $A = $ conv(K). In other words, we wish to find a set K such that every element of A is expressible as a convex combination of elements in K. Moreover, elements of K should be *irreducible* in the sense that they themselves should not be expressible as non-trivial convex combinations in A. This latter idea leads us to the definition of an extreme point.

6.8.1 DEFINITION Let A be a convex set in a real vector space **E**. A point $x \in A$ is said to be an ***extreme point*** of A if, whenever $x = ty + (1 - t)z$ for any two points $y, z \in A$ and $t \in [0, 1]$, it must happen that either $x = y$ or $x = z$.

We write ext(A) for the set of all extreme points in A (which may be empty). For brevity of notation, we will write $[x, y]$ for the line $\{tx + (1 - t)y : 0 \le t \le 1\}$.

6.8.2 EXAMPLE

(i) For $n \in \mathbb{N}$, the n-simplex Δ_n is the convex hull in $\mathbb{R}^{n+1}$ of the standard basis $\{e_1, e_2, \ldots, e_{n+1}\}$ (see Figure 6.2). Geometrically, it is easy to see that ext$(\Delta_n) = \{e_1, e_2, \ldots, e_{n+1}\}$.

(ii) For $n \in \mathbb{N}$, the cube $C_n := \{x \in \mathbb{R}^n : \|x\|_\infty \le 1\}$ is a convex set. Once again, one may verify that ext$(C_n) = \{(\pm 1, \pm 1, \ldots, \pm 1)\}$.

(iii) For $n \in \mathbb{N}$, the ball

$$B = \{x \in \mathbb{R}^n : \|x\|_2 \le 1\}$$

is a convex set with ext$(B) = S^{n-1}$, the sphere in $\mathbb{R}^n$. Therefore there are convex sets with infinitely many extreme points.

(iv) At the other extreme, there are convex sets with no extreme points! Let A denote the closed unit ball in $L^1[0, 1]$. We claim that ext$(A) = \emptyset$. To see this,

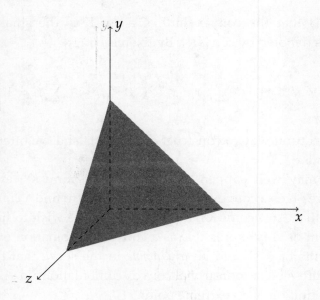

FIGURE 6.2   2-Simplex in $\mathbb{R}^3$

suppose $f \in A$ with $f \neq 0$, then define $F : [0,1] \to \mathbb{R}$ by

$$F(s) := \int_0^s |f(t)| dt.$$

This is a continuous function (by Lemma 4.1.7), $F(0) = 0$ and $F(1) = \|f\|_1$. Hence there exists $s \in [0,1]$ such that $F(s) = \|f\|_1/2$. Set

$$g := 2f\chi_{[0,s]} \text{ and } h := 2f\chi_{[s,1]}.$$

Then $\|g\|_1 = \|h\|_1 = \|f\|_1 \leq 1$ and

$$f = \frac{1}{2}g + \frac{1}{2}h.$$

Since $h \neq g$, f is not an extreme point of A. Thus $\mathrm{ext}(A) = \varnothing$.

An extreme point is a special case of a more general notion and one that we will need below.

6.8.3 DEFINITION   Let A be a convex set in a vector space **E**. A *face* of A is a subset $F \subset A$ such that

(i) F is convex and

(ii) if $[x,y] \cap F \neq \varnothing$ for any two points $x, y \in A$, then $[x,y] \subset F$.

A face F of A is said to be *proper* if $F \neq A$.

6.8.4 EXAMPLE

(i) A singleton set $\{x\}$ is a face if and only if x is an extreme point.

(ii) For $n \in \mathbb{N}$, let Δ_n denote the n-simplex in $\mathbb{R}^{n+1}$. Then

   (a) Each of the $(n+1)$ extreme points constitutes a singleton set that is a face,

   (b) Each line $L_{i,j}$ connecting e_i to e_j intersects with Δ_n in a face,

   (c) Each plane spanned by sets of the form $\{e_i, e_j, e_k\}$ intersects with Δ_n in a face, and so on.

   Continuing this process, one concludes that Δ_n has at least

   $$\sum_{k=0}^{n} \binom{n+1}{k} = 2^{n+1} - 1$$

   faces (in fact, it has exactly these many faces).

(iii) Let $C_n = \{x \in \mathbb{R}^n : \|x\|_\infty \leq 1\}$. Then

   (a) Each of the 2^n extreme points constitutes a singleton set that is a face,

   (b) Each line joining two extreme points intersects with C_n in a face,

   (c) There are $2n$ faces of C_n that are copies of C_{n-1}, and so on.

   It is an interesting exercise to determine the exact number of such faces.

(iv) If $\mathbf{E}$ is a normed linear space and $B_{\mathbf{E}}$ is the closed unit ball of $\mathbf{E}$, then the sphere $S_{\mathbf{E}} = \{x \in \mathbf{E} : \|x\| = 1\}$ is a face in $B_{\mathbf{E}}$ (do verify this).

The next lemma tells us where to look for faces.

6.8.5 LEMMA
Let A be a convex set in a vector space $\mathbf{E}$ and let $\varphi : \mathbf{E} \to \mathbb{K}$ be a linear functional. Suppose that $\alpha := \sup_{x \in A} \mathrm{Re}(\varphi)(x) < \infty$, then the set

$$F := \{a \in A : \mathrm{Re}(\varphi)(a) = \alpha\}$$

is a face of A.

Proof. Since φ is linear, F is a convex subset of A. Now suppose $x, y \in A$ and $t \in [0,1]$ are such that $tx + (1-t)y \in F$. Then

$$\alpha = \mathrm{Re}(\varphi)(tx + (1-t)y) = t\,\mathrm{Re}(\varphi)(x) + (1-t)\,\mathrm{Re}(\varphi)(y) \leq t\alpha + (1-t)\alpha = \alpha.$$

Therefore $\mathrm{Re}(\varphi)(x) = \mathrm{Re}(\varphi)(y) = \alpha$ and F is a face of A. $\qquad\square$

6.8.6 PROPOSITION *Let A be a convex subset of a vector space* **E** *and let* $F \subset A$ *be a face of A.*

(i) *If* $B \subset F$, *then B is a face of F if and only if B is a face of A.*

(ii) $\text{ext}(F) = F \cap \text{ext}(A)$.

Proof. Part (ii) is a direct consequence of part (i), so we only prove part (i).

If B is a face of A, then B is trivially a face of F because $F \subset A$. Conversely, suppose B is a face of F. Let $x, y \in A$ be such that $[x, y] \cap B \neq \emptyset$. Then $[x, y] \cap F \neq \emptyset$. Since F is a face of A, $[x, y] \subset F$. Since B is a face of F, $[x, y] \subset B$. This proves that B is a face of A. $\qquad\square$

We have already seen an example of a convex set without any extreme points. Our first goal, therefore, is to determine conditions under which extreme points exist. As it turns out, we need a number of things to hold!

6.8.7 PROPOSITION *Let A be a compact, convex subset of a locally convex space. Then* $\text{ext}(A) \neq \emptyset$.

Proof. Extreme points correspond to singleton faces, so it seems reasonable to look for a *minimal* face. Since compactness is involved, it is prudent to work with *closed* faces. Therefore we set

$$\mathcal{F} := \{F \subset A : F \text{ is a non-empty, closed face of } A\}.$$

Note that $\mathcal{F}$ is non-empty because $A \in \mathcal{F}$. For $F_1, F_2 \in \mathcal{F}$, we write $F_1 \leq F_2$ if $F_2 \subset F_1$. This is clearly a partial order on $\mathcal{F}$.

(i) Let $\mathcal{C}$ be a totally ordered subset of $\mathcal{F}$. Then any finite subcollection of $\mathcal{C}$ has a non-empty intersection. Since A is compact,

$$K := \bigcap_{F \in \mathcal{C}} F$$

is a non-empty, closed subset of A. Since it is the intersection of convex sets, K is also convex. Finally, suppose $[x, y] \cap K \neq \emptyset$ for some $x, y \in A$. For any $F \in \mathcal{C}$, $[x, y] \cap F \neq \emptyset$. Since F is a face, $[x, y] \subset F$. This is true for each $F \in \mathcal{C}$ and thus $[x, y] \subset K$. Hence $K \in \mathcal{F}$. With this in hand, it is clear that K is an upper bound for $\mathcal{C}$.

(ii) By Zorn's Lemma, $\mathcal{F}$ has a maximal element, which we denote by F_0. We claim that F_0 is a singleton set. Suppose not, then there would exist two distinct points $x, y \in F_0$. By Corollary 6.7.11 there is a continuous linear functional ψ such that

$$\mathrm{Re}(\psi)(x) \neq \mathrm{Re}(\psi)(y).$$

Since F_0 is compact and ψ is continuous, $\alpha := \sup_{z \in F_0} \mathrm{Re}(\psi)(z)$ is finite. Define

$$F := \{z \in F_0 : \mathrm{Re}(\psi)(z) = \alpha\}.$$

Then F is non-empty, closed and is a face of F_0 by Lemma 6.8.5. By Proposition 6.8.6, F is a face of A and is thus a member of $\mathcal{F}$. By construction, it cannot happen that both x and y are in F. Therefore $F \neq F_0$. Since $F_0 \leq F$, this contradicts the maximality of F_0. Hence F_0 must be a singleton set and that point is in $\mathrm{ext}(A)$.

$\square$

A comment on the proof: We needed A to be compact to ensure that K is non-empty and that α is finite. Perforce, this required us to work with *topological* vector spaces. Furthermore, we needed local convexity to ensure that the dual space had enough linear functionals to separate points. It is not hard to imagine that all this goes horribly wrong if the space is not locally convex (see Exercise 6.51).

We are now able to answer the question we started the section with. Given a convex set A, we would like to find a minimal set K such that $A = \mathrm{conv}(K)$. We now have a natural candidate for K, namely $\mathrm{ext}(A)$. While it is not always true that $A = \mathrm{conv}(\mathrm{ext}(A))$, the Krein–Milman Theorem gives us the best possible result.

6.8.8 THEOREM (KREIN AND MILMAN, 1940) *Let A be a compact, convex subset of a locally convex space. Then*

$$A = \overline{\mathrm{conv}(\mathrm{ext}(A))}.$$

Proof. Note that $\mathrm{ext}(A) \subset A$. Since A is both closed and convex, it follows that

$$B := \overline{\mathrm{conv}(\mathrm{ext}(A))} \subset A.$$

Seeking a contradiction, we assume that $B \neq A$ and choose $x_0 \in A \setminus B$. Now B is closed, convex and non-empty by Proposition 6.8.7 and $\{x_0\}$ is compact. By the Hahn–Banach Separation Theorem (Theorem 6.7.10), there is a continuous linear functional ψ such that

$$\operatorname{Re}(\psi)(x_0) > \sup_{b \in B} \operatorname{Re}(\psi)(b).$$

Since ψ is continuous and A is compact, $\alpha := \sup_{a \in A} \operatorname{Re}(\psi)(a)$ is finite. Define

$$F := \{a \in A : \operatorname{Re}(\psi)(a) = \alpha\}.$$

By Lemma 6.8.5, F is a face of A and is both non-empty and compact. By Proposition 6.8.7,

$$\operatorname{ext}(F) \neq \emptyset.$$

By Proposition 6.8.6, $\operatorname{ext}(A) \cap F \neq \emptyset$. However, if $y \in \operatorname{ext}(A) \cap F$, then $y \notin B$. This contradicts the fact that $\operatorname{ext}(A) \subset B$. Hence our assumption that $B \neq A$ must have been false and the theorem is proved. $\qquad\square$

If $\mathbf{E}$ is a normed linear space, then the dual space $\mathbf{E}^*$ is locally convex when equipped with the weak-$*$ topology. Also, the closed unit ball $B_{\mathbf{E}^*}$ of $\mathbf{E}^*$ is both convex and compact (by the Banach–Alaoglu Theorem). The next corollary is thus an immediate application of the Krein–Milman Theorem.

6.8.9 COROLLARY *Let $\mathbf{E}$ be a normed linear space and $B_{\mathbf{E}^*}$ denote the closed unit ball of $\mathbf{E}^*$. Then*

$$B_{\mathbf{E}^*} = \overline{\operatorname{conv}(\operatorname{ext}(B_{\mathbf{E}^*}))},$$

where the closure is taken in the weak-$*$ topology. In particular, $B_{\mathbf{E}^*}$ has extreme points.

Corollary 6.8.9 lends itself to a variety of interesting applications. For instance, we have just seen that the closed unit ball in $L^1[0,1]$ does not have any extreme points. Therefore not only is it not the dual of $L^\infty[0,1]$, $L^1[0,1]$ cannot be the dual space of *any* normed linear space!

6.9 Uniformly Convex Spaces

Through this chapter, we have had a growing realization that convexity is intimately linked to the weak topologies. Now we take a closer look at it in the norm topology and focus on the *geometry* of the Banach space (specifically, the shape of the unit ball). To begin with, recall the following definition from Definition 4.6.1.

6.9.1 DEFINITION A normed linear space **E** is said to be *strictly convex* if, for any distinct $x, y \in \mathbf{E}$ with $\|x\| = \|y\| = 1$, we have

$$\left\| \frac{x+y}{2} \right\| < 1.$$

Geometrically, this means that the unit ball of **E** is 'round' (if you draw a line connecting two points on the sphere, then the mid-point of that line lies in the interior of the ball). We have already seen some consequences of strict convexity in the exercises at the end of Chapter 4. Now we will explore a stronger version of convexity; one that will have important *topological* consequences for the normed linear space.

6.9.2 DEFINITION (CLARKSON, 1936) A normed linear space **E** is said to be *uniformly convex* if, for any $0 < \epsilon \le 2$, there is a $\delta > 0$ such that, for any two vectors $x, y \in \mathbf{E}$ with $\|x\| = \|y\| = 1$,

$$\|x - y\| \ge \epsilon \Rightarrow \left\| \frac{x+y}{2} \right\| \le 1 - \delta.$$

Note that any uniformly convex space is strictly convex. In fact, you might wonder what the difference is! Unfortunately, it is impossible to visualize the difference because the two notions coincide on finite dimensional spaces. However, as we will see below, a uniformly convex space is *reflexive* and there exist strictly convex spaces that are not reflexive. Therefore the two notions are indeed distinct.

6.9.3 EXAMPLE

(i) Any inner product space is uniformly convex.

   Proof. This follows from the Parallelogram law. If $x, y \in \mathbf{E}$ with $\|x\| = \|y\| = 1$, then

   $$\left\| \frac{x+y}{2} \right\|^2 = \frac{1}{2} \left(\|x\|^2 + \|y\|^2 \right) - \left\| \frac{x-y}{2} \right\|^2 = 1 - \left\| \frac{x-y}{2} \right\|^2.$$

   Therefore for $0 < \epsilon \le 2$, one can see that $\delta := 1 - \sqrt{1 - \frac{\epsilon^2}{4}} > 0$ works. □

(ii) As mentioned above, a finite dimensional space need not be uniformly convex. For instance, $\mathbb{R}^2$ equipped with the supremum norm is not even strictly convex. However, a finite dimensional space is uniformly convex if and only if it is strictly convex.

Proof. Suppose $\mathbf{E}$ is a finite dimensional space that is not uniformly convex, then there exists $\epsilon > 0$ for which no $\delta > 0$ satisfies the requirement of Definition 6.9.2. In particular, for each $n \geq 2$, there exist $x_n, y_n \in \mathbf{E}$ such that $\|x_n\| = \|y_n\| = 1, \|x_n - y_n\| \geq \epsilon$, but

$$\left\| \frac{x_n + y_n}{2} \right\| > 1 - \frac{1}{n}.$$

By finite dimensionality of $\mathbf{E}$, we may choose limit points $x \in \overline{\{x_n\}}$ and $y \in \overline{\{y_n\}}$. Then it would follow that $\|x\| = \|y\| = 1$. Furthermore, since

$$1 - \frac{1}{n} < \left\| \frac{x_n + y_n}{2} \right\| \leq 1,$$

for all $n \in \mathbb{N}$, it follows that $\|\frac{x+y}{2}\| = 1$. This proves that $\mathbf{E}$ cannot be strictly convex. □

(iii) Equip ℓ^1 with the norm

$$\|x\|_s := \|x\|_1 + \|x\|_2.$$

Then $(\ell^1, \| \cdot \|_s)$ is strictly convex, but not uniformly convex.

Proof. That $\| \cdot \|_s$ is well-defined follows from the fact that $\ell^1 \subset \ell^2$ (Exercise 2.12). That $\| \cdot \|_s$ is strictly convex follows from the fact that $\| \cdot \|_2$ is strictly convex (by Example (i) above). To see that it is not uniformly convex, choose

$$x_n := e_1 + e_2 + \ldots + e_n \text{ and } y_n := e_{n+1} + e_{n+2} + \ldots + e_{2n}.$$

Then $\|x_n\|_s = n + \sqrt{n} = \|y_n\|_s$. Furthermore, $\|x_n - y_n\|_s = n + \sqrt{2n}$ and

$$\left\| \frac{x_n + y_n}{2} \right\|_s = n + \sqrt{\frac{n}{2}}.$$

Therefore if $u_n := \frac{x_n}{n+\sqrt{n}}$ and $v_n := \frac{y_n}{n+\sqrt{n}}$, then $\|u_n\|_s = \|v_n\|_s = 1$ and

$$\lim_{n \to \infty} \|u_n - v_n\|_s = \lim_{n \to \infty} \frac{n + \sqrt{2n}}{n + \sqrt{n}} = 1,$$

$$\lim_{n \to \infty} \left\| \frac{u_n + v_n}{2} \right\|_s = \lim_{n \to \infty} \frac{n + \sqrt{\frac{n}{2}}}{n + \sqrt{n}} = 1.$$

Therefore for $\epsilon = \frac{1}{2}$, there is no $\delta > 0$ that satisfies the requirement of Definition 6.9.2. As we will see later, there is *no norm* on ℓ^1 that could make it uniformly convex. □

Our goal is to show that $L^p[a, b]$ is uniformly convex whenever $1 < p < \infty$. The original proof is due to Clarkson [8], but we present a more recent proof due to Shioji [56].

6.9.4 DEFINITION Let A be a convex subset of a real vector space and let $f : A \to [-\infty, +\infty]$ be a function. We say that f is **strictly convex** if, whenever $x, y \in A$ are two distinct points and $0 < t < 1$, then

$$f(tx + (1 - t)y) < tf(x) + (1 - t)f(y).$$

If $1 < p < \infty$, then the function $t \mapsto t^p$ defined on $[0, \infty)$ is strictly convex. We will use this fact below.

6.9.5 LEMMA *Fix $1 < p < \infty$. Let $\alpha : \mathbb{K}^2 \to [0, \infty)$ be the function*

$$\alpha(z_1, z_2) = \frac{|z_1|^p + |z_2|^p}{2} - \left|\frac{z_1 + z_2}{2}\right|^p.$$

Then $\alpha(z_1, z_2) > 0$ for any pair $(z_1, z_2) \in \mathbb{K}^2$ such that $z_1 \neq z_2$.

Proof. Suppose $(z_1, z_2) \in \mathbb{K}^2$ with $z_1 \neq z_2$, then by the convexity of $t \mapsto t^p$,

$$\left|\frac{z_1 + z_2}{2}\right|^p \leq \left(\frac{|z_1| + |z_2|}{2}\right)^p \leq \frac{|z_1|^p + |z_2|^p}{2}.$$

If $|z_1| = |z_2|$ and $z_1 \neq z_2$, then the first inequality is strict. If $|z_1| \neq |z_2|$, then the second inequality is strict because $t \mapsto t^p$ is strictly convex. In either case, $\alpha(z_1, z_2) > 0$. □

6.9.6 LEMMA *Let $\alpha : \mathbb{K}^2 \to [0, \infty)$ be the function defined in Lemma 6.9.5. For $\eta > 0$, define*

$$D := \{(z_1, z_2) \in \mathbb{K}^2 : |z_1|^p + |z_2|^p = 1 \text{ and } |z_1 - z_2|^p \geq \eta\}.$$

Then $\theta := \inf\{\alpha(z_1, z_2) : (z_1, z_2) \in D\} > 0$. Furthermore, for any pair $(z_1, z_2) \in \mathbb{K}^2$ with $|z_1 - z_2|^p \geq \eta(|z_1|^p + |z_2|^p)$,

$$|z_1|^p + |z_2|^p \leq \frac{\alpha(z_1, z_2)}{\theta}.$$

Proof. D is closed and bounded in $\mathbb{K}^2$ and thus compact. Furthermore, α is continuous and strictly positive on D by Lemma 6.9.5. Therefore α attains its infimum on D, which must be strictly positive. Now if $(z_1, z_2) \in \mathbb{K}^2$ with $|z_1 - z_2|^p \geq \eta(|z_1|^p + |z_2|^p)$, then set

$$t := (|z_1|^p + |z_2|^p)^{1/p}.$$

If $w_1 := z_1/t$ and $w_2 = z_2/t$, then $(w_1, w_2) \in D$. Therefore $\alpha(w_1, w_2) \geq \theta$. This implies that

$$\theta \leq \frac{\alpha(z_1, z_2)}{t^p},$$

which implies the result. $\qquad\square$

6.9.7 THEOREM (CLARKSON, 1936) *If* $1 < p < \infty$, *then* $L^p[a, b]$ *is uniformly convex.*

Proof. Let $0 < \epsilon \leq 2$ be fixed. We wish to find a $\delta = \delta(\epsilon) > 0$ such that

$$\left\| \frac{f + g}{2} \right\|_p < 1 - \delta.$$

for any pair of functions $f, g \in L^p[a, b]$ such that $\|f\|_p = \|g\|_p = 1$ and $\|f - g\|_p \geq \epsilon$. To that end, let $f, g \in L^p[a, b]$ be two such functions. Let $\eta := \frac{\epsilon^p}{4}$,

$$D := \{(z_1, z_2) \in \mathbb{K}^2 : |z_1|^p + |z_2|^p = 1 \text{ and } |z_1 - z_2|^p \geq \eta\}$$

and let $\theta > 0$ be as in Lemma 6.9.6. Set

$$M := \{x \in [a, b] : |f(x) - g(x)|^p \geq \eta(|f(x)|^p + |g(x)|^p)\}.$$

Then for any $x \in M$,

$$|f(x)|^p + |g(x)|^p \leq \frac{1}{\theta}\left(\frac{|f(x)|^p + |g(x)|^p}{2} - \left| \frac{f(x) + g(x)}{2} \right|^p \right)$$

by Lemma 6.9.6. Integrating over M, we get

$$\begin{aligned}
\int_M (|f|^p + |g|^p) &\leq \frac{1}{\theta}\left(\int_M \frac{|f|^p + |g|^p}{2} - \int_M \left| \frac{f + g}{2} \right|^p \right) \\
&\leq \frac{1}{\theta}\left(\int_a^b \frac{|f|^p + |g|^p}{2} - \int_a^b \left| \frac{f + g}{2} \right|^p \right) \\
&\leq \frac{1}{\theta}\left(\frac{\|f\|_p^p + \|g\|_p^p}{2} - \left\| \frac{f + g}{2} \right\|_p^p \right) \\
&= \frac{1}{\theta}\left(1 - \left\| \frac{f + g}{2} \right\|_p^p \right)
\end{aligned} \qquad (6.4)$$

where the second inequality holds because α is an non-negative function. Therefore we have

$$
\begin{aligned}
\|f - g\|_p^p &= \int_{[a,b]\setminus M} |f - g|^p + \int_M |f - g|^p \\
&= \int_{[a,b]\setminus M} |f - g|^p + 2^p \int_M \left| \frac{f - g}{2} \right|^p \\
&\leq \eta \int_{[a,b]\setminus M} (|f|^p + |g|^p) + 2^p \int_M \frac{|f|^p + |g|^p}{2}
\end{aligned}
$$

where the last inequality holds, once again, because α is non-negative. Combining this with Equation 6.4, we conclude that

$$
\begin{aligned}
\epsilon^p \leq \|f - g\|_p^p &\leq \eta(\|f\|_p^p + \|g\|_p^p) + \frac{2^{p-1}}{\theta} \left(1 - \left\| \frac{f + g}{2} \right\|_p^p \right) \\
&\leq 2\eta + \frac{2^{p-1}}{\theta} \left(1 - \left\| \frac{f + g}{2} \right\|_p^p \right).
\end{aligned}
$$

Since $\eta = \epsilon^p / 4$, we conclude that

$$
\delta = \frac{\epsilon^p \theta}{2^p}
$$

satisfies the requirement of Definition 6.9.2 (note that θ and therefore δ only depends on ϵ). $\qquad\square$

There are many interesting consequences of uniform convexity, some of which we explore in the exercises given at the end of the chapter. For now, though, we prove that uniform convexity of the norm implies that the space is reflexive. Along the way, we will also prove an important density theorem due to Goldstine.

6.9.8 LEMMA *Let* $\mathbf{E}$ *be a normed linear space and* $\{\varphi_1, \varphi_2, \ldots, \varphi_n\} \subset \mathbf{E}^*$ *be a finite set of bounded linear functionals on* $\mathbf{E}$. *If* $T \in \mathbf{E}^{**}$, *then there exists* $x \in \mathbf{E}$ *such that*

$$
T(\varphi_i) = \varphi_i(x)
$$

for all $1 \leq i \leq n$.

Proof. The proof works along similar lines to Lemma 6.5.9. Set $\mathbf{F} := \mathbb{K}^n$ equipped with the Euclidean norm and define $S : \mathbf{E} \to \mathbf{F}$ by

$$
S(x) := (\varphi_1(x), \varphi_2(x), \ldots, \varphi_n(x)).
$$

Define $w := (T(\varphi_1), T(\varphi_2), \ldots, T(\varphi_n)) \in \mathbf{F}$. We then need to prove that $w \in \text{Range}(S)$. Suppose for argument's sake that $w \notin \text{Range}(S)$. Since $\mathbf{F}$ is finite dimensional, Range(S) is closed. By Proposition 4.3.3, there is a linear functional $\psi \in \mathbf{F}^*$ such that $\psi(w) = 1$ and

$$\psi(S(x)) = 0$$

for all $x \in \mathbf{E}$. By the Riesz Representation Theorem, there is a tuple $z = (\alpha_1, \ldots, \alpha_n) \in \mathbf{F}$ such that $\psi(y) = \langle y, z \rangle$ for all $y \in \mathbf{F}$. In particular,

$$\sum_{i=1}^{n} \alpha_i \varphi_i(x) = 0$$

for all $x \in \mathbf{E}$. Therefore if $\varphi := \sum_{i=1}^{n} \alpha_i \varphi_i$, then φ is the zero linear functional on $\mathbf{E}$. However,

$$T(\varphi) = \sum_{i=1}^{n} \alpha_i T(\varphi_i) = \langle w, z \rangle = 1.$$

This is clearly impossible. Therefore w must belong to Range(S). $\qquad\square$

6.9.9 THEOREM (GOLDSTINE, 1938) *Let* $\mathbf{E}$ *be a normed linear space and* $J : \mathbf{E} \to \mathbf{E}^{**}$ *be the natural inclusion. Then*

(i) *$J(\mathbf{E})$ is dense in $\mathbf{E}^{**}$ with respect to the weak-$*$ topology and*

(ii) *$J(B_{\mathbf{E}})$ is dense in $B_{\mathbf{E}^{**}}$ with respect to the weak-$*$ topology.*

Proof. Since J is an isometry, it clearly suffices to prove part (i). Let V be a weak-$*$ open set in $\mathbf{E}^{**}$ and we wish to prove that $V \cap J(\mathbf{E}) \neq \varnothing$. As before, we may assume that

$$V = \{T \in \mathbf{E}^{**} : |T(\varphi_i) - T_0(\varphi_i)| < \epsilon \text{ for all } 1 \leq i \leq n\}$$

for some $T_0 \in \mathbf{E}^{**}$, a finite set $\{\varphi_1, \varphi_2, \ldots, \varphi_n\} \subset \mathbf{E}^*$ and some $\epsilon > 0$. Now choose $x \in \mathbf{E}$ by Lemma 6.9.8 such that

$$T_0(\varphi_i) = \varphi_i(x)$$

for all $1 \leq i \leq n$. Then $J(x)$ lies in V and thus $V \cap J(\mathbf{E}) \neq \varnothing$. This is true for any weak-$*$ open set V, which proves that $J(\mathbf{E})$ is weak-$*$ dense in $\mathbf{E}^{**}$. $\qquad\square$

We now come to the main result of this section and the final one of this chapter.

6.9.10 THEOREM (MILMAN, 1938, AND PETTIS, 1939) *A uniformly convex Banach space is reflexive.*

Proof. Let $\mathbf{E}$ be a uniformly convex Banach space. To prove that $\mathbf{E}$ is reflexive, it suffices to show that

$$J(B_{\mathbf{E}}) = B_{\mathbf{E}^{**}}.$$

However, $J(B_{\mathbf{E}})$ is closed in the norm-topology, since J is isometric. Therefore it suffices to show that $J(B_{\mathbf{E}})$ is norm-dense in $B_{\mathbf{E}^{**}}$. By Goldstine's Theorem, we know that $J(B_{\mathbf{E}})$ is weak-$*$ dense in $B_{\mathbf{E}^{**}}$. The point of the proof is to use uniform convexity to translate weak-$*$ density to norm-density.

Fix $2 \geq \epsilon > 0$ and $T_0 \in B_{\mathbf{E}^{**}}$ with $\|T_0\| = 1$. Choose $\delta > 0$ such that the condition in Definition 6.9.2 is satisfied. Since $\|T_0\| = 1$, there exists $\varphi \in \mathbf{E}^*$ such that $\|\varphi\| = 1$ and

$$|T_0(\varphi)| > 1 - \frac{\delta}{2}.$$

Set $V := \{T \in B_{\mathbf{E}^{**}} : |T(\varphi) - T_0(\varphi)| < \frac{\delta}{2}\}$. Then V is weak-$*$ open, so by Goldstine's Theorem, there is a point $x \in B_{\mathbf{E}}$ such that $J(x) \in V$. We claim that

$$\|T_0 - J(x)\| \leq \epsilon.$$

Suppose not, then there exists $\psi \in \mathbf{E}^*$ such that $\|\psi\| = 1$ and $|T_0(\psi) - \psi(x)| > \epsilon$. Now consider

$$W := \{S \in B_{\mathbf{E}^{**}} : |S(\psi) - \psi(x)| > \epsilon\} \cap V.$$

Then W is a weak-$*$ open set and $T_0 \in W$. By Goldstine's Theorem, there is $y \in B_{\mathbf{E}}$ such that $J(y) \in W$. Observe that

$$\|y - x\| = \|J(y) - J(x)\| \geq |\psi(y) - \psi(x)| > \epsilon. \tag{6.5}$$

Since $J(y) \in V$, $|\varphi(y) - T_0(\varphi)| < \frac{\delta}{2}$. Therefore

$$\left| \varphi\left(\frac{x+y}{2}\right) - T_0(\varphi) \right| < \frac{\delta}{2}.$$

Since $|T_0(\varphi)| > 1 - \frac{\delta}{2}$, we conclude that

$$\left\| \frac{x+y}{2} \right\| \geq \left| \varphi\left(\frac{x+y}{2}\right) \right| > 1 - \delta.$$

Together with Equation 6.5, this contradicts our choice of δ. Therefore it must happen that $\|T_0 - J(x)\| \leq \epsilon$. Such a vector exists for any $\epsilon > 0$ and thus $J(B_{\mathbf{E}}) = B_{\mathbf{E}^{**}}$. $\qquad\square$

We should mention that the converse of the Milman–Pettis Theorem is false. There is an example (due to M.M. Day [11]) of a reflexive Banach space that is not topologically isomorphic to a uniformly convex space.

We now come to the end of our long discussion on weak topologies. The highlights were the Hahn–Banach Separation Theorem(s), Helley's Theorem, the Banach–Alaoglu Theorem and the Krein–Milman Theorem. You will find many of the exercises built around these results.

6.10 Exercises

Weak Convergence

EXERCISE 6.1 Let $\mathbf{E} := c$, equipped with the supremum norm. For a sequence $(x^k) \subset \mathbf{E}$, prove that $x^k \xrightarrow{w} x$ if and only if the following three conditions hold.

(a) (x^k) is norm-bounded,

(b) $x_j^k \to x_j$ for each $j \in \mathbb{N}$ and

(c) $\lim_{k\to\infty}(\lim_{j\to\infty} x_j^k) = \lim_{j\to\infty} x_j$.

Hint: Use Exercise 4.19 and Corollary 6.1.7.

EXERCISE 6.2 (⚒) Prove Lemma 6.1.10.

EXERCISE 6.3 The aim of this exercise is to show that the conclusion of Lemma 6.1.10 does not hold for $p = 1$. For each $n \in \mathbb{N}$, define $f_n \in L^1[0,1]$ by

$$f_n(t) = \begin{cases} 1 & : \text{ if } \frac{k}{2^n} + \frac{1}{2^{2n+1}} \leq t < \frac{k+1}{2^n} \text{ and } 0 \leq k \leq 2^n - 1, \\ 1 - 2^{n+1} & : \text{otherwise.} \end{cases}$$

In other words, $f_n = 1$ on 2^n disjoint intervals, each of length $(\frac{1}{2^n} - \frac{1}{2^{2n+1}})$ and $f_n = 1 - 2^{n+1}$ on the remainder of $[0,1]$.

(i) Prove that (f_n) is a bounded sequence in $L^1[0,1]$ and that

$$\left| \int_0^x f_n(t)dt \right| \leq \frac{1}{2^n}$$

for all $x \in [0,1]$. Conclude that

$$\lim_{n\to\infty} \int_0^x f_n(t)dt = 0$$

for each $x \in [0,1]$.

(ii) Let E_n denote the set on which $f_n = 1$. Prove that $m(E_n) = 1 - \frac{1}{2^{n+1}}$.

(iii) If $E = \cap_{n=1}^{\infty} E_n$, then prove that $m(E) \geq \frac{1}{2}$ and that

$$\int_E f_n(t)dt = m(E)$$

for all $n \in \mathbb{N}$. Conclude that (f_n) does not converge weakly to 0.

EXERCISE 6.4 Let $(f_n) \subset L^1[0,1]$ be the sequence of functions constructed in Exercise 6.3. If $1 < p < \infty$, then $(f_n) \subset L^p[0,1]$ as well. Show that (f_n) does not converge weakly to 0 in $L^p[0,1]$ either. Why does this not contradict Lemma 6.1.10?

EXERCISE 6.5 ($\maltese$) Give an example of a sequence $(f_n) \subset L^1[0,1]$ such that $(\|f_n\|_1)$ is uniformly bounded, $\lim_{n \to \infty} f_n(x) = 0$ a.e., but (f_n) does not converge weakly in $L^1[0,1]$. (This shows that Proposition 6.1.12 does not hold for $p = 1$.)

EXERCISE 6.6 Give an example of a norm bounded sequence in ℓ^1 that does not have a weakly convergent subsequence.

EXERCISE 6.7 ($\maltese$) Recall from Theorem 5.3.9 that there is a quotient map $p : \ell^1 \to \ell^2$. Let $\mathbf{F} := \ker(p)$.

(i) Prove that ℓ^2 is not isomorphic to a closed subspace of ℓ^1.

(ii) Prove that $\mathbf{F}$ is not complemented in ℓ^1 (in the sense of Definition 5.7.3).

For some context to this question, see Exercise 5.48 and the problems after it.

EXERCISE 6.8 Let $\mathbf{E}$ be a normed linear space and $(x_n) \subset \mathbf{E}$ be a sequence such that $x_n \xrightarrow{w} x$. Suppose $(\varphi_n) \subset \mathbf{E}^*$ is a sequence such that $\lim_{n \to \infty} \varphi_n = \varphi$ in the norm of $\mathbf{E}^*$. Prove that $\lim_{n \to \infty} \varphi_n(x_n) = \varphi(x)$.

Hahn–Banach Separation Theorem

EXERCISE 6.9 Let C be a convex, open subset of a normed linear space $\mathbf{E}$ such that $0 \in C$. Let $p : \mathbf{E} \to \mathbb{R}_+$ be the associated Minkowski functional (Theorem 6.2.2). Prove that p is a seminorm if C is *balanced* ($\alpha x \in C$ whenever $|\alpha| = 1$ and $x \in C$).

EXERCISE 6.10 ($\maltese$) Let $\mathbf{E}$ be a normed linear space.

(i) If $\{A_\alpha : \alpha \in J\}$ are a family of convex subsets of $\mathbf{E}$, then prove that $\cap_{\alpha \in J} A_\alpha$ is convex.

(ii) If A is a convex set, then prove that $\overline{A}^{\|\cdot\|}$ is a convex set.

EXERCISE 6.11 (✗) Let **E** be a real vector space.

(i) Prove that a subset A of **E** is a convex set if and only if, whenever $\{x_1, x_2, \ldots, x_n\} \subset A$ and $\{t_1, t_2, \ldots, t_n\} \subset [0, 1]$ are such that $\sum_{i=1}^{n} t_i = 1$, then

$$\sum_{i=1}^{n} t_i x_i \in A.$$

(ii) If $K \subset \mathbf{E}$, then prove that

$$\operatorname{conv}(K) = \bigcup_{n=1}^{\infty} \left\{ \sum_{i=1}^{n} t_i x_i : x_i \in K, 0 \leq t_i \leq 1 \text{ and } \sum_{i=1}^{n} t_i = 1 \right\}.$$

EXERCISE 6.12 Let **E** be a normed linear space and K be a subset of **E**.

(i) If K is finite, then prove that $\operatorname{conv}(K)$ is compact.

(ii) If K is compact and **E** is a Banach space, then prove that $\overline{\operatorname{conv}(K)}$ is compact.

(iii) Give an example of a compact subset K of a Banach space such that $\operatorname{conv}(K)$ is not closed.

EXERCISE 6.13 The goal of this problem is to show that Proposition 6.2.3 holds in a finite dimensional space without any assumptions on the convex set. Let **E** be a finite dimensional normed linear space over $\mathbb{R}$ and let $C \subset \mathbf{E}$ be a non-empty convex set such that $0 \notin C$.

(i) Let $\{x_n\}_{n=1}^{\infty}$ be a countable set that is dense in C (which exists because **E** is separable). For each $n \in \mathbb{N}$, define

$$C_n := \operatorname{conv}(\{x_1, x_2, \ldots, x_n\}).$$

Then prove that C_n is compact and that $\bigcup_{n=1}^{\infty} C_n$ is dense in C.

(ii) For each $n \in \mathbb{N}$, prove that there is a linear functional $\varphi_n \in \mathbf{E}^*$ such that $\|\varphi_n\| = 1$ and

$$\varphi_n(x) \geq 0$$

for all $x \in C_n$.

(iii) Conclude that there is a linear functional $\varphi \in \mathbf{E}^*$ such that $\|\varphi\| = 1$ and $\varphi(x) \geq 0$ for all $x \in C$.

(iv) If A and B are two non-empty, disjoint, convex sets in $\mathbf{E}$, prove that there is a closed hyperplane that separates A from B.

EXERCISE 6.14 (⚒) Prove Theorem 6.2.8.

Hint: Use Lemma 4.2.3 along with the real versions of the theorem.

The Weak Topology

EXERCISE 6.15 (⚒) Let (x_n) be a sequence in a normed linear space $\mathbf{E}$. Prove that $\lim_{n\to\infty} x_n = x$ with respect to $\sigma(\mathbf{E}, \mathbf{E}^*)$ if and only if $x_n \xrightarrow{w} x$.

EXERCISE 6.16 Let (x_n) be a sequence in normed linear space $\mathbf{E}$ such that $x_n \xrightarrow{w} x_0$. For each $n \in \mathbb{N}$, set

$$K_n := \overline{\operatorname{conv}(\{x_n, x_{n+1}, x_{n+2}, \ldots\})}.$$

(i) Prove that $x_0 \in \bigcap_{n=1}^{\infty} K_n$.

(ii) Let $\varphi \in \mathbf{E}^*$ and $\epsilon > 0$. Let

$$A := \{x \in \mathbf{E} : |\varphi(x) - \varphi(x_0)| \leq \epsilon\}.$$

Prove that A is a closed convex set and that there is $N \in \mathbb{N}$ such that $K_n \subset A$ for all $n \geq N$.

(iii) Prove that $\bigcap_{n=1}^{\infty} K_n = \{x_0\}$.

EXERCISE 6.17 (⚒) Let $\mathbf{E}$ be an infinite dimensional normed linear space. Prove that any weakly open set containing 0 must also contain a non-zero subspace. Conclude that the open unit ball $B_{\mathbf{E}}(0,1) = \{x \in \mathbf{E} : \|x\| < 1\}$ is not weakly open.

Hint: See the proof of Theorem 6.3.13.

6.10.1 DEFINITION Let X be a topological space. A function $f : X \to (-\infty, +\infty]$ is said to be ***lower semi-continuous*** if the set $f^{-1}((-\infty, \alpha]) = \{x \in X : f(x) \leq \alpha\}$ is closed in X for each $\alpha \in (-\infty, \infty]$.

EXERCISE 6.18 Let $\mathbf{E}$ be a normed linear space and $f : \mathbf{E} \to (-\infty, +\infty]$ be a convex function.

(i) If f is lower semi-continuous with respect to the norm topology, then prove that f is lower semi-continuous with respect to the weak topology.

(ii) Give an example of a convex function that is continuous with respect to the norm topology, but not continuous with respect to the weak topology.

EXERCISE 6.19 Let $\mathbf{E}$ be a normed linear space. Prove that the norm $\|\cdot\| : \mathbf{E} \to [0,\infty)$ is a convex function that is lower semi-continuous with respect to the norm topology. Conclude that if (x_n) is a sequence in $\mathbf{E}$ such that $x_n \xrightarrow{w} x$ in $\mathbf{E}$, then

$$\|x\| \leq \liminf_{n\to\infty} \|x_n\|.$$

EXERCISE 6.20 (⚒) Let $(\mathbf{E}_1, \|\cdot\|_{\mathbf{E}_1})$ and $(\mathbf{E}_2, \|\cdot\|_{\mathbf{E}_2})$ be two normed linear spaces. On the vector space $\mathbf{E}_1 \times \mathbf{E}_2$, define the norm

$$\|(x_1, x_2)\|_1 := \|x_1\|_{\mathbf{E}_1} + \|x_2\|_{\mathbf{E}_2}.$$

We had seen in Exercise 2.5 that $(\mathbf{E}_1 \times \mathbf{E}_2, \|\cdot\|_1)$ is a normed linear space, which we denoted by $\mathbf{E}_1 \oplus_1 \mathbf{E}_2$. Consider two topologies on $\mathbf{E}_1 \times \mathbf{E}_2$:

$$\begin{aligned} \tau_w &:= \text{the weak topology } \sigma(\mathbf{E}_1 \oplus_1 \mathbf{E}_2, (\mathbf{E}_1 \oplus_1 \mathbf{E}_2)^*) \text{ and} \\ \tau_p &:= \text{the product topology } \sigma(\mathbf{E}_1, \mathbf{E}_1^*) \times \sigma(\mathbf{E}_2, \mathbf{E}_2^*). \end{aligned}$$

Prove that $\tau_w = \tau_p$.

6.10.2 DEFINITION Given a linear operator $T : \mathbf{E} \to \mathbf{F}$ between two normed linear spaces, we say that T is ***norm-weak*** continuous if T is continuous as a map

$$T : (\mathbf{E}, \sigma(\mathbf{E}, \|\cdot\|)) \to (\mathbf{F}, \sigma(\mathbf{F}, \mathbf{F}^*)).$$

The notions of ***norm–norm***, ***weak–weak*** and ***weak–norm*** continuity are defined analogously. Note that *norm–norm* continuity is the usual notion of continuity (boundedness) of the operator.

EXERCISE 6.21 (⚒) Let $T : \mathbf{E} \to \mathbf{F}$ be a linear operator between two normed linear spaces.

(i) If T is norm–norm continuous, then prove that T is weak–weak continuous. In other words, prove that

$$T : (\mathbf{E}, \sigma(\mathbf{E}, \mathbf{E}^*)) \to (\mathbf{F}, \sigma(\mathbf{F}, \mathbf{F}^*))$$

is continuous.

(ii) If **E** and **F** are both Banach spaces and T is norm–weak continuous, then prove that T is norm–norm continuous.

Hint: Use Exercise 6.20.

(iii) If **E** and **F** are both Banach spaces and T is weak–weak continuous, then prove that T is norm–norm continuous.

EXERCISE 6.22 Let $T : \mathbf{E} \to \mathbf{F}$ be a linear operator between two normed linear spaces.

(i) If T is norm–norm continuous (bounded) and Range(T) is finite dimensional, then prove that T is weak–norm continuous. In other words, prove that

$$T : (\mathbf{E}, \sigma(\mathbf{E}, \mathbf{E}^*)) \to (\mathbf{F}, \| \cdot \|)$$

is continuous.

Hint: Use Exercise 4.6 on an appropriate finite, linearly independent subset of **E**.

(ii) If T is weak–norm continuous, then prove that Range(T) is finite dimensional.

Hint: Construct finitely many linear functionals $\{\varphi_1, \varphi_2, \ldots, \varphi_n\} \subset \mathbf{E}^*$ such that $\bigcap_{i=1}^{n} \ker(\varphi_i) \subset \ker(T)$.

EXERCISE 6.23 Let **E** be a Banach space and $\{e_n\}_{n=1}^{\infty}$ be a countable subset of **E** such that for each $x \in \mathbf{E}$, there are unique scalars $\{\alpha_1, \alpha_2, \ldots\}$ such that

$$\sum_{i=1}^{n} \alpha_i x_i \overset{w}{\to} x$$

as $n \to \infty$. Then prove that $\{e_n\}$ is a Schauder basis for **E**.

Hint: Use Corollary 6.3.12 together with Exercise 6.11.

Weak Sequential Compactness

EXERCISE 6.24 Let **E** be a reflexive space and **F** be any Banach space. If $T \in \mathcal{B}(\mathbf{E}, \mathbf{F})$ is a bounded linear operator, then prove that $T(B_{\mathbf{E}})$ is a closed subset of **F**.

Hint: Use Theorem 6.4.4 and Exercise 6.21.

EXERCISE 6.25 Use Theorem 6.4.4 to give an alternate solution to Exercise 4.28: Let $\mathbf{E}$ be a reflexive Banach space and $\varphi \in \mathbf{E}^*$ be a bounded linear functional. Prove that there is a point $x_0 \in \mathbf{E}$ such that $\|x_0\| = 1$ and

$$\|\varphi\| = |\varphi(x_0)|.$$

In other words, every bounded linear functional attains its norm on the unit sphere.

6.10.3 DEFINITION Let $\mathbf{E}$ be a normed linear space. A sequence $(x_n) \subset \mathbf{E}$ is said to be *weakly Cauchy* if, for each $\varphi \in \mathbf{E}^*$, $(\varphi(x_n))$ is a Cauchy sequence in $\mathbb{K}$.

EXERCISE 6.26 If $\mathbf{E}$ is a reflexive Banach space, then prove that every weakly Cauchy sequence in $\mathbf{E}$ is weakly convergent. (In other words, $\mathbf{E}$ is *weakly sequentially complete*).

EXERCISE 6.27 Let $\mathbf{E}$ be a reflexive Banach space and (K_n) be a sequence of non-empty norm-bounded, closed, convex subsets of $\mathbf{E}$ such that $K_{n+1} \subset K_n$ for all $n \in \mathbb{N}$. Prove that

$$\bigcap_{n=1}^{\infty} K_n \neq \varnothing.$$

Hint: Use Theorem 6.4.4 and Exercise 6.16.

EXERCISE 6.28 The example given in this problem shows that reflexivity is necessary for Exercise 6.27 to work. Let $\mathbf{E} := c_0$. For $n \in \mathbb{N}$, define

$$K_n := \{x = (x_k) \in \mathbf{E} : \|x\| \leq 1 \text{ and } x_1 = x_2 = \ldots = x_n = 1\}.$$

Prove that each K_n is a closed, bounded, convex set. Prove that $K_{n+1} \subset K_n$ and that

$$\bigcap_{n=1}^{\infty} K_n = \varnothing.$$

6.10.4 DEFINITION An operator $T : \mathbf{E} \to \mathbf{F}$ between two normed linear spaces is said to be *completely continuous* if, for any sequence $(x_n) \subset \mathbf{E}$,

$$x_n \xrightarrow{w} x \Rightarrow T(x_n) \xrightarrow{s} T(x).$$

EXERCISE 6.29 If $T : \mathbf{E} \to \mathbf{F}$ is a compact operator, then prove that it is completely continuous.

Hint: Use Lemma 6.1.4, Remark 5.6.2 and Exercise 6.21.

EXERCISE 6.30 Let $\mathbf{E}$ be a reflexive Banach space and $T : \mathbf{E} \to \mathbf{F}$ be a completely continuous operator. Then prove that T is compact.

Hint: First assume $\mathbf{E}$ is separable and appeal to Theorem 6.4.4.

EXERCISE 6.31 Let X be a compact, Hausdorff space. Prove that $C(X)$ is reflexive if and only if X is a finite set.

Hint: Use Theorem 6.4.4.

The Weak-∗ Topology

EXERCISE 6.32

(i) Let $\mathbf{E}$ be a Banach space and $(\varphi_n) \subset \mathbf{E}^*$ be a sequence of bounded linear functionals on $\mathbf{E}$. If $\varphi_n \xrightarrow{w^*} \varphi$, then prove that (φ_n) is norm-bounded and that

$$\|\varphi\| \leq \liminf_{n \to \infty} \|\varphi_n\|.$$

(ii) Give an example to show that the conclusion of Part (i) does not hold if $\mathbf{E}$ is not assumed to be complete.

EXERCISE 6.33 Let $1 < q \leq \infty$ and let p denote the conjugate exponent of q. Let $\Delta : L^q[a,b] \to (L^p[a,b])^*$ be the natural inclusion from Definition 4.1.1. For a sequence $(g_n) \subset L^q[a,b]$, prove that

$$\Delta(g_n) \xrightarrow{w^*} \Delta(g)$$

if and only if $(\|g_n\|_q)$ is a bounded sequence and

$$\lim_{n \to \infty} \int_a^x g_n(t)dt = \int_a^x g(t)dt$$

for all $x \in [a,b]$. (Compare with Lemma 6.1.10).

EXERCISE 6.34 (✖) Complete the proof of Proposition 6.5.6.

EXERCISE 6.35 Let $\Delta_{c_0} : \ell^1 \to (c_0)^*$ and $\Delta_c : \ell^1 \to c^*$ be the isomorphisms described in Exercise 4.13 and Exercise 4.19, respectively. For $n \in \mathbb{N}$, let $e_n \in \ell^1$ denote the nth standard basis vector. Prove that

$$\Delta_{c_0}(e_n) \xrightarrow{w^*} 0, \text{ but } \Delta_c(e_n) \xrightarrow{w^*}\!\!\!\!\!/\ \ 0.$$

EXERCISE 6.36 Let $\mathbf{E} := c_0$ equipped with the supremum norm. For $n \in \mathbb{N}$, define $\varphi_n : \mathbf{E} \to \mathbb{K}$ by $\varphi_n((x_k)) := x_n$. Let

$$C := \mathrm{conv}(\{\varphi_n\}) \subset \mathbf{E}^*$$

be the convex hull of the set $\{\varphi_n\}$. Prove that $\varphi_n \xrightarrow{w^*} 0$, but that $0 \notin \overline{C}^{\|\cdot\|}$ (compare this with Corollary 6.3.12).

EXERCISE 6.37 Let $\mathbf{E}$ be a normed linear space and $\mathbf{W} \subset \mathbf{E}^*$ be a subspace. Let $\mathbf{W}^\perp \subset \mathbf{E}$ be defined as in Definition 4.5.8. Prove that $(\mathbf{W}^\perp)^\perp$ is the weak-$*$ closure of $\mathbf{W}$.

Hint: If $\varphi \in (\mathbf{W}^\perp)^\perp$, choose a basic weak-$*$ open set V containing φ. Assume $V \cap \mathbf{W} = \emptyset$ and follow the ideas in Lemma 6.9.8.

Weak-$*$ Compactness

EXERCISE 6.38 (✘) Let $\mathbf{E} = c_0$ and set $B_\mathbf{E} := \{x \in \mathbf{E} : \|x\| \leq 1\}$. Define

$$U_j := \{(x_n) \in B_\mathbf{E} : |x_j| < 1\}.$$

Prove that each U_j is weakly open in $B_\mathbf{E}$ (with the subspace topology) and that $\{U_j : j \in \mathbb{N}\}$ is an open cover for $B_\mathbf{E}$ which does not have a finite subcover.

EXERCISE 6.39 (✘) Prove that the function d defined in Lemma 6.6.9 is indeed a metric.

EXERCISE 6.40 Let $\{(X_n, d_n) : n \in \mathbb{N}\}$ be a countable family of metric spaces. Let $X := \prod_{n=1}^\infty X_n$ be the product space and define $d : X \times X \to [0, \infty)$ by

$$d((x_n), (y_n)) := \sum_{n=1}^\infty \frac{d_n(x_n, y_n)}{2^n(1 + d_n(x_n, y_n))}.$$

Prove that d is a metric on X and that the associated metric topology coincides with the product topology on X.

EXERCISE 6.41 Let $\mathbf{E}$ be a separable normed linear space and let $\{x_1, x_2, \ldots\} \subset B_\mathbf{E}$ be a countable dense set. Define

$$B_0 := \prod_{n=1}^\infty B_\mathbb{K}[0, \|x_n\|],$$

and equip it with the product topology. Define $\Theta_0 : B_{\mathbf{E}^*} \to B_0$ by

$$\Theta_0(\varphi) := (\varphi(x_1), \varphi(x_2), \ldots).$$

(i) Prove that Θ_0 induces a homeomorphism $B_{\mathbf{E}^*} \to \Theta_0(B_{\mathbf{E}^*})$, when $B_{\mathbf{E}^*}$ is equipped with the weak-$*$ topology.

(ii) Use Exercise 6.40 to conclude that $B_{\mathbf{E}^*}$ is metrizable in the weak-$*$ topology.

This gives an alternate proof of Lemma 6.6.9.

EXERCISE 6.42 Prove the 'dual' of Lemma 6.6.9: Let $\mathbf{E}$ be a normed linear space such that $\mathbf{E}^*$ is separable. Prove that $B_{\mathbf{E}}$ is metrizable in the weak topology.

EXERCISE 6.43 Let $\mathbf{E}$ be a normed linear space.

(i) Prove that there is a compact, Hausdorff topological space $X_{\mathbf{E}}$ and an isometric linear map

$$\Gamma : \mathbf{E} \to C(X_{\mathbf{E}}),$$

where $C(X_{\mathbf{E}})$ is equipped with the supremum norm.

(ii) If $\mathbf{E}$ is separable, prove that there is an isometric linear map $\Gamma : \mathbf{E} \to \ell^\infty$.

Locally Convex Spaces

EXERCISE 6.44 ($\maltese$) Prove that the following spaces are topological vector spaces.

(i) Any normed linear space.

(ii) $\mathfrak{S}$, the space of all real sequences, equipped with the product topology (described in Exercise 2.10).

(iii) For $0 < p < 1$, ℓ^p equipped with the metric topology described in Exercise 2.11.

EXERCISE 6.45 ($\maltese$) If $0 < p < 1$, prove that ℓ^p is not locally convex.

Hint: Use the hint from Exercise 2.11 to prove that, for any $\epsilon > 0$, the convex hull of $\{x \in \ell^p : d(x,0) < \epsilon\}$ is not bounded.

EXERCISE 6.46 For $0 < p < 1$.

(i) For $y = (y_n) \in \ell^\infty$, define $\varphi_y : \ell^p \to \mathbb{K}$ by

$$\varphi_y((x_n)) := \sum_{i=1}^{\infty} x_i y_i.$$

Prove that $\varphi_y \in (\ell^p)^*$.

(ii) If $\varphi \in (\ell^p)^*$, let $y = (y_n)$ be given by

$$y_n := \varphi(e_n).$$

Prove that $y \in \ell^\infty$ and that $\varphi = \varphi_y$.

Hint: If $y \notin \ell^\infty$, then there is a subsequence (y_{n_k}) such that $|y_{n_k}| > 2^k$ for all $k \in \mathbb{N}$.

(iii) Conclude that $(\ell^p)^*$ separates points of ℓ^p.

EXERCISE 6.47 (⚒) If $\mathbf{E}$ is a normed linear space, then prove that $(\mathbf{E}^*, \sigma(\mathbf{E}^*, \mathbf{E}))$ is a locally convex space.

EXERCISE 6.48 (⚒) Prove Lemma 6.7.7.

The Krein–Milman Theorem

EXERCISE 6.49 Let $\mathbf{E}$ be a normed linear space. Let $B_\mathbf{E}$ denote the closed unit ball in $\mathbf{E}$ and let $S_\mathbf{E} := \{x \in E : \|x\| = 1\}$.

(i) Prove that $\text{ext}(B_\mathbf{E}) \subset S_\mathbf{E}$.

(ii) If $\mathbf{E}$ is an inner product space, then prove that $\text{ext}(B_\mathbf{E}) = S_\mathbf{E}$.

Hint: If $x, y \in \mathbf{E}$ are such that $\text{Re}(\langle x, y\rangle) = \|x\|\|y\|$, then $\{x, y\}$ must be linearly dependent.

EXERCISE 6.50 Let $\mathbf{E} = \ell^1$ and let $\{e_1, e_2, \ldots\}$ denotes the standard Schauder basis for $\mathbf{E}$. If $B_\mathbf{E}$ denote the closed unit ball of $\mathbf{E}$, then prove that $\text{ext}(B_\mathbf{E}) = \{\lambda e_n : n \in \mathbb{N}, \lambda \in \mathbb{K} \text{ with } |\lambda| = 1\}$.

EXERCISE 6.51 Let $\mathbf{E} = L^1[0,1]$ and let $B_\mathbf{E}$ denote the closed unit ball of $\mathbf{E}$. For each $\alpha \in (0,1]$, set

$$F_\alpha := \{f \in B_\mathbf{E} : f|_{(0,\alpha)} \equiv 0\}.$$

Prove that $\{F_\alpha : \alpha \in (0,1]\}$ is a totally ordered collection of faces in $B_\mathbf{E}$ such that

$$\bigcap_{\alpha \in (0,1]} F_\alpha = \varnothing.$$

(Compare this with the proof of Proposition 6.8.7).

EXERCISE 6.52 Let $\mathbf{E} := C([0,1], \mathbb{R})$ denote the real normed linear space of continuous, real-valued functions on $[0,1]$, equipped with the supremum norm. Let $B_\mathbf{E}$ denote the unit ball in $\mathbf{E}$.

(i) Suppose $f \in B_\mathbf{E}$ and $x \in [0,1]$ are such that $|f(x)| < 1$, then prove that there is a function $g \in \mathbf{E}$ such that $(f+g)$ and $(f-g)$ are both in $B_\mathbf{E}$.

(ii) Conclude that $\text{ext}(B_\mathbf{E}) = \{\pm 1\}$.

(iii) Prove that $\mathbf{E}$ is not the dual of any normed linear space.

EXERCISE 6.53 Prove that c_0 is not the dual of any normed linear space.

Hint: See Exercise 2.38.

6.10.5 DEFINITION Let $\mathbf{E}$ be a topological vector space. A subset H of $\mathbf{E}$ is said to be a *closed half-space* if there exists $\varphi \in \mathbf{E}^*$ and $\alpha \in \mathbb{R}$ such that $H = \{x \in \mathbf{E} : \text{Re}(\varphi)(x) \leq \alpha\}$.

EXERCISE 6.54 Let K be a subset of a locally convex space. Prove that $\overline{\text{conv}(K)}$ is the intersection of all closed half-spaces containing K.

Uniformly Convex Spaces

EXERCISE 6.55 If $1 < p < \infty$, then prove that the function $f : [0,\infty) \to [0,\infty)$ given by $f(t) = t^p$ is strictly convex.

EXERCISE 6.56 If $\mathbf{E}$ is a uniformly convex normed linear space and (x_n) is a sequence in $B_\mathbf{E}$ such that

$$\lim_{n,m\to\infty} \left\| \frac{x_n + x_m}{2} \right\| = 1,$$

then prove that (x_n) is a Cauchy sequence.

EXERCISE 6.57 If $\mathbf{E}$ is a uniformly convex Banach space, then prove that for every $\varphi \in \mathbf{E}^*$, there is a unique point $x \in \mathbf{E}$ such that $\|x\| = 1$ and $\|\varphi\| = \varphi(x)$ (note that we have already seen this fact in Exercise 4.28).

Hint: Use Exercise 6.56.

EXERCISE 6.58 Let $\mathbf{E}$ be a uniformly convex Banach space and $(x_n) \subset \mathbf{E}$ be a sequence such that $x_n \xrightarrow{w} x$. If $\|x_n\| \to \|x\|$ in $\mathbb{R}$, then prove that $x_n \xrightarrow{s} x$.

EXERCISE 6.59 For $n \in \mathbb{N}$, let $f_n \in L^1[0, 2\pi]$ be the function $f_n(t) = \sin(nt) + 1$. If $f \equiv 1$, then

(i) Prove that $f_n \xrightarrow{w} f$ and that $\|f_n\| \to \|f\|$ in $\mathbb{R}$.

(ii) Prove that (f_n) does not converge strongly to f.

Thus the conclusion of Exercise 6.58 does not hold if the space is not uniformly convex.

EXERCISE 6.60 (✖) Use Goldstine's Theorem to prove the converse of Corollary 6.6.7: If $\mathbf{E}$ is a normed linear space such that $B_{\mathbf{E}}$ is weakly compact, then prove that $\mathbf{E}$ is reflexive.

EXERCISE 6.61 In Exercise 6.60, we have indirectly proved the following: If $\mathbf{E}$ is a normed linear space such that $B_{\mathbf{E}}$ is weakly compact, then $\mathbf{E}$ is complete. Can you prove this fact directly from first principles? (In other words, start with a Cauchy sequence and prove that it converges in $\mathbf{E}$).

Additional Reading

- In Exercise 6.7, we used a quotient map $p : \ell^1 \to \ell^2$ and observed that $\ker(p)$ does not have a complementary subspace in ℓ^1. The same proof also works if we replace ℓ^2 by ℓ^p for $1 < p < \infty$, or even c_0. It is an interesting fact, however, that all these kernels are mutually isomorphic! For more on this and a related open problem, see Section 2.f of this book by Lindenstrauss and Tzafriri.

  Lindenstrauss, Joram, and Tzafriri, Lior (1977). *Classical Banach Spaces*. I. Ergebnisse der Mathematik und ihrer Grenzgebiete, Band 92. Sequence spaces. Berlin–New York: Springer-Verlag. ISBN: 3-540-08072-4.

- In Exercise 6.57, it was shown that every bounded linear functional on a uniformly convex space attains its norm at a (unique) point on the unit sphere. In fact, we know from Exercise 4.28 that this property holds for any reflexive space. What is remarkable, though, is that this property characterizes reflexive spaces!

James, Robert C. (1972 and 1973). Reflexivity and the sup of linear functionals. *Israel Journal of Mathematics*, 13, 289–300. ISSN: 0021-2172. DOI: 10.1007/BF02762803.

Chapter 7

Spectral Theory

7.1 Banach Algebras

In a first course in linear algebra, one encounters a variety of interesting ideas surrounding linear operators on finite dimensional spaces. The central object of this discussion is the concept of an *eigenvalue* or *characteristic value*. The eigenvalues of an operator together with their eigenvectors help us build a collection of ideas such as the Cayley–Hamilton Theorem, the question of diagonalizability of an operator, the theory of canonical forms and much much more.

In this chapter, we will revisit the idea of an eigenvalue in the context of operators on infinite dimensional spaces. Here, the set of eigenvalues needs to be replaced by the *spectrum* of an operator. This is a compact subset of the complex plane that carries a great deal of information about the operator and is the object we wish to study.

> **Note.** Before we get going, we make one important assumption. Henceforth all vector spaces will be over $\mathbb{C}$. The precise reason for this will be explained in due course, but it is related to the Fundamental Theorem of Algebra.

7.1.1 Definition

(i) An *algebra* over $\mathbb{C}$ is a complex vector space $\mathbf{A}$ together with a multiplication map $m : \mathbf{A} \times \mathbf{A} \to \mathbf{A}$ given by $(x, y) \mapsto xy$ satisfying two properties.

(a) Bilinearity: For all scalars $\alpha, \beta \in \mathbb{C}$ and vectors $a, b, c \in \mathbf{A}$,

$$(\alpha a + \beta b)c = \alpha(ac) + \beta(bc) \text{ and } a(\alpha b + \beta c) = \alpha(ab) + \beta(ac).$$

(b) Associativity: $a(bc) = (ab)c$ for all $a, b, c \in \mathbf{A}$.

(ii) An algebra $\mathbf{A}$ is said to be a ***normed algebra*** if there is a norm on $\mathbf{A}$ such that $(\mathbf{A}, \| \cdot \|)$ is a normed linear space and

$$\|ab\| \leq \|a\|\|b\| \tag{7.1}$$

for all $a, b \in \mathbf{A}$.

(iii) A ***Banach algebra*** is a normed algebra that is complete with respect to the induced metric.

The property of Equation 7.1 is called *submultiplicativity* of the norm. We saw in Remark 2.1.2 that the triangle inequality ensures that the addition map $a : \mathbf{A} \times \mathbf{A} \to \mathbf{A}$ is continuous. Likewise, submultiplicativity ensures that the multiplication map $m : \mathbf{A} \times \mathbf{A} \to \mathbf{A}$ is continuous. If $x_n \to x$ and $y_n \to y$, then $x_n y_n \to xy$.

7.1.2 EXAMPLE

(i) $\mathbb{C}$ is a Banach algebra when equipped with the usual multiplication and the absolute value norm.

(ii) If $\mathbb{C}^n$ is equipped with the supremum norm, then we may define multiplication componentwise. If $x = (x_i)$ and $y = (y_i) \in \mathbb{C}^n$, then

$$xy := (x_1 y_1, x_2 y_2, \ldots, x_n y_n).$$

It is entirely obvious that $\mathbb{C}^n$ is a Banach algebra under this operation.

(iii) If X is a compact Hausdorff space, consider $C(X)$ equipped with the operations of pointwise addition and scalar multiplication. Then $C(X)$ is a Banach space under the supremum norm. For $f, g \in C(X)$, we define $fg : X \to \mathbb{C}$ by

$$(fg)(x) := f(x)g(x).$$

$C(X)$ is clearly closed under this operation and is a Banach algebra.

(iv) If X is a locally compact, Hausdorff space, then

$$C_b(X) := \{f : X \to \mathbb{C} : f \text{ is continuous and bounded}\}$$

is a Banach algebra (once again, it is equipped with the supremum norm and pointwise operations).

(v) If X is a locally compact Hausdorff space, let $C_0(X)$ denote the space of continuous functions vanishing at infinity. Then $C_0(X)$ is a Banach algebra when equipped with the supremum norm (Exercise 2.45).

(vi) As a special case of the previous examples with $X = \mathbb{N}$, $\ell^\infty = C_b(\mathbb{N})$ and $c_0 = C_0(\mathbb{N})$ are both Banach algebras.

(vii) If $a, b \in \mathbb{R}$ with $a < b$, then $L^\infty[a, b]$ is a Banach space. When equipped with pointwise multiplication, it forms a Banach algebra.

(viii) Let $\mathbf{E}$ be a Banach space and let $\mathcal{B}(\mathbf{E})$ denote the Banach space of bounded operators from $\mathbf{E}$ to itself. For $S, T \in \mathcal{B}(\mathbf{E})$, define $ST \in \mathcal{B}(\mathbf{E})$ by

$$ST := S \circ T.$$

This makes $\mathcal{B}(\mathbf{E})$ into an algebra. If $x \in \mathbf{E}$, then

$$\|ST(x)\| = \|S(T(x))\| \le \|S\|\|T(x)\| \le \|S\|\|T\|\|x\|.$$

Therefore $\|ST\| \le \|S\|\|T\|$ and $\mathcal{B}(\mathbf{E})$ is a Banach algebra.

(ix) Let $\mathbf{E} = \mathbb{C}^n$ (equipped with any norm) and fix a basis of $\mathbf{E}$. From the previous example, the space $M_n(\mathbb{C})$ of $n \times n$ complex matrices is a Banach algebra under the usual operations of matrix addition and multiplication, when equipped with the operator norm. However, one should be careful here. Not every norm on $M_n(\mathbb{C})$ is submultiplicative! See Exercise 7.3 for such an example.

(x) Consider $\ell^1(\mathbb{Z})$, which is a Banach space under the 1-norm. We define a convolution multiplication on $\ell^1(\mathbb{Z})$ by

$$(x \cdot y)_n := \sum_{k=-\infty}^{\infty} x_k y_{n-k}$$

where $x = (x_j)$ and $y = (y_j)$ are in $\ell^1(\mathbb{Z})$. We need to prove that this operation is well-defined. If $x, y \in \ell^1(\mathbb{Z})$, then

$$\sum_{n=-\infty}^{\infty} |(x \cdot y)_n| = \sum_{n=-\infty}^{\infty} \left| \sum_{k=-\infty}^{\infty} x_k y_{n-k} \right|$$

$$\leq \sum_{n=-\infty}^{\infty} \sum_{k=-\infty}^{\infty} |x_k||y_{n-k}|.$$

Since both x and y are in $\ell^1(\mathbb{Z})$, we may interchange the summation signs (by Proposition 1.2.13) and conclude that

$$\sum_{n=-\infty}^{\infty} |(x \cdot y)_n| \leq \sum_{k=-\infty}^{\infty} |x_k| \left(\sum_{n=-\infty}^{\infty} |y_{n-k}| \right) = \|x\|_1 \|y\|_1.$$

Hence $x \cdot y \in \ell^1(\mathbb{Z})$ and $\|x \cdot y\|_1 \leq \|x\|_1 \|y\|_1$. Thus $\ell^1(\mathbb{Z})$ is a Banach algebra.

(xi) Consider $L^1(\mathbb{R})$ equipped with the 1-norm. By the argument in Theorem 2.3.9, $L^1(\mathbb{R})$ is a Banach space. If $f, g \in L^1(\mathbb{R})$, we define

$$f * g(x) := \int_{\mathbb{R}} f(t)g(x-t)dt.$$

Then by the same ideas as above, $f * g \in L^1(\mathbb{R})$ and $\|f * g\|_1 \leq \|f\|_1 \|g\|_1$. Therefore $L^1(\mathbb{R})$ is a Banach algebra (Exercise 7.4).

7.1.3 DEFINITION Let $\mathbf{A}$ be a Banach algebra and $\mathbf{I}$ be a vector subspace of $\mathbf{A}$. $\mathbf{I}$ is said to be

  (i) a *subalgebra* of $\mathbf{A}$ if $ab \in \mathbf{I}$ whenever $a \in \mathbf{I}$ and $b \in \mathbf{I}$.

 (ii) a *left ideal* of $\mathbf{A}$ if $ab \in \mathbf{I}$ whenever $a \in \mathbf{A}$ and $b \in \mathbf{I}$.

(iii) a *right ideal* of $\mathbf{A}$ if $ba \in \mathbf{I}$ whenever $a \in \mathbf{A}$ and $b \in \mathbf{I}$.

 (iv) a (two-sided) *ideal* if it is both a left and a right ideal.

  (v) a *maximal ideal* if it is an ideal that is maximal in the collection of all proper ideals of $\mathbf{A}$.

If $\mathbf{I}$ is a two-sided ideal of $\mathbf{A}$, we write $\mathbf{I} \lhd \mathbf{A}$.

7.1.4 EXAMPLE

(i) Let $\mathbf{A} = C[0,1]$ and $\Phi : \mathbf{A} \to \mathbb{C}$ be the surjective ring homomorphism

$$f \mapsto f(1).$$

Then there is a ring isomorphism $\mathbf{A}/\ker(\Phi) \to \mathbb{C}$ and $\ker(\Phi)$ must therefore be a maximal ideal of $\mathbf{A}$.

(ii) If $\mathbf{A} = M_n(\mathbb{C})$, then $\mathbf{A}$ has no non-trivial ideals.

Proof. Suppose $\mathbf{J}$ is an ideal of $\mathbf{A}$ such that $\mathbf{J} \neq \{0\}$, then choose $0 \neq T = (T_{i,j}) \in \mathbf{J}$. Then there exists $1 \leq i, j \leq n$ such that $T_{i,j} =: a \neq 0$. Let $E_{k,l}$ be the permutation matrix obtained by switching the kth row of the identity matrix with the lth row. Since $\mathbf{J}$ is an ideal,

$$T' := E_{1,j} T E_{i,1} \in \mathbf{J}$$

and $T'_{1,1} = a \neq 0$. Now let $F_{i,i}$ be the matrix with 1 in the $(i,i)^{th}$ entry and zero elsewhere. Then

$$\frac{1}{a} F_{1,1} T' F_{1,1} = F_{1,1} \in \mathbf{J}.$$

Similarly, $F_{2,2}, F_{3,3}, \ldots, F_{n,n} \in \mathbf{J}$. Adding them up, we conclude that the identity matrix I is in $\mathbf{J}$. Since $\mathbf{J}$ is an ideal, it must happen that $\mathbf{J} = \mathbf{A}$. $\square$

(iii) If X is a locally compact Hausdorff space, then $C_0(X)$ is a vector subspace of $C_b(X)$. Furthermore, if $f \in C_0(X)$ and $g \in C_b(X)$, we wish to check that $fg \in C_0(X)$. To do this, we may assume that $g \neq 0$. So for $\epsilon > 0$, choose a compact set $K \subset X$ such that

$$|f(x)| < \frac{\epsilon}{\|g\|}$$

whenever $x \in X \setminus K$. It follows that $|f(x)g(x)| < \epsilon$ if $x \in X \setminus K$. This is true for any $\epsilon > 0$ and thus $gf = fg \in C_0(X)$. We have proved that $C_0(X)$ is an ideal in $C_b(X)$.

(iv) Let $\mathbf{E}$ be a Banach space and let $\mathcal{F}(\mathbf{E})$ be the subspace of $\mathcal{B}(\mathbf{E})$ consisting of all operators of finite rank. Then $\mathcal{F}(\mathbf{E})$ is an ideal in $\mathcal{B}(\mathbf{E})$ (do try Exercise 7.5).

The final example is so important that it warrants a proposition. This originally appeared as an exercise (Exercise 5.46), but we prove it here as well. If $\mathbf{E}$ is a normed linear space, we write $\mathcal{K}(\mathbf{E})$ for the set of all compact operators on $\mathbf{E}$.

7.1.5 PROPOSITION   *If $\mathbf{E}$ is a Banach space, then $\mathcal{K}(\mathbf{E})$ is a closed ideal in $\mathcal{B}(\mathbf{E})$.*

Proof. Since **E** is a Banach space, $\mathcal{K}(\mathbf{E})$ is a closed subspace of $\mathcal{B}(\mathbf{E})$ by Proposition 5.6.4. Let $S \in \mathcal{K}(\mathbf{E})$ and $T \in \mathcal{B}(\mathbf{E})$, we wish to prove that ST and TS are both in $\mathcal{K}(\mathbf{E})$.

Let $B_{\mathbf{E}}$ denote the closed unit ball in **E**. Since T is bounded, $T(B_{\mathbf{E}})$ is a bounded subset of **E**. Since S is compact, $S(T(B_{\mathbf{E}}))$ has compact closure. Thus $ST \in \mathcal{K}(\mathbf{E})$. On the other hand, since S is compact, $S(B_{\mathbf{E}})$ is totally bounded. Since T is uniformly continuous, $T(S(B_{\mathbf{E}}))$ is totally bounded. Since **E** is complete, $T(S(B_{\mathbf{E}}))$ has compact closure. Thus $TS \in \mathcal{K}(\mathbf{E})$ as well.   □

The next result tells us that we may quotient a Banach algebra by an ideal provided that it is *closed* (just as we did for normed linear spaces in Proposition 2.5.2).

7.1.6 PROPOSITION   *If **A** is a Banach algebra and* $\mathbf{I} \lhd \mathbf{A}$ *is a closed ideal, then* $\mathbf{A}/\mathbf{I}$ *is a Banach algebra.*

Proof. Since **I** is a closed subspace of **A**, we know that $\mathbf{A}/\mathbf{I}$ is a normed linear space when equipped with the norm

$$\|a + \mathbf{I}\| = \inf\{\|a + b\| : b \in \mathbf{I}\}.$$

It is complete by Theorem 2.5.8. Since **A** is a ring and **I** is an ideal, $\mathbf{A}/\mathbf{I}$ carries a ring structure, where multiplication is given by

$$(a + \mathbf{I}) \cdot (b + \mathbf{I}) = ab + \mathbf{I}.$$

Observe that the quotient map $\pi : \mathbf{A} \to \mathbf{A}/\mathbf{I}$ is thus a ring homomorphism and a linear map. An easy calculation now shows that $\mathbf{A}/\mathbf{I}$ is an algebra. To verify submultiplicativity of the norm, let $a, b \in \mathbf{A}$ and $c, d \in \mathbf{I}$. Then $x := cb + ad + cd \in \mathbf{I}$ and thus

$$\|ab + \mathbf{I}\| \le \|ab + x\| = \|(a + c)(b + d)\| \le \|a + c\|\|b + d\|.$$

This is true for any $c, d \in \mathbf{I}$, so $\|ab + \mathbf{I}\| \le \|a + \mathbf{I}\|\|b + \mathbf{I}\|$.   □

Israel Moiseevic Gelfand (1913–2009) was a prolific Soviet mathematician and one of the leading lights of Mathematics in the 20th century. After working odd jobs in Moscow, Gelfand started his research work under Kolmogorov in 1932. Two degrees and six years later, Gelfand had laid the foundations of the theory of Banach algebras.

After groundbreaking work on C*-algebras (which we will encounter in Chapter 8), he went on to study many disparate fields from the representation theory of non-compact groups to cell biology. He also had a deep passion for education. He ran a weekly seminar for 44 years (1946–1989), established a Mathematics School by Correspondence in the Soviet Union and founded the Gelfand Outreach Program in the U.S. in 1992 for high school students.

7.1.7 DEFINITION Let **A** and **B** be Banach algebras.

(i) A map $\Phi : \mathbf{A} \to \mathbf{B}$ is called a *homomorphism* of Banach algebras if Φ is a continuous (bounded) linear operator such that

$$\Phi(ab) = \Phi(a)\Phi(b)$$

for all $a, b \in \mathbf{A}$.

(ii) A bijective homomorphism is called a *topological isomorphism* of Banach algebras. Note that the inverse of such a homomorphism is a (continuous) homomorphism by the Bounded Inverse Theorem.

(iii) If a topological isomorphism is also isometric, then we say that **A** and **B** are isomorphic *as Banach algebras*. In symbols, we denote this by $\mathbf{A} \cong \mathbf{B}$.

7.1.8 EXAMPLE

(i) If $\mathbf{I} \lhd \mathbf{A}$ is a closed ideal, then the natural quotient map $\pi : \mathbf{A} \to \mathbf{A}/\mathbf{I}$ is a bounded linear map by Proposition 2.5.3. Since π is multiplicative, it is a homomorphism of Banach algebras.

(ii) Let X be a compact, Hausdorff space and $\mathbf{A} = C(X)$. For $x_0 \in X$, the evaluation map $\Phi : \mathbf{A} \to \mathbb{C}$ given by

$$f \mapsto f(x_0)$$

homomorphism of Banach algebras since $\mathbf{A}$ carries the supremum norm.

(iii) Let $\mathbf{A}$ be any Banach algebra and $\mathcal{B}(\mathbf{A})$ be the space of bounded linear operators on $\mathbf{A}$ (where we think of $\mathbf{A}$ merely as a Banach space). For each $a \in \mathbf{A}$, define $L_a : \mathbf{A} \to \mathbf{A}$ by

$$L_a(b) := ab.$$

For any $b \in \mathbf{A}$, $\|L_a(b)\| = \|ab\| \leq \|a\|\|b\|$. Thus $L_a \in \mathcal{B}(\mathbf{A})$ and $\|L_a\| \leq \|a\|$. We now get a function $L : \mathbf{A} \to \mathcal{B}(\mathbf{A})$ given by

$$L(a) := L_a.$$

It is easy to see that this map is linear and it is multiplicative because the multiplication on $\mathbf{A}$ is associative. Thus L is a homomorphism of Banach algebras. In analogy with group theory, it is called the *left regular representation* of $\mathbf{A}$.

(iv) Let $\mathbf{A} := L^\infty[0,1]$ and $\mathbf{H} := L^2[0,1]$. Then $\mathcal{B}(\mathbf{H})$ is a Banach algebra. For each $\phi \in \mathbf{A}$, define $M_\phi : \mathbf{H} \to \mathbf{H}$ by

$$M_\phi(g) := \phi g.$$

Note that this is well-defined because $\|\phi g\|_2 \leq \|\phi\|_\infty \|g\|_2$ for any $g \in \mathbf{H}$. Furthermore, it is clear that

$$M_{\phi_1 + \phi_2} = M_{\phi_1} + M_{\phi_2}, \, M_{\alpha\phi} = \alpha M_\phi \text{ and } M_{\phi_1 \phi_2} = M_{\phi_1} M_{\phi_2}$$

for any $\phi_1, \phi_2 \in \mathbf{A}$ and $\alpha \in \mathbb{C}$. Thus we may define $\Phi : \mathbf{A} \to \mathcal{B}(\mathbf{H})$ by

$$\Phi(f) := M_\phi.$$

This is now a homomorphism of Banach algebras. The operator M_ϕ defined above is called a *multiplication operator* and we will discuss these operators in great detail in Chapter 10.

The First Isomorphism Theorem for Banach algebras should now come as no surprise. Between Theorem 2.5.9 and Proposition 7.1.6, it should be routine to prove.

7.1.9 THEOREM (FIRST ISOMORPHISM THEOREM) *Let* $\Phi : \mathbf{A} \to \mathbf{B}$ *be a homomorphism of Banach algebras. Then* $\ker(\Phi)$ *is a closed ideal in* $\mathbf{A}$ *and there is a unique injective homomorphism*

$$\widehat{\Phi} : \mathbf{A}/\ker(\Phi) \to \mathbf{B}$$

such that $\widehat{\Phi} \circ \pi = \Phi$. *Furthermore,* $\|\widehat{\Phi}\| = \|\Phi\|$.

7.2 Invertible Elements

The main reason to study Banach algebras is that it introduces ring-theoretic notions into operator theory. In particular, we may talk about invertible elements (provided the algebra has a multiplicative unit). Perhaps the most interesting idea in this section is that the set of all invertible elements (which forms a group under multiplication) is *open* in the ambient algebra.

7.2.1 DEFINITION A Banach algebra $\mathbf{A}$ is said to be **unital** if there exists $e \in \mathbf{A}$ such that $ae = ea = a$ for all $a \in \mathbf{A}$.

If such a unit exists, we will denote it by 1 or $1_\mathbf{A}$. Also, if $\mathbf{A}$ is unital, we will think of $\mathbb{C}$ as a subset of $\mathbf{A}$ via the inclusion $\alpha \mapsto \alpha 1_\mathbf{A}$. Therefore if you see a dangling scalar 'λ' in $\mathbf{A}$, then you must think of it as '$\lambda 1_\mathbf{A}$'. The next lemma states that we may assume that $\|1_\mathbf{A}\| = 1$ without any loss of generality, which is what we will do henceforth.

7.2.2 LEMMA *Let* $(\mathbf{A}, \|\cdot\|)$ *be a unital Banach algebra. Then there is an equivalent norm* $\|\cdot\|_L$ *on* $\mathbf{A}$ *under which it remains a Banach algebra such that* $\|1_\mathbf{A}\|_L = 1$.

Proof. Consider the left regular representation $L : \mathbf{A} \to \mathcal{B}(\mathbf{A})$ as described in Example 7.1.8. Note that L is injective because if L_a is the identity map on $\mathcal{B}(\mathbf{A})$, then $a = L_a(1_\mathbf{A}) = 1_\mathbf{A}$. Therefore we may define $\|\cdot\|_L : \mathbf{A} \to \mathbb{R}_+$ by

$$\|a\|_L := \|L(a)\|_{\mathcal{B}(\mathbf{A})}.$$

This is clearly a norm on $\mathbf{A}$.

(i) If $a, b \in \mathbf{A}$, then

$$\|ab\|_L := \|L(ab)\|_{\mathcal{B}(\mathbf{A})} = \|L(a)L(b)\|_{\mathcal{B}(\mathbf{A})} \le \|L(a)\|_{\mathcal{B}(\mathbf{A})}\|L(b)\|_{\mathcal{B}(\mathbf{A})} = \|a\|_L\|b\|_L.$$

Hence $(\mathbf{A}, \|\cdot\|_L)$ is a normed algebra.

(ii) Suppose $(a_n) \subset \mathbf{A}$ is a sequence that is Cauchy with respect to $\|\cdot\|_L$, then this implies that the sequence $(L(a_n))$ is Cauchy in $\mathcal{B}(\mathbf{A})$. In particular, the sequence $a_n = L(a_n)(1_\mathbf{A})$ is Cauchy in $(\mathbf{A}, \|\cdot\|)$. Therefore there exists $a \in \mathbf{A}$ such that

$$\lim_{n\to\infty} \|a_n - a\| = 0.$$

However, $L : \mathbf{A} \to \mathcal{B}(\mathbf{A})$ is a bounded operator with respect to $\|\cdot\|$. Therefore

$$\lim_{n\to\infty} \|a_n - a\|_L = \lim_{n\to\infty} \|L(a_n) - L(a)\|_{\mathcal{B}(\mathbf{A})} \le \|L\| \lim_{n\to\infty} \|a_n - a\| = 0.$$

Thus $(\mathbf{A}, \|\cdot\|_L)$ is a Banach algebra.

(iii) We know that $\|a\|_L = \|L(a)\|_{\mathcal{B}(\mathbf{A})} \le \|L\|\|a\|$ for any $a \in \mathbf{A}$. Since $\mathbf{A}$ is complete with respect to each of the norms, Lemma 5.3.13 implies that they are equivalent norms.

To complete the proof, observe that $\|1_\mathbf{A}\|_L = \|L(1_\mathbf{A})\|_{\mathcal{B}(\mathbf{A})} = \|I\|_{\mathcal{B}(\mathbf{A})} = 1$. $\square$

Note. Henceforth if $\mathbf{A}$ is a unital Banach algebra, we will assume that $\|1_\mathbf{A}\| = 1$.

Before we proceed to the examples, we need to recall Urysohn's Lemma for locally compact Hausdorff spaces. We had mentioned it in Remark 2.3.10 for metric spaces, but here is the version we now need (For a proof, you may refer to Rudin [51]).

7.2.3 THEOREM (URYSOHN'S LEMMA) *Let X be a locally compact, Hausdorff space. Let E and F be two disjoint closed subsets of X. If E is compact, then there is a continuous function $f : X \to [0,1]$ such that*

$$f|_E \equiv 1 \text{ and } f|_F \equiv 0.$$

7.2.4 EXAMPLE

(i) If X is a compact Hausdorff space, then $C(X)$ is unital whose unit is the constant function **1**.

(ii) More generally, if X is a locally compact, Hausdorff space, then $C_b(X)$ is unital. In particular, $\ell^\infty = C_b(\mathbb{N})$ is unital.

(iii) If X is a non-compact, locally compact, Hausdorff space, then $C_0(X)$ is non-unital. In particular, $c_0 = C_0(\mathbb{N})$ is non-unital.

Proof. If $e \in C_0(X)$ is a unit, then there must exist a compact set $K \subset X$ such that $|e(x)| < \frac{1}{2}$ whenever $x \notin K$. Since X is not compact, there is a point $x_0 \in X \setminus K$. Let $U \subset X \setminus K$ be an open neighbourhood of x_0 with compact closure. By Urysohn's Lemma, there is a continuous function $f : X \to [0,1]$ such that $f(x_0) = 1$ and

$$f|_{X \setminus U} \equiv 0.$$

Since f is supported on $\overline{U}$, $f \in C_0(X)$ and

$$1 = |f(x_0)| = |f(x_0)e(x_0)| < \frac{1}{2}.$$

This contradiction proves that no such element $e \in C_0(X)$ can exist. $\square$

(iv) $M_n(\mathbb{C})$ is unital, where the unit is given by the identity matrix. More generally, $\mathcal{B}(\mathbf{E})$ is unital for any Banach space $\mathbf{E}$, where the unit is given by the identity operator.

(v) $L^1(\mathbb{R})$ is non-unital.

Proof. Suppose $e \in L^1(\mathbb{R})$ is a unit, then we apply Lemma 4.1.7. For $\epsilon = \frac{1}{2}$, there is a $\delta > 0$ such that, for any measurable $V \subset \mathbb{R}$,

$$m(V) < \delta \Rightarrow \int_V |e(x)| dx < \frac{1}{2}.$$

Let $V = (-\delta/4, \delta/4)$ and $f = \chi_V$, the characteristic function of V. Then for any $x \in V$, $m(x - V) = m(V) < \delta$. Therefore

$$1 = |f(x)| = |(e * f)(x)| = \left| \int_{\mathbb{R}} e(t) f(x - t) dt \right| = \left| \int_{x-V} e(t) dt \right| < \frac{1}{2}.$$

This contradiction proves that $L^1(\mathbb{R})$ does not have a unit. $\square$

(vi) $\ell^1(\mathbb{Z})$ is unital with unit $e = (e_n)$ given by

$$e_n = \begin{cases} 1 & : n = 0, \\ 0 & : n \neq 0. \end{cases}$$

(vii) $L^\infty[a,b]$ is unital, where the unit is the constant function **1**.

7.2.5 DEFINITION   Let **A** be a unital Banach algebra.

(i) An element $a \in \mathbf{A}$ is said to be **invertible** if there exists $b \in \mathbf{A}$ such that $ab = ba = 1_{\mathbf{A}}$. Such an element b is unique. It is called the **inverse** of a and is denoted by a^{-1}.

(ii) The set of all invertible elements in **A** is denoted by $GL(\mathbf{A})$.

7.2.6 EXAMPLE

(i) If $\mathbf{A} = \mathbb{C}$, then $GL(\mathbf{A}) = \mathbb{C}^\times = \mathbb{C} \setminus \{0\}$.

(ii) Let X be a compact Hausdorff space and $\mathbf{A} = C(X)$. If $f \in C(X)$ is invertible, then there exists $g \in C(X)$ such that $f(x)g(x) = 1$ for all $x \in X$. In particular, it must happen that $f(x) \neq 0$ for all $x \in X$. Conversely, if $f \in C(X)$ is such that $f(x) \neq 0$ for all $x \in X$, then $g : X \to \mathbb{C}$ given by $g(x) := \frac{1}{f(x)}$ is well-defined and continuous. Since $fg = \mathbf{1}$, f is invertible in $C(X)$. Therefore

$$GL(C(X)) = \{f \in C(X) : f(x) \neq 0 \text{ for all } x \in X\}.$$

(iii) Let **E** be a Banach space, then $GL(\mathcal{B}(\mathbf{E}))$ is merely the set of all invertible operators from **E** to itself.

(iv) As a special case, consider $M_n(\mathbb{C}) \cong \mathcal{B}(\mathbb{C}^n)$. Then the set of invertible $n \times n$ matrices is often written as $GL_n(\mathbb{C})$. Note that

$$GL_n(\mathbb{C}) = \{T \in M_n(\mathbb{C}) : \det(T) \neq 0\}$$

where 'det' is the determinant function.

We now give a general recipe to construct an invertible element. Suppose **A** is a unital Banach algebra and $x \in \mathbf{A}$. Then

$$\left\| \sum_{k=0}^{\infty} \frac{x^k}{k!} \right\| \leq \sum_{k=0}^{\infty} \frac{\|x^k\|}{k!} \leq \sum_{k=0}^{\infty} \frac{\|x\|^k}{k!} < \infty.$$

Since **A** is a Banach space, Proposition 2.3.8 allows us to make the following definition.

7.2.7 DEFINITION For $x \in \mathbf{A}$, the *exponential* of x is defined as

$$\exp(x) := \sum_{k=0}^{\infty} \frac{x^k}{k!}.$$

If $x, y \in A$ commute, then one can show that $\exp(x) \exp(y) = \exp(x + y)$. It then follows that $\exp(x)$ is invertible for any $x \in \mathbf{A}$. The proofs of these facts are omitted here (do try Exercise 7.18).

The next theorem is crucial for us going forward. It gives a clean condition under which you can expect an element to be invertible. Moreover, it proves that $GL(\mathbf{A})$ is an open set.

7.2.8 THEOREM *Let* **A** *be a unital Banach algebra. If* $a \in \mathbf{A}$ *is such that* $\|1 - a\| < 1$, *then* $a \in GL(\mathbf{A})$. *Furthermore,* a^{-1} *is given by the* ***Neumann series***

$$a^{-1} = 1 + (1 - a) + (1 - a)^2 + \ldots = \sum_{k=0}^{\infty} (1 - a)^k. \tag{7.2}$$

Also,

$$\|1 - a^{-1}\| \leq \frac{\|1 - a\|}{1 - \|1 - a\|}.$$

Proof. Observe that $\|(1 - a)^k\| \leq \|(1 - a)\|^k$. Since $\|1 - a\| < 1$, the series appearing on the right hand side of Equation 7.2 is absolutely convergent. Since

A is a Banach space, the series

$$b := \sum_{n=0}^{\infty} (1-a)^n$$

converges in **A** (Proposition 2.3.8). If $x := (1-a)$, then $\|x\| < 1$, so $\lim_{n\to\infty} x^{n+1} = 0$. By the continuity of multiplication,

$$ab = \lim_{n\to\infty} \sum_{k=0}^{n} a(1-a)^k = \lim_{n\to\infty} \sum_{k=0}^{n} (1-x)x^k = \lim_{n\to\infty} (1 - x^{n+1}) = 1.$$

Similarly, $ba = 1$ as well and we have proved that a is invertible. Now observe that

$$\|1-b\| \le \sum_{n=1}^{\infty} \|(1-a)^n\| \le \sum_{n=1}^{\infty} \|1-a\|^n \le \frac{\|1-a\|}{1-\|1-a\|}.$$

$\square$

If **A** is a unital Banach algebra and $x, y \in GL(\mathbf{A})$, then it is clear that xy is also invertible and that $(xy)^{-1} = y^{-1}x^{-1}$. Therefore $GL(\mathbf{A})$ is a group under multiplication. Now $GL(\mathbf{A})$ also inherits the norm topology as a subset of **A**. We know that the multiplication map $m : GL(\mathbf{A}) \times GL(\mathbf{A}) \to GL(\mathbf{A})$ is continuous by the submultiplicativity of the norm. The next result shows that the inverse map $i : GL(\mathbf{A}) \to GL(\mathbf{A})$ is also continuous, making $GL(\mathbf{A})$ a topological group.

7.2.9 DEFINITION A *topological group* is a group G equipped with a topology such that the multiplication operation $m : G \times G \to G$ and the inverse operation $i : G \to G$ are both continuous.

7.2.10 COROLLARY *Let **A** be a unital Banach algebra. Then $GL(\mathbf{A})$ is an open subset of **A** and a topological group.*

Proof.

(i) If $a \in GL(\mathbf{A})$, set $r := \frac{1}{\|a^{-1}\|^{-1}} > 0$. If $b \in \mathbf{A}$ is such that $\|a - b\| < r$, then

$$\|1 - a^{-1}b\| < 1.$$

By Theorem 7.2.8, $a^{-1}b$ is invertible which implies that b is invertible. Thus $B_\mathbf{A}(a, r) = \{b \in \mathbf{A} : \|a - b\| < r\} \subset GL(\mathbf{A})$. This is true for each $a \in GL(\mathbf{A})$, proving that $GL(\mathbf{A})$ is open.

(ii) We show that the map $x \mapsto x^{-1}$ is continuous. Suppose $a_n \to a$ in $GL(\mathbf{A})$, then we wish to prove that $a_n^{-1} \to a^{-1}$. Since multiplication by a is a homeomorphism, it suffices to prove that

$$a_n^{-1}a \to 1.$$

Let $b_n := a_n a^{-1}$, then $b_n \to 1$ and we wish to prove that $b_n^{-1} \to 1$. So fix $\epsilon > 0$ and choose $1 > \delta > 0$ such that

$$\frac{\delta}{1-\delta} < \epsilon.$$

Since $b_n \to 1$, there exists $N \in \mathbb{N}$ such that $\|b_n - 1\| < \delta$ for all $n \geq N$. By Theorem 7.2.8,

$$\|b_n^{-1} - 1\| \leq \frac{\delta}{(1-\delta)} < \epsilon$$

for all $n \geq N$. Therefore $b_n^{-1} \to 1$. $\square$

We end this section with two nice applications of Corollary 7.2.10 which we will need in Chapter 8. The next result is merely a restatement of a fact you would know from ring theory and a proof may be found in Dummit & Foote [15].

7.2.11 PROPOSITION *Let $\mathbf{A}$ be a unital Banach algebra, then every proper ideal $\mathbf{I} \lhd \mathbf{A}$ is contained in a maximal ideal. In particular, every unital Banach algebra has a maximal ideal.*

7.2.12 PROPOSITION *Let $\mathbf{A}$ be a unital Banach algebra.*

(i) *If $\mathbf{I} \lhd \mathbf{A}$ is a proper ideal, then $\bar{\mathbf{I}}$ is a proper ideal.*

(ii) *Every maximal ideal in $\mathbf{A}$ is closed.*

Proof. Observe that Part (ii) follows directly from Part (i). To prove Part (i), note that if $\mathbf{I}$ is an ideal, then so is $\bar{\mathbf{I}}$ (do verify this). If $\mathbf{I} \lhd \mathbf{A}$ is a proper ideal, then $\mathbf{I} \cap GL(\mathbf{A}) = \emptyset$. Since $GL(\mathbf{A})$ is open,

$$\bar{\mathbf{I}} \cap GL(\mathbf{A}) = \emptyset.$$

In particular, $1_{\mathbf{A}} \notin \bar{\mathbf{I}}$, which proves that $\bar{\mathbf{I}} \neq \mathbf{A}$. $\square$

Simple examples show that Proposition 7.2.12 is no longer true in non-unital Banach algebras (see Exercise 7.24). If **A** is a Banach algebra and $\mathbf{I} \lhd \mathbf{A}$, we had proved in Proposition 2.5.3 that the quotient map $\pi : \mathbf{A} \to \mathbf{A}/\mathbf{I}$ has norm 1. In the unital case, we now can say more.

7.2.13 PROPOSITION *Let* **A** *be a unital Banach algebra and* $\mathbf{I} \lhd \mathbf{A}$ *be a proper closed ideal. Let* $\pi : \mathbf{A} \to \mathbf{A}/\mathbf{I}$ *be the natural homomorphism. Then* π *is continuous and* $\|\pi\| = \|\pi(1_\mathbf{A})\| = 1$.

Proof. By Proposition 2.5.3, it suffices to show that $\|\pi\| = \|\pi(1_\mathbf{A})\|$. Observe that since **I** is a proper closed ideal, $\mathbf{I} \cap GL(\mathbf{A}) = \emptyset$. In particular, **I** does not contain any elements $a \in \mathbf{A}$ such that $\|a - 1_\mathbf{A}\| < 1$. Therefore

$$\|\pi(1_\mathbf{A})\| = \inf\{\|1_\mathbf{A} - x\| : x \in \mathbf{I}\} \geq 1.$$

Conversely, since $\|\pi\| = 1$ and $\|1_\mathbf{A}\| = 1$ (by assumption), we have $\|\pi(1_\mathbf{A})\| \leq \|\pi\|\|1_\mathbf{A}\| \leq 1$. Therefore $\|\pi\| = \|\pi(1_\mathbf{A})\|$. $\qquad \square$

7.3 Spectrum of an Element of a Banach Algebra

We now arrive at the central notion of this chapter. The spectrum of an operator is a generalization of the set of all eigenvalues of a matrix. However, this is a very poor description. When defined in the context of Banach algebras, we may not only talk about spectra of operators, but also of continuous (or Borel) functions. This allows us to *compare* the spectra of elements when there is a homomorphism of Banach algebras. This simple device will prove to be of utmost importance to us in Chapter 8.

For now, we begin with a definition and establish the basic properties and examples. Throughout this section, **A** will denote a unital Banach algebra with unit $1_\mathbf{A}$.

7.3.1 DEFINITION Let $a \in \mathbf{A}$.

(i) The *spectrum of a* is defined as

$$\sigma_\mathbf{A}(a) = \sigma(a) := \{\lambda \in \mathbb{C} : (a - \lambda 1_\mathbf{A}) \notin GL(\mathbf{A})\}.$$

An element of $\sigma(a)$ is called a *spectral value* of a.

(ii) The *resolvent of a* is defined as

$$\rho_\mathbf{A}(a) = \rho(a) := \mathbb{C} \setminus \sigma(a).$$

7.3.2 EXAMPLE

(i) If $\mathbf{E} := \mathbb{C}^n$, then an operator in $\mathcal{B}(\mathbf{E})$ is invertible if and only if it is injective (since $\mathbf{E}$ is finite dimensional). Therefore if $T \in \mathcal{B}(\mathbf{E})$, then $\sigma(T)$ is the set of eigenvalues of T.

(ii) If $\mathbf{E}$ is a Banach space, then an operator is invertible if and only if it is bijective (by the Bounded Inverse Theorem). Therefore if $T \in \mathcal{B}(\mathbf{E})$, then

$$\sigma(T) = \{\lambda \in \mathbb{C} : (T - \lambda I) \text{ is not bijective}\}.$$

(iii) Let $T \in \mathcal{B}(C[0,1])$ be the operator

$$T(f)(x) := \int_0^x f(t)dt.$$

Then T is not surjective because if $g \in \text{Range}(T)$, then g is a differentiable function. Hence $0 \in \sigma(T)$. However, 0 is not an eigenvalue of T (Exercise 7.28).

(iv) Let X be a compact Hausdorff space and $\mathbf{A} := C(X)$. By the discussion in Example 7.2.6, $g \in C(X)$ is invertible if and only if $0 \notin g(X)$. Hence if $f \in C(X)$, then

$$\sigma(f) = f(X).$$

(v) Let $\mathbf{A} = \ell^\infty(I)$ for some set I. Then for any $f \in \mathbf{A}$, $\sigma(f) = \overline{f(I)}$ is the closure of the range of f in $\mathbb{C}$. The argument is exactly as above and I invite you to try it (Exercise 7.16).

7.3.3 PROPOSITION *For any $a \in \mathbf{A}$, $\sigma(a)$ is a compact subset of the disc*

$$\{z \in \mathbb{C} : |z| \leq \|a\|\} \subset \mathbb{C}.$$

Proof. If $|\lambda| > \|a\|$, then $\|a/\lambda\| < 1$ and $(\lambda 1_{\mathbf{A}} - a) = \lambda(1_{\mathbf{A}} - a/\lambda)$ is invertible by Theorem 7.2.8. Therefore

$$\sigma(a) \subset \{z \in \mathbb{C} : |z| \leq \|a\|\}.$$

In particular, $\sigma(a)$ is bounded. Now consider the function $f : \mathbb{C} \to \mathbf{A}$ given by $\lambda \mapsto (\lambda 1_{\mathbf{A}} - a)$. Observe that f is continuous and $\lambda \in \rho(a)$ if and only if $f(\lambda) \in GL(\mathbf{A})$. Thus

$$\rho(a) = f^{-1}(GL(\mathbf{A})).$$

Since $GL(\mathbf{A})$ is open in $\mathbf{A}$, $\rho(a)$ is open in $\mathbb{C}$. Hence $\sigma(a)$ is closed in $\mathbb{C}$ and thus compact. $\qquad\square$

Our immediate goal is to prove that the spectrum is non-empty. The next definition serves as a useful tool.

7.3.4 DEFINITION Let $\mathbf{A}$ be a Banach algebra and $\Omega \subset \mathbb{C}$ be an open set. A function $F : \Omega \to \mathbf{A}$ is said to be *analytic* if there is a continuous function $G : \Omega \to \mathbf{A}$ such that

$$\lim_{h \to 0} \left\| \frac{F(z+h) - F(z)}{h} - G(z) \right\| = 0$$

for all $z \in \Omega$.

If such a function exists, it is clearly unique, so we may write $F' = G$ on Ω. Suppose F is analytic and $\tau \in \mathbf{A}^*$ is a bounded linear functional, then $H : \Omega \to \mathbb{C}$ given by $H = \tau \circ F$ is analytic (in the usual sense) and $H' = \tau \circ G$ (Exercise 7.25).

7.3.5 LEMMA *Let $\mathbf{A}$ be a unital Banach algebra and $a \in \mathbf{A}$. Define $F : \rho(a) \to \mathbf{A}$ by*

$$F(z) = (z - a)^{-1}.$$

Then F is analytic and $F'(z) = -(z - a)^{-2}$.

Proof. We know that the inverse map is continuous on $GL(\mathbf{A})$ (by Corollary 7.2.10) and the map $b \mapsto b^2$ is continuous on $\mathbf{A}$. Therefore if $G : \rho(a) \to \mathbf{A}$ is given by $z \mapsto -(z - a)^{-2}$, then G is continuous. Now for any $x, y \in GL(\mathbf{A})$, we have the identity

$$x^{-1} - y^{-1} = x^{-1}(y - x)y^{-1}.$$

Fix $z \in \rho(a)$ and choose $h \in \mathbb{C}$ of small enough modulus so that $x = (z + h - a)$ and $y = (z - a)$ are both invertible. By the above identity, we have

$$\left\| \frac{F(z+h) - F(z)}{h} - G(z) \right\| = \left\| \left((z+h-a)^{-1} - (z-a)^{-1} \right) (z-a)^{-1} \right\|.$$

Since the map $z \mapsto (z - a)^{-1}$ is continuous, we conclude that

$$\lim_{h \to 0} \left\| \left((z+h-a)^{-1} - (z-a)^{-1} \right) (z-a)^{-1} \right\| = 0.$$

This proves the result. $\qquad\square$

From the Fundamental Theorem of Algebra, we know that an operator on a finite dimensional complex vector space must have an eigenvalue. The next result generalizes this fact to a possibly infinite dimensional Banach algebra. This result is central to the subject and this is one of the principal reasons that we chose to work over the field of complex numbers. In fact, if you are familiar with one of the standard proofs of the Fundamental Theorem of Algebra, it should come as no surprise that the main tool used in the proof is Liouville's Theorem.

7.3.6 THEOREM *If* $\mathbf{A}$ *is a unital Banach algebra and* $a \in \mathbf{A}$, *then* $\sigma(a)$ *is nonempty.*

Proof. Let $a \in \mathbf{A}$. If $a = 0$, then $0 \in \sigma(a)$, so we may assume that $a \neq 0$. Suppose $\sigma(a) = \varnothing$, then $\rho(a) = \mathbb{C}$, so we consider $F : \mathbb{C} \to \mathbf{A}$ given by

$$F(z) = (z - a)^{-1}.$$

By Lemma 7.3.5, F is analytic and $F'(z) = -(z - a)^{-2}$. If $z \in \mathbb{C}$ is such that $|z| > R := 2\|a\|$, then

$$\left\| \left(1 - \frac{a}{z}\right) - 1 \right\| = \left\| \frac{a}{z} \right\| < \frac{1}{2}.$$

By Corollary 7.2.10, $\left(1 - \frac{a}{z}\right) \in GL(\mathbf{A})$ and

$$\left\| \left(1 - \frac{a}{z}\right)^{-1} - 1 \right\| \leq 1.$$

If $|z| > R$, then since $\|1\| = 1$,

$$\|F(z)\| = \left\| z^{-1} \left(1 - \frac{a}{z}\right)^{-1} \right\| \leq \frac{2}{R}. \tag{7.3}$$

Let $\tau \in \mathbf{A}^*$ be a fixed bounded linear functional and consider $H := \tau \circ F$. Then H is an entire function and it is bounded by Equation 7.3. Therefore H is constant by Liouville's Theorem. However, we know from Lemma 7.3.5 that

$$H'(z) = -\tau((z - a)^{-2}).$$

We conclude that $\tau(a^{-2}) = -H'(0) = 0$. This is true for all $\tau \in \mathbf{A}^*$. Since $\mathbf{A}^*$ separates points of $\mathbf{A}$ (Corollary 4.2.8), $a^{-2} = 0$. However, we had assumed that $a \neq 0$ and this contradiction proves that $\sigma(a)$ could not have been empty. $\square$

Recall that a *division ring* is a ring in which every non-zero element has an inverse. The next result merely states that there are no non-trivial Banach algebras that can be division rings.

7.3.7 COROLLARY (GELFAND, 1941, AND MAZUR, 1938) *If* **A** *is a unital Banach algebra in which every non-zero element is invertible, then* **A** $= \mathbb{C}1_{\mathbf{A}}$.

Proof. Let $a \in \mathbf{A}$, then there is $\lambda \in \mathbb{C}$ such that $a - \lambda 1_{\mathbf{A}}$ is not invertible (by Theorem 7.3.6). By hypothesis, $a - \lambda 1_{\mathbf{A}} = 0$. Therefore to each $a \in \mathbf{A}$, there is a $\lambda \in \mathbb{C}$ such that $a = \lambda 1_{\mathbf{A}}$. This proves the result. $\square$

7.3.8 DEFINITION For an element a in a unital Banach algebra **A**, the *spectral radius of a* is defined by

$$r(a) := \sup\{|\lambda| : \lambda \in \sigma(a)\}.$$

By Proposition 7.3.3, $r(a) \leq \|a\|$. Furthermore, since $\sigma(a)$ is compact, there is a point $\lambda_0 \in \sigma(a)$ such that $r(a) = |\lambda_0|$. As we will see below, it quite possible that $r(a) < \|a\|$.

7.3.9 EXAMPLE

(i) Let X be a compact Hausdorff space and $\mathbf{A} := C(X)$. For $f \in \mathbf{A}$, we had seen that $\sigma(f) = f(X)$. Therefore

$$r(f) = \|f\|_{\infty}.$$

(ii) If $T = \begin{bmatrix} 0 & 1 \\ 0 & 0 \end{bmatrix} \in M_2(\mathbb{C})$, then $T^2 = 0$. Therefore the only eigenvalue of T is 0. Thus $\sigma(T) = \{0\}$ and

$$r(T) = 0 < \|T\|.$$

More generally, if $T \in M_n(\mathbb{C})$ is such that $r(T) = 0$, then the only eigenvalue of T is zero. Therefore, the minimal polynomial of T is of the form $p(x) = x^k$ for some $k \in \mathbb{N}$. This means that $T^k = 0$. Thus the only operators on $\mathbb{C}^n$ with $r(T) = 0$ are the *nilpotent* operators.

(iii) Let $1 \leq p \leq \infty$ and $T : \ell^p \to \ell^p$ be the left-shift operator given by

$$T((x_j)) = (x_2, x_3, \ldots).$$

Then we wish to determine $\sigma(T)$.

If $\lambda \in \mathbb{C}$ such that $|\lambda| < 1$, then the sequence $x := (\lambda, \lambda^2, \lambda^3, \ldots)$ is an element of ℓ^p and $Tx = \lambda x$. Therefore $\lambda \in \sigma(T)$ and we see that

$$\{z \in \mathbb{C} : |z| < 1\} \subset \sigma(T).$$

Now observe that $\|T(x)\| \leq \|x\|$ for all $x \in \ell^p$ so that $\|T\| \leq 1$. By Proposition 7.3.3, we know that $\sigma(T)$ is a closed set that is contained in $\mathbb{D} := \{z \in \mathbb{C} : |z| \leq 1\}$. Therefore the only possibility is that $\|T\| = 1$ and

$$\sigma(T) = \mathbb{D}.$$

In particular, $r(T) = 1 = \|T\|$.

Our next goal in this section is to explicitly compute the spectral radius. To that end, we need the following theorem which is extremely useful in analyzing the spectra of operators. As is customary in algebra, we will henceforth write $\mathbb{C}[z]$ for the set of all polynomials with complex coefficients.

7.3.10 DEFINITION Let $\mathbf{A}$ be a unital Banach algebra, $a \in \mathbf{A}$ and $p(z) \in \mathbb{C}[z]$ be a polynomial given explicitly as $p(z) = a_0 + a_1 z + \ldots + a_n z^n$. The element $p(a) \in \mathbf{A}$ is then defined as

$$p(a) := a_0 1_{\mathbf{A}} + a_1 a + \ldots + a_n a^n.$$

Note that if $\mathbf{A}$ is non-unital, then we may apply polynomials $p(z) \in \mathbb{C}[z]$ to elements of $\mathbf{A}$ provided $p(0) = 0$.

7.3.11 THEOREM (SPECTRAL MAPPING THEOREM) *Let $\mathbf{A}$ be a unital Banach algebra, $a \in \mathbf{A}$ and $p(z) \in \mathbb{C}[z]$. Then*

$$\sigma(p(a)) = p(\sigma(a)) = \{p(\lambda) : \lambda \in \sigma(a)\}.$$

Proof. If $p(z) = a_0 + a_1 z + \ldots + a_n z^n$, then $p(a) = a_0 1_{\mathbf{A}} + a_1 a + \ldots + a_n a^n$. Now if $\alpha \in \mathbb{C}$, then by the Fundamental Theorem of Algebra, there exist scalars $\gamma, \beta_1, \beta_2, \ldots, \beta_n \in \mathbb{C}$ such that $p(z) - \alpha = \gamma(z - \beta_1)(z - \beta_2) \ldots (z - \beta_n)$. In $\mathbf{A}$ this translates to

$$p(a) - \alpha 1_{\mathbf{A}} = \gamma(a - \beta_1 1_{\mathbf{A}})(a - \beta_2 1_{\mathbf{A}}) \ldots (a - \beta_n 1_{\mathbf{A}}).$$

Now if $\beta_i \notin \sigma(a)$ for all $1 \le i \le n$, then each element $(a - \beta_i 1_{\mathbf{A}})$ is invertible. This implies that $(p(a) - \alpha 1_{\mathbf{A}})$ is invertible. Therefore we have the implication

$$\alpha \in \sigma(p(a)) \Rightarrow \beta_i \in \sigma(a) \text{ for some } 1 \le i \le n.$$

Conversely, if $(p(a) - \alpha 1_{\mathbf{A}})$ were invertible (with inverse $b \in \mathbf{A}$), then

$$1_{\mathbf{A}} = b\gamma \prod_{k=1}^{n} (a - \beta_k 1_{\mathbf{A}}) = \gamma \prod_{k=1}^{n} (a - \beta_k 1_{\mathbf{A}}) b.$$

For any fixed $i \in \{1, 2, \dots, n\}$, $(a - \beta_i 1_{\mathbf{A}})$ commutes with $(a - \beta_j 1_{\mathbf{A}})$ for all $1 \le j \le n$. Thus there exist $c, d \in \mathbf{A}$ such that

$$1_{\mathbf{A}} = c(a - \beta_i 1_{\mathbf{A}}) = (a - \beta_i 1_{\mathbf{A}}) d.$$

This means that $(a - \beta_i 1_{\mathbf{A}})$ is invertible, so $\beta_i \notin \sigma(a)$. We conclude that

$$\alpha \notin \sigma(p(a)) \Rightarrow \beta_i \notin \sigma(a) \text{ for all } 1 \le i \le n.$$

Therefore

$$\alpha \in \sigma(p(a)) \Leftrightarrow \beta_i \in \sigma(a) \quad \text{for some } 1 \le i \le n$$
$$\Leftrightarrow p(\lambda) - \alpha = 0 \quad \text{for some } \lambda \in \sigma(a)$$
$$\Leftrightarrow \alpha \in p(\sigma(a)). \qquad \qquad \square$$

The Spectral Mapping Theorem is extremely useful. In fact, one of the major themes in subsequent chapters will be to extend this theorem to other classes of functions (such as continuous functions). For now, we use this theorem to explicitly determine the spectral radius.

7.3.12 THEOREM (SPECTRAL RADIUS FORMULA) *For any $a \in \mathbf{A}$,*

$$r(a) = \lim_{n \to \infty} \|a^n\|^{1/n}.$$

In particular, this limit exists.

Proof.

(i) By Proposition 7.3.3, $r(a) \le \|a\|$. In fact, if $\lambda \in \sigma(a)$, then $\lambda^n \in \sigma(a^n)$ by the Spectral Mapping Theorem. Hence $|\lambda^n| \le \|a^n\|$ and thus $|\lambda| \le \|a^n\|^{1/n}$. This is true for each $\lambda \in \sigma(a)$ and each $n \in \mathbb{N}$, which proves that

$$r(a) \le \liminf_{n \to \infty} \|a^n\|^{1/n}.$$

(ii) We now wish to prove that $r(a) \geq \limsup_{n\to\infty} \|a^n\|^{1/n}$. To that end, let D be the open disc in $\mathbb{C}$ centered at 0 of radius $1/r(a)$ (If $r(a) = 0$, then we take D to be all of $\mathbb{C}$). If $\lambda \in D$, then $|\frac{1}{\lambda}| > r(a)$. Therefore

$$1_{\mathbf{A}} - \lambda a = \lambda \left(\frac{1_{\mathbf{A}}}{\lambda} - a \right) \in GL(\mathbf{A}).$$

So if $\tau \in \mathbf{A}^*$ is a fixed bounded linear functional, consider the map $g : D \to \mathbb{C}$ given by

$$g(\lambda) := \tau((1_{\mathbf{A}} - \lambda a)^{-1}).$$

As in Theorem 7.3.6, g is analytic. Therefore there are unique scalars $(\alpha_n) \subset \mathbb{C}$ such that

$$g(\lambda) = \sum_{n=0}^{\infty} \alpha_n \lambda^n.$$

Now if $|\lambda| < 1/\|a\|$, then $\lambda \in D$ and $\|\lambda a\| < 1$, so by Theorem 7.2.8,

$$(1_{\mathbf{A}} - \lambda a)^{-1} = \sum_{n=0}^{\infty} \lambda^n a^n.$$

Hence

$$g(\lambda) = \sum_{n=0}^{\infty} \tau(\lambda^n a^n) = \sum_{n=0}^{\infty} \lambda^n \tau(a^n).$$

So by uniqueness of the (α_n),

$$g(\lambda) = \sum_{n=0}^{\infty} \tau(a^n) \lambda^n \tag{7.4}$$

for all $\lambda \in D$. In particular, for fixed $\lambda \in D$, the series in Equation 7.4 converges in $\mathbb{C}$ and so the sequence $(\tau(a^n)\lambda^n)$ converges to 0. This is true for all $\tau \in \mathbf{A}^*$. By the Principle of Uniform Boundedness, there exists $M > 0$

such that

$$\|\lambda^n a^n\| \leq M$$

for all $n \geq 0$. Therefore

$$\|a^n\|^{1/n} \leq \frac{M^{1/n}}{|\lambda|}$$

for all $n \geq 1$. Taking $\limsup$ on both sides, we get

$$\limsup_{n \to \infty} \|a^n\|^{1/n} \leq \frac{1}{|\lambda|}.$$

This is true for all $\lambda \in \mathbb{C}$ such that $|\lambda| < 1/r(a)$. Therefore $\limsup \|a^n\|^{1/n} \leq r(a)$, which proves the theorem. $\qquad \square$

In Example 7.3.9, we had seen that an operator on a finite dimensional Banach space has spectrum $\{0\}$ if and only if it is nilpotent. The next example shows that this is no longer true for operators on infinite dimensional spaces.

7.3.13 EXAMPLE Let $\mathbf{E} := C[0,1]$ equipped with the supremum norm, $\mathbf{A} := \mathcal{B}(C[0,1])$ and $T \in \mathbf{A}$ be the operator

$$T(f)(x) = \int_0^x f(t)dt.$$

We had seen in Example 2.2.6 that T is a bounded operator with $\|T\| \leq 1$. We now wish to compute the spectral radius of T. Note that

$$T^2(f)(x) = \int_0^x \int_0^t f(s)dsdt.$$

Hence

$$|T^2(f)(x)| \leq \|f\|_\infty \int_0^x \int_0^t dsdt = \|f\|_\infty \frac{x^2}{2}.$$

Therefore $\|T^2\| \leq \frac{1}{2}$. Proceeding in this fashion, we may show that $\|T^n\| \leq \frac{1}{n!}$ for all $n \geq 1$. Hence

$$r(T) \leq \lim_{n \to \infty} \left(\frac{1}{n!}\right)^{1/n} = 0.$$

Thus $\sigma(T) = \{0\}$ even though T is not nilpotent.

An element a of a unital Banach algebra **A** with the property that $\sigma(a) = \{0\}$ is said to be **quasinilpotent**. We give another example of such an element in Exercise 7.49.

The notion of a *linear metric ring* was first introduced by Nagumo in 1936 [44], who used the submultiplicativity of the norm to prove some basic results, including the fact that the group of invertibles is open. In 1941, I. M. Gelfand (1913–2009) developed the theory more fully and termed these objects *normed rings* [20]. He proved the Spectral Radius Formula and also discussed the key role played by maximal ideals (which we will explore in Chapter 8). To highlight the applicability of his new methods, he also gave a simple proof of a theorem due to Wiener regarding functions with absolutely convergent Fourier series. This was considered a resounding success for the nascent subject at the time.

In 1945, W. Ambrose christened these rings **Banach Algebras** during a period of vigorous growth for the subject. This period culminated in M.A. Naimark's book *Normed Rings* (1956), which would go on to have a profound impact on the subject in the years to come.

7.4 Spectrum of an Operator

We now specialize some of the ideas from the previous section to the case of a bounded operator on a Banach space. Given a linear operator T on a Banach space **E**, we had observed in Example 7.3.2 that

$$\sigma(T) = \sigma_{\mathcal{B}(\mathbf{E})}(T) = \{\lambda \in \mathbb{C} : (T - \lambda I) \text{ is not bijective}\}.$$

We now discuss the relationship between spectral values and eigenvalues.

7.4.1 DEFINITION If $T \in \mathcal{B}(\mathbf{E})$, the **point spectrum** of T is the set

$$\sigma_p(T) := \{\lambda \in \mathbb{C} : (T - \lambda I) \text{ is not injective}\}.$$

Elements in $\sigma_p(T)$ are called **eigenvalues** of T. If $\lambda \in \sigma_p(T)$, non-zero vectors in $\ker(T - \lambda I)$ are called **eigenvectors** of T associated to λ.

By definition, $\sigma_p(T) \subset \sigma(T)$. If **E** is finite dimensional, then we had seen in Example 7.3.2 that $\sigma_p(T) = \sigma(T)$. However, in the infinite dimensional case this may not be true.

7.4.2 EXAMPLE Let $T : C[0,1] \to C[0,1]$ be the operator given by

$$T(f)(x) := \int_0^x f(t)dt.$$

We had seen in Example 7.3.13 that $\sigma(T) = \{0\}$. However, by Exercise 7.28, 0 is not an eigenvalue of T so, $\sigma_p(T) = \emptyset$.

In order to get a better understanding of the spectral values that may not be eigenvalues, we introduce the following definition (we had seen this earlier in Exercise 5.16).

7.4.3 DEFINITION We say that an operator $T \in \mathcal{B}(\mathbf{E})$ is **bounded below** if there exists $c > 0$ such that $\|T(x)\| \geq c\|x\|$ for all $x \in \mathbf{E}$.

7.4.4 PROPOSITION *If $T \in \mathcal{B}(\mathbf{E})$ is bounded below if and only if T is injective and Range(T) is closed in* **E**.

Proof.

(i) If T is bounded below, then clearly T is injective. To show that Range(T) is closed in **E**, choose a sequence (y_n) in Range(T) such that $y_n \to y$. Write $y_n = T(x_n)$ for some $x_n \in \mathbf{E}$. If $c > 0$ is such that $\|T(x)\| \geq c\|x\|$ for all $x \in \mathbf{E}$, then it follows that

$$\|x_n - x_m\| \leq \frac{1}{c}\|y_n - y_m\|.$$

Since (y_n) is Cauchy, (x_n) is Cauchy in **E**. Therefore there is a vector $x \in \mathbf{E}$ such that $x_n \to x$. Since T is bounded, $y = T(x) \in$ Range(T). Therefore Range(T) is closed.

(ii) Now suppose T is injective and Range(T) is closed. Then $T : \mathbf{E} \to$ Range(T) defines a bijective, bounded linear operator between Banach spaces. By the Bounded Inverse Theorem, $T^{-1} :$ Range$(T) \to \mathbf{E}$ is bounded. Therefore there exists $M > 0$ such that $\|T^{-1}(y)\| \leq M\|y\|$ for all $y \in$ Range(T). So if $c := \frac{1}{M}$, then $\|T(x)\| \geq c\|x\|$ holds for all $x \in \mathbf{E}$. □

7.4.5 DEFINITION If $T \in \mathcal{B}(\mathbf{E})$, the *approximate point spectrum* of T is the set

$$\sigma_{ap}(T) := \{\lambda \in \mathbb{C} : (T - \lambda I) \text{ is not bounded below}\}.$$

If $(T - \lambda I)$ is not injective, then it clearly cannot be bounded below. Also, if $(T - \lambda I)$ is not bounded below, then it cannot be invertible (by the Bounded Inverse Theorem). Therefore

$$\sigma_p(T) \subset \sigma_{ap}(T) \subset \sigma(T). \tag{7.5}$$

We will shortly give examples to show that these three sets may be distinct. Before that, we make the following observation.

7.4.6 PROPOSITION $\sigma_{ap}(T)$ *is a compact set.*

Proof. Since $\sigma_{ap}(T) \subset \sigma(T)$, it is bounded. We now show that $\sigma_{ap}(T)$ is closed. Suppose $\lambda \notin \sigma_{ap}(T)$, then there is a $c > 0$ such that

$$\|(T - \lambda I)(x)\| \geq c\|x\|$$

for all $x \in E$. If $\omega \in \mathbb{C}$ is such that $|\lambda - \omega| < \frac{c}{2}$, then by the triangle inequality,

$$\|(T - \omega I)(x)\| \geq \|(T - \lambda I)(x)\| - |\lambda - \omega|\|x\| \geq \frac{c}{2}\|x\|.$$

Therefore $\omega \notin \sigma_{ap}(T)$. We have thus proved that the complement of $\sigma_{ap}(T)$ is open. $\square$

Notice that $\lambda \in \sigma_{ap}(T)$ if and only if there is a sequence $(x_n) \subset E$ such that $\|x_n\| = 1$ for all $n \in \mathbb{N}$ and

$$\lim_{n \to \infty} \|Tx_n - \lambda x_n\| = 0.$$

The next proposition refines the relationships in Equation 7.5. Before we get there, we need the following observation that was buried in Corollary 7.2.10. Note that if $X \subset \mathbb{C}$ and $\lambda \in \mathbb{C}$, we will use the notation $d(\lambda, X)$ to denote $\inf\{|\lambda - z| : z \in X\}$.

7.4.7 LEMMA *Let* $\mathbf{A}$ *be a unital Banach algebra and* $a \in \mathbf{A}$. *For any* $\lambda \in \rho(a)$,

$$d(\lambda, \sigma(a)) \geq \frac{1}{\|(a - \lambda 1_{\mathbf{A}})^{-1}\|}.$$

Proof. Suppose $\omega \in \mathbb{C}$ is such that

$$|\lambda - \omega| = \|(a - \lambda 1_{\mathbf{A}}) - (a - \omega 1_{\mathbf{A}})\| < \frac{1}{\|(a - \lambda 1_{\mathbf{A}})^{-1}\|},$$

then the proof of Corollary 7.2.10 shows that $(a - \omega 1_{\mathbf{A}})$ is also invertible. Therefore if $\omega \in \sigma(a)$, then it must happen that

$$|\lambda - \omega| \geq \frac{1}{\|(a - \lambda 1_{\mathbf{A}})^{-1}\|}.$$

Taking an infimum over all $\omega \in \sigma(a)$ proves the result. $\square$

Note that if X is a subset of $\mathbb{C}$, then the boundary of X (denoted by ∂X) is defined by $\overline{X} \cap \overline{\mathbb{C} \setminus X}$.

7.4.8 PROPOSITION *If $T \in \mathcal{B}(\mathbf{E})$, then $\partial \sigma(T) \subset \sigma_{ap}(T)$. In particular, $\sigma_{ap}(T)$ is non-empty.*

Proof. Let $\lambda \in \partial \sigma(T)$. By definition, there is a sequence $(\lambda_n) \subset \mathbb{C} \setminus \sigma(T)$ such that $\lambda_n \to \lambda$. Since $\sigma(T)$ is closed, $\lambda \in \sigma(T)$. Therefore

$$\lim_{n \to \infty} d(\lambda_n, \sigma(T)) = 0.$$

By Lemma 7.4.7, it follows that $\lim_{n \to \infty} \|(T - \lambda_n I)^{-1}\| = \infty$. By passing to a subsequence if need be, we may assume that $\|(T - \lambda_n I)^{-1}\| > n$ for all $n \in \mathbb{N}$. Thus for each $n \in \mathbb{N}$, there is a vector $x_n \in \mathbf{E}$ such that $\|x_n\| = 1$ and

$$\|(T - \lambda_n I)^{-1}(x_n)\| > n.$$

So if

$$y_n := \frac{(T - \lambda_n I)^{-1}(x_n)}{\|(T - \lambda_n I)^{-1}(x_n)\|},$$

then $\|y_n\| = 1$ and

$$\|(T - \lambda I)(y_n)\| \leq \|(T - \lambda_n I)(y_n)\| + \|(\lambda_n - \lambda)(y_n)\|$$

$$= \frac{1}{\|(T - \lambda_n I)^{-1}(x_n)\|} + |\lambda_n - \lambda| \|y_n\|$$

$$\leq \frac{1}{n} + |\lambda_n - \lambda|.$$

This last term converges to 0 as $n \to \infty$, proving that $\lambda \in \sigma_{ap}(T)$. $\square$

Given an operator $T \in \mathcal{B}(\mathbf{E})$, recall that the transpose of T is the operator $T' \in \mathcal{B}(\mathbf{E}^*)$ defined by $T'(\varphi)(x) = \varphi(T(x))$. The next proposition relates the spectrum of an operator to that of its transpose. Once again, this result will be important to us in the context of Hilbert spaces in Chapter 8.

7.4.9 PROPOSITION *Let $T \in \mathcal{B}(\mathbf{E})$ and let $T' \in \mathcal{B}(\mathbf{E}^*)$ denote the transpose of T. Then $\sigma(T) = \sigma(T')$.*

Proof. Write $I_{\mathbf{E}}$ and $I_{\mathbf{E}^*}$ for the identity operators on $\mathbf{E}$ and $\mathbf{E}^*$ respectively. If $\lambda \in \mathbb{C}$, then

$$(T - \lambda I_{\mathbf{E}})' = T' - \lambda I_{\mathbf{E}^*}.$$

(i) Suppose $\lambda \in \mathbb{C}$ is such that $(T - \lambda I_\mathbf{E})$ is invertible. Let $R \in \mathcal{B}(\mathbf{E})$ be such that $R(T - \lambda I_\mathbf{E}) = (T - \lambda I_\mathbf{E})R = I_\mathbf{E}$. Then by taking a transpose (Proposition 4.5.7), we see that

$$(T' - \lambda I_{\mathbf{E}^*})R' = R'(T' - \lambda I_{\mathbf{E}^*}) = I_{\mathbf{E}^*}.$$

Thus $(T' - \lambda I_{\mathbf{E}^*})$ is invertible. This proves that $\sigma(T') \subset \sigma(T)$.

(ii) For the reverse inclusion, suppose $\lambda \notin \sigma(T')$. We prove that $\lambda \notin \sigma(T)$ by proving that $(T - \lambda I_\mathbf{E})$ is bijective. Write $S := (T - \lambda I_\mathbf{E})$ and assume that S' is invertible. We claim that S is bounded below. For $x \in \mathbf{E}$, there exists $\varphi \in \mathbf{E}^*$ such that $\|\varphi\| = 1$ and $|\varphi(x)| = \|x\|$ (by Corollary 4.2.7). Write $\psi := (S')^{-1}(\varphi) \in \mathbf{E}^*$, then $\|\psi\| \le \|(S')^{-1}\|\|\varphi\| = \|(S')^{-1}\|$ and

$$\|x\| = |\varphi(x)| = |\psi(S(x))| \le \|\psi\|\|S(x)\| \le \|(S')^{-1}\|\|S(x)\|.$$

Therefore if $c := \frac{1}{\|(S')^{-1}\|}$, then the inequality

$$\|S(x)\| \ge c\|x\|$$

holds for all $x \in \mathbf{E}$. By Proposition 7.4.4, $\mathrm{Range}(S)$ is closed and S is injective. However, by Proposition 4.5.9, we have

$$\mathrm{Range}(S)^\perp = \ker(S') = \{0\}.$$

Therefore $\mathrm{Range}(S)$ is dense in $\mathbf{E}$. We conclude that $\mathrm{Range}(S) = \mathbf{E}$ and S is bijective. Since $\mathbf{E}$ is a Banach space, S is invertible, proving that $\lambda \notin \sigma(T)$. We have thus proved that $\sigma(T) \subset \sigma(T')$. $\qquad\square$

We end with some examples which illustrate the ideas developed in this section.

7.4.10 EXAMPLE Let $1 \le p < \infty$ and $T : \ell^p \to \ell^p$ be the left-shift operator given by

$$T((x_j)) = (x_2, x_3, \ldots).$$

(i) In Example 7.3.9, we had seen that $\sigma(T) = \mathbb{D} = \{z \in \mathbb{C} : |z| \le 1\}$. In fact, we had even shown that $\{z \in \mathbb{C} : |z| < 1\} \subset \sigma_p(T)$. Moreover, if $\lambda \in \sigma_p(T)$ and $x = (x_j) \in \ell^p$ is an associated eigenvector, then it must follow that

$$x_n = \lambda^{n-1}x_1$$

for all $n \ge 2$. Since $x \in \ell^p$ and is non-zero, it follows that $|\lambda| < 1$. Hence

$$\sigma_p(T) = \{z \in \mathbb{C} : |z| < 1\}.$$

(ii) Now $\sigma_p(T) \subset \sigma_{ap}(T)$ and $S^1 \subset \sigma_{ap}(T)$ by Proposition 7.4.8. We conclude that

$$\sigma_{ap}(T) = \sigma(T) = \mathbb{D}.$$

We now consider the transpose $T' : (\ell^p)^* \to (\ell^p)^*$. Let $1 < q \leq \infty$ be the conjugate exponent of p and let $\Delta : \ell^q \to (\ell^p)^*$ be the isomorphism from Theorem 4.1.3. Then it is easy to see that the map

$$S := \Delta^{-1} \circ T' \circ \Delta : \ell^q \to \ell^q$$

is the right-shift operator $S((x_j)) = (0, x_1, x_2, \ldots)$ (Exercise 7.35). We wish to understand the spectrum of S.

(i) By Proposition 7.4.9, we know that $\sigma(S) = \mathbb{D}$.

(ii) We claim that $\sigma_p(S) = \emptyset$. To see this, fix $\lambda \in \mathbb{C}$ and suppose $x = (x_j) \in \ell^q$ such that $S(x) = \lambda x$. It follows that

$$\lambda x_1 = 0 \text{ and } \lambda x_n = x_{n-1}$$

for all $n \geq 2$. Hence $x = 0$ must hold and λ cannot be an eigenvalue of S.

(iii) By Proposition 7.4.8, we know that $S^1 \subset \sigma_{ap}(S)$. If $\lambda \in \mathbb{D}$ with $|\lambda| < 1$, then for any $x \in \ell^q$,

$$\|(S - \lambda I)x\| \geq |\|S(x)\| - |\lambda|\|x\|| \geq (1 - |\lambda|)\|x\|.$$

Thus, $(S - \lambda I)$ is bounded below, proving that $\lambda \notin \sigma_{ap}(T)$. Thus

$$\sigma_{ap}(S) = S^1.$$

7.4.11 EXAMPLE Fix a sequence $\alpha = (\alpha_n) \in \ell^\infty$. Define $T : \ell^2 \to \ell^2$ by

$$T((x_j)) := (\alpha_1 x_1, \alpha_2 x_2, \ldots).$$

We had seen in Exercise 2.15 that T is a bounded operator with $\|T\| = \|\alpha\|_\infty$.

(i) For each j, let e_j denote the jth standard basis vector in ℓ^2. Then $T(e_j) = \alpha_j e_j$. Hence $\alpha_j \in \sigma_p(T)$ for each $j \in \mathbb{N}$. Conversely, if $\lambda \in \sigma_p(T)$, let $x = (x_j)$ be an eigenvector corresponding to λ so that

$$(\alpha_1 x_1, \alpha_2 x_2, \ldots) = (\lambda x_1, \lambda x_2, \ldots).$$

Since $x \neq 0$, there is a $j \in \mathbb{N}$ such that $x_j \neq 0$. It follows that $\lambda = \alpha_j$. Therefore

$$\sigma_p(T) = \{\alpha_j : j \in \mathbb{N}\}.$$

(ii) We claim that $\sigma(T) = \overline{\{\alpha_j : j \in \mathbb{N}\}}$. To see this, first note that $\sigma_p(T) \subset \sigma(T)$ and $\sigma(T)$ is closed. Therefore $\overline{\{\alpha_j : j \in \mathbb{N}\}} \subset \sigma(T)$. Conversely, if $\lambda \notin \overline{\{\alpha_j : j \in \mathbb{N}\}}$, then there is a $\delta > 0$ such that $|\lambda - \alpha_j| \geq \delta$ for all $j \in \mathbb{N}$. If $x = (x_j) \in \ell^2$, then

$$\|T(x) - \lambda x\|_2 = \left(\sum_{j=1}^{\infty} |\alpha_j x_j - \lambda x_j|^2 \right)^2 \geq \delta \|x\|_2.$$

Therefore $(T - \lambda I)$ is bounded below. In particular, it is injective. Also, if $y = (y_j) \in \ell^2$, we define $x_j := \frac{y_j}{\alpha_j - \lambda}$. Then

$$\sum_{j=1}^{\infty} |x_j|^2 \leq \frac{1}{\delta^2} \sum_{j=1}^{\infty} |y_j|^2 < \infty.$$

Thus $x = (x_j) \in \ell^2$ and clearly $(T - \lambda I)(x) = y$. Thus $(T - \lambda I)$ is surjective as well, proving that $\lambda \notin \sigma(T)$. We conclude that

$$\sigma(T) = \overline{\{\alpha_j : j \in \mathbb{N}\}}.$$

(iii) Finally, $\sigma_{ap}(T)$ is a closed subset of $\sigma(T)$ containing $\sigma_p(T)$. Therefore

$$\sigma_{ap}(T) = \sigma(T).$$

7.4.12 EXAMPLE Fix $1 \leq p < \infty$ and let $T : L^p[0,1] \to L^p[0,1]$ be given by

$$T(f)(t) := tf(t).$$

Then T is a well-defined linear operator with

$$\|T(f)\|_p = \left(\int_0^1 |tf(t)|^p dt \right)^{1/p} \leq \|f\|_p.$$

Hence T is bounded with $\|T\| \leq 1$.

(i) We claim that $\sigma_p(T) = \emptyset$. To see this, suppose $\lambda \in \sigma_p(T)$, then there is a non-zero function $f \in L^p[0,1]$ such that $T(f) = \lambda f$. Then it must happen that

$$(t - \lambda)f(t) = 0$$

for all $t \in [0,1]$. If $t \neq \lambda$, then it follows that $f(t) = 0$. Therefore f is zero on the set $[0,1] \setminus \{\lambda\}$. In other words, f must be the zero function in $L^p[0,1]$. This contradicts the assumption that f is an eigenvector. Therefore

$$\sigma_p(T) = \emptyset.$$

(ii) We claim that $\sigma(T) = [0,1]$. To see this, suppose that $\lambda \notin [0,1]$, then there is a $\delta > 0$ such that $|\lambda - t| \geq \delta$ for all $t \in [0,1]$. Then for any $f \in L^p[0,1]$, we have

$$\|T(f) - \lambda f\|_p = \left(\int_0^1 |tf(t) - \lambda f(t)|^p \right)^{1/p} \geq \delta \|f\|_p.$$

Hence $(T - \lambda I)$ is bounded below. In particular, it is injective. For surjectivity, suppose $g \in L^p[0,1]$, define $f : [0,1] \to \mathbb{C}$ by

$$f(t) = \frac{g(t)}{t - \lambda}.$$

Then f is well-defined, measurable (since it is the quotient of measurable functions) and

$$\int_0^1 |f(t)|^p dt \leq \frac{1}{\delta^p} \int_0^1 |g(t)|^p dt < \infty.$$

Thus $f \in L^p[0,1]$ and $(T - \lambda I)(f) = g$. Therefore $(T - \lambda I)$ is bijective and $\lambda \notin \sigma(T)$. Thus

$$\sigma(T) \subset [0,1].$$

Conversely, if $\lambda \in [0,1]$, then define $f_n : [0,1] \to [0,1]$ by

$$f_n := \frac{1}{m(F_n)^{1/p}} \chi_{F_n},$$

where $F_n = \left[\lambda - \frac{1}{n}, \lambda + \frac{1}{n}\right] \cap [0,1]$. Then $f_n \in L^p[0,1]$ with $\|f_n\|_p = 1$ and

$$\|T(f_n) - \lambda f_n\|_p \leq \left(\int_{F_n} |t - \lambda|^p dt \right)^{1/p} \|f_n\|_p \leq \frac{2}{n}.$$

Therefore $\|T(f_n) - \lambda f_n\| \to 0$, so that $\lambda \in \sigma_{ap}(T)$. We conclude that

$$\sigma(T) = \sigma_{ap}(T) = [0,1].$$

The operator T here is an example of a multiplication operator (see Example 7.1.8). We will discuss these operators and their spectra in great detail in Chapter 10.

7.5 Spectrum of a Compact Operator

In Section 5.6, we had seen that compact operators provide a useful generalization of finite rank operators. In this section, we explore the spectrum of such an operator. The results of this section are not only interesting in and of themselves, they will also guide our intuition later in the book when we study the spectra of non-compact operators. As before, $\mathbf{E}$ will denote a Banach space in what follows.

7.5.1 PROPOSITION *If $T \in \mathcal{B}(\mathbf{E})$ has finite rank, then $\sigma_p(T) = \sigma_{ap}(T) = \sigma(T)$.*

Proof. Since $\sigma_p(T) \subset \sigma_{ap}(T) \subset \sigma(T)$, it suffices to prove that $\sigma(T) \subset \sigma_p(T)$. Suppose $\lambda \notin \sigma_p(T)$, then $(T - \lambda I)$ is injective.

(i) If $\mathbf{E}$ is finite dimensional, then $(T - \lambda I)$ is surjective and thus invertible in $\mathcal{B}(\mathbf{E})$.

(ii) If $\mathbf{E}$ is infinite dimensional, then it must happen that $\lambda \neq 0$, otherwise $T :$ $\mathbf{E} \to \text{Range}(T)$ would be an injective map from an infinite dimensional space to a finite dimensional space. Now consider the map

$$S := (T - \lambda I)|_{\text{Range}(T)} \colon \text{Range}(T) \to \text{Range}(T).$$

This is an injective map on a finite dimensional space, which must therefore be surjective. Now if $y \in \mathbf{E}$, then $T(y) \in \text{Range}(T)$. Therefore, there exists $z \in \text{Range}(T)$ such that $(T - \lambda I)(z) = S(z) = T(y)$. Thus

$$y = (T - \lambda I)\left(\frac{z - y}{\lambda}\right) \in \text{Range}(T - \lambda I).$$

We have proved that $\text{Range}(T - \lambda I) = \mathbf{E}$ and thus $(T - \lambda I)$ is invertible in $\mathcal{B}(\mathbf{E})$.

In either case, we conclude that $\lambda \notin \sigma(T)$. Thus $\sigma(T) \subset \sigma_p(T)$. $\square$

Henceforth we write $\mathcal{K}(\mathbf{E})$ for the set of all compact operators on $\mathbf{E}$.

7.5.2 LEMMA *If $T \in \mathcal{K}(\mathbf{E})$ and $\lambda \neq 0$, then $\ker(T - \lambda I)$ is finite dimensional.*

Proof. Let $\mathbf{F} := \ker(T - \lambda I)$ and observe that $T(\mathbf{F}) \subset \mathbf{F}$. Hence we may consider the restriction

$$S := T|_{\mathbf{F}} \colon \mathbf{F} \to \mathbf{F}.$$

Since T is compact, so is S (verify this!). Furthermore, S is a non-zero scalar multiple of the identity map. This implies that the closed unit ball in $\mathbf{F}$ is compact and thus $\mathbf{F}$ is finite dimensional. $\square$

For the next proposition, we need the following lemma which was an exercise in Chapter 5 (Exercise 5.49).

7.5.3 LEMMA *Let* $\mathbf{E}$ *be a normed linear space and* $\mathbf{F} < \mathbf{E}$ *be a finite dimensional subspace. Then there is a closed subspace* $\mathbf{F}'$ *of* $\mathbf{E}$ *such that* $\mathbf{F}' \cap \mathbf{F} = \{0\}$ *and* $\mathbf{E} = \mathbf{F} + \mathbf{F}'$.

Proof. Let $\{e_1, e_2, \ldots, e_n\}$ be a basis for $\mathbf{F}$ and let $\{\varphi_1, \varphi_2, \ldots, \varphi_n\}$ denote the corresponding dual basis. Since $\mathbf{F}$ is finite dimensional, each φ_i is bounded. By the Hahn-Banach Theorem, there exist $\{\psi_1, \psi_2, \ldots, \psi_n\} \subset \mathbf{E}^*$ such that $\psi_i|_{\mathbf{F}} = \varphi_i$. Define

$$\mathbf{F}' := \bigcap_{i=1}^{n} \ker(\psi_i).$$

Then $\mathbf{F}'$ is clearly a closed subspace of $\mathbf{E}$. Also, if $x \in \mathbf{F}' \cap \mathbf{F}$, then $\varphi_i(x) = \psi_i(x) = 0$ for all $1 \leq i \leq n$. Since $\{\varphi_1, \varphi_2, \ldots, \varphi_n\}$ is a basis for $\mathbf{F}^*$, it follows that $x = 0$. Hence

$$\mathbf{F}' \cap \mathbf{F} = \{0\}.$$

Finally, if $x \in \mathbf{E}$, then $y := \sum_{i=1}^{n} \psi_i(x) e_i \in \mathbf{F}$. Also, if $z := x - y$, then

$$\psi_j(z) = \psi_j(x) - \psi_j\left(\sum_{i=1}^{n} \psi_i(x) e_i\right) = 0$$

for each $1 \leq j \leq n$. Thus $z \in \mathbf{F}'$. Since $x = y + z$, we have proved that $\mathbf{E} = \mathbf{F} + \mathbf{F}'$. $\qquad \square$

7.5.4 PROPOSITION *Let* $T \in \mathcal{K}(\mathbf{E})$ *and* $\mathbf{W}$ *be a closed subspace of* $\mathbf{E}$. *If* $\lambda \neq 0$, *then* $(T - \lambda I)(\mathbf{W})$ *is a closed subspace of* $\mathbf{E}$.

Proof. Since $\lambda \neq 0$, we may divide by λ throughout to assume that $\lambda = 1$. Therefore we begin with a closed subspace $\mathbf{W}$ of $\mathbf{E}$ and prove that $(T - I)(\mathbf{W})$ is closed.

(i) First assume that $(T - I)$ is injective on $\mathbf{W}$. Equivalently, we assume that

$$\mathbf{W} \cap \ker(T - I) = \{0\}.$$

To prove that $(T - I)(\mathbf{W})$ is closed, we choose $y \in \overline{(T - I)(\mathbf{W})}$ and we wish to prove that $y \in (T - I)(\mathbf{W})$. To do that, we begin with a sequence $(x_n) \subset \mathbf{W}$ such that

$$\lim_{n \to \infty} (T(x_n) - x_n) = y. \tag{7.6}$$

(a) We claim that (x_n) is a bounded sequence. Suppose not, then by passing to a subsequence, we may assume that $\|x_n\| \geq n$ for all $n \in \mathbb{N}$. Set $z_n := \frac{x_n}{\|x_n\|}$. Then (z_n) is a sequence of unit vectors such that

$$\lim_{n\to\infty} \|Tz_n - z_n\| = \lim_{n\to\infty} \frac{\|Tx_n - x_n\|}{\|x_n\|} = 0, \tag{7.7}$$

since $\|T(x_n) - x_n\| \to \|y\|$ as $n \to \infty$. However, since T is compact and (z_n) is a bounded sequence, there is a subsequence (z_{n_k}) and a vector $z \in \mathbf{E}$ such that

$$\lim_{k\to\infty} T(z_{n_k}) = z.$$

From Equation 7.7 we conclude that

$$\|T(z) - z\| = \lim_{k\to\infty} \|T(z_{n_k}) - z_{n_k}\| = 0.$$

Since $z_{n_k} \in \mathbf{W}$ for all $k \in \mathbb{N}$ and $\mathbf{W}$ is closed, $z \in \mathbf{W}$. Since $\mathbf{W} \cap \ker(T - I) = \{0\}$, we conclude that $z = 0$. However, by Equation 7.7 and the continuity of the norm (Remark 2.1.2), we have

$$\|z\| = \lim_{k\to\infty} \|T(z_{n_k})\| = \lim_{n\to\infty} \|z_{n_k}\| = 1.$$

This contradiction proves that (x_n) must be a bounded sequence.

(b) Now we appeal to the fact that T is compact once again. Since (x_n) is bounded, there is a subsequence (x_{n_k}) and a vector $x \in \mathbf{E}$ such that $\lim_{k\to\infty} T(x_{n_k}) = x$. From Equation 7.6, we see that

$$\lim_{k\to\infty} x_{n_k} = x - y =: u$$

which is in $\mathbf{W}$ because $\mathbf{W}$ is closed. Therefore

$$y = \lim_{k\to\infty} (T - I)(x_{n_k}) = (T - I)(u) \in (T - I)(\mathbf{W})$$

which is what we had set out to prove.

(ii) Now suppose $(T - I)$ is not injective on $\mathbf{W}$. Since T is compact, $\ker(T - I)$ is finite dimensional by Lemma 7.5.2. By Lemma 7.5.3, there is a closed subspace $\mathbf{F}'$ of $\mathbf{E}$ such that $\mathbf{F}' \cap \ker(T - I) = \{0\}$ and $\mathbf{F}' + \ker(T - I) = \mathbf{E}$. Therefore

$$(T - I)(\mathbf{W}) = (T - I)(\mathbf{F}' \cap \mathbf{W}).$$

By construction, $\mathbf{F}' \cap \mathbf{W}$ is a closed subspace and $(T - I)$ is injective on $\mathbf{F}' \cap \mathbf{W}$. By part (i), $(T - I)(\mathbf{F}' \cap \mathbf{W})$ is closed in $\mathbf{E}$. This completes the proof. $\square$

The next theorem is an important one and was first proved by Ivar Fredholm in the context of integral operators.

7.5.5 THEOREM (FREDHOLM ALTERNATIVE) *Let* $T \in \mathcal{K}(E)$ *and* $\lambda \neq 0$. *Then* $(T - \lambda I)$ *is injective if and only if it is surjective.*

Proof. As before, we may assume that $\lambda = 1$.

$\Rightarrow$: Suppose $(T - I)$ is injective, but not surjective.

(a) For each $n \geq 0$, set

$$\mathbf{F}_n := (T - I)^n(\mathbf{E}).$$

Observe that each $\mathbf{F}_n$ satisfies $T(\mathbf{F}_n) \subset \mathbf{F}_n$, $\mathbf{E} = \mathbf{F}_0$ and $\mathbf{F}_{n+1} \subset \mathbf{F}_n$ for all $n \geq 0$. By Proposition 7.5.4, $\mathbf{F}_1$ is closed in $\mathbf{F}_0$. By induction, it now follows that $\mathbf{F}_{n+1}$ is closed in $\mathbf{F}_n$ for each $n \geq 0$.

(b) We claim that $\mathbf{F}_{n+1} \neq \mathbf{F}_n$ for all $n \geq 0$. To see this, fix $x_0 \in \mathbf{E}$ which is not in the range of $(T - I)$. We claim that

$$(T - I)^n(x_0) \notin \mathbf{F}_{n+1}.$$

Suppose not, then there is $x \in \mathbf{E}$ such that $(T - I)^n(x_0) = (T - I)^{n+1}(x)$. Then

$$0 = (T - I)^n \left(x_0 - (T - I)(x) \right).$$

Since $(T - I)$ is injective, so is $(T - I)^n$. It follows that $x_0 = (T - I)(x)$, which contradicts the assumption that $x_0 \notin \text{Range}(T - I)$. Therefore $\mathbf{F}_{n+1} \neq \mathbf{F}_n$ for all $n \geq 0$.

(c) Since each $\mathbf{F}_{n+1}$ is closed in $\mathbf{F}_n$, we may apply Riesz' Lemma (Lemma 2.4.14). For each $n \geq 0$, there is a unit vector $x_n \in \mathbf{F}_n$ such that

$$d(x_n, \mathbf{F}_{n+1}) \geq \frac{1}{2}.$$

Now $(I - T)(x_n) \in (I - T)(\mathbf{F}_n) \subset \mathbf{F}_{n+1}$ and if $m > n$, then $T(x_m) \in \mathbf{F}_m \subset \mathbf{F}_{n+1}$. Therefore

$$\|T(x_n) - T(x_m)\| = \|x_n - ((I - T)(x_n) + T(x_m))\| \geq d(x_n, \mathbf{F}_{n+1}) \geq \frac{1}{2}.$$

Therefore the sequence $(T(x_n))$ cannot have a convergent subsequence. However, each x_n is a unit vector, so this contradicts the fact that T is a compact operator.

The only conclusion therefore is that $(T - I)$ is surjective.

$\Leftarrow$: Suppose $(T - I)$ is surjective. We wish to prove that $(T - I)$ is injective. To do this, we consider the transpose $T' \in \mathcal{B}(\mathbf{E}^*)$. Since T is compact, T' is a compact operator by Theorem 5.6.14. Furthermore,

$$\ker(T' - I_{\mathbf{E}^*}) = \text{Range}(T - I_{\mathbf{E}})^{\perp} = \{0\}$$

by Proposition 4.5.9. Hence $(T' - I_{\mathbf{E}^*})$ is injective and therefore surjective by the first part of the proof.

Now if $x \in \ker(T - I)$, then by the Hahn–Banach Theorem (Corollary 4.2.7), there is a bounded linear functional $\varphi \in \mathbf{E}^*$ such that $|\varphi(x)| = \|x\|$. Also, there is a $\psi \in \mathbf{E}^*$ such that

$$T'(\psi) - \psi = \varphi.$$

Since $(T - I)(x) = 0$, we see that

$$\|x\| = |\varphi(x)| = |T'(\psi)(x) - \psi(x)| = |\psi((T - I)(x))| = |\psi(0)| = 0.$$

Therefore $x = 0$, proving that $(T - I)$ is injective. $\qquad\square$

It is important to point out that the Fredholm Alternative holds only for non-zero scalars. For instance, let $T : \ell^2 \to \ell^2$ be the operator given by

$$T((x_j)) := \left(x_1, \frac{x_2}{2}, \frac{x_3}{3}, \dots\right).$$

Then T is a compact operator by Example 5.6.6 and it is clearly injective. But it cannot be surjective because that would imply that T is invertible in $\mathcal{B}(\ell^2)$, which would violate the fact that $\mathcal{K}(\ell^2)$ is a proper ideal in $\mathcal{B}(\ell^2)$.

Our next theorem is the final step before we are able to completely describe the spectrum of a compact operator.

7.5.6 LEMMA *Let $T \in \mathcal{K}(\mathbf{E})$ and $r > 0$. Then the set $A_r := \{z \in \sigma(T) : |z| \geq r\}$ is finite.*

Proof. Seeking a contradiction, we assume that A_r is infinite.

(i) If $\lambda \in A_r$, then $(T - \lambda I)$ is not invertible in $\mathcal{B}(\mathbf{E})$. By the Bounded Inverse Theorem, this means that $(T - \lambda I)$ is not bijective on $\mathbf{E}$. By Theorem 7.5.5, it follows that $(T - \lambda I)$ is not injective. Hence every element of A_r is an eigenvalue of T.

(ii) Since A_r is assumed to be infinite, we may choose a countable set $\{\lambda_1, \lambda_2, \ldots\}$ in A_r consisting of distinct scalars. To each λ_i, we choose an eigenvector $x_i \in E$. We claim that the set $\{x_1, x_2, \ldots\}$ is linearly independent. To see this, suppose we have an expression of the form

$$\sum_{i=1}^{k} \alpha_i x_i = 0. \tag{7.8}$$

with non-zero coefficients. Then we may choose such an expression with the least number of non-zero coefficients. Having chosen such an expression, we apply $(T - \lambda_k I)$ to it to see that

$$\sum_{i=1}^{k} \alpha_i \lambda_i x_i - \alpha_i \lambda_k x_i = \sum_{i=1}^{k-1} \alpha_i (\lambda_i - \lambda_k) x_i = 0.$$

Since the expression in Equation 7.8 had the least number of non-zero coefficients, it follows that $\alpha_i(\lambda_i - \lambda_k) = 0$ for all $1 \leq i \leq k-1$. Since the $\{\lambda_j\}$ are all distinct, we conclude that $\alpha_i = 0$ for all $1 \leq i \leq k-1$. In turn, this implies that $\alpha_k = 0$ as well, which proves that the set $\{x_1, x_2, \ldots\}$ is linearly independent.

(iii) For each $n \in \mathbb{N}$, consider the subspace

$$\mathbf{F}_n := \operatorname{span}\{x_1, x_2, \ldots, x_n\}.$$

Clearly, $\mathbf{F}_n \subset \mathbf{F}_{n+1}$, each $\mathbf{F}_n$ is closed (since it is finite dimensional) and $\mathbf{F}_n \neq \mathbf{F}_{n+1}$ by part (ii). By Riesz' Lemma, we may choose $y_n \in \mathbf{F}_n$ such that $\|y_n\| = 1$ and

$$d(y_n, \mathbf{F}_{n+1}) \geq \frac{1}{2}.$$

Now if $n > m$, then

$$T(y_n) - T(y_m) = \lambda_n y_n + (T - \lambda_n I)(y_n) - T(y_m).$$

Since x_n is an eigenvector associated to λ_n, $(T - \lambda_n I)(y_n) \in \mathbf{F}_{n-1}$. Also, $T(y_m) \in \mathbf{F}_m \subset \mathbf{F}_{n-1}$. Therefore

$$\|T(y_n) - T(y_m)\| \geq d(\lambda_n y_n, \mathbf{F}_{n-1}) \geq \frac{|\lambda_n|}{2} \geq \frac{r}{2}.$$

Once again, this proves that the sequence $(T(y_n))$ cannot have a convergent subsequence. Since each y_n is a unit vector, this violates the assumption that T is compact.

We are forced to conclude that A_r is a finite set. $\qquad\square$

We are now in a position to completely describe the spectrum of a compact operator. Note that if **E** is finite dimensional, then every operator in $\mathcal{B}(\mathbf{E})$ is compact and we have completely described the spectrum in this case in Proposition 7.5.1. It is the infinite dimensional case that we now tackle.

7.5.7 THEOREM *Let* **E** *be a Banach space and* $T \in \mathcal{K}(\mathbf{E})$. *Then*

    (i) Each non-zero point in $\sigma(T)$ *is an eigenvalue of* T *and an isolated point of* $\sigma(T)$.

    (ii) $\sigma(T)$ *is a countable set.*

    (iii) If **E** *is infinite dimensional, then* $0 \in \sigma(T)$.

    (iv) $\sigma(T) = \sigma_{ap}(T)$.

Proof.

    (i) We know from the Fredholm Alternative that every non-zero element in $\sigma(T)$ is an eigenvalue of T. Now for each $r > 0$, let

$$A_r := \{z \in \sigma(T) : |z| \geq r\}.$$

By Lemma 7.5.6, A_r is a finite set. If $\lambda \in \sigma(T)$ is non-zero, then for $r := \frac{|\lambda|}{3}$, we have

$$\{z \in \mathbf{C} : |\lambda - z| < r\} \cap \sigma(T) \subset A_r.$$

Therefore there are atmost finitely many spectral values in an open set containing λ. This proves that λ is an isolated point of $\sigma(T)$.

    (ii) For each $n \in \mathbb{N}$, the set $A_{\frac{1}{n}}$ is a finite set and

$$\sigma(T) \setminus \{0\} = \bigcup_{n=1}^{\infty} A_{\frac{1}{n}}.$$

This proves that $\sigma(T)$ is a countable set.

    (iii) If **E** is infinite dimensional, the identity operator I is not compact. Therefore the ideal $\mathcal{K}(\mathbf{E})$ cannot contain any invertible operators. Hence if $T \in \mathcal{K}(\mathbf{E})$, then 0 must be a spectral value of T.

    (iv) By the previous steps, it suffices to prove that $0 \in \sigma_{ap}(T)$ when **E** is infinite dimensional. To do this, we consider two cases: If 0 is an isolated point in $\sigma(T)$, then $0 \in \partial\sigma(T) \subset \sigma_{ap}(T)$ by Proposition 7.4.8. If 0 is not an isolated point in $\sigma(T)$, then 0 must be a limit point of $\sigma_p(T)$ by part (i).

However, $\sigma_p(T) \subset \sigma_{ap}(T)$ and $\sigma_{ap}(T)$ is closed (Proposition 7.4.6). Therefore $0 \in \sigma_{ap}(T)$.

□

Given a sequence (α_n) of scalars that converge to 0, we can construct a compact operator whose spectrum is precisely $\overline{\{\alpha_j : j \in \mathbb{N}\}}$. Indeed, we may define $T : \ell^2 \to \ell^2$ by

$$T((x_j)) = (\alpha_1 x_1, \alpha_2 x_2, \ldots).$$

That this operator is compact was proved in Example 5.6.6 and we computed its spectrum in Example 7.4.11. At the other extreme, there exist compact operators without any eigenvalues at all. Of course, any such operator must be quasinilpotent. One such example is described in Exercise 7.49.

7.6 Exercises

Banach Algebras

EXERCISE 7.1 (⚒) Let $\mathbb{D} := \{z \in \mathbb{C} : |z| \leq 1\}$ and let $A(\mathbb{D})$ denote the space of all continuous functions on $\mathbb{D}$ which are holomorphic in the interior of $\mathbb{D}$. When equipped with the supremum norm, prove that $A(\mathbb{D})$ is a Banach algebra. This is called the ***disc algebra***.

EXERCISE 7.2 (⚒) Let $a, b \in \mathbb{R}$ with $a < b$ and $n \in \mathbb{N}$. Let $C^n[a, b]$ be the space of all complex-valued functions on $[a, b]$ that are n-times continuously differentiable. Define a norm on $C^n[a, b]$ by

$$\|f\| := \sum_{k=0}^{n} \frac{1}{k!} \|f^{(k)}\|_\infty$$

where $f^{(k)}$ denotes the kth derivative of f. Prove that $C^n[a, b]$ is a Banach algebra with this norm.

Hint: For completeness, use the completeness of $(C[a, b], \|\cdot\|_\infty)$, the Fundamental Theorem of Calculus and induction.

EXERCISE 7.3 (⚒) Let $\mathbf{A} := M_n(\mathbb{C})$ under the usual algebraic operations of matrix addition and multiplication. Define $\|\cdot\|_\infty : \mathbf{A} \to \mathbb{R}_+$ by

$$\|(a_{i,j})\|_\infty := \sup\{|a_{i,j}| : 1 \leq i, j \leq n\}.$$

Prove that $\|\cdot\|_\infty$ is not submultiplicative.

EXERCISE 7.4 (✗) Let $f, g \in L^1(\mathbb{R})$.

(i) Define $h : \mathbb{R} \times \mathbb{R} \to \mathbb{R}$ by

$$h(x, t) := f(x)g(x - t).$$

Prove that h is Lebesgue measurable and that

$$\int_{\mathbb{R}^2} |h(x, t)| dx dt < \infty.$$

(ii) Conclude that $f * g \in L^1(\mathbb{R})$ and that $\|f * g\|_1 \leq \|f\|_1 \|g\|_1$.

EXERCISE 7.5 (✗) Let $\mathbf{E}$ be a Banach space and set

$$\mathcal{F}(\mathbf{E}) := \{T \in \mathcal{B}(\mathbf{E}) : T \text{ has finite rank}\}.$$

Prove that $\mathcal{F}(\mathbf{E})$ is a two-sided ideal of $\mathcal{B}(\mathbf{E})$.

EXERCISE 7.6 Prove Theorem 7.1.9.

EXERCISE 7.7 Prove the following generalization of Lemma 7.2.2: Let $(\mathbf{A}, \| \cdot \|)$ be a Banach space over $\mathbb{C}$. Suppose that $\mathbf{A}$ has the structure of an algebra with a unit $e \neq 0$ such that the multiplication map $m : \mathbf{A} \times \mathbf{A} \to \mathbf{A}$ is continuous. Then prove that there is a norm $\| \cdot \|_L$ on $\mathbf{A}$ which is equivalent to $\| \cdot \|$ such that $(\mathbf{A}, \| \cdot \|_L)$ is a Banach algebra.

EXERCISE 7.8 Let $\mathbf{A}$ be a normed algebra. Prove that the completion of $\mathbf{A}$ defined in Exercise 2.29 is a Banach algebra.

EXERCISE 7.9 For $n \in \mathbb{N}$, let $g_n \in L^1(\mathbb{R})$ be a non-negative function satisfying two conditions

(a) g_n vanishes outside the interval $[-1/n, +1/n]$.

(b)

$$\int_{-\infty}^{\infty} g_n(t) dt = 1.$$

For any $f \in L^1(\mathbb{R})$, prove that $\lim_{n \to \infty} \|f * g_n - f\|_1 = 0$ (In other words, $L^1(\mathbb{R})$ has an *approximate unit*, even though it does not have a unit).

EXERCISE 7.10 Let $\{\mathbf{A}_\lambda : \lambda \in J\}$ be a family of Banach algebras. Define $\prod_{\lambda \in J} \mathbf{A}_\lambda$ to be the set of all functions $a : J \to \bigcup_{\lambda \in J} \mathbf{A}_\lambda$ for which $a(\lambda) \in \mathbf{A}_\lambda$ for all $\lambda \in J$ and

$$\|a\|_\infty := \sup\{\|a(\lambda)\| : \lambda \in J\} < \infty.$$

Prove that $\prod_{\lambda \in J} \mathbf{A}_\lambda$ is a Banach algebra.

EXERCISE 7.11 Let $\{\mathbf{A}_\lambda : \lambda \in J\}$ be a family of Banach algebras and let $\prod_{\lambda \in J} \mathbf{A}_\lambda$ be as in Exercise 7.10. Let $\mathbf{I} \subset \prod_{\lambda \in J} \mathbf{A}_\lambda$ consist of all elements $a \in \prod_{\lambda \in J} \mathbf{A}_\lambda$ such that $a(\lambda) = 0$ for all but finitely many $\lambda \in J$. Prove that $\mathbf{I}$ is a two-sided ideal of $\prod_{\lambda \in J} \mathbf{A}_\lambda$.

Thus the closure of $\mathbf{I}$, denoted by $\sum_{\lambda \in J} \mathbf{A}_\lambda$, is an ideal in $\prod_{\lambda \in J} \mathbf{A}_\lambda$.

EXERCISE 7.12 Let X be a locally compact Hausdorff space and let $\mathbf{A}$ be a Banach algebra. Define $C_b(X, \mathbf{A})$ to be the space of all continuous functions from X to $\mathbf{A}$ that are bounded. In other words, $f \in C_b(X, \mathbf{A})$ if $f : X \to \mathbf{A}$ is continuous and

$$\|f\|_\infty := \sup\{\|f(x)\| : x \in X\} < \infty.$$

Prove that $C_b(X, \mathbf{A})$ is a Banach algebra under this norm and the pointwise operations.

EXERCISE 7.13 (⚒) Let X be a locally compact Hausdorff space and let $\mathbf{A}$ be a Banach algebra. Let $C_0(X, \mathbf{A})$ denote the space of all continuous functions from X to $\mathbf{A}$ that vanish at infinity. In other words, $f \in C_0(X, \mathbf{A})$ if $f : X \to \mathbf{A}$ is continuous and for any $\epsilon > 0$, there is a compact set $K \subset X$ such that $\|f(x)\| < \epsilon$ for all $x \notin K$.

Prove that $C_0(X, \mathbf{A})$ is a closed ideal in $C_b(X, \mathbf{A})$.

EXERCISE 7.14 Let X and Y be compact Hausdorff spaces. Prove that

$$C(X, C(Y)) \cong C(X \times Y).$$

EXERCISE 7.15 Let $\mathbf{A}$ be a non-unital Banach algebra and let $\widetilde{\mathbf{A}} := \mathbf{A} \times \mathbb{C}$. Define addition and scalar multiplication on $\widetilde{\mathbf{A}}$ componentwise and define multiplication by the formula

$$(a, \alpha) \cdot (b, \beta) := (ab + \alpha b + \beta a, \alpha\beta).$$

Define a norm on $\widetilde{\mathbf{A}}$ by $\|(a, \alpha)\| := \|a\| + |\alpha|$. Then prove that

(i) $\widetilde{\mathbf{A}}$ is a unital Banach algebra.

(ii) The map $\iota : \mathbf{A} \to \widetilde{\mathbf{A}}$ given by $a \mapsto (a,0)$ is an injective homomorphism such that $\iota(\mathbf{A}) \lhd \widetilde{\mathbf{A}}$ and $\iota(\mathbf{A})$ has codimension 1 in $\widetilde{\mathbf{A}}$.

(iii) If $\mathbf{B}$ is a unital Banach algebra and $\Phi : \mathbf{A} \to \mathbf{B}$ is a homomorphism, then there is a unique homomorphism $\widetilde{\Phi} : \widetilde{\mathbf{A}} \to \mathbf{B}$ such that $\widetilde{\Phi}(1_{\widetilde{\mathbf{A}}}) = 1_{\mathbf{B}}$ and $\widetilde{\Phi} \circ \iota = \Phi$.

(iv) If $(\widehat{\mathbf{A}}, \widehat{\iota})$ is a pair satisfying the properties $(i) - (iii)$, then there is an isomorphism $\Psi : \widetilde{\mathbf{A}} \to \widehat{\mathbf{A}}$ such that $\Psi(1_{\widetilde{\mathbf{A}}}) = 1_{\widehat{\mathbf{A}}}$ and $\Psi \circ \iota = \widehat{\iota}$.

The pair $(\widetilde{\mathbf{A}}, \iota)$ is thus unique and is called the ***unitalization*** of $\mathbf{A}$. (Compare this with Theorem 8.2.11, where we choose a different norm on the unitalization).

Invertible Elements

EXERCISE 7.16 (✖) Let I be any set and $\mathbf{A} := \ell^\infty(I)$ be the space of bounded functions from I to $\mathbb{C}$ equipped with the supremum norm. For an element $f \in \mathbf{A}$, prove that $f \in GL(\mathbf{A})$ if and only if $0 \notin \overline{f(I)}$, the closure of the range of f.

EXERCISE 7.17 Let $\mathbf{A}$ be a Banach algebra and let (a_n) and (b_n) be two sequences in $\mathbf{A}$ such that $\sum_{n=0}^\infty \|a_n\| < \infty$ and $\sum_{n=0}^\infty \|b_n\| < \infty$. By Proposition 2.3.8, we may write

$$a := \sum_{n=0}^\infty a_n \text{ and } b := \sum_{n=0}^\infty b_n.$$

If $c_n := \sum_{k=0}^n a_k b_{n-k}$, then prove that $\sum_{n=0}^\infty \|c_n\| < \infty$ and that $ab = \sum_{n=1}^\infty c_n$.

EXERCISE 7.18 (✖) Let $\mathbf{A}$ be a unital Banach algebra and $x \in \mathbf{A}$. Let $\exp(x)$ be the element defined in Definition 7.2.7.

(i) If $y \in \mathbf{A}$ is such that $xy = yx$, then prove that $\exp(x + y) = \exp(x) \exp(y)$.

   Hint: Use Exercise 7.17.

(ii) Prove that $\exp(x) \in GL(\mathbf{A})$ for any $x \in \mathbf{A}$.

EXERCISE 7.19 Let $\Phi : \mathbf{A} \to \mathbf{B}$ be a surjective homomorphism of unital Banach algebras. For $y \in \mathbf{B}$, let $b := \exp(y) \in GL(\mathbf{B})$. Prove that there is an element $a \in GL(\mathbf{A})$ such that $\Phi(a) = b$.

EXERCISE 7.20 Let G be a topological group and let G_0 denote the connected component of the identity in G. Prove that G_0 is a subgroup of G and that it is normal in G.

EXERCISE 7.21 Let $\mathbf{A}$ be a unital Banach algebra and $GL(\mathbf{A})$ be the group of invertible elements in $\mathbf{A}$. Let $GL_0(\mathbf{A})$ denote the connected component of the identity in $GL(\mathbf{A})$.

(i) Prove that $GL_0(\mathbf{A})$ is the path component of the identity.

(ii) For any $x \in \mathbf{A}$, prove that $\exp(x) \in GL_0(\mathbf{A})$.

EXERCISE 7.22 (⚒) Let $S^1 := \{z \in \mathbb{C} : |z| = 1\}$ and let $\mathbf{A}$ denote the Banach algebra $C(S^1)$.

(i) If $f \in GL(\mathbf{A})$, then think of f as a loop in $\mathbb{C}^\times = \mathbb{C} \setminus \{0\}$ (by part (ii) of Example 7.2.6). Prove that $f \in GL_0(\mathbf{A})$ if and only if f is a *contractible* loop.

(ii) Construct an element $f \in GL(\mathbf{A})$ such that $f \notin GL_0(\mathbf{A})$. In particular, such an element cannot be of the form $\exp(g)$ for any $g \in \mathbf{A}$.

EXERCISE 7.23 Let $\Phi : \mathbf{A} \to \mathbf{B}$ be a homomorphism of unital Banach algebras such that $\Phi(1_{\mathbf{A}}) = 1_{\mathbf{B}}$.

(i) If $a \in GL(\mathbf{A})$, then prove that $\Phi(a) \in GL(\mathbf{B})$.

(ii) Give an example of a homomorphism as above and an element $a \in \mathbf{A}$ such that $\Phi(a) \in GL(\mathbf{B})$, but $a \notin GL(\mathbf{A})$.

EXERCISE 7.24 (⚒) Let $\mathbf{A} := c_0$, equipped with the supremum norm and pointwise algebraic operations. If $\mathbf{I} := c_{00}$, then prove that $\mathbf{I}$ is a proper ideal of $\mathbf{A}$ that is dense in $\mathbf{A}$.

The Spectrum of an Element of a Banach Algebra

EXERCISE 7.25 (⚒) Let $\mathbf{A}$ be a Banach algebra, Ω an open subset of $\mathbb{C}$ and $F : \Omega \to \mathbf{A}$ be a function that is analytic in the sense of Definition 7.3.4. If $\tau \in \mathbf{A}^*$ is a bounded linear functional, define $H : \Omega \to \mathbb{C}$ by $H := \tau \circ F$. Prove that H is analytic (in the usual sense) and that $H' = \tau \circ F'$.

EXERCISE 7.26 Let $\mathbf{A}$ be a Banach algebra, Ω an open, connected subset of $\mathbb{C}$ and $F : \Omega \to \mathbf{A}$ a function that is analytic in the sense of Definition 7.3.4. If $F' = 0$ on Ω, then prove that F is a constant function.

EXERCISE 7.27 Let $\mathbf{A}$ be a unital Banach algebra and $a \in \mathbf{A}$. If U is an open set in $\mathbb{C}$ such that $\sigma(a) \subset U$, then prove that there is a $\delta > 0$ such that

$$\sigma(b) \subset U$$

for any $b \in \mathbf{A}$ with $\|b - a\| < \delta$. This property is called the *upper semi-continuity* of the spectrum.

Hint: Use the proof of Theorem 7.3.6 to show that

$$\inf\left\{\frac{1}{\|(a - \lambda 1_{\mathbf{A}})^{-1}\|} : \lambda \in \mathbb{C} \setminus U\right\} > 0.$$

EXERCISE 7.28 (⚒) Let $\mathbf{E} := C[0,1]$ and $T \in \mathcal{B}(\mathbf{E})$ be the operator given by

$$T(f)(x) = \int_0^x f(t)dt.$$

Prove that 0 is a spectral value of T but not an eigenvalue of T.

EXERCISE 7.29 Let $\mathbf{A}$ be a unital Banach algebra and $a \in \mathbf{A}$.

(i) If $b \in GL(\mathbf{A})$, then prove that $\sigma(bab^{-1}) = \sigma(a)$.

(ii) If $\alpha \in \mathbb{C}$, then prove that $\sigma(\alpha a) = \alpha\sigma(a)$.

(iii) If $a \in GL(\mathbf{A})$, then prove that $\sigma(a^{-1}) = \{\lambda^{-1} : \lambda \in \sigma(a)\}$.

EXERCISE 7.30 (⚒) Let $\mathbf{A}$ be a unital Banach algebra and $a, b \in \mathbf{A}$.

(i) Prove that $\sigma(ab) \cup \{0\} = \sigma(ba) \cup \{0\}$.

Hint: If $1 \notin \sigma(ab)$ and $c := (1_{\mathbf{A}} - ab)^{-1}$, then prove that $(1_{\mathbf{A}} + bca) = (1_{\mathbf{A}} - ba)^{-1}$.

(ii) Give an example where $\sigma(ab) \neq \sigma(ba)$.

EXERCISE 7.31 Let $\Phi : \mathbf{A} \to \mathbf{B}$ be a homomorphism between two unital Banach algebras such that $\Phi(1_{\mathbf{A}}) = 1_{\mathbf{B}}$. For any $a \in \mathbf{A}$, prove that

$$\sigma_{\mathbf{B}}(\Phi(a)) \subset \sigma_{\mathbf{A}}(a).$$

7.6.1 DEFINITION Let $\mathbf{A}$ be a unital Banach algebra. An element $p \in \mathbf{A}$ is said to be an ***idempotent*** if $p^2 = p$. Furthermore, p is said to be a *non-trivial* idempotent if p is neither 0 nor $1_{\mathbf{A}}$.

EXERCISE 7.32 Let p be a non-trivial idempotent in a unital Banach algebra $\mathbf{A}$. Prove that $\sigma(p) = \{0, 1\}$.

EXERCISE 7.33 If $\mathbf{E}$ is a Banach space and $P \in \mathcal{B}(\mathbf{E})$ is an idempotent, then prove that

$$\mathbf{E} = \text{Range}(P) \oplus \ker(P).$$

EXERCISE 7.34 Let $\mathbf{A} = C^1[0,1]$ be the space of all continuously differentiable functions on $[0,1]$, equipped with the norm

$$\|f\| := \|f\|_\infty + \|f'\|_\infty.$$

We had seen in Exercise 7.2 that $\mathbf{A}$ is a Banach algebra. If $f \in \mathbf{A}$ is the identity function $f(x) = x$, then prove that

$$\|f\| = 2 \text{ and } r(f) = 1.$$

The Spectrum of an Operator

EXERCISE 7.35 (⚒) Let $1 \leq p < \infty$ and $T : \ell^p \to \ell^p$ be the left-shift operator given by

$$T((x_j)) = (x_2, x_3, \ldots).$$

If q is the conjugate exponent of p, let $\Delta : \ell^q \to (\ell^p)^*$ be the isomorphism described in Theorem 4.1.3. Prove that the map $\Delta^{-1} \circ T' \circ \Delta : \ell^q \to \ell^q$ is the right-shift operator

$$(x_j) \mapsto (0, x_1, x_2, \ldots).$$

EXERCISE 7.36 Let $\alpha = (\alpha_n) \in \ell^\infty$ and $T : \ell^2 \to \ell^2$ be given by

$$T((x_j)) := (\alpha_1 x_1, \alpha_2 x_2, \ldots).$$

We had seen in Example 7.4.11 that $\sigma(T) = \overline{\{\alpha_j : j \in \mathbb{N}\}}$. If $\lambda \notin \sigma(T)$, then prove that there is a sequence $\beta = (\beta_n) \in \ell^\infty$ such that

$$(T - \lambda I)^{-1}((y_j)) = (\beta_1 y_1, \beta_2 y_2, \ldots)$$

for all $(y_j) \in \ell^2$.

EXERCISE 7.37 Let K be a non-empty compact subset of $\mathbb{C}$. Prove that there is a bounded operator $T \in \mathcal{B}(\ell^2)$ such that

$$\sigma(T) = K = \sigma_{ap}(T).$$

EXERCISE 7.38 Let $\mathbf{E}$ be a Banach space and $T \in \mathcal{B}(\mathbf{E})$ be an isometry.

(i) Prove that $\sigma_{ap}(T) \subset S^1$.

(ii) If T is not invertible, then prove that $\sigma(T) = \mathbb{D} = \{z \in \mathbb{C} : |z| \leq 1\}$.

Hint: If $\sigma(T) \neq \mathbb{D}$, then examine $\partial\sigma(T)$.

(iii) If T is invertible, then prove that $\sigma(T) = \sigma_{ap}(T) \subset S^1$.

The Spectrum of a Compact Operator

EXERCISE 7.39 ($\maltese$) Let $\mathbf{E}$ be a Banach space and $P \in \mathcal{K}(\mathbf{E})$ a compact operator that is idempotent in $\mathcal{B}(\mathbf{E})$ (see Definition 7.6.1). Prove that P has finite rank.

EXERCISE 7.40 For $n \in \mathbb{N}$, define $P_n : \ell^2 \to \ell^2$ by

$$P_n((x_j)) := (x_1, x_2, \ldots, x_n, 0, 0, \ldots).$$

Prove that an operator $T \in \mathcal{B}(\ell^2)$ is compact if and only if $\lim_{n\to\infty} \|P_n T - T\| = 0$.

Hint: Use the proof of Theorem 5.6.9.

EXERCISE 7.41 Let $\mathbf{E}$ be a Banach space and $T \in \mathcal{B}(\mathbf{E})$ be a bounded operator. If there is a polynomial $p(z) \in \mathbb{C}[z]$ such that $p(T) \in \mathcal{K}(\mathbf{E})$, then prove that $\sigma(T)$ is a countable set.

EXERCISE 7.42 Let $\mathbf{H} = L^2[0,1]$ and $K \in L^2([0,1]^2)$ be the function

$$K(x,y) = \begin{cases} (1-x)y & : \text{ if } 0 \leq y \leq x \leq 1, \\ (1-y)x & : \text{ if } 0 \leq x \leq y \leq 1. \end{cases}$$

Let $T \in \mathcal{B}(\mathbf{H})$ be the associated integral operator given by

$$T(f)(x) = \int_0^1 K(x,y) f(y) dy.$$

We had seen in Example 5.6.8 that T is a compact operator. The aim of this exercise is to determine the spectrum of T. Let λ be a non-zero eigenvalue of T with eigenvector $f \in L^2[0,1]$.

(i) Prove that $f(0) = f(1) = 0$.

(ii) Prove that f is absolutely continuous (and thus continuous).

(iii) Prove that f is continuously differentiable and that f' satisfies the equation

$$\lambda f'(x) = -\int_0^x y f(y)\,dy + \int_x^1 (1-y)f(y)\,dy.$$

(iv) Prove that f' is continuously differentiable and that f'' satisfies the equation

$$\lambda f''(x) = -f(x).$$

(v) Prove that $\lambda = \frac{1}{n^2\pi^2}$ for some $n \in \mathbb{N}$. Conclude that

$$\sigma(T) = \left\{ 0, \frac{1}{\pi^2}, \frac{1}{4\pi^2}, \ldots \right\}.$$

Weighted Shift Operator

7.6.2 DEFINITION For $\alpha = (\alpha_n) \in \ell^\infty$, define $T_\alpha : \ell^2 \to \ell^2$ by

$$T_\alpha((x_j)) := (0, \alpha_1 x_1, \alpha_2 x_2, \ldots).$$

This operator is called a (unilateral) **weighted shift operator**.

EXERCISE 7.43

(i) For $n \in \mathbb{N}$, prove that

$$\|T_\alpha^n\| = \sup\{|\alpha_i \alpha_{i+1} \ldots \alpha_{i+n-1}| : i \in \mathbb{N}\}.$$

(ii) If $\alpha \in c$ is a convergent sequence of non-negative real numbers, then prove that

$$r(T_\alpha) \le \lim_{n\to\infty} \alpha_n.$$

EXERCISE 7.44 Let $\alpha = (\alpha_n)$ and $\beta = (\beta_n)$ be two bounded sequences. If there exists $\epsilon > 0$ and $M > 0$ such that

$$\epsilon \le \left| \frac{\alpha_1 \alpha_2 \ldots \alpha_n}{\beta_1 \beta_2 \ldots \beta_n} \right| \le M$$

for all $n \in \mathbb{N}$, then construct an invertible operator $S \in \mathcal{B}(\ell^2)$ such that

$$S T_\alpha S^{-1} = T_\beta.$$

Conclude that $\sigma(T_\alpha) = \sigma(T_\beta)$.

Hint: Choose S of the form $S((x_j)) = (\gamma_1 x_1, \gamma_2 x_2, \ldots)$ for some sequence $(\gamma_n) \in \ell^\infty$.

EXERCISE 7.45 Let $\alpha = (\alpha_n)$ and $\beta = (\beta_n)$ be two bounded sequences. If $|\alpha_n| = |\beta_n|$ for all $n \in \mathbb{N}$, then construct a unitary operator $U \in \mathcal{B}(\ell^2)$ such that

$$UT_\alpha U^{-1} = T_\beta.$$

EXERCISE 7.46

(i) If $\alpha_n \neq 0$ for all $n \in \mathbb{N}$, then prove that $\sigma_p(T_\alpha) = \emptyset$.

(ii) If $\alpha_n = 0$ for some $n \in \mathbb{N}$, then prove that $\sigma_p(T_\alpha) = \{0\}$.

EXERCISE 7.47

(i) If $\lambda \in S^1$, then prove that there is a unitary operator $U \in \mathcal{B}(\ell^2)$ such that $UT_\alpha U^{-1} = \lambda T_\alpha$. Conclude that $\sigma(T)$ is a union of concentric circles around 0.

(ii) Prove that $\sigma_{ap}(T)$ contains the circle $\{z \in \mathbb{C} : |z| = r(T_\alpha)\}$.

EXERCISE 7.48 Prove that the weighted shift operator $T_\alpha : \ell^2 \to \ell^2$ is compact if and only if $\alpha \in c_0$.

Hint: Use Exercise 7.40.

EXERCISE 7.49 (⚒) Define $T : \ell^2 \to \ell^2$ by

$$T((x_j)) := \left(0, x_1, \frac{x_2}{2}, \frac{x_3}{3}, \ldots\right).$$

(i) Prove that T is a compact operator.

(ii) Use Exercise 7.43 to prove that $\sigma(T) = \{0\}$.

(iii) Prove that $\sigma_p(T) = \emptyset$ and that $\sigma_{ap}(T) = \{0\}$.

Additional Reading

- Perhaps the best introduction to the study of Operator theory through the lens of Banach algebras is this aptly titled book by Douglas.

  Douglas, Ronald G. (1998). *Banach Algebra Techniques in Operator Theory*. Second. Vol. 179. Graduate Texts in Mathematics. xvi+194. New York: Springer-Verlag. ISBN: 0-387-98377-5. DOI: 10.1007/978-1-4612-1656-8.

- My first encounter with Banach algebra theory came through Fourier analysis (specifically, chapter 9 of Rudin [51]). Very soon, I found myself immersed in this excellent book by Loomis which remains fresh and relevant to this day.

Loomis, Harold Loomis (1956). *Introduction to Abstract Harmonic Analysis*. Izdat. Inostran. Lit., Moscow, 251.

Chapter 8

C*-Algebras

8.1 Operators on Hilbert Spaces

In the previous chapter, we studied operators on Banach spaces and their spectra and we were particularly successful in describing the spectrum of a compact operator. However, in order to say more, one has to impose further restrictions on the operator and the underlying Banach space.

Given a Banach space $\mathbf{E}$ and a bounded operator $T \in \mathcal{B}(\mathbf{E})$, we have defined a transpose operator $T' \in \mathcal{B}(\mathbf{E}^*)$. If $\mathbf{E}$ is a Hilbert space, then there is a conjugate-linear isomorphism $\Delta : \mathbf{E} \to \mathbf{E}^*$. Therefore the transpose of the operator may be thought of as an operator on $\mathbf{E}$ itself. This operator is called the *adjoint* of T. This adjoint operation introduces a further structure on $\mathcal{B}(\mathbf{E})$ and is the focus of our attention in this section. An abstraction of this structure leads us to the notion of a C*-algebra, which we will study through the remainder of the chapter.

Throughout this section, $\mathbf{H}$ and $\mathbf{K}$ will denote complex Hilbert spaces and we will use $\langle \cdot, \cdot \rangle$ to denote the inner product in either space. As before, $\mathcal{B}(\mathbf{H}, \mathbf{K})$ will denote the collection of bounded operators from $\mathbf{H}$ to $\mathbf{K}$ and $\mathcal{B}(\mathbf{H})$ will denote $\mathcal{B}(\mathbf{H}, \mathbf{H})$.

8.1.1 LEMMA *If $T \in \mathcal{B}(\mathbf{H}, \mathbf{K})$, then*

$$\|T\| = \sup\{|\langle T(x), y \rangle| : x \in \mathbf{H}, y \in \mathbf{K}, \|x\| = \|y\| = 1\}.$$

Proof. Since $\|T\| = \sup\{\|T(x)\| : x \in \mathbf{H}, \|x\| = 1\}$, Corollary 4.2.7 implies that

$$\|T\| = \sup\{|\psi(T(x))| : x \in \mathbf{H}, \psi \in \mathbf{K}^*, \|x\| = \|\psi\| = 1\}.$$

The result now follows from the Riesz Representation Theorem. $\qquad\square$

The next result is an immediate corollary and it is used implicitly in many results that follow.

8.1.2 COROLLARY *If $S_1, S_2 \in \mathcal{B}(\mathbf{H}, \mathbf{K})$ are such that*

$$\langle S_1(x), y \rangle = \langle S_2(x), y \rangle$$

for all $x \in \mathbf{H}$ and $y \in \mathbf{K}$, then $S_1 = S_2$.

We now arrive at the central concept of this section, that of the adjoint of an operator.

8.1.3 PROPOSITION *Let $T : \mathbf{H} \to \mathbf{K}$ be a bounded linear operator. Then there is a unique bounded operator $S : \mathbf{K} \to \mathbf{H}$ such that*

$$\langle T(x), y \rangle = \langle x, S(y) \rangle \tag{8.1}$$

*for all $x \in \mathbf{H}$ and $y \in \mathbf{K}$. This operator is called the **adjoint** of T and is denoted by T^*.*

Proof. The uniqueness of this operator follows directly from Corollary 8.1.2, therefore we only prove existence. Let $T' : \mathbf{K}^* \to \mathbf{H}^*$ be the transpose of T, given explicitly by the formula

$$T'(\psi)(x) = \psi(T(x))$$

for $x \in \mathbf{H}$ and $\psi \in \mathbf{K}^*$. Since $\mathbf{H}$ is a Hilbert space, there is a conjugate linear isometry $\Delta_{\mathbf{H}} : \mathbf{H} \to \mathbf{H}^*$ given by $\Delta_{\mathbf{H}}(y)(x) = \langle x, y \rangle$. The map $\Delta_{\mathbf{K}} : \mathbf{K} \to \mathbf{K}^*$ is defined similarly. Let $S : \mathbf{K} \to \mathbf{H}$ be given by

$$S = \Delta_{\mathbf{H}}^{-1} \circ T' \circ \Delta_{\mathbf{K}}.$$

It is clear that S is a bounded linear operator. Now if $x \in \mathbf{H}$ and $y \in \mathbf{K}$, we have

$$\langle x, S(y) \rangle = \Delta_{\mathbf{H}}(S(y))(x) = T'(\Delta_{\mathbf{K}}(y))(x) = \Delta_{\mathbf{K}}(y)(T(x)) = \langle T(x), y \rangle.$$

$\square$

In many ways, the adjoint of an operator behaves like the complex conjugate of a complex number. Let us look at some examples.

8.1.4 EXAMPLE

(i) Let $\mathbf{H} = \mathbb{C}^n$ and $T \in \mathcal{B}(\mathbf{H})$. If $\Lambda = \{e_1, e_2, \ldots, e_n\}$ is an orthonormal basis for $\mathbf{H}$, let $A = (a_{i,j})$ denote the matrix of T with respect to Λ. In other words,

$$a_{i,j} = \langle T(e_j), e_i \rangle.$$

Then

$$\langle T^*(e_j), e_i \rangle = \langle e_j, T(e_i) \rangle = \overline{\langle T(e_i), e_j \rangle} = \overline{a_{j,i}}.$$

Hence the matrix of T^* with respect to Λ is $(\overline{a_{j,i}})$, the conjugate transpose of A.

(ii) If $\mathbf{H} = L^2[0,1]$ and $K \in L^2([0,1]^2)$, we define $T : \mathbf{H} \to \mathbf{H}$ to be the associated integral operator

$$T(f)(x) = \int_0^1 K(x,y)f(y)dy.$$

We had proved in Example 2.2.17 that T is a well-defined bounded operator. Now for any $f, g \in \mathbf{H}$, let $h := T^*(g)$. Then the equation $\langle T(f), g \rangle = \langle f, T^*(g) \rangle$ translates to

$$\int_0^1 \int_0^1 K(x,y)f(y)\overline{g(x)}dydx = \int_0^1 f(x)\overline{h(x)}dx.$$

By taking conjugates and using the Fubini–Tonelli Theorem, we have

$$\int_0^1 \int_0^1 \overline{K(x,y)}g(x)dx\overline{f(y)}dy = \int_0^1 h(y)\overline{f(y)}dy.$$

This must be true for any $f \in \mathbf{H}$, so

$$T^*(g)(y) = h(y) = \int_0^1 \overline{K(x,y)}g(x)dx.$$

In other words, T^* is an integral operator whose kernel is the 'conjugate transpose' of K.

(iii) If $\mathbf{H} = \ell^2$ and $S \in \mathcal{B}(\mathbf{H})$ be the right shift operator

$$S(x_1, x_2, \ldots) = (0, x_1, x_2, \ldots).$$

If $x = (x_j)$ and $y = (y_j) \in \mathbf{H}$, then the equation $\langle S(x), y \rangle = \langle x, S^*(y) \rangle$ translates to

$$\langle x, S^*(y) \rangle = x_1 \overline{y_2} + x_2 \overline{y_3} + \ldots.$$

Therefore S^* is the left shift operator given by $S^*(y) = (y_2, y_3, \ldots)$.

(iv) Let $\mathbf{H} = L^2[0,1]$ and $\phi \in L^\infty[0,1]$. Let $M_\phi \in \mathcal{B}(\mathbf{H})$ be the multiplication operator given by

$$M_\phi(g) := \phi g.$$

We had seen in Example 7.1.8 that M_ϕ is a bounded operator. Now if $g, h \in \mathbf{H}$, then

$$\langle g, (M_\phi)^*(h) \rangle = \langle M_\phi(g), h \rangle = \int_0^1 \phi(x)g(x)\overline{h(x)}dx = \int_0^1 g(x)\overline{\overline{\phi(x)}h(x)}dx.$$

Therefore $(M_\phi)^*$ is the multiplication operator $M_{\overline{\phi}}$, where $\overline{\phi} \in L^\infty[0,1]$ is the function $\overline{\phi}(x) = \overline{\phi(x)}$.

We now establish some basic properties of the adjoint. The first result is an easy consequence of the defining equation (Equation 8.1), so we omit its proof here.

8.1.5 PROPOSITION *For $T, S \in \mathcal{B}(\mathbf{H})$ and $\alpha \in \mathbb{C}$,*

  (i) $(\alpha T + S)^* = \overline{\alpha}T^* + S^*$

  (ii) $(TS)^* = S^*T^*$

  (iii) $(T^*)^* = T$

  (iv) If $T \in GL(\mathcal{B}(\mathbf{H}))$, then $T^* \in GL(\mathcal{B}(\mathbf{H}))$ and $(T^{-1})^* = (T^*)^{-1}$

The next result might seem innocuous at the moment. However, this identity is a crucial ingredient in the definition of an abstract C*-algebra and we will see consequences of it throughout this chapter.

8.1.6 PROPOSITION *If $T \in \mathcal{B}(\mathbf{H})$, then $\|T\| = \|T^*\| = \|T^*T\|^{1/2}$.*

Proof.   For $x \in \mathbf{H}$ with $\|x\| \leq 1$, we have

$$\|Tx\|^2 = \langle Tx, Tx \rangle = \langle T^*Tx, x \rangle \leq \|T^*Tx\|\|x\| \leq \|T^*T\| \leq \|T^*\|\|T\|$$

and thus $\|T\| \leq \|T^*\|$. The reverse inequality is true since $T^{**} = T$. Hence $\|T\| = \|T^*\|$. Now the inequalities above show that

$$\|T\|^2 \leq \|T^*T\| \leq \|T^*\|\|T\|.$$

Since these are now equalities, we conclude that $\|T^*T\| = \|T\|^2$.   □

As we will soon see, operators that commute with their adjoints are easier to understand than those that do not. We now isolate such operators and their brethren.

8.1.7 DEFINITION Let $T \in \mathcal{B}(\mathbf{H})$. We say that T is

(i) *normal* if $TT^* = T^*T$.

(ii) *self-adjoint* if $T = T^*$.

8.1.8 EXAMPLE

(i) Let $\mathbf{H} = \mathbb{C}^n$ and $T \in \mathcal{B}(\mathbf{H})$. If $\Lambda = \{e_1, e_2, \ldots, e_n\}$ is an orthonormal basis for $\mathbf{H}$, let $A = (a_{i,j})$ denote the matrix of T with respect to Λ. Then T is normal if and only if A commutes with A^*, the conjugate transpose of A. Also, T is self-adjoint if and only if

$$a_{i,j} = \overline{a_{j,i}}$$

for all $1 \leq i, j \leq n$. In particular, if A is a real symmetric matrix, then T is self-adjoint. The astute reader would remember that any real symmetric matrix can be diagonalized (this is one of the highlights of a first course in linear algebra). In fact, we are just beginning a long quest for an analogous result for operators on a Hilbert space.

(ii) If $\mathbf{H} = \ell^2$ and $S \in \mathcal{B}(\mathbf{H})$ is the right shift operator given by

$$S((x_j)) = (0, x_1, x_2, \ldots).$$

Then S^* is the left-shift operator $S^*((y_j)) = (y_2, y_3, \ldots)$. Thus $S^*S = I$ while $SS^* \neq I$. Therefore S is not normal.

(iii) If $\mathbf{H} = L^2[0, 1]$, $\phi \in L^\infty[0, 1]$ and $M_\phi \in \mathcal{B}(\mathbf{H})$ be the multiplication operator, then $M_\phi^* = M_{\overline{\phi}}$. Thus M_ϕ is normal because

$$M_\phi^* M_\phi = M_{\overline{\phi}} M_\phi = M_{\overline{\phi}\phi} = M_{\phi\overline{\phi}} = M_\phi M_\phi^*.$$

Notice that the normality of M_ϕ is really a reflection of the fact that ϕ and $\overline{\phi}$ commute in $L^\infty[0, 1]$.

The next few results are here to help us recognize these classes of operators. They will also be useful when we try to understand their spectra.

8.1.9 PROPOSITION $T \in \mathcal{B}(\mathbf{H})$ *is self-adjoint if and only if* $\langle Tx, x \rangle \in \mathbb{R}$ *for all* $x \in \mathbf{H}$

Proof. If T is self-adjoint, then for any $x \in \mathbf{H}$,

$$\langle Tx, x \rangle = \langle x, T^*x \rangle = \langle x, Tx \rangle = \overline{\langle Tx, x \rangle},$$

so that $\langle Tx, x \rangle \in \mathbb{R}$. Conversely, suppose $\langle Tx, x \rangle \in \mathbb{R}$ for all $x \in \mathbf{H}$. Then for any $x \in \mathbf{H}$, we have

$$\langle Tx, x \rangle = \overline{\langle Tx, x \rangle} = \overline{\langle x, T^*x \rangle} = \langle T^*x, x \rangle.$$

If $S := (T - T^*)$, then $\langle Sz, z \rangle = 0$ for all $z \in \mathbf{H}$. Therefore for any $x, y \in \mathbf{H}$ and $\alpha \in \mathbb{C}$,

$$0 = \langle S(x + \alpha y), x + \alpha y \rangle = \langle Sx, x \rangle + \overline{\alpha}\langle Sx, y \rangle + \alpha\langle Sy, x \rangle + |\alpha|^2\langle Sy, y \rangle$$
$$= \overline{\alpha}\langle Tx, y \rangle - \overline{\alpha}\langle T^*x, y \rangle + \alpha\langle Ty, x \rangle - \alpha\langle T^*y, x \rangle.$$

We conclude that $\overline{\alpha}\langle Tx, y \rangle + \alpha\langle Ty, x \rangle = \overline{\alpha}\langle T^*x, y \rangle + \alpha\langle T^*y, x \rangle$ for any $\alpha \in \mathbb{C}$. Set $\alpha = 1$ and then $\alpha = i$ to see that

$$\langle Tx, y \rangle + \langle Ty, x \rangle = \langle T^*x, y \rangle + \langle T^*y, x \rangle \text{ and}$$
$$-i\langle Tx, y \rangle + i\langle Ty, x \rangle = -i\langle T^*x, y \rangle + i\langle T^*y, x \rangle$$

Taking linear combinations, we conclude that

$$\langle Tx, y \rangle = \langle T^*x, y \rangle$$

holds for all $x, y \in \mathbf{H}$. Therefore $T = T^*$. $\square$

8.1.10 THEOREM *If $T \in \mathcal{B}(\mathbf{H})$ is self-adjoint, then*

$$\|T\| = \sup\{|\langle Tx, x \rangle| : x \in \mathbf{H}, \|x\| = 1\}.$$

Proof. Let $\beta := \sup\{|\langle Tx, x \rangle| : x \in \mathbf{H}, \|x\| = 1\}$. If $x \in \mathbf{H}$ with $\|x\| = 1$, we have

$$|\langle T(x), x \rangle| \leq \|T(x)\|\|x\| \leq \|T\|\|x\|^2 = \|T\|$$

by the Cauchy–Schwarz Inequality. Therefore $\beta \leq \|T\|$.

Conversely, since $T = T^*$, we have

$$\langle T(x \pm y), x \pm y \rangle = \langle Tx, x \rangle \pm 2 \operatorname{Re} \langle Tx, y \rangle + \langle Ty, y \rangle$$

for any $x, y \in \mathbf{H}$ with $\|x\| = \|y\| = 1$. By the parallelogram law,

$$\begin{aligned}
4 \operatorname{Re} \langle Tx, y \rangle &= \langle T(x+y), x+y \rangle - \langle T(x-y), x-y \rangle \\
&\leq \beta(\|x+y\|^2 + \|x-y\|^2) \\
&= 2\beta(\|x\|^2 + \|y\|^2) \\
&= 4\beta.
\end{aligned}$$

Now if $\lambda \in \mathbb{C}$ is chosen so that $|\lambda| = 1$ and $\langle Tx, y \rangle = |\langle Tx, y \rangle|$, then we may replace x by λx in the previous argument to conclude that

$$4 |\langle T(x), y \rangle| = 4 \operatorname{Re} \langle T(\lambda x), y \rangle \leq 4\beta.$$

By Lemma 8.1.1, $\|T\| \leq \beta$. □

The corollaries now come thick and fast.

8.1.11 COROLLARY *If $T \in \mathcal{B}(\mathbf{H})$ and $\langle Tx, x \rangle = 0$ for all $x \in \mathbf{H}$, then $T = 0$.*

Proof. Since $\langle Tx, x \rangle \in \mathbb{R}$ for all $x \in \mathbf{H}$, T is self-adjoint by Proposition 8.1.9. Hence $T = 0$ by Theorem 8.1.10. □

Recall that $T \in \mathcal{B}(\mathbf{H})$ is an isometry if $\|T(x)\| = \|x\|$ for all $x \in \mathbf{H}$ and T is a unitary if T is a surjective isometry (Definition 3.4.5). The next two corollaries help us identify such operators through algebraic properties.

8.1.12 COROLLARY $T \in \mathcal{B}(\mathbf{H})$ *is an isometry if and only if $T^*T = I$.*

Proof. T is an isometry if and only if $\langle Tx, Tx \rangle = \langle x, x \rangle$ for all $x \in \mathbf{H}$. This is equivalent to $\langle (T^*T - I)x, x \rangle = 0$, so the result now follows from the previous corollary. □

8.1.13 COROLLARY $T \in \mathcal{B}(\mathbf{H})$ *is a unitary if and only if $TT^* = T^*T = I$.*

Proof. If T is a unitary, then it is an isometry and thus $T^*T = I$ must hold. Furthermore, T is invertible in $\mathcal{B}(\mathbf{H})$ and thus has a right inverse as well. It must follow (as it does in ring theory) that the left inverse and right inverse must coincide. In other words,

$$TT^* = I$$

must also hold.

Conversely, if $TT^* = T^*T = I$, then T is an isometry by Corollary 8.1.12. Furthermore, since T has a right inverse, it is also surjective. This proves that T is a unitary. $\qquad\square$

The previous two corollaries are worth staring at for a minute. An isometry is one that satisfies the equation $T^*T = I$, while a unitary is one that satisfies $TT^* = I = T^*T$ as well. This subtle distinction is best captured by the right shift operator $S \in \mathcal{B}(\ell^2)$ given by

$$S((x_j)) = (0, x_1, x_2, \ldots).$$

As described in Example 8.1.8, $S^*S = I$ but $SS^* \neq I$. This is corroborated by the fact that S is quite obviously isometric but not surjective. Furthermore,

$$I - SS^* = P_{e_1},$$

the projection of ℓ^2 onto the one-dimensional subspace spanned by e_1. In other words, S fails to be a unitary by a rather small margin, to wit, a compact operator. Such an operator is termed *essentially normal*. Let us now end with a result which will be very useful to us when we study normal operators in Chapter 10.

8.1.14 THEOREM $T \in \mathcal{B}(\mathbf{H})$ *is normal if and only if* $\|Tx\| = \|T^*x\|$ *for all* $x \in \mathbf{H}$.

Proof. Suppose that $\|T(x)\| = \|T^*(x)\|$ holds for all $x \in \mathbf{H}$. Then we may unwrap the equation $\|T(x)\|^2 = \|T^*(x)\|^2$ to find that

$$\langle T^*Tx, x \rangle = \langle TT^*x, x \rangle$$

holds. Hence $\langle (T^*T - TT^*)x, x \rangle = 0$. This last equation holds for all $x \in \mathbf{H}$ if and only if $TT^* - T^*T$ is the zero operator (by Corollary 8.1.11). Therefore T must be normal.

For the converse, we assume that T is normal and simply follow the implications in the reverse direction to conclude that $\|T(x)\| = \|T^*(x)\|$ holds for all $x \in \mathbf{H}$. $\quad\square$

Before we give an application of this theorem, let us prove an analogue of Proposition 4.5.9 in the setting of Hilbert spaces.

8.1.15 LEMMA *Let* $S \in \mathcal{B}(\mathbf{H})$. *Then*

  (i) $\ker(S) = \operatorname{Range}(S^*)^{\perp}$.

 (ii) $\overline{\operatorname{Range}(S^*)} = \ker(S)^{\perp}$.

(iii) $\overline{\operatorname{Range}(S^*S)} = \ker(S)^{\perp}$.

Proof.

(i) For any $x \in \mathbf{H}$, note that

$$x \in \ker(S) \Leftrightarrow \langle S(x), y \rangle = 0 \text{ for all } y \in \mathbf{H}$$

$$\Leftrightarrow \langle x, S^*(y) \rangle = 0 \text{ for all } y \in \mathbf{H}$$

$$\Leftrightarrow x \in \mathrm{Range}(S^*)^\perp.$$

(ii) For any subspace $\mathbf{M}$ of $\mathbf{H}$, we know that $\overline{\mathbf{M}} = (\mathbf{M}^\perp)^\perp$ from Proposition 3.1.10. Together with part (i), this proves that $\overline{\mathrm{Range}(S^*)} = (\mathrm{Range}(S^*)^\perp)^\perp = \ker(S)^\perp$.

(iii) From part (ii), we know that $\overline{\mathrm{Range}(S^*)} = \ker(S)^\perp$. Since $\mathrm{Range}(S^*S) \subset \mathrm{Range}(S^*)$, it suffices to prove that $\mathrm{Range}(S^*) \subset \overline{\mathrm{Range}(S^*S)}$. To that end, fix $y \in \mathrm{Range}(S^*)$ and $\epsilon > 0$ and choose $x \in \mathbf{H}$ such that $y = S^*(x)$. Express x in the form

$$x = u + v,$$

where $u \in \ker(S^*)$ and $v \in \ker(S^*)^\perp$ (by Proposition 3.1.11). Then $y = S^*(v)$. By part (ii) applied to S^*, there exists $w \in \mathrm{Range}(S)$ such that $\|v - w\| < \epsilon$. Write $w = S(z)$ for some $z \in \mathbf{H}$. Then

$$\|y - S^*S(z)\| = \|S^*(v) - S^*(w)\| \leq \epsilon \|S\|.$$

This is true for any $\epsilon > 0$, so $y \in \overline{\mathrm{Range}(S^*S)}$. We have thus proved that $\mathrm{Range}(S^*) \subset \overline{\mathrm{Range}(S^*S)}$. $\qquad\square$

8.1.16 COROLLARY *If $T \in \mathcal{B}(\mathbf{H})$ is a normal operator, then $\sigma(T) = \sigma_{ap}(T)$.*

Proof. We know that $\sigma_{ap}(T) \subset \sigma(T)$, so it suffices to show that if $\lambda \notin \sigma_{ap}(T)$, then $\lambda \notin \sigma(T)$. Now if $\lambda \notin \sigma_{ap}(T)$, then $S := (T - \lambda I)$ is bounded below. However, S is a normal operator since T is normal. By Theorem 8.1.14,

$$\|S^*(x)\| = \|S(x)\|$$

for all $x \in \mathbf{H}$. In particular, S^* is bounded below and is thus injective. By Lemma 8.1.15,

$$\mathrm{Range}(S)^\perp = \ker(S^*) = \{0\}.$$

Hence Range(S) is dense in **H**. By Proposition 7.4.4, Range(S) is also closed. In other words,

$$\text{Range}(S) = \mathbf{H}.$$

Since S is bounded below, it is injective. We conclude that S is bijective and thus $\lambda \notin \sigma(T)$. $\square$

This concludes our preliminary discussion on the adjoint of an operator on a Hilbert space and we are now ready to encounter C*-algebras.

8.2   C*-Algebras

As we had seen in Chapter 7, the notion of a Banach algebra is well suited to study operators on Banach spaces and their spectra. A C*-algebra plays an analogous role when studying operators on Hilbert spaces. The axioms given below are thus modelled on the defining properties of $\mathcal{B}(\mathbf{H})$.

8.2.1 DEFINITION   Let **A** be a Banach algebra.

(i) An *involution* on **A** is a map $\delta : \mathbf{A} \to \mathbf{A}$ such that for all $a, b \in \mathbf{A}$ and $\alpha \in \mathbb{C}$,

   (a) $\delta(\delta(a)) = a$.

   (b) $\delta(ab) = \delta(b)\delta(a)$.

   (c) $\delta(\alpha a + b) = \overline{\alpha}\delta(a) + \delta(b)$.

   If such an involution exists, we write $a^* := \delta(a)$.

(ii) **A** is said to be a *C*-algebra* if there is an involution $a \mapsto a^*$ on **A** such that

$$\|a^*a\| = \|a\|^2 \tag{8.2}$$

   holds for all $a \in \mathbf{A}$.

Equation 8.2 is called the *C*-identity*. There are Banach algebras with involutions that do not satisfy this identity (see Part (vi) of Example 8.2.4). While these objects are certainly useful, the rigidity offered by Equation 8.2 will prove to be crucial while proving some important theorems concerning C*-algebras.

8.2.2 REMARK Let $\mathbf{A}$ be a C*-algebra.

(i) Since $(a^*)^* = a$, the map $a \mapsto a^*$ is bijective.

(ii) If $\mathbf{A}$ is unital, then for any $a \in \mathbf{A}$,

$$a^* = a^* \cdot 1_{\mathbf{A}} = (1_{\mathbf{A}}^* \cdot a)^*.$$

Hence $a = 1_{\mathbf{A}}^* \cdot a$ and, similarly, $a = a \cdot 1_{\mathbf{A}}^*$. By the uniqueness of the identity, it follows that $1_{\mathbf{A}} = 1_{\mathbf{A}}^*$.

(iii) Moreover, $\|1_{\mathbf{A}}\|^2 = \|1_{\mathbf{A}}^* \cdot 1_{\mathbf{A}}\| = \|1_{\mathbf{A}} \cdot 1_{\mathbf{A}}\| = \|1_{\mathbf{A}}\|$. Assuming that $\mathbf{A} \neq \{0\}$, we conclude that $\|1_{\mathbf{A}}\| = 1$. Recall that, for a Banach algebra, we had to resort to Lemma 7.2.2 and choose a norm on $\mathbf{A}$ such that $\|1_{\mathbf{A}}\| = 1$. For a C*-algebra, however, the norm automatically satisfies this condition.

(iv) If $\mathbf{A}$ is unital, then for any $\alpha \in \mathbb{C}$, $\alpha^* := (\alpha \cdot 1_{\mathbf{A}})^* = \overline{\alpha}$.

The next lemma is a useful way to verify the C*-identity.

8.2.3 LEMMA *If $\mathbf{A}$ is a Banach algebra and $a \mapsto a^*$ is an involution such that*

$$\|a\|^2 \leq \|a^*a\|$$

for all $a \in \mathbf{A}$, then $\mathbf{A}$ is a C*-algebra.

Proof. We need to prove that $\|a^*a\| \leq \|a\|^2$. Since $\|a^*a\| \leq \|a^*\|\|a\|$, it suffices to prove that $\|a^*\| \leq \|a\|$. However,

$$\|a^*\|^2 \leq \|(a^*)^*a^*\| = \|aa^*\| \leq \|a\|\|a^*\|.$$

Therefore $\|a^*\| \leq \|a\|$ must hold. $\square$

8.2.4 EXAMPLE

(i) Let $\mathbf{A} = \mathbb{C}$ equipped with the usual norm. Then $z \mapsto \overline{z}$ is an involution on $\mathbb{C}$ that makes it a C*-algebra.

(ii) If $\mathbf{H}$ is a Hilbert space, then $\mathcal{B}(\mathbf{H})$ is a C*-algebra by Proposition 8.1.6.

(iii) If $\mathbf{H} = \mathbb{C}^n$, then $\mathcal{B}(\mathbf{H}) \cong M_n(\mathbb{C})$. Therefore $M_n(\mathbb{C})$ is a C*-algebra. Under this identification, Example 8.1.4 shows that

$$(a_{i,j})^* := (\overline{a_{j,i}}).$$

(iv) For any Hilbert space $\mathbf{H}$, $\mathcal{K}(\mathbf{H})$ is a closed ideal in $\mathcal{B}(\mathbf{H})$ and thus a Banach algebra in its own right. Also, if $T \in \mathcal{K}(\mathbf{H})$, then $T' \in \mathcal{K}(\mathbf{H}^*)$ by Theorem 5.6.14. Since T^* is obtained from T' by composing with bounded operators, T^* is also a compact operator (see Proposition 7.1.5). Hence the adjoint operation on $\mathcal{B}(\mathbf{H})$ restricts to $\mathcal{K}(\mathbf{H})$, making $\mathcal{K}(\mathbf{H})$ a C*-algebra.

(v) If X is a locally compact Hausdorff space, then $C_0(X)$ is a C*-algebra with involution $f \mapsto f^*$, where $f^*(x) = \overline{f(x)}$.

(vi) $L^\infty[0,1]$ is a C*-algebra with the same involution as above.

(vii) Let $\mathbf{A} = C^1[0,1]$ be the Banach algebra with norm

$$\|f\| := \|f\|_\infty + \|f'\|_\infty$$

(see Exercise 7.2). The map $f \mapsto f^* := \overline{f}$ is an involution in $\mathbf{A}$. However, if $f(x) = x$, then $\|f\|^2 = (1+1)^2 = 4$, while

$$\|f^*f\| = \|f^2\| = 1 + 2 = 3.$$

Therefore $\mathbf{A}$ is not a C*-algebra with respect to this involution and norm.

8.2.5 DEFINITION Let $\mathbf{A}$ be a C*-algebra.

(i) A subset $\mathbf{B}$ of $\mathbf{A}$ is said to be *self-adjoint* if $a^* \in \mathbf{B}$ whenever $a \in \mathbf{B}$.

(ii) A subset $\mathbf{B}$ of $\mathbf{A}$ is said to be a *C*-subalgebra* if $\mathbf{B}$ is a closed, self-adjoint subalgebra of $\mathbf{A}$ (as in Definition 7.1.3).

(iii) When $\mathbf{A}$ is unital, a C*-subalgebra $\mathbf{B}$ is said to *unital in* $\mathbf{A}$ if $\mathbf{B}$ contains the unit of $\mathbf{A}$.

(iv) If $\{\mathbf{B}_\alpha\}_{\alpha \in J}$ is a family of C*-subalgebras of $\mathbf{A}$, then

$$\bigcap_{\alpha \in J} \mathbf{B}_\alpha$$

is also a C*-subalgebra of $\mathbf{A}$ (do verify this). Therefore if $\mathcal{S}$ is a subset of $\mathbf{A}$, then there is a unique *smallest* C*-subalgebra of $\mathbf{A}$ containing $\mathcal{S}$. This is called the *C*-subalgebra generated by* $\mathcal{S}$ and is denoted by $C^*(\mathcal{S})$.

The next example is fundamental to almost everything we are going to learn in this chapter and Chapter 10.

8.2.6 EXAMPLE Let $T \in \mathcal{B}(\mathbf{H})$ be a normal operator, define

$$\mathcal{A}(T) := C^*(\{I, T\})$$

to be the smallest C*-algebra containing both I and T. We examine the elements in $\mathcal{A}(T)$ more closely.

(i) $T^n \in \mathcal{A}(T)$ and $(T^*)^m \in \mathcal{A}(T)$ for any $n, m \geq 0$.

(ii) Let $\mathbb{C}[x, y]$ denote the set of all polynomials in two commuting variables with complex coefficients. If $p(x, y) \in \mathbb{C}[x, y]$ is given by $p(x, y) = \sum_{i,j} \alpha_{i,j} x^i y^j$, then we may define

$$p(T, T^*) := \sum_{i,j} \alpha_{i,j} T^i (T^*)^j.$$

There is no ambiguity in this expression since T and T^* commute. Furthermore, if $p_1(x, y), p_2(x, y) \in \mathbb{C}[x, y]$, then

$$(p_1 p_2)(T, T^*) = p_1(T, T^*) p_2(T, T^*).$$

(iii) The collection $\mathcal{C} = \{p(T, T^*) : p(x, y) \in \mathbb{C}[x, y]\}$ is a self-adjoint subalgebra of $\mathcal{B}(\mathbf{H})$ and any self-adjoint subalgebra containing $\{I, T\}$ must contain $\mathcal{C}$. Furthermore, it is easy to verify that the closure of such a subalgebra is a C*-subalgebra. Therefore

$$\mathcal{A}(T) = \overline{\{p(T, T^*) : p(x, y) \in \mathbb{C}[x, y]\}}.$$

(iv) Since $\mathbb{C}[x, y]$ is a commutative ring, it follows that $\mathcal{C}$ is a commutative subalgebra. Thus $\mathcal{A}(T)$ is a commutative C*-algebra that is unital in $\mathcal{B}(\mathbf{H})$.

8.2.7 LEMMA *If* $\mathbf{A}$ *is a C*-algebra, then for any* $a \in \mathbf{A}$,

(i) $\|a\| = \|a^*\|$

(ii) $\|aa^*\| = \|a\|^2$

Proof. Assume without loss of generality that $a \neq 0$.

(i) Note that $\|a\|^2 = \|a^*a\| \leq \|a^*\|\|a\|$. Hence $\|a\| \leq \|a^*\|$. The reverse inequality follows from the fact that $(a^*)^* = a$.

(ii) By part (i), $\|aa^*\| = \|(a^*)^*a\| = \|a^*\|^2 = \|a\|^2.$ $\qquad\qquad\square$

8.2.8 PROPOSITION *If* **A** *is a C*-algebra, then for any* $a \in$ **A**,

$$\|a\| = \sup\{\|ax\| : x \in \mathbf{A}, \|x\| \le 1\}$$
$$= \sup\{\|xa\| : x \in \mathbf{A}, \|x\| \le 1\}$$
$$= \sup\{\|x^*ay\| : x, y \in \mathbf{A}, \|x\| \le 1, \|y\| \le 1\}.$$

Proof. Assume without loss of generality that $a \ne 0$ and let $\beta := \sup\{\|ax\| : x \in \mathbf{A}, \|x\| \le 1\}$. Since **A** is a Banach algebra, $\|ax\| \le \|a\|$ for all $x \in \mathbf{A}$ such that $\|x\| \le 1$. Thus $\beta \le \|a\|$.

Furthermore, if $x = a^*/\|a\|$, then $\|x\| = 1$ by Lemma 8.2.7 and $\|ax\| = \|a\|$. This proves that $\beta = \|a\|$. The second equality is similar and the third follows from the first two. We leave the details to the reader. $\qquad \square$

8.2.9 DEFINITION

(i) A function $\Phi : \mathbf{A} \to \mathbf{B}$ between two C*-algebras is called a **-homomorphism* if Φ is a homomorphism of algebras such that

$$\Phi(a^*) = \Phi(a)^*$$

for all $a \in \mathbf{A}$. Note that a *-homomorphism is not required to be continuous by definition; they are automatically continuous (see Proposition 8.2.15).

(ii) A bijective *-homomorphism is called an **-isomorphism* of C*-algebras. If there is a *-isomorphism from **A** to **B**, we write $\mathbf{A} \cong \mathbf{B}$. (While we have used the same symbol for an isomorphism of Banach algebras, the meaning will be clear from the context in which it is stated).

8.2.10 EXAMPLE

(i) Let X be a compact, Hausdorff space, $\mathbf{A} = C(X)$ and $x_0 \in X$. Then $\Phi : \mathbf{A} \to \mathbb{C}$ given by $f \mapsto f(x_0)$ is a *-homomorphism.

(ii) Let $\mathbf{A} = C(X)$ and $x_1, x_2, \ldots, x_n$ be points in X (with possible repeats). Define $\Phi : \mathbf{A} \to M_n(\mathbb{C})$ by

$$f \mapsto \begin{pmatrix} f(x_1) & 0 & \ldots & 0 \\ 0 & f(x_2) & \ldots & 0 \\ \vdots & & & \vdots \\ 0 & 0 & \ldots & f(x_n) \end{pmatrix}.$$

This is a *-homomorphism from **A** to $M_n(\mathbb{C})$.

(iii) Conversely, let $T = \text{diag}(\lambda_1, \lambda_2, \ldots, \lambda_n) \in M_n(\mathbb{C})$ be a diagonal matrix and let $X = \{\lambda_1, \lambda_2, \ldots, \lambda_n\} = \sigma(T) \subset \mathbb{C}$. Define $\Phi : C(X) \to M_n(\mathbb{C})$ by

$$f \mapsto \begin{pmatrix} f(\lambda_1) & 0 & \ldots & 0 \\ 0 & f(\lambda_2) & \ldots & 0 \\ \vdots & & & \vdots \\ 0 & 0 & \ldots & f(\lambda_n) \end{pmatrix}.$$

This is a $*$-homomorphism, with the property that $\Phi(\zeta) = T$, where $\zeta \in C(X)$ is the identity function $\zeta(z) = z$.

(iv) If $\mathbf{A} = L^\infty[0,1]$ and $\mathbf{B} = \mathcal{B}(L^2[0,1])$, then define $\Psi : \mathbf{A} \to \mathbf{B}$ by

$$\phi \mapsto M_\phi.$$

Then Ψ is a homomorphism of Banach algebras such that $\Psi(\phi^*) = \Psi(\phi)^*$ (see Example 7.1.8 and Example 8.1.4) and is thus a $*$-homomorphism.

Let $\mathbf{A}$ be a C*-algebra. A (left, right, or two-sided) ideal of $\mathbf{A}$ is defined exactly as we did for Banach algebras in Definition 7.1.3. If $\mathbf{I}$ is a two-sided ideal of $\mathbf{A}$, then we write $\mathbf{I} \lhd \mathbf{A}$. We will study quotients of C*-algebras later in the chapter. For now, we only need this notation in order to describe the unitalization of a C*-algebra.

8.2.11 THEOREM *Let $\mathbf{A}$ be a non-unital C*-algebra, then there is a pair $(\mathbf{A}^u, \iota)$ satisfying the following properties:*

(a) *$\mathbf{A}^u$ is a unital C*-algebra.*

(b) *$\iota : \mathbf{A} \to \mathbf{A}^u$ is an isometric $*$-homomorphism such that $\iota(\mathbf{A}) \lhd \mathbf{A}^u$ and $\iota(\mathbf{A})$ has codimension 1 in $\mathbf{A}^u$.*

(c) *If $\mathbf{B}$ is any unital C*-algebra and $\Phi : \mathbf{A} \to \mathbf{B}$ a $*$-homomorphism, then there is a unique $*$-homomorphism $\widetilde{\Phi} : \mathbf{A}^u \to \mathbf{B}$ such that $\widetilde{\Phi}(1_{\mathbf{A}^u}) = 1_{\mathbf{B}}$ and $\widetilde{\Phi} \circ \iota = \Phi$.*

(d) *If $(\widehat{\mathbf{A}}, \widehat{\iota})$ is a pair satisfying properties (a)-(c), then there is a $*$-isomorphism $\Psi : \mathbf{A}^u \to \widehat{\mathbf{A}}$ such that $\Psi \circ \iota = \widehat{\iota}$.*

*The pair $(\mathbf{A}^u, \iota)$ is thus unique and is called the **unitalization** of $\mathbf{A}$.*

Proof. Let $\mathcal{B}(\mathbf{A})$ denote the space of bounded operators on $\mathbf{A}$ (treated as Banach space) and let $\iota : \mathbf{A} \to \mathcal{B}(\mathbf{A})$ be the left-regular representation given by $a \mapsto L_a$, where

$$L_a(b) := ab.$$

Let $\mathbf{A}^u := \{L_a + \lambda \cdot 1_{\mathcal{B}(\mathbf{A})} : a \in \mathbf{A}, \lambda \in \mathbb{C}\}$ and define an involution on $\mathbf{A}^u$ by

$$(L_a + \lambda \cdot 1_{\mathcal{B}(\mathbf{A})})^* := L_{a^*} + \overline{\lambda} \cdot 1_{\mathcal{B}(\mathbf{A})}.$$

We show that this algebra $\mathbf{A}^u$ satisfies all the required conditions.

(a) If $a \in \mathbf{A}$, then $\|L_a\| = \sup\{\|ab\| : \|b\| \le 1\} = \|a\|$ by Proposition 8.2.8. Therefore ι is isometric. In particular, its image is closed in $\mathcal{B}(\mathbf{A})$. Since $\mathbf{A}^u$ is obtained by adding a finite dimensional space to $\iota(\mathbf{A})$, it follows from Corollary 2.5.7 that $\mathbf{A}^u$ is a closed subspace of $\mathcal{B}(\mathbf{A})$. Since it is clearly an algebra, $\mathbf{A}^u$ is a unital Banach algebra. In order to prove that $\mathbf{A}^u$ is a C*-algebra, we appeal to Lemma 8.2.3 and simply verify that

$$\|x\|^2 \le \|x^*x\|$$

for all $x \in \mathbf{A}^u$. If $x = L_a + \lambda 1_{\mathcal{B}(\mathbf{A})}$ and $b \in \mathbf{A}$ is such that $\|b\| \le 1$, then

$$\begin{aligned}
\|x(b)\|^2 &= \|ab + \lambda b\|^2 \\
&= \|(ab + \lambda b)^*(ab + \lambda b)\| \\
&= \|b^*(x^*x(b))\| \\
&\le \|x^*x(b)\| \\
&\le \|x^*x\|.
\end{aligned}$$

It follows that $\|x\|^2 \le \|x^*x\|$.

(b) By definition, ι is a *-homomorphism and $\iota(\mathbf{A}) = \{L_a : a \in \mathbf{A}\}$ is an ideal in $\mathbf{A}^u$ (verify this). We now need to prove that $\mathbf{A}^u/\iota(\mathbf{A})$ is one-dimensional. Clearly, $\mathbf{A}^u/\iota(\mathbf{A})$ has dimension atmost 1. If it had dimension zero, then there would exist $a \in \mathbf{A}$ such that $1_{\mathcal{B}(\mathbf{A})} = L_a$. This would imply that $ab = b$ for all $b \in \mathbf{A}$; taking adjoints, it would follow that $b^*a^* = b^*$, and so $ca = c$ for all $c \in \mathbf{A}$. We would then be forced to conclude that a is a unit of $\mathbf{A}$, which contradicts the assumption that $\mathbf{A}$ is non-unital. Hence $\mathbf{A}^u/\iota(\mathbf{A})$ is one-dimensional.

(c) If $\Phi : \mathbf{A} \to \mathbf{B}$ is a *-homomorphism with $\mathbf{B}$ unital, then define $\widetilde{\Phi} : \mathbf{A}^u \to \mathbf{B}$ by

$$L_a + \lambda 1_{\mathcal{B}(\mathbf{A})} \mapsto \Phi(a) + \lambda 1_{\mathbf{B}}.$$

Then $\widetilde{\Phi}$ is well-defined (since ι is injective) and satisfies the required conditions.

(d) This is left as an exercise.

□

Observe that if $\mathbf{A}$ were a unital C*-algebra and $\iota : \mathbf{A} \to \mathcal{B}(\mathbf{A})$ were the left-regular representation, then the algebra $\mathbf{A}^u$ defined above is precisely $\iota(\mathbf{A})$ and is thus isomorphic to $\mathbf{A}$. This allows us to make the following definition.

8.2.12 DEFINITION Let $\mathbf{A}$ be a C*-algebra with unitalization $(\mathbf{A}^u, \iota)$. If $a \in \mathbf{A}$, we define the spectrum of a to be

$$\sigma_{\mathbf{A}}(a) := \sigma_{\mathbf{A}^u}(\iota(a)).$$

This allows us to work with both unital and non-unital C*-algebras on a (somewhat) equal footing. However, do notice that if a is an element of a non-unital C*-algebra $\mathbf{A}$, then $0 \in \sigma_{\mathbf{A}}(a)$ (Exercise 8.11). Armed with this definition, we now prove a result that shows how the C*-identity imposes strong constraints on the algebra. To begin with, we identify normal and self-adjoint elements as we did for operators on Hilbert spaces.

8.2.13 DEFINITION Let $\mathbf{A}$ be a C*-algebra, then $a \in \mathbf{A}$ is said to be

(i) *normal* if $aa^* = a^*a$.

(ii) *self-adjoint* if $a = a^*$.

Observe that if b is any element of a C*-algebra, then b^*b is self-adjoint. It is this simple fact that is repeatedly exploited.

8.2.14 LEMMA *Let $\mathbf{A}$ be a C*-algebra.*

(i) *For any $a \in \mathbf{A}$, $\|a\| = r(a^*a)^{1/2}$.*

(ii) *If $a \in \mathbf{A}$ is self-adjoint, then $\|a\| = r(a)$.*

Proof. Since $\|a\|^2 = \|a^*a\|$ and a^*a is self-adjoint, part (i) follows from part(ii).

To prove part (ii), let $a \in \mathbf{A}$ be self-adjoint. Then $\|a^2\| = \|a^*a\| = \|a\|^2$. By induction, it follows that $\|a^{2^n}\| = \|a\|^{2^n}$ for all $n \geq 1$. By the Spectral Radius Formula,

$$r(a) = \lim_{k \to \infty} \|a^k\|^{1/k} = \lim_{n \to \infty} \|a^{2^n}\|^{1/2^n} = \|a\|.$$

$\square$

Let us take a moment to understand Lemma 8.2.14. On a unital C*-algebra, the norm is a *topological* quantity; one that induces a metric on it. However, the spectral radius is defined in terms of the invertibility (or lack thereof) of certain elements. Therefore it is an *algebraic* quantity; one that ought not to be related to the metric. That these two seemingly disparate notions are equal for self-adjoint elements is another remarkable feature of C*-algebra theory.

This lemma leads to the next (equally remarkable) result that every ∗-homomorphism is automatically continuous.

8.2.15 PROPOSITION *Let* $\Phi : \mathbf{A} \to \mathbf{B}$ *be a ∗-homomorphism between two C*-algebras. Then* Φ *is bounded and* $\|\Phi\| \leq 1$.

Proof.

(i) Assume first that $\mathbf{A}$ is unital. Let $\mathbf{C} := \overline{\text{Range}(\Phi)}$, then $\mathbf{C}$ is a C*-subalgebra of $\mathbf{B}$ and $e := \Phi(1_{\mathbf{A}})$ satisfies

$$e\Phi(a) = \Phi(a) = \Phi(a)e$$

for all $a \in \mathbf{A}$. It follows that $\mathbf{C}$ is unital and $e = 1_{\mathbf{C}}$. Since Φ is multiplicative,

$$\Phi(GL(\mathbf{A})) \subset GL(\mathbf{C}).$$

Hence if $b \in \mathbf{A}$ and $\lambda \in \sigma_{\mathbf{C}}(\Phi(b))$, then $(b - \lambda 1_{\mathbf{A}})$ cannot be invertible in $\mathbf{A}$. In other words,

$$\sigma_{\mathbf{C}}(\Phi(b)) \subset \sigma_{\mathbf{A}}(b).$$

Therefore if $b \in \mathbf{A}$ is self-adjoint, we may apply Lemma 8.2.14 to conclude that

$$\|\Phi(b)\| = \sup\{|\lambda| : (\Phi(b) - \lambda 1_{\mathbf{C}}) \notin GL(\mathbf{C})\}$$

$$\leq \sup\{|\lambda| : (b - \lambda 1_{\mathbf{A}}) \notin GL(\mathbf{A})\}$$

$$= \|b\|.$$

Finally, if $a \in \mathbf{A}$, then $b = a^*a$ is self-adjoint. Therefore

$$\|\Phi(a)\|^2 = \|\Phi(a^*a)\| \leq \|a^*a\| = \|a\|^2.$$

This proves that $\|\Phi(a)\| \leq \|a\|$ for all $a \in \mathbf{A}$.

(ii) Now suppose **A** is not unital. Replacing **B** by its unitalization if need be, we may assume that **B** is unital. Let $(\mathbf{A}^u, \iota)$ be the unitalization of **A**. By Theorem 8.2.11, we may extend Φ to a $*$-homomorphism $\Psi : \mathbf{A}^u \to \mathbf{B}$ such that $\Psi \circ \iota = \Phi$. By part (i), $\|\Psi\| \leq 1$. Since ι is isometric, it follows that

$$\|\Phi\| = \|\Psi \circ \iota\| \leq \|\Psi\| \|\iota\| = \|\Psi\| \leq 1. \qquad \square$$

We record one fact that was used in the previous result and is something we will refer back to repeatedly.

8.2.16 PROPOSITION *Let* $\Phi : \mathbf{A} \to \mathbf{B}$ *be a homomorphism between two unital Banach algebras such that* $\Phi(1_{\mathbf{A}}) = 1_{\mathbf{B}}$. *If* $a \in \mathbf{A}$, *then* $\sigma_{\mathbf{B}}(\Phi(a)) \subset \sigma_{\mathbf{A}}(a)$.

We end the section with a uniqueness result that is, once again, a consequence of the C*-identity.

8.2.17 COROLLARY *Let* **A** *be a complex algebra equipped with an involution. Then there is atmost one norm on* **A** *that makes it a C*-algebra.*

Proof. Suppose $\| \cdot \|_1$ and $\| \cdot \|_2$ are two norms on **A** that make it a C*-algebra. Then the identity map

$$\iota : (\mathbf{A}, \| \cdot \|_1) \to (\mathbf{A}, \| \cdot \|_2)$$

is a $*$-homomorphism. By Proposition 8.2.15, $\|a\|_2 \leq \|a\|_1$ for all $a \in \mathbf{A}$. By symmetry, it follows that $\|a\|_1 = \|a\|_2$ for all $a \in \mathbf{A}$. $\qquad \square$

It is worth mentioning that there are complex algebras with involution which do not admit *any* norm that satisfies the C*-identity. The preceding result merely states that there cannot be *two* such norms.

We end this section with an example that has been a long time coming. Let $\mathbf{H} := L^2[0,1]$ and let $K \in L^2([0,1]^2)$. In Example 2.2.17, we had defined an operator $T \in \mathcal{B}(\mathbf{H})$ by

$$T(f)(x) = \int_0^1 K(x,y)f(y)dy.$$

We had proved in that example that $\|T\| \leq \|K\|_2$ and we had mentioned that this inequality is not an equality in general. We now use Lemma 8.2.14 to give one such example.

8.2.18 EXAMPLE Let $K : [0,1]^2 \to \mathbb{R}$ be given by

$$K(x,y) = \begin{cases} 1 & : \text{ if } x \geq y, \\ 0 & : \text{ otherwise.} \end{cases}$$

so that

$$T(f)(x) = \int_0^x f(y)dy.$$

By Example 8.1.4, $T^* \in \mathcal{B}(\mathbf{H})$ is given by

$$T^*(g)(y) = \int_y^1 g(t)dt.$$

Therefore $S := T^*T \in \mathcal{B}(\mathbf{H})$ is given by

$$S(f)(x) = \int_x^1 \int_0^y f(t)dtdy.$$

By Lemma 8.2.14, $\|T\| = r(S)^{1/2}$, so we compute the spectral radius of S. By Example 5.6.8, T is a compact operator. Since $\mathcal{K}(\mathbf{H})$ is an ideal in $\mathcal{B}(\mathbf{H})$, S is a compact operator. By Theorem 7.5.7,

$$r(S) = \max\{|\lambda| : \lambda \text{ is an eigenvalue of } S\}.$$

We now compute the eigenvalues of S. Suppose λ is an eigenvalue with corresponding eigenvector f, then

$$\lambda f(x) = \int_x^1 \int_0^y f(t)dtdy. \tag{8.3}$$

It follows that f is twice continuously differentiable and

$$\lambda f''(x) = -f(x).$$

Therefore f is of the form $\alpha e^{i\omega x} + \beta e^{-i\omega x}$ for some $\alpha, \beta \in \mathbb{C}$, where $\omega^2 = \frac{1}{\lambda}$. One may then explicitly compute

$$S(f)(x) = \lambda f(x) + \frac{1}{i\omega}(\alpha - \beta)x - \lambda\left(\alpha e^{i\omega} - \beta e^{-i\omega}\right) - \frac{1}{i\omega}(\alpha - \beta).$$

Therefore $\alpha = \beta$ must hold and we have $f(x) = 2\alpha \cos(\omega x)$. Since $f \neq 0$, it follows that $\alpha \neq 0$. Furthermore, Equation 8.3 shows that $f(1) = 0$. Therefore

$$\omega = \frac{(2n+1)\pi}{2}$$

for some $n \in \mathbb{N}$. Thus the eigenvalues of S are of the form

$$\lambda_n = \frac{4}{(2n+1)^2\pi^2}.$$

In particular, the largest eigenvalue is $4/\pi^2$. Hence

$$\|T\| = \frac{2}{\pi}.$$

Finally, it is easy to see that

$$\|K\|_2 = \left(\int_0^1 \int_0^1 |K(x,y)|^2 dy dx \right)^{1/2} = \frac{1}{\sqrt{2}}.$$

Therefore $\|T\| < \|K\|_2$.

8.3 Commutative C*-Algebras

As mentioned above, C*-algebras are best suited to understand operators on Hilbert spaces. The first such application is to the study of normal operators. In Example 8.2.6, we had seen that a normal operator generates a commutative C*-algebra. The main goal of this section is to completely characterize commutative C*-algebras. This result (Theorem 8.3.12) is a fundamental representation theorem and one that has deep implications throughout the subject.

We will begin our journey with the study of unital commutative Banach algebras. The first major result is a connection between maximal ideals and linear functionals that respect the multiplicative structure of the algebra.

8.3.1 DEFINITION Let **A** be a Banach algebra.

(i) A linear functional $\tau : \mathbf{A} \to \mathbb{C}$ is said to be *multiplicative* if $\tau(ab) = \tau(a)\tau(b)$ for all $a, b \in \mathbf{A}$ (A priori, we do not require such a linear functional to be continuous).

(ii) The *Gelfand spectrum of* **A** is defined as

$$\Omega(\mathbf{A}) = \{\tau : \mathbf{A} \to \mathbb{C} : \tau \text{ is a non-zero multiplicative linear functional}\}$$

8.3.2 LEMMA *Let* **A** *be a unital commutative Banach algebra.*

(i) *If* $\tau \in \Omega(\mathbf{A})$, *then* $\tau(1_\mathbf{A}) = 1$.

(ii) *If* $\tau \in \Omega(\mathbf{A})$, *then* $\ker(\tau)$ *is a maximal ideal.*

(iii) *The map*

$$\vartheta : \tau \mapsto \ker(\tau)$$

defines a bijection between $\Omega(\mathbf{A})$ *and the set of all maximal ideals of* **A**.

(iv) $\Omega(\mathbf{A})$ *is non-empty.*

(v) *Every* $\tau \in \Omega(\mathbf{A})$ *is continuous and* $\|\tau\| = \tau(1_{\mathbf{A}}) = 1$.

Proof.

(i) For all $a \in \mathbf{A}, \tau(a) = \tau(a \cdot 1_{\mathbf{A}}) = \tau(a)\tau(1_{\mathbf{A}})$. Choose $a \in \mathbf{A}$ such that $\tau(a) \neq 0$, then it follows that $\tau(1_{\mathbf{A}}) = 1$.

(ii) If $\tau \in \Omega(\mathbf{A})$, then τ is surjective (because it is non-zero) and so τ induces an ring isomorphism $\widehat{\tau} : \mathbf{A}/\ker(\tau) \to \mathbb{C}$. Since $\mathbb{C}$ is a field, $\ker(\tau)$ is a maximal ideal.

(iii) If $\tau \in \Omega(\mathbf{A})$, then $\ker(\tau)$ is maximal, so ϑ is well-defined. We prove that ϑ is bijective.

  (a) If $\tau_1 \neq \tau_2$, then choose $a \in \mathbf{A}$ such that $\tau_1(a) \neq \tau_2(a)$. Then $a - \tau_2(a) \cdot 1_{\mathbf{A}} \in \ker(\tau_2) \setminus \ker(\tau_1)$, so the map ϑ is injective.

  (b) If $\mathbf{I} \lhd \mathbf{A}$ is a maximal ideal, then $\mathbf{I}$ is closed by Proposition 7.2.12 and therefore $\mathbf{A}/\mathbf{I}$ is a Banach algebra. Furthermore, if $a + \mathbf{I} \neq \mathbf{I}$, then define

  $$\mathbf{J} := \{ab + x : x \in \mathbf{I}, b \in \mathbf{A}\}.$$

  Observe that $\mathbf{J}$ is a vector subspace of $\mathbf{A}$ and it contains $\mathbf{I}$. Furthermore, if $z \in \mathbf{J}$ and $c \in \mathbf{A}$, then cz and zc are both in $\mathbf{J}$ because $\mathbf{I}$ is an ideal. Thus $\mathbf{J}$ is an ideal of $\mathbf{A}$. Since $\mathbf{J} \neq \mathbf{I}$, it must happen that $\mathbf{J} = \mathbf{A}$. Hence there exists $x \in \mathbf{I}$ and $b \in \mathbf{A}$ such that

  $$ab + x = 1_{\mathbf{A}}.$$

  Therefore $(a + \mathbf{I})(b + \mathbf{I}) = 1_{\mathbf{A}} + \mathbf{I}$. Thus every non-zero element in $\mathbf{A}/\mathbf{I}$ is invertible. By Corollary 7.3.7, there is a Banach algebra isomorphism

  $$\Phi : \mathbf{A}/\mathbf{I} \to \mathbb{C}.$$

  Let $\pi : \mathbf{A} \to \mathbf{A}/\mathbf{I}$ be the quotient map, then $\tau := \Phi \circ \pi$ is a multiplicative linear functional. Furthermore, τ is non-zero because Φ is an isomorphism and π is non-zero. Therefore $\tau \in \Omega(\mathbf{A})$ and $\ker(\tau) = \mathbf{I}$. We have proved that ϑ is surjective.

(iv) By Proposition 7.2.11, $\mathbf{A}$ contains maximal ideals and thus $\Omega(\mathbf{A})$ is non-empty by part (iii).

(v) Let $\tau \in \Omega(\mathbf{A})$.

(a) By part (ii), $\mathbf{I} := \ker(\tau)$ is a maximal ideal. By Proposition 7.2.12, $\mathbf{I}$ is closed in $\mathbf{A}$. Hence $\pi : \mathbf{A} \to \mathbf{A}/\mathbf{I}$ is continuous by Proposition 7.2.13. Furthermore, the First Isomorphism Theorem allows us to define an injective homomorphism

$$\widehat{\tau} : \mathbf{A}/\mathbf{I} \to \mathbb{C}$$

such that $\widehat{\tau} \circ \pi = \tau$. Since both $\mathbf{A}/\mathbf{I}$ and $\mathbb{C}$ are finite dimensional, $\widehat{\tau} : \mathbf{A}/\mathbf{I} \to \mathbb{C}$ is a bounded linear map. Since $\tau = \widehat{\tau} \circ \pi$, it follows that τ is continuous.

(b) Now for any $a \in \mathbf{A}$, there is $\alpha \in \mathbb{C}$ such that $a + \mathbf{I} = \alpha + \mathbf{I}$ and

$$|\widehat{\tau}(a + \mathbf{I})| = |\widehat{\tau}(\alpha + \mathbf{I})| = |\alpha|.$$

By Proposition 7.1.6, $|\alpha| = \|\alpha + \mathbf{I}\|$ and so $\|\widehat{\tau}\| = 1$. Hence

$$\|\tau\| \leq \|\widehat{\tau}\| \|\pi\| \leq 1.$$

Since $\tau(1_{\mathbf{A}}) = 1$ and $\|1_{\mathbf{A}}\|$ is assumed to be 1, it follows that $\|\tau\| = 1$. $\square$

8.3.3 PROPOSITION *Let $\mathbf{A}$ be a unital commutative Banach algebra and $a \in \mathbf{A}$. Then*

$$\sigma(a) = \{\tau(a) : \tau \in \Omega(\mathbf{A})\}$$

Proof. Suppose $\lambda \in \sigma(a)$, then $x := (a - \lambda \cdot 1_{\mathbf{A}})$ is not invertible. Let $\mathbf{I}$ be the principal ideal generated by x. Then $\mathbf{I}$ is contained in a maximal ideal $\mathbf{J}$ by Proposition 7.2.11. Let τ be the corresponding element of $\Omega(\mathbf{A})$ given by Lemma 8.3.2. Then $\mathbf{J} = \ker(\tau)$, so $\tau(x) = 0$. Since $\tau(1_{\mathbf{A}}) = 1$, we conclude that $\lambda = \tau(a)$.

Conversely, if $\tau \in \Omega(\mathbf{A})$, then $x := a - \tau(a) \cdot 1_{\mathbf{A}}$ is in $\ker(\tau)$, which is a proper ideal of $\mathbf{A}$. Hence x cannot be invertible and thus $\tau(a) \in \sigma(a)$. $\square$

Let $\mathbf{A}$ be a unital Banach algebra. If

$$B_{\mathbf{A}^*} := \{\psi \in \mathbf{A}^* : \|\psi\| \leq 1\},$$

then $B_{\mathbf{A}^*}$ is compact in the weak-* topology by the Banach–Alaoglu Theorem (Theorem 6.6.3). By Lemma 8.3.2, $\Omega(\mathbf{A})$ is a subset of $B_{\mathbf{A}^*}$ and therefore inherits the weak-* topology. The next result just begs to be proved.

8.3.4 PROPOSITION *If $\mathbf{A}$ is a unital Banach algebra, then $\Omega(\mathbf{A})$ is a compact Hausdorff space in the weak-* topology.*

Proof. It suffices to show that $\Omega(\mathbf{A})$ is closed in $B_{\mathbf{A}^*}$. The proof is similar to the proof of the Banach–Alaoglu Theorem itself. Suppose $\rho \in \overline{\Omega(\mathbf{A})}$, then we wish to prove that ρ is non-zero and multiplicative.

(i) To prove that ρ is non-zero, we simply show that $\rho(1_{\mathbf{A}}) = 1$. To that end, consider the evaluation map $\pi : B_{\mathbf{A}^*} \to \mathbb{C}$ given by

$$\pi(\psi) := \psi(1_{\mathbf{A}}).$$

By definition of the weak-$*$ topology, π is continuous. If $\epsilon > 0$ and $z_0 \in \mathbb{C}$, set $B_{\mathbb{C}}(z_0, \epsilon) := \{z \in \mathbb{C} : |z - z_0| < \epsilon\}$. Then

$$V := \pi^{-1}(B_{\mathbb{C}}(\rho(1_{\mathbf{A}}), \epsilon))$$

is an open subset of $B_{\mathbf{A}^*}$ which contains ρ. Therefore there must exist $\tau \in \Omega(\mathbf{A})$ such that $\tau \in V$. It then follows that

$$|1 - \rho(1_{\mathbf{A}})| = |\tau(1_{\mathbf{A}}) - \rho(1_{\mathbf{A}})| < \epsilon.$$

This is true for any $\epsilon > 0$, so $\rho(1_{\mathbf{A}}) = 1$.

(ii) To prove that ρ is multiplicative, fix $a, b \in \mathbf{A}$ and $\epsilon > 0$. For $x \in \mathbf{A}$, let $\pi_x : B_{\mathbf{A}^*} \to \mathbb{C}$ denote the evaluation map $\psi \mapsto \psi(x)$. Now consider

$$V := \pi_a^{-1}(B_{\mathbb{C}}(\rho(a), \epsilon)) \cap \pi_b^{-1}(B_{\mathbb{C}}(\rho(b), \epsilon)) \cap \pi_{ab}^{-1}(B_{\mathbb{C}}(\rho(ab), \epsilon)).$$

Once again, since each π_x is continuous, V is an open set in $B_{\mathbf{A}^*}$ containing ρ and must therefore contain a point $\tau \in \Omega(\mathbf{A})$. Since $\|\tau\| = 1$ and $\|\rho\| \leq 1$, we may estimate

$$|\rho(ab) - \rho(a)\rho(b)| \leq |\rho(ab) - \tau(ab)| + |\tau(ab) - \tau(a)\tau(b)|$$
$$+ |\tau(a)\tau(b) - \rho(a)\tau(b)| + |\rho(a)\tau(b) - \rho(a)\rho(b)|$$
$$< \epsilon(1 + \|b\| + \|a\|).$$

Since this is true for every $\epsilon > 0$, it follows that $\rho(ab) = \rho(a)\rho(b)$. $\qquad\square$

Henceforth we will think of $\Omega(\mathbf{A})$ as a compact Hausdorff space when equipped with the weak-$*$ topology. Given a normed linear space $\mathbf{E}$ and an element $x \in \mathbf{E}$, we found it profitable to study the linear functional $\widehat{x} : \mathbf{E}^* \to \mathbb{C}$ given by

$$\widehat{x}(\psi) = \psi(x).$$

If $\mathbf{E}$ were replaced by a unital Banach algebra $\mathbf{A}$, then it seems reasonable to restrict this function to those linear functionals that respect the Banach algebra structure of **A**. That such a restriction yields a weak-∗ continuous function is built in to the definition.

8.3.5 DEFINITION Given $a \in \mathbf{A}$, the *Gelfand transform* of a is the map $\widehat{a} : \Omega(\mathbf{A}) \to \mathbb{C}$ given by

$$\widehat{a}(\tau) := \tau(a).$$

As mentioned above, $\widehat{a} \in C(\Omega(\mathbf{A}))$. The map $\Gamma_\mathbf{A} : \mathbf{A} \to C(\Omega(\mathbf{A}))$ given by

$$\Gamma_\mathbf{A}(a) = \widehat{a}$$

is called the *Gelfand representation* of **A**.

8.3.6 THEOREM *Let **A** be a unital commutative Banach algebra.*

(i) *For each $a \in \mathbf{A}$, $\|\widehat{a}\|_\infty = r(a)$, the spectral radius of a.*

(ii) *The Gelfand representation $\Gamma_\mathbf{A} : \mathbf{A} \to C(\Omega(\mathbf{A}))$ is a homomorphism of Banach algebras.*

Proof.

(i) In Proposition 8.3.3, we proved that

$$\sigma(a) = \{\tau(a) : \tau \in \Omega(\mathbf{A})\}.$$

This is precisely the range of $\widehat{a}$ and hence the result.

(ii) For any $a, b \in \mathbf{A}$ and $\tau \in \Omega(\mathbf{A})$, we have

$$\widehat{ab}(\tau) = \tau(ab) = \tau(a)\tau(b) = \widehat{a}(\tau)\widehat{b}(\tau).$$

This is true for all $\tau \in \Omega(\mathbf{A})$ and so $\widehat{ab} = \widehat{a}\widehat{b}$. Hence $\Gamma_\mathbf{A}$ is multiplicative. Similarly, we see that $\Gamma_\mathbf{A}$ is also linear. Finally, $\Gamma_\mathbf{A}$ is continuous since

$$\|\Gamma_\mathbf{A}(a)\| = \|\widehat{a}\|_\infty = r(a) \leq \|a\|$$

by Proposition 7.3.3. □

We now identify $\Omega(\mathbf{A})$ for one important Banach algebra. Let X be a compact Hausdorff space and $\mathbf{A} = C(X)$. For any $x \in X$, the map $\tau_x : \mathbf{A} \to \mathbb{C}$ given by

$f \mapsto f(x)$ is a non-zero multiplicative linear functional. Therefore we get a function $\eta_X : X \to \Omega(\mathbf{A})$ given by

$$\eta_X(x) = \tau_x.$$

We wish to prove that η_X is a homeomorphism.

8.3.7 LEMMA *Let* $\mathbf{I} \lhd C(X)$ *be a maximal ideal. Then there exists* $x_0 \in X$ *such that*

$$\mathbf{I} = \ker(\tau_{x_0}).$$

Proof. Since $\mathbf{I}$ is a maximal ideal and $\ker(\tau_{x_0})$ is a proper ideal, it suffices to prove that there is a point $x_0 \in X$ such that

$$\mathbf{I} \subset \ker(\tau_{x_0}).$$

Suppose there is no such point, then for all $x \in X$, there exists $f_x \in \mathbf{I}$ such that $f_x(x) \neq 0$. Then there is an open neighbourhood V_x of X such that $f_x(y) \neq 0$ for all $y \in V_x$. Now the family $\{V_x : x \in X\}$ forms an open cover of X and so must have a finite subcover, say $\{V_{x_1}, V_{x_2}, \dots, V_{x_n}\}$. Define

$$h := \sum_{i=1}^{n} f_{x_i} \overline{f_{x_i}}.$$

Then $h \in \mathbf{I}$ since $\mathbf{I}$ is an ideal. If $x \in X$ then there is $1 \leq i \leq n$ such that $x \in V_{x_i}$. Hence $f_{x_i}(x) \neq 0$ and so $h(x) > 0$. Thus h is strictly positive on X, which implies that $h \in GL(C(X))$ (see Example 7.2.6). However, a proper ideal cannot contain invertible elements. This contradiction proves that the claim is true. $\qquad\square$

8.3.8 THEOREM *Let* X *be a compact Hausdorff space. Then the map* $\eta_X : X \to \Omega(C(X))$ *given by*

$$x \mapsto \tau_x$$

is a homeomorphism.

Proof.

(i) If $x, y \in X$ are two distinct points, then by Uyrsohn's Lemma, there is a function $f \in C(X)$ such that $f(x) = 0$ and $f(y) = 1$ (see Theorem 7.2.3). Therefore $\tau_x(f) \neq \tau_y(f)$. This proves that η_X is injective.

(ii) Let $\tau \in \Omega(C(X))$ be a multiplicative linear functional. By Lemma 8.3.2, $\ker(\tau)$ is a maximal ideal in $C(X)$. By Lemma 8.3.7, there is a point $x_0 \in X$ such that

$$\ker(\tau) = \ker(\tau_{x_0}).$$

By Lemma 6.5.9, there must exist $\alpha \in \mathbb{C}$ such that $\tau = \alpha \tau_{x_0}$. Since $\tau(1_{C(X)}) = 1 = \tau_{x_0}(1_{C(X)})$, we see that $\alpha = 1$. Therefore $\tau = \tau_{x_0}$ and η_X is surjective.

(iii) Recall that $\Omega(C(X))$ carries the weak-$*$ topology. Therefore to prove that η_X is continuous, it suffices to prove that

$$\widehat{f} \circ \eta_X : X \to \mathbb{C}$$

is continuous for each $f \in C(X)$ (by Proposition 6.3.3). However, if $f \in C(X)$, then $\widehat{f} \circ \eta_X = f$, which is continuous. Therefore η_X is continuous.

Since X is compact and $\Omega(C(X))$ is Hausdorff, η_X is a homeomorphism. $\qquad \square$

8.3.9 REMARK If X is compact and Hausdorff, then consider the map Φ : $C(\Omega(C(X))) \to C(X)$ given by

$$g \mapsto g \circ \eta_X.$$

Since η_X is continuous, this is well-defined. Since η_X is a homeomorphism, we get a map $\Psi : C(X) \to C(\Omega(C(X)))$ given by

$$h \mapsto h \circ \eta_X^{-1}.$$

It is easy to see that Ψ and Φ are both Banach algebra homomorphisms and they satisfy

$$\Psi \circ \Phi = \mathrm{id}_{C(\Omega(C(X)))} \text{ and } \Phi \circ \Psi = \mathrm{id}_{C(X)}.$$

Therefore Φ is an isomorphism of Banach algebras. Now consider the map $\Phi \circ \Gamma_{C(X)} : C(X) \to C(X)$. If $f \in C(X)$, then

$$\Phi \circ \Gamma_{C(X)}(f) = \Phi(\widehat{f}) = \widehat{f} \circ \eta_X = f.$$

Hence $\Phi \circ \Gamma_{C(X)}$ is the identity map on $C(X)$. Since Φ is an isomorphism, we identify $C(\Omega(C(X)))$ with $C(X)$ and treat $\Gamma_{C(X)}$ as the identity map.

We now consider the Gelfand representation in the case of commutative C*-algebras. Recall (Definition 8.2.13) that an element a in a C*-algebra is said to be self-adjoint if $a = a^*$.

8.3.10 REMARK Let **A** be a C*-algebra and $a \in \mathbf{A}$, then we may set

$$b = \frac{a + a^*}{2} \text{ and } c = \frac{i(a^* - a)}{2}.$$

Then b and c are self-adjoint elements such that $a = b + ic$. Moreover, this expression is unique (Exercise 8.16).

8.3.11 LEMMA *Let* **A** *be a unital C*-algebra and* $\tau \in \Omega(\mathbf{A})$. *For any* $a \in \mathbf{A}$,

$$\tau(a^*) = \overline{\tau(a)}.$$

In particular, $\tau(a) \in \mathbb{R}$ *if* a *is self-adjoint.*

Proof.

(i) We begin with the special case when a is self-adjoint. By Lemma 8.3.2, $\|\tau\| = 1$. Hence if $t \in \mathbb{R}$, then

$$\begin{aligned}
|\tau(a + it)|^2 &\leq \|a + it\|^2 = \|(a + it)^*(a + it)\| \\
&= \|(a - it)(a + it)\| = \|a^2 + t^2\| \\
&\leq \|a^2\| + t^2.
\end{aligned}$$

Therefore if $\tau(a) = \alpha + i\beta$ with $\alpha, \beta \in \mathbb{R}$, then

$$|\alpha|^2 + (\beta + t)^2 \leq \|a^2\| + t^2.$$

Expanding this out, we see that $|\alpha|^2 + 2t\beta \leq \|a^2\|$ holds for all $t \in \mathbb{R}$. For this to hold, it must happen that $\beta = 0$ and so $\tau(a) \in \mathbb{R}$.

(ii) Now if $a \in \mathbf{A}$ is arbitrary, then write $a = b + ic$, where $b, c \in \mathbf{A}$ are self-adjoint as in Remark 8.3.10. Then $\tau(b), \tau(c) \in \mathbb{R}$ by part (i) and

$$a^* = b - ic.$$

Hence $\tau(a) = \tau(b) + i\tau(c)$ and $\tau(a^*) = \tau(b) - i\tau(c) = \overline{\tau(a)}$. $\square$

We now arrive at the Gelfand–Naimark characterization of unital commutative C*-algebras.

8.3.12 THEOREM (GELFAND AND NAIMARK, 1943) *Let **A** be a unital commutative C*-algebra and let $\Omega(\mathbf{A})$ denote its Gelfand spectrum. Then the Gelfand representation*

$$\Gamma_{\mathbf{A}} : \mathbf{A} \to C(\Omega(\mathbf{A}))$$

is an isometric isomorphism of C*-algebras.

Proof.

(i) We start by showing that $\Gamma_{\mathbf{A}}$ is isometric. If $a \in \mathbf{A}$, we wish to prove that

$$\|a\| = \|\widehat{a}\|_\infty = r(a).$$

To do that, we appeal to the Spectral Radius Formula. Since **A** is commutative and satisfies the C*-identity, we have

$$\|a^2\| = \|(a^2)^* a^2\|^{1/2} = \|(a^* a)(a^* a)\|^{1/2} = (\|a^* a\|^2)^{1/2} = (\|a\|^4)^{1/2} = \|a\|^2.$$

By induction, it follows that

$$\|a^{2^n}\| = \|a\|^{2^n}$$

for all $n \in \mathbb{N}$. Therefore

$$r(a) = \lim_{k \to \infty} \|a^k\|^{1/k} = \lim_{n \to \infty} \|a^{2^n}\|^{1/2^n} = \|a\|.$$

Hence $\Gamma_{\mathbf{A}}$ is isometric.

(ii) To prove that $\Gamma_{\mathbf{A}}$ is surjective, consider $\mathcal{A} := \text{Range}(\Gamma_{\mathbf{A}})$. We show that $\mathcal{A}$ satisfies the hypotheses of the Stone–Weierstrass Theorem (see Section A.1).

(a) Since **A** is a Banach space and $\Gamma_{\mathbf{A}}$ is isometric, it follows that $\mathcal{A}$ is complete. Therefore $\mathcal{A}$ is closed in $C(\Omega(\mathbf{A}))$.

(b) Since **A** is unital,

$$\Gamma_{\mathbf{A}}(1_{\mathbf{A}})(\tau) = \tau(1_{\mathbf{A}}) = 1$$

for all $\tau \in \Omega(\mathbf{A})$. Thus $\mathcal{A}$ contains the constant functions in $C(\Omega(\mathbf{A}))$.

(c) If $\tau, \rho \in \Omega(\mathbf{A})$ are two different element, then there exists $a \in \mathbf{A}$ such that $\tau(a) \neq \rho(a)$. Therefore

$$\widehat{a}(\tau) \neq \widehat{a}(\rho)$$

and $\mathcal{A}$ separates points of $\Omega(\mathbf{A})$.

(d) Suppose $\widehat{a} \in \mathcal{A}$, then

$$\widehat{a}^*(\tau) = \tau(a^*) = \overline{\tau(a)} = \overline{\widehat{a}}(\tau)$$

by Lemma 8.3.11. Therefore $\mathcal{A}$ is closed under complex conjugation, and satisfies all the hypotheses of the Stone–Weierstrass Theorem. Therefore $\mathcal{A} = C(\Omega(\mathbf{A}))$.

□

We end the section with a result which describes the *functoriality* of the Gelfand–Naimark isomorphism. In other words, not only does the theorem provide an isomorphism of algebras, it also provides a dictionary through which continuous functions between compact Hausdorff spaces may be interpreted as ∗-homomorphisms between unital commutative C*-algebras.

8.3.13 PROPOSITION *Let* $\mathbf{A}$ *and* $\mathbf{B}$ *be unital commutative C*-algebras and* $\Gamma_{\mathbf{A}} : \mathbf{A} \to C(\Omega(\mathbf{A}))$ *and* $\Gamma_{\mathbf{B}} : \mathbf{B} \to C(\Omega(\mathbf{B}))$ *be the corresponding Gelfand representations.*

(i) *Let* $\Phi : \mathbf{A} \to \mathbf{B}$ *be a ∗-homomorphism. Then the transpose* $\Phi' : \mathbf{B}^* \to \mathbf{A}^*$ *induces a continuous function*

$$\widetilde{\Phi} : \Omega(\mathbf{B}) \to \Omega(\mathbf{A}).$$

(ii) *If* $f : \Omega(\mathbf{B}) \to \Omega(\mathbf{A})$ *is a continuous function, then the map* $\Phi_f : \mathbf{A} \to \mathbf{B}$ *given by*

$$\Phi_f(a) = \Gamma_{\mathbf{B}}^{-1}(\Gamma_{\mathbf{A}}(a) \circ f)$$

is a ∗-homomorphism.

(iii) *If* $\Phi : \mathbf{A} \to \mathbf{B}$ *is a ∗-homomorphism and* $f = \widetilde{\Phi}$ *as in part (i), then* $\Phi_f = \Phi$.

(iv) *Conversely, if* $f : \Omega(\mathbf{B}) \to \Omega(\mathbf{A})$ *is a continuous function and* $\Phi_f : \mathbf{A} \to \mathbf{B}$ *is as in part (ii), then* $\widetilde{\Phi_f} = f$.

Therefore there is a bijection between the set of all continuous functions $\Omega(\mathbf{B}) \to \Omega(\mathbf{A})$ *and the set of all ∗-homomorphism* $\mathbf{A} \to \mathbf{B}$.

Proof.

(i) If $\tau \in \Omega(\mathbf{B})$, then $\Phi'(\tau) = \tau \circ \Phi \in \Omega(\mathbf{A})$ since both τ and Φ are multiplicative. Therefore $\widetilde{\Phi}$ is a well-defined map. To verify that Φ' is continuous, we once again appeal to Proposition 6.3.3. If $a \in \mathbf{A}$, then $\hat{a} \circ \Phi' : \mathbf{B}^* \to \mathbb{C}$ is given by

$$\hat{a} \circ \Phi'(\psi) = \Phi'(\psi)(a) = \psi(\Phi(a)) = \widehat{\Phi(a)}(\psi).$$

The map $\hat{a} \circ \Phi' : \mathbf{B}^* \to \mathbb{C}$ is thus continuous with respect to the weak-$*$ topology. This is true for every $a \in \mathbf{A}$ and so Φ' is continuous. It then follows that $\tilde{\Phi}$ is also continuous.

(ii) This is a routine verification.

(iii) If $\Phi : \mathbf{A} \to \mathbf{B}$ is a $*$-homomorphism and $f = \tilde{\Phi}$, then for any $\tau \in \Omega(\mathbf{B})$,

$$\begin{aligned}
\Gamma_{\mathbf{B}}(\Phi_f(a))(\tau) &= \Gamma_{\mathbf{A}}(a) \circ \tilde{\Phi}(\tau) \\
&= \Gamma_{\mathbf{A}}(a)(\tau \circ \Phi) \\
&= \tau(\Phi(a)) \\
&= \Gamma_{\mathbf{B}}(\Phi(a))(\tau).
\end{aligned}$$

Since $\Gamma_{\mathbf{B}}$ is injective, it follows that $\Phi_f = \Phi$.

(iv) Let $f : \Omega(\mathbf{B}) \to \Omega(\mathbf{A})$ be a continuous function. Then for any $\tau \in \Omega(\mathbf{B})$ and $a \in \mathbf{A}$, we have

$$\begin{aligned}
\widetilde{\Phi_f}(\tau)(a) &= \tau(\Phi_f(a)) \\
&= \tau(\Gamma_{\mathbf{B}}^{-1}(\Gamma_{\mathbf{A}}(a) \circ f)) \\
&= (\Gamma_{\mathbf{A}}(a) \circ f)(\tau) \\
&= f(\tau)(a).
\end{aligned}$$

Hence $\widetilde{\Phi_f} = f$. □

Let us now look at the most important special case of this result.

8.3.14 REMARK Let X and Y be compact Hausdorff spaces. Given a continuous function $g : Y \to X$, the map $\Phi : C(X) \to C(Y)$ given by

$$\Phi(f) = f \circ g$$

is clearly a $*$-homomorphism. Conversely, $\mathbf{A} := C(X)$ and $\mathbf{B} := C(Y)$ are unital, commutative C*-algebras. Therefore if $\Phi : C(X) \to C(Y)$ is a $*$-homomorphism, then there is a continuous function $h : \Omega(C(Y)) \to \Omega(C(X))$ such that

$$\Phi(a) = \Gamma_{\mathbf{B}}^{-1}(\Gamma_{\mathbf{A}}(a) \circ h)$$

for all $a \in \mathbf{A}$. By Theorem 8.3.8, there are natural homeomorphisms $\eta_X : X \to \Omega(C(X))$ and $\eta_Y : Y \to \Omega(C(Y))$. If $g := \eta_X^{-1} \circ h \circ \eta_Y$, then $g : Y \to X$ is continuous. Furthermore, if $f \in C(X)$ and $y \in Y$,

$$\begin{aligned}
\Phi(f)(y) &= \Gamma_{\mathbf{B}}^{-1}(\Gamma_{\mathbf{A}}(f) \circ h)(y) \\
&= (\Gamma_{\mathbf{A}}(f) \circ h)(\eta_Y(y)) \\
&= f(\eta_X^{-1}(h(\eta_Y(y)))) \\
&= f \circ g(y).
\end{aligned}$$

Thus to every $*$-homomorphism $\Phi : C(X) \to C(Y)$ there is a continuous function $g : Y \to X$ such that $\Phi(f) = f \circ g$ for all $f \in C(X)$.

The Gelfand–Naimark Theorem is a cornerstone in the theory of C*-algebras. We will use this theorem in the context of normal operators, but its influence on the subject goes far deeper than that. Oftentimes, a conceptual breakthrough for non-commutative C*-algebras is preceded by a deep understanding of the corresponding notion for topological spaces. This has led to some of the most important theorems in the subject and has given rise to new areas of mathematics (such as K-theory and non-commutative geometry) in the process.

> **Aside.** There is another theorem that goes by the name of the *Gelfand–Naimark Theorem*. It is a representation theorem for (possibly non-commutative) C*-algebras and is as important to the subject as the one described here. Taken together, these two theorems are the bedrock on which C*-algebra theory has been developed. We will discuss it in some detail in the final section of the book, Section 10.8.

8.4   Spectral Permanence Theorem

Let $\mathbf{A}$ be a unital Banach algebra and let $b \in \mathbf{A}$. We had defined the spectrum of b to be the set

$$\sigma_{\mathbf{A}}(b) = \{\lambda \in \mathbb{C} : (b - \lambda 1_{\mathbf{A}}) \notin GL(\mathbf{A})\}.$$

What we did not emphasize earlier is how this set depends on the ambient algebra. In particular, suppose $\mathbf{B} \subset \mathbf{A}$ is a closed subalgebra of $\mathbf{A}$ such that $1_{\mathbf{A}} \in \mathbf{B}$ and

$b \in \mathbf{B}$, then we wish to understand the set

$$\sigma_{\mathbf{B}}(b) = \{\lambda \in \mathbb{C} : (b - \lambda 1_{\mathbf{A}}) \notin GL(\mathbf{B})\}$$

and its relationship with $\sigma_{\mathbf{A}}(b)$. Naturally, this depends on the relationship between $GL(\mathbf{B})$ and $GL(\mathbf{A})$. Clearly, $GL(\mathbf{B}) \subset GL(\mathbf{A})$. However, if an element in **B** is invertible in **A** then its inverse may not be in **B**, as the next example shows.

8.4.1 EXAMPLE Let $\mathbf{A} = C(S^1)$ and $\mathbf{B} \subset \mathbf{A}$ be the subalgebra generated by $\{1_{\mathbf{A}}, \zeta\}$, where $\zeta(z) = z$. In other words,

$$\mathbf{B} = \overline{\{p(z) : p(x) \in \mathbb{C}[x]\}}.$$

Then ζ is invertible in **A** because $\zeta^{-1}(z) = \frac{1}{z}$ is a continuous function on S^1. However, we claim that ζ is not invertible in **B**. If it were, then ζ^{-1} would be the limit of a sequence (p_n) of polynomials in z. Furthermore, the polynomials would be uniformly bounded, being a Cauchy sequence in **A**. By the Dominated Convergence Theorem, it would follow that

$$\int_{S^1} \frac{dz}{z} = \lim_{n \to \infty} \int_{S^1} p_n(z) dz = 0.$$

This is patently false and thus $\zeta \notin GL(\mathbf{B})$.

8.4.2 REMARK Let **B** be a closed subalgebra of a unital Banach algebra **A** such that $1_{\mathbf{A}} \in B$. For $b \in \mathbf{B}$, we now have four sets to consider simultaneously; $\sigma_{\mathbf{B}}(b)$, $\sigma_{\mathbf{A}}(b)$ and their respective complements $\rho_{\mathbf{B}}(b)$ and $\rho_{\mathbf{A}}(b)$. Since $GL(\mathbf{B}) \subset GL(\mathbf{A})$,

$$\sigma_{\mathbf{A}}(b) \subset \sigma_{\mathbf{B}}(b).$$

We now wish to understand the relationship between these two sets, first for Banach algebras and then for C*-algebras. Once again, we draw our inspiration from Example 8.4.1.

8.4.3 EXAMPLE Let $\mathbf{A} = C(S^1)$ and $\mathbf{B} \subset \mathbf{A}$ be the subalgebra generated by $\{1_{\mathbf{A}}, \zeta\}$.

(i) For any $f \in C(S^1)$, the spectrum of f is given by the range of f by Example 7.3.2. Hence

$$\sigma_{\mathbf{A}}(\zeta) = \zeta(S^1) = S^1.$$

(ii) We wish to determine $\sigma_{\mathbf{B}}(\zeta)$. By Proposition 8.3.3,

$$\sigma_{\mathbf{B}}(\zeta) = \{\tau(\zeta) : \tau \in \Omega(\mathbf{B})\},$$

so we now determine $\Omega(\mathbf{B})$.

(a) For each $\lambda \in \mathbb{D} := \{z \in \mathbb{C} : |z| \leq 1\}$, define τ_λ for a polynomial $p(z) \in \mathbf{B}$ by

$$\tau_\lambda(p(z)) = p(\lambda).$$

By the maximum modulus principle,

$$|p(\lambda)| \leq \sup_{|z|=1} |p(z)| = \|p\|_\infty.$$

Hence τ_λ extends to a bounded linear functional on $\mathbf{B}$ and is clearly multiplicative (because it is multiplicative on a dense subalgebra).

(b) We claim that *every* element of $\Omega(\mathbf{B})$ is of this form. Given $\tau \in \Omega(\mathbf{B})$, let $\lambda = \tau(\zeta)$. Then $|\lambda| \leq \|\zeta\| = 1$. Also, for any $p(x) \in \mathbb{C}[x]$,

$$\tau(p(z)) = p(\tau(\zeta)) = p(\lambda) = \tau_\lambda(p(z)).$$

Since $\tau = \tau_\lambda$ on a dense set, it follows that $\tau = \tau_\lambda$ on $\mathbf{B}$.

Hence the map $\lambda \mapsto \tau_\lambda$ defines a bijective function $\eta_{\mathbf{B}} : \mathbb{D} \to \Omega(\mathbf{B})$. Therefore

$$\sigma_{\mathbf{B}}(\zeta) = \{\tau_\lambda(\zeta) : \lambda \in \mathbb{D}\} = \mathbb{D}.$$

This proves that the spectra $\sigma_{\mathbf{B}}(b)$ and $\sigma_{\mathbf{A}}(b)$ can be different. However, as we will show below, they share an interesting relationship that would not seem obvious at first. As we did before, we will write ∂X for the boundary of set $X \subset \mathbb{C}$.

8.4.4 THEOREM *Let* $\mathbf{B}$ *be a closed subalgebra of a unital Banach algebra* $\mathbf{A}$ *containing the unit of* $\mathbf{A}$. *If* $b \in \mathbf{B}$, *then* $\partial \sigma_{\mathbf{B}}(b) \subset \sigma_{\mathbf{A}}(b)$

Proof. Seeking a contradiction, suppose there is a point $\lambda \in \partial \sigma_{\mathbf{B}}(b) \setminus \sigma_{\mathbf{A}}(b)$. Then $(b - \lambda 1_{\mathbf{A}}) \in GL(\mathbf{A})$ and there is a sequence $(\lambda_n) \subset \rho_{\mathbf{B}}(b)$ such that $\lambda_n \to \lambda$. Hence $(b - \lambda_n 1_{\mathbf{A}}) \in GL(\mathbf{B}) \subset GL(\mathbf{A})$. By the continuity of the inverse map in $GL(\mathbf{A})$, we have

$$(b - \lambda_n 1_{\mathbf{A}})^{-1} \to (b - \lambda 1_{\mathbf{A}})^{-1}$$

in $GL(\mathbf{A})$. But, $(b - \lambda_n 1_{\mathbf{A}})^{-1} \in \mathbf{B}$ for all $n \in \mathbb{N}$ and so $(b - \lambda 1_{\mathbf{A}})^{-1} \in \mathbf{B}$, whence $\lambda \notin \sigma_{\mathbf{B}}(b)$. This contradiction proves that no such λ can exist. $\square$

To understand the relationship between $\sigma_{\mathbf{A}}(b)$ and $\sigma_{\mathbf{B}}(b)$, we must introduce the notion of a *hole* in a compact set.

8.4.5 DEFINITION Let $K \subset \mathbb{C}$ be a compact set, then $\mathbb{C} \setminus K$ has exactly one unbounded connected component, which we denote by X_∞. We list the remaining (finitely many) bounded components as $X_1, X_2, \ldots, X_n$, so that

$$\mathbb{C} \setminus K = X_\infty \sqcup X_1 \sqcup X_2 \sqcup \ldots \sqcup X_n.$$

Each set X_i (for $i \neq \infty$) is called a **hole** in K.

8.4.6 LEMMA *Let X be a connected topological space and $K \subset X$ be a closed set such that $\partial K = \emptyset$. Then either $K = X$ or $K = \emptyset$.*

Proof. If $\partial K = \emptyset$, then $X = \text{int}(K) \sqcup X \setminus K$, where $\text{int}(K)$ denotes the interior of K. Since both these sets are open and X is connected, either $\text{int}(K) = \emptyset$ or $X \setminus K = \emptyset$. If $K \neq X$, then $\text{int}(K) = \emptyset$ and $K = \text{int}(K) \sqcup \partial K = \emptyset$. Hence the result. □

The next result is a partial version of the Spectral Permanence Theorem that applies for Banach algebras.

8.4.7 THEOREM *Let $\mathbf{B}$ be a closed subalgebra of a unital Banach algebra $\mathbf{A}$ containing the unit of $\mathbf{A}$. If $b \in \mathbf{B}$, then $\sigma_{\mathbf{B}}(b)$ is obtained from $\sigma_{\mathbf{A}}(b)$ by adjoining to it some (and perhaps none) of its holes.*

Proof. We know that $\sigma_{\mathbf{A}}(b) \subset \sigma_{\mathbf{B}}(b)$, so we wish to understand the set $\sigma_{\mathbf{B}}(b) \cap \rho_{\mathbf{A}}(b)$. As mentioned above, we may write $\rho_{\mathbf{A}}(b)$ as

$$\rho_{\mathbf{A}}(b) = X_\infty \sqcup X_1 \sqcup X_2 \sqcup \ldots \sqcup X_n,$$

where X_∞ denotes the unbounded component of $\rho_{\mathbf{A}}(b)$.

(i) Since $\partial \sigma_{\mathbf{B}}(b) \subset \sigma_{\mathbf{A}}(b)$, it follows that X_∞ must intersect $\sigma_{\mathbf{B}}(b)$ trivially.

(ii) Now let X be a hole in $\sigma_{\mathbf{A}}(b)$. Let $K = X \cap \sigma_{\mathbf{B}}(b)$, thought of as a closed subspace of X. The boundary $\partial_X(K)$ of K relative to X is $\partial_X(K) = \overline{K} \cap \overline{X \setminus K}$. Now $K \subset \sigma_{\mathbf{B}}(b)$ and $X \setminus K \subset \rho_{\mathbf{B}}(b)$. By Theorem 8.4.4, it follows that

$$\partial_X(K) \subset \sigma_{\mathbf{B}}(b) \cap \overline{\rho_{\mathbf{B}}(b)} = \partial \sigma_{\mathbf{B}}(b) \subset \sigma_{\mathbf{A}}(b) \subset \mathbb{C} \setminus X.$$

However, $\partial_X(K) \subset X$ and thus $\partial_X(K) = \emptyset$. Lemma 8.4.6 now implies that either $K = \emptyset$ or $K = X$. In other words, either $X \cap \sigma_{\mathbf{B}}(b) = \emptyset$ or $X \subset \sigma_{\mathbf{B}}(b)$.

This applies to each hole in $\sigma_{\mathbf{A}}(b)$ and hence the result. □

For instance, if $\sigma_{\mathbf{A}}(b) = S^1$, then $\sigma_{\mathbf{B}}(b)$ must be either S^1 or $\mathbb{D}$. That both cases are possible is evident from Example 8.4.3. The next corollary is a crucial step in our transition from Banach algebras to C*-algebras.

8.4.8 COROLLARY *Let* $\mathbf{B}$ *be a closed subalgebra of a unital Banach algebra* $\mathbf{A}$ *containing the unit of* $\mathbf{A}$. *If* $b \in \mathbf{B}$ *is such that* $\sigma_{\mathbf{B}}(b) \subset \mathbb{R}$, *then* $\sigma_{\mathbf{B}}(b) = \sigma_{\mathbf{A}}(b)$.

Proof. As before, we know that $\sigma_{\mathbf{A}}(b) \subset \sigma_{\mathbf{B}}(b)$. Now if $\sigma_{\mathbf{B}}(b) \subset \mathbb{R}$, then $\partial\sigma_{\mathbf{B}}(b) = \sigma_{\mathbf{B}}(b)$. Therefore it follows from Theorem 8.4.4 that $\sigma_{\mathbf{B}}(b) \subset \sigma_{\mathbf{A}}(b)$ as well. □

Now let us pass from Banach algebras to C*-algebras. Let $\mathbf{A}$ be a unital C*-algebra and $\mathbf{B} \subset \mathbf{A}$ be a closed Banach subalgebra of $\mathbf{A}$ that contains $1_{\mathbf{A}}$. If $b \in \mathbf{B}$, then we know that $\sigma_{\mathbf{B}}(b)$ may be obtained from $\sigma_{\mathbf{A}}(b)$ by adjoining some holes. However, if we also require that $\mathbf{B}$ is closed under the involution operation, then the two sets must coincide! This theorem is a stark reminder of how different C*-algebras can be from more general Banach algebras.

We should hasten to point out that the result applies to closed subalgebras of a C*-algebra that are closed under the involution operation. If $C(S^1)$ is equipped with the usual involution, then the algebra $\mathbf{B}$ of Example 8.4.3 does not satisfy this hypothesis.

8.4.9 THEOREM (SPECTRAL PERMANENCE THEOREM) *Let* $\mathbf{A}$ *be a unital C*-algebra and let* $\mathbf{B}$ *be a C*-subalgebra of* $\mathbf{A}$ *that contains the unit of* $\mathbf{A}$. *For any* $b \in \mathbf{B}$,

$$\sigma_{\mathbf{B}}(b) = \sigma_{\mathbf{A}}(b).$$

Proof. The proof relies on Corollary 8.4.8 and the fact that the spectrum of a self-adjoint element is contained in $\mathbb{R}$. Let us see how this works.

(i) Let us first consider the case when $b \in \mathbf{B}$ is self-adjoint. Set $\mathbf{C}$ to be the C*-algebra generated by $\{1_{\mathbf{A}}, b\}$. Then $\mathbf{C}$ is commutative as in Example 8.2.6. Therefore

$$\sigma_{\mathbf{C}}(b) = \{\tau(b) : \tau \in \Omega(\mathbf{C})\}$$

by Proposition 8.3.3. However, if $\tau \in \Omega(\mathbf{C})$, then $\tau(b) \in \mathbb{R}$ since b is self-adjoint. Hence $\sigma_{\mathbf{C}}(b) \subset \mathbb{R}$ and Corollary 8.4.8 implies that

$$\sigma_{\mathbf{A}}(b) = \sigma_{\mathbf{C}}(b).$$

For the same reason, $\sigma_{\mathbf{B}}(b) = \sigma_{\mathbf{C}}(b)$, proving that $\sigma_{\mathbf{B}}(b) = \sigma_{\mathbf{A}}(b)$.

(ii) Now suppose $b \in \mathbf{B}$ is an arbitrary element. Since $\sigma_\mathbf{A}(b) \subset \sigma_\mathbf{B}(b)$, it suffices to show that $\sigma_\mathbf{B}(b) \subset \sigma_\mathbf{A}(b)$. To that end, let $\lambda \in \sigma_\mathbf{B}(b)$ and let $c := b - \lambda 1_\mathbf{A}$. We know that $c \notin GL(\mathbf{B})$ and we wish to prove that $c \notin GL(\mathbf{A})$. If c were invertible in $\mathbf{A}$, then there would exist $d \in \mathbf{A}$ such that $cd = 1_\mathbf{A} = dc$. Applying the involution, we see that $c^* d^* = d^* c^* = 1_\mathbf{A}$. Hence

$$(d^* d)(cc^*) = (cc^*)(d^* d) = 1_\mathbf{A}.$$

Therefore (cc^*) is invertible in $\mathbf{A}$. Since cc^* is self-adjoint, it follows from part (i) that cc^* is invertible in $\mathbf{B}$. Hence there exists $c' \in \mathbf{B}$ such that

$$cc^* c' = 1_\mathbf{A}$$

and c is has a right inverse in $\mathbf{B}$. Similarly, $c^* c$ is invertible in $\mathbf{B}$, proving that c has a left inverse in $\mathbf{B}$. Thus c is invertible in $\mathbf{B}$ and $\lambda \notin \sigma_\mathbf{B}(b)$. This is a contradiction, which proves that $\sigma_\mathbf{B}(b) \subset \sigma_\mathbf{A}(b)$. $\qquad\square$

We record an important observation from part (i) of the previous proof. As usual, it works for non-unital C*-algebras just as well as unital ones by passing to the unitalization if needed.

8.4.10 PROPOSITION *Let* $\mathbf{A}$ *be a* C*-*algebra and* $a \in \mathbf{A}$ *be a self-adjoint element. Then* $\sigma_\mathbf{A}(a) \subset \mathbb{R}$.

8.5 Continuous Functional Calculus

In Example 8.2.6, we had defined the C*-algebra generated by a normal operator. We now make the obvious definition for normal elements.

8.5.1 EXAMPLE Let $\mathbf{A}$ be a unital C*-algebra and $a \in \mathbf{A}$ be a normal element. Define $\mathcal{A}(a)$ to be the C*-subalgebra of $\mathbf{A}$ generated by the set $\{1_\mathbf{A}, a\}$. If $p(x, y) \in \mathbb{C}[x, y]$ is a polynomial in two commuting variables given by $p(x, y) = \sum_{i,j} \alpha_{i,j} x^i y^j$, then we may define

$$p(a, a^*) := \sum_{i,j} \alpha_{i,j} a^i (a^*)^j.$$

Once again, this is well-defined since a commutes with a^*. Furthermore,

$$\mathcal{A}(a) = \overline{\{p(a, a^*) : p(x, y) \in \mathbb{C}[x, y]\}}.$$

Finally, since $\mathbb{C}[x, y]$ is commutative, it follows that $\mathcal{A}(a)$ is a commutative C*-subalgebra of $\mathbf{A}$ that is unital in $\mathbf{A}$.

The next theorem is the most important application of the Gelfand–Naimark Theorem. At heart, it merely determines the Gelfand spectrum of the C*-algebra $\mathcal{A}(a)$.

8.5.2 THEOREM *Let* **A** *be a unital C*-algebra and* $a \in$ **A** *be a normal element. Then there is an isometric *-isomorphism*

$$\Theta : C(\sigma_{\mathbf{A}}(a)) \to \mathcal{A}(a)$$

such that $\Theta(p(z, \bar{z})) = p(a, a^*)$ *for any polynomial* $p(x, y) \in \mathbb{C}[x, y]$.

Proof.   By the Spectral Permanence Theorem and Proposition 8.3.3,

$$\sigma_{\mathbf{A}}(a) = \sigma_{\mathcal{A}(a)}(a) = \{\tau(a) : \tau \in \Omega(\mathcal{A}(a))\}.$$

Therefore if $\tau \in \Omega(\mathcal{A}(a))$, then $\tau(a) \in \sigma_{\mathbf{A}}(a)$. Hence we have a well-defined map $\widehat{a} : \Omega(\mathcal{A}(a)) \to \sigma_{\mathbf{A}}(a)$ given by

$$\widehat{a}(\tau) := \tau(a).$$

Furthermore, $\widehat{a}$ is continuous because $\Omega(\mathcal{A}(a))$ is equipped with the weak-* topology. We claim that $\widehat{a}$ is a homeomorphism.

(i) If $\tau, \rho \in \Omega(\mathcal{A}(a))$ are two multiplicative linear functionals such that $\widehat{a}(\tau) = \widehat{a}(\rho)$, then $\tau(a) = \rho(a)$. Also, $\tau(1_{\mathbf{A}}) = \rho(1_{\mathbf{A}})$, and by Lemma 8.3.11 we know that $\tau(a^*) = \rho(a^*)$. Therefore if $p(x, y) \in \mathbb{C}[x, y]$ is any polynomial, then $\tau(p(a, a^*)) = \rho(p(a, a^*))$. Since both τ and ρ are continuous, we conclude that $\tau = \rho$ in $\Omega(\mathcal{A}(a))$. Hence $\widehat{a}$ is injective.

(ii) If $\lambda \in \sigma_{\mathbf{A}}(a) = \sigma_{\mathcal{A}(a)}(a)$, then by Proposition 8.3.3, there is a $\tau \in \Omega(\mathcal{A}(a))$ such that $\tau(a) = \lambda$. Therefore $\widehat{a}$ is also surjective.

Since $\Omega(\mathcal{A}(a))$ is compact and $\sigma_{\mathbf{A}}(a)$ is Hausdorff, $\widehat{a}$ is a homeomorphism. Therefore we get a *-homomorphism $\Psi : C(\sigma_{\mathbf{A}}(a)) \to C(\Omega(\mathcal{A}(a)))$ given by

$$\Psi(f) := f \circ \widehat{a}.$$

Since $\widehat{a}$ is a homeomorphism, Ψ is an isometric *-isomorphism (see Exercise 2.50). Consider the Gelfand–Naimark isomorphism $\Gamma : \mathcal{A}(a) \to C(\Omega(\mathcal{A}(a)))$ and define

$\Theta : C(\sigma_{\mathbf{A}}(a)) \to \mathcal{A}(a)$ by

$$\Theta := \Gamma^{-1} \circ \Psi.$$

Then Θ is clearly an isometric $*$-isomorphism. Furthermore, consider $\zeta \in C(\sigma_{\mathbf{A}}(a))$ to be the identity function $\zeta(z) = z$. Then

$$\Theta(\zeta) = \Gamma^{-1}(\zeta \circ \hat{a}) = \Gamma^{-1}(\hat{a}) = a.$$

Since Θ respects the adjoint operation and $\Theta(\mathbf{1}) = 1_{\mathbf{A}}$, it follows that $\Theta(p(z, \bar{z})) = p(a, a^*)$ for any polynomial $p(x, y) \in \mathbb{C}[x, y]$. This proves the theorem. □

If $\mathbf{A}$ is a C*-algebra and $a \in \mathbf{A}$, we will henceforth write $\sigma(a)$ to denote $\sigma_{\mathbf{A}}(a)$. The Spectral Permanence Theorem allows us to use this symbol with impunity even when we are dealing with a unital C*-subalgebra of $\mathbf{A}$ that contains a.

If $\mathbf{A}$ is a unital C*-algebra and $a \in \mathbf{A}$ is a normal element, then we know what $p(a, a^*)$ means for any polynomial $p(x, y) \in \mathbb{C}[x, y]$. The above isomorphism allows us to extend this definition to include *continuous* functions that are not necessarily polynomials. In the process, it provides us with an important tool to understand the spectrum of a normal operator.

8.5.3 DEFINITION Let $\mathbf{A}$ be a unital C*-algebra, $a \in \mathbf{A}$ be a normal element and let $\Theta : C(\sigma(a)) \to \mathcal{A}(a)$ be the isomorphism from Theorem 8.5.2. For any continuous function $f \in C(\sigma(a))$, we define

$$f(a) := \Theta(f).$$

This map $f \mapsto f(a)$ is called the ***continuous functional calculus***.

8.5.4 REMARK If $\mathbf{A}$ is a non-unital C*-algebra and $a \in \mathbf{A}$ is a normal element, then the continuous functional calculus applies in the unitalization $\mathbf{A}^u$. In other words, if $f \in C(\sigma(a))$, then $f(a) \in \mathbf{A}^u$. However, $0 \in \sigma(a)$ by Exercise 8.11 and if $f(0) = 0$, then f may be approximated by polynomials $p(z, \bar{z})$ which do not have a constant term. For such a polynomial, $p(a, a^*)$ lies in the non-unital C*-algebra generated by a and thus $f(a) \in \mathbf{A}$ as well.

Given a polynomial $p(x) \in \mathbb{C}[x]$ and an element a of a Banach algebra, we had proved the Spectral Mapping Theorem to determine the spectrum of $p(a)$. The next result extends the Spectral Mapping Theorem to understand the spectrum of $f(a)$ when a is a normal element in a unital C*-algebra $\mathbf{A}$ and $f \in C(\sigma(a))$.

8.5.5 THEOREM (SPECTRAL MAPPING THEOREM) *Let* **A** *be a unital C*-algebra and* $a \in \mathbf{A}$ *be a normal element. Then for any* $f \in C(\sigma(a))$,

$$\sigma(f(a)) = f(\sigma(a)).$$

Proof.   Note that $f \mapsto f(a)$ is an isometric $*$-isomorphism from $\mathbf{C} := C(\sigma(a))$ to $\mathbf{B} := \mathcal{A}(a)$. Hence

$$\sigma_{\mathbf{B}}(f(a)) = \sigma_{\mathbf{C}}(f).$$

By the Spectral Permanence Theorem, $\sigma_{\mathbf{B}}(f(a)) = \sigma_{\mathbf{A}}(f(a))$ and by Example 7.3.2, $\sigma_{\mathbf{C}}(f) = f(\sigma(a))$. Hence the result.   □

Under the continuous functional calculus, the identity function $\zeta \in C(\sigma(a))$ is mapped to a. Now the norm of ζ in $C(\sigma(a))$ is merely the spectral radius $r(a)$. Since the continuous functional calculus is isometric, we get

8.5.6 COROLLARY *Let* a *be a normal element in a C*-algebra, then* $\|a\| = r(a)$.

Notice that we do not require the ambient C*-algebra to be unital for this theorem. If it is not, then we simply replace it with its unitalization and apply the same argument.

8.5.7 DEFINITION Let **A** be a C*-algebra. An element $a \in \mathbf{A}$ is said to be

(i) a **projection** if $a = a^* = a^2$.

(ii) a **unitary** if $aa^* = a^*a = 1_{\mathbf{A}}$ (this definition only makes sense if **A** is unital).

8.5.8 REMARK Let **H** be a Hilbert space and $\mathbf{A} = \mathcal{B}(\mathbf{H})$.

(i) If **M** is a closed subspace of **H** and $P \in \mathcal{B}(\mathbf{H})$ is the orthogonal projection onto **M**, then $P = P^2$ by Proposition 3.1.9. Furthermore, if $x, y \in \mathbf{H}$, then we may write $x = x_1 + x_2$ and $y = y_1 + y_2$, where $x_1, y_1 \in \mathbf{M}$ and $x_2, y_2 \in \mathbf{M}^{\perp}$. Therefore

$$\langle P(x), y \rangle = \langle x_1, y \rangle = \langle x_1, y_1 \rangle = \langle x_1, P(y) \rangle = \langle x, P(y) \rangle.$$

Therefore $P = P^*$ holds. Conversely, if $P \in \mathcal{B}(\mathbf{H})$ is an operator such that $P = P^* = P^2$, then $\mathbf{M} = \text{Range}(P)$ is a subspace of **H**. In Exercise 8.30, we ask you to show that this is indeed a closed subspace of **H** whose orthogonal projection is P. Therefore there is a bijection between the set of all closed subspaces of **H** and the set of all projections in $\mathcal{B}(\mathbf{H})$.

(ii) $T \in \mathcal{B}(\mathbf{H})$ is a unitary if it is a Hilbert space isomorphism that preserves the inner product. However, we had shown in Corollary 8.1.13 that this is equivalent to the condition $TT^* = T^*T = I$. Hence our definition agrees with Definition 3.4.5.

The next lemma is an easy one given what we know about multiplicative linear functionals. Notice that part (i) of this lemma was proved earlier in Proposition 8.4.10.

8.5.9 LEMMA *Let $\mathbf{A}$ be a $C*$-algebra and $a \in \mathbf{A}$.*

(i) *If $a \in \mathbf{A}$ is self-adjoint, then $\sigma(a) \subset \mathbb{R}$.*

(ii) *If $a \in \mathbf{A}$ is unitary, then $\sigma(a) \subset S^1$.*

(iii) *If $a \in \mathbf{A}$ is a projection, then $\sigma(a) \subset \{0,1\}$.*

Proof. Replacing $\mathbf{A}$ by its unitalization if need be, we may assume that $\mathbf{A}$ is unital. In each of the three cases, a is a normal element. Therefore $\mathcal{A}(a)$ is a commutative $C*$-algebra. By the Spectral Permanence Theorem,

$$\sigma_{\mathbf{A}}(a) = \sigma_{\mathcal{A}(a)}(a).$$

Therefore we may replace $\mathbf{A}$ by $\mathcal{A}(a)$ and assume without loss of generality that $\mathbf{A}$ is commutative. In that case,

$$\sigma(a) = \{\tau(a) : \tau \in \Omega(\mathbf{A})\}.$$

Therefore we fix $\tau \in \Omega(\mathbf{A})$ and ascertain what $\tau(a)$ looks like.

(i) If $a = a^*$, then $\tau(a) \in \mathbb{R}$ by Lemma 8.3.11.

(ii) Since $\tau(1_{\mathbf{A}}) = 1$, we have

$$1 = \tau(1_{\mathbf{A}}) = \tau(a^*a) = \tau(a^*)\tau(a) = \overline{\tau(a)}\tau(a) = |\tau(a)|^2.$$

Therefore $|\tau(a)| = 1$.

(iii) If a is a projection, then $\tau(a) = \overline{\tau(a)} = \tau(a)^2$. Hence $\tau(a) \in \{0,1\}$.

Therefore we have determined the form of $\tau(a)$ for each $\tau \in \Omega(\mathbf{A})$ and this completes the proof. $\square$

The converse of Lemma 8.5.9 gives us our first glimpse into the power of the continuous functional calculus.

8.5.10 THEOREM *Let* $\mathbf{A}$ *be a* C^*-*algebra and* $a \in \mathbf{A}$ *be a normal element.*

  (i) *If* $\sigma(a) \subset \mathbb{R}$, *then* $a = a^*$.

  (ii) *If* $\sigma(a) \subset S^1$, *then* a *is unitary.*

  (iii) *If* $\sigma(a) \subset \{0, 1\}$, *then* a *is a projection.*

Proof.   Once again, replacing $\mathbf{A}$ by its unitalization, we may assume that $\mathbf{A}$ is unital. Let $f \mapsto f(a)$ denote the functional calculus from $C(\sigma(a)) \to \mathcal{A}(a) \subset \mathbf{A}$. In particular, if $\zeta(z) = z$, then

$$a = \zeta(a) \text{ and } a^* = \overline{\zeta}(a).$$

Furthermore, the constant function $\mathbf{1} \in C(\sigma(a))$ is mapped to $1_{\mathbf{A}}$ and the map is isometric.

  (i) If $\sigma(a) \subset \mathbb{R}$, then $\zeta = \overline{\zeta}$ in $C(\sigma(a))$. Therefore $a = a^*$.

  (ii) If $\sigma(a) \subset S^1$, then

$$\zeta(z)\overline{\zeta}(z) = \overline{\zeta}(z)\zeta(z) = 1$$

   for all $z \in \sigma(a)$. Hence $aa^* = a^*a = 1_{\mathbf{A}}$.

  (iii) If $\sigma(a) \in \{0, 1\}$, then $\zeta = \zeta^2 = \overline{\zeta}$ in $C(\sigma(a))$. Once again, these properties transfer to a through the continuous functional calculus and we have $a = a^2 = a^*$. $\square$

We end the section with another rigidity results concerning C^*-algebras that we now have access to. In Proposition 8.2.15, we proved that any $*$-homomorphism is norm decreasing. Now we show that injective $*$-homomorphisms are norm preserving. The next lemma will help us get there and is a useful result in and of itself.

8.5.11 LEMMA *Let* $\Phi : \mathbf{A} \to \mathbf{B}$ *be a* $*$-*homomorphism between two unital* C^*-*algebras such that* $\Phi(1_{\mathbf{A}}) = 1_{\mathbf{B}}$. *If* $a \in \mathbf{A}$ *is normal and* $f \in C(\sigma_{\mathbf{A}}(a))$ *is a continuous function, then* $f \in C(\sigma_{\mathbf{B}}(\Phi(a)))$ *and*

$$f(\Phi(a)) = \Phi(f(a)).$$

Proof.   By Proposition 8.2.16, $\sigma_{\mathbf{B}}(\Phi(a)) \subset \sigma_{\mathbf{A}}(a)$. Therefore every $f \in C(\sigma_{\mathbf{A}}(a))$ also belongs to $C(\sigma_{\mathbf{B}}(\Phi(a)))$. Let $\mathcal{A}$ be the collection of functions in $C(\sigma_{\mathbf{A}}(a))$ that satisfy the conclusion.

(i) Since $\Phi(1_{\mathbf{A}}) = 1_{\mathbf{B}}$, the constant function **1** belongs to $\mathcal{A}$.

(ii) If $\zeta \in C(\sigma(a))$ is the identity function $\zeta(z) = z$, then ζ is clearly a member of $\mathcal{A}$. Hence $\mathcal{A}$ separates points of $\sigma_{\mathbf{A}}(a)$.

(iii) Since the continuous functional calculus is an algebra homomorphism, it is easy to see that $\mathcal{A}$ is a subalgebra of $C(\sigma(a))$. Furthermore, if $f \in \mathcal{A}$, then

$$\overline{f}(\Phi(a)) = f(\Phi(a))^* = \Phi(f(a))^* = \Phi(f(a)^*) = \Phi(\overline{f}(a)).$$

Hence $\overline{f} \in \mathcal{A}$.

(iv) Since the continuous functional calculus is isometric and Φ is continuous, it follows that $\mathcal{A}$ is closed in $C(\sigma_{\mathbf{A}}(a))$.

By the Stone–Weierstrass Theorem, $\mathcal{A} = C(\sigma_{\mathbf{A}}(a))$, which is what we needed to prove. $\qquad\qquad\square$

8.5.12 PROPOSITION *If* $\Phi : \mathbf{A} \to \mathbf{B}$ *is an injective $*$-homomorphism between two C^*-algebras, then Φ is isometric.*

Proof. As in the proof of Proposition 8.2.15, we break the proof into two cases.

(i) Suppose first that **A** is unital. Let $\mathbf{C} := \overline{\mathrm{Range}(\Phi)}$, then **C** is a C*-subalgebra of **B**. As in Proposition 8.2.15,

$$\Phi(1_{\mathbf{A}}) = 1_{\mathbf{C}}.$$

By the C*-identity, it suffices to show that $\|\Phi(a)\| = \|a\|$ for all self-adjoint element $a \in \mathbf{A}$. So fix a self-adjoint element $a \in \mathbf{A}$. Since $\|\Phi(a)\| = r(\Phi(a))$ and $\|a\| = r(a)$, it suffices to show that $\sigma_{\mathbf{C}}(\Phi(a)) = \sigma_{\mathbf{A}}(a)$. By Proposition 8.2.16, we know that

$$\sigma_{\mathbf{C}}(\Phi(a)) \subset \sigma_{\mathbf{A}}(a).$$

Therefore we now show that $\sigma_{\mathbf{A}}(a) \subset \sigma_{\mathbf{C}}(\Phi(a))$. Suppose not, then there is a point $\lambda \in \sigma_{\mathbf{A}}(a)$ with $\lambda \notin \sigma_{\mathbf{C}}(\Phi(a))$. Since $\sigma_{\mathbf{C}}(\Phi(a))$ is closed, there is a continuous function $f \in C(\sigma_{\mathbf{A}}(a))$ such that $f(\lambda) = 1$ and

$$f|_{\sigma_{\mathbf{C}}(\Phi(a))} = 0$$

by Urysohn's Lemma. Therefore $f(\Phi(a)) = 0$. Since the continuous functional calculus is injective, $f(a) \neq 0$. However, by Lemma 8.5.11, we have

$$\Phi(f(a)) = f(\Phi(a)) = 0.$$

This contradicts the assumption that Φ is injective. Therefore it must be that $\sigma_{\mathbf{C}}(\Phi(a)) = \sigma_{\mathbf{A}}(a)$, which implies that $\|\Phi(a)\| = \|a\|$.

(ii) Now suppose $\mathbf{A}$ is not unital. Replacing $\mathbf{B}$ by its unitalization if necessary, we may assume that $\mathbf{B}$ is unital. If $(\mathbf{A}^u, \iota)$ denotes the unitalization of $\mathbf{A}$, then there is a $*$-homomorphism $\Psi : \mathbf{A}^u \to \mathbf{B}$ such that $\Psi \circ \iota = \Phi$. We claim that Ψ is injective. To see this, suppose $x \in \mathbf{A}^u$ is such that $\Psi(x) = 0$. By the description of $\mathbf{A}^u$ in Theorem 8.2.11, we may express x in the form $x = \iota(a) + \lambda 1_{\mathbf{A}^u}$ for some $a \in \mathbf{A}$ and $\lambda \in \mathbb{C}$. Then $\Phi(a) + \lambda 1_{\mathbf{B}} = \Psi(x) = 0$. Hence

$$\Phi(a) = -\lambda 1_{\mathbf{B}}.$$

If $\lambda \neq 0$, then $b := \frac{-a}{\lambda}$ has the property that $\Phi(b) = 1_{\mathbf{B}}$. For any $c \in \mathbf{A}$, we have

$$\Phi(cb) = \Phi(c).$$

Since Φ is injective, $cb = c$. Similarly, $bc = c$ and it follows that $b = 1_{\mathbf{A}}$. This contradicts the assumption that $\mathbf{A}$ is not unital. Therefore it must happen that $\lambda = 0$. However, if that happens, then $\Phi(a) = 0$. Since Φ is injective, $a = 0$, which implies that $x = 0$. We conclude that Ψ is injective. Hence Ψ is isometric by part (i). Since ι is isometric, we conclude that Φ is isometric as well.

$\square$

8.6 Ideals and Quotients of C*-Algebras

It may be somewhat surprising that we have not yet discussed quotients of C*-algebras. These are natural constructions and one would expect to encounter them early in the theory. However, we needed the continuous functional calculus in order to do so correctly. As you might imagine, our goal is to prove a First Isomorphism Theorem in this context. There is one more reason to go down this road, though. We wish to prove (as we will in Theorem 8.6.7) that the range of any $*$-homomorphism is a closed C*-subalgebra of the codomain. Once again, this is a consequence of the rigidity provided by the C*-identity.

Let **A** be a C*-algebra and **I** be a closed two-sided ideal (in the sense of Definition 7.1.3). Then **A**/**I** is a Banach algebra (Proposition 7.1.6) with the quotient norm given by

$$\|a + \mathbf{I}\| = \inf\{\|a + b\| : b \in \mathbf{I}\}.$$

We now wish to define an involution on **A**/**I** by

$$(a + \mathbf{I})^* := a^* + \mathbf{I}$$

and show that **A**/**I** is a C*-algebra. However, for this to be well-defined we need **I** to be closed under the involution operation. The next lemma helps us prove this fact.

8.6.1 LEMMA *Let* **A** *be a C*-algebra and* **I** $\lhd$ **A** *be a closed ideal in* **A**. *For any* $a \in \mathbf{I}$, *there is a sequence of self-adjoint elements* $(e_n) \subset \mathbf{I}$ *such that* $\sigma(e_n) \subset [0,1]$ *for all* $n \in \mathbb{N}$ *and*

$$\lim_{n \to \infty} \|a - ae_n\| = 0.$$

Proof. Replacing **A** by its unitalization if need be, we may assume that **A** is unital.

(i) Suppose first that $a = a^*$, so that $\sigma(a) \subset \mathbb{R}$. For $n \in \mathbb{N}$, define $f_n \in C(\sigma(a))$ by

$$f_n(t) := \frac{nt^2}{1 + nt^2}.$$

We then define $e_n := f_n(a)$ by the continuous functional calculus.

(a) Since $f_n = \overline{f_n}$, it follows that $e_n = e_n^*$.

(b) Now $f_n \in C(\sigma(a))$ is a limit of polynomials in $\{p_{n,k} : k \in \mathbb{N}\}$ since $\sigma(a)$ is a compact subset of $\mathbb{R}$. Furthermore, since $f_n(0) = 0$, we may choose these polynomials $p_{n,k}$ such that $p_{n,k}(0) = 0$ (do verify this). Since **I** is an ideal and $a \in \mathbf{I}$,

$$p_{n,k}(a) \in \mathbf{I}$$

for all k. Since **I** is closed, $e_n \in \mathbf{I}$ for all $n \in \mathbb{N}$.

(c) Since $f_n(t) \in [0,1]$ for all $t \in \sigma(a)$, it follows from the Spectral Mapping Theorem that $\sigma(e_n) \subset [0,1]$. By Corollary 8.5.6, we conclude that $\|e_n\| = r(e_n) \leq 1$. Also,

$$1 - f_n(t) = \frac{1}{1 + nt^2}.$$

By the same argument, $\|1_{\mathbf{A}} - e_n\| \leq 1$.

(d) Hence

$$\|a - ae_n\|^2 = \|a(1_\mathbf{A} - e_n)\|^2$$
$$= \|(1_\mathbf{A} - e_n)a^2(1_\mathbf{A} - e_n)\|$$
$$\leq \|a^2(1_\mathbf{A} - e_n)\|$$
$$= \frac{1}{n}\|na^2(1_\mathbf{A} - e_n)\|$$
$$= \frac{1}{n}\|e_n\| \leq \frac{1}{n}.$$

Therefore $\|a - ae_n\| \to 0$ as $n \to \infty$.

(ii) Now suppose $a \in \mathbf{A}$ is an arbitrary element, then a^*a is self-adjoint and in **I**. By part (i), there is sequence (e_n) of self-adjoint elements in **I** such that $\sigma(e_n) \subset [0,1]$ and

$$\lim_{n\to\infty} \|a^*a(1_\mathbf{A} - e_n)\| = 0.$$

Since $\|1_\mathbf{A} - e_n\| \leq 1$,

$$\|ae_n - a\|^2 = \|(1_\mathbf{A} - e_n)a^*a(1_\mathbf{A} - e_n)\| \leq \|a^*a(1_\mathbf{A} - e_n)\| \to 0$$

as $n \to \infty$. □

Aside. We should mention here that the conclusion of Lemma 8.6.1 may be strengthened. If **I** is a separable, closed ideal in a C*-algebra **A**, one may find a sequence (e_n) of self-adjoint elements in **I** such that $\sigma(e_n) \subset [0,1]$ and

$$\lim_{n\to\infty} \|ae_n - a\| = 0$$

for all $a \in \mathbf{I}$. Such a sequence is termed an *approximate unit* for the ideal. In fact, even if **I** is not separable, the conclusion holds provided one is willing to replace the sequence by a (possibly uncountable) net. For more on this, do have a look at Murphy [43, Chapter 3].

8.6.2 EXAMPLE

(i) Let $\mathbf{A} = \mathcal{B}(\ell^2)$ and $\mathbf{I} = \mathcal{K}(\ell^2)$. Then $\mathbf{I}$ is a closed ideal in $\mathbf{A}$ by Proposition 7.1.5. Let P_n be the projection onto the subspace spanned by $\{e_1, e_2, \ldots, e_n\}$. Then for any $T \in \mathbf{I}$,

$$\lim_{n \to \infty} \|T - P_n T\| = 0,$$

by Theorem 5.6.9.

(ii) Let $\mathbf{A} = C[0,1]$ and $\mathbf{I} = C_0((0,1/2))$, then we may choose $e_n \in \mathbf{I}$ such that $e_n(x) = 1$ if $1/n \le x \le 1/2 - 1/n$ and $0 \le e_n \le 1$. Then for any $f \in \mathbf{I}$,

$$\lim_{n \to \infty} \|f - f e_n\| \to 0.$$

(See Exercise 8.39.)

8.6.3 PROPOSITION *Every closed ideal of a C*-algebra is self-adjoint.*

Proof. Let $\mathbf{A}$ be a C*-algebra and $\mathbf{I}$ be a closed ideal in $\mathbf{A}$. Let $a \in \mathbf{I}$ and choose a sequence $(e_n) \subset \mathbf{I}$ of self-adjoint elements as in Lemma 8.6.1 such that $\lim_{n \to \infty} \|a e_n - a\| = 0$. Then $e_n a^* \in \mathbf{I}$ for all $n \in \mathbb{N}$ and

$$\lim_{n \to \infty} \|a^* - e_n a^*\| = \lim_{n \to \infty} \|a - a e_n\| = 0.$$

Since $\mathbf{I}$ is closed, $a^* \in \mathbf{I}$. $\qquad\square$

8.6.4 LEMMA *Let $\mathbf{A}$ be a C*-algebra and $\mathbf{I} \lhd \mathbf{A}$ a closed ideal. Then for any $a \in \mathbf{A}$,*

$$\|a + \mathbf{I}\| = \inf\{\|a - ax\| : x \in \mathbf{I}, x = x^* \text{ and } \sigma(x) \subset [0,1]\}.$$

Proof. Once again, replacing $\mathbf{A}$ by its unitalization if need be, we may assume that $\mathbf{A}$ is itself unital. Fix $a \in \mathbf{A}$ and recall that

$$\|a + \mathbf{I}\| = \inf\{\|a - b\| : b \in \mathbf{I}\}.$$

Let $S = \{x \in \mathbf{I} : x = x^*, \sigma(x) \subset [0,1]\}$ and set $\beta := \inf\{\|a - ax\| : x \in S\}$. Since $ax \in \mathbf{I}$ for each $x \in S$, it follows that

$$\|a + \mathbf{I}\| \le \beta.$$

Now suppose $b \in I$, then choose a sequence $(e_n) \in S$ such that $\lim_{n \to \infty} \|b - be_n\| = 0$ by Lemma 8.6.1. Then $\|(1_A - e_n)\| \leq 1$ and

$$\begin{aligned} \|a + b\| &\geq \|(a + b)(1_A - e_n)\| \\ &= \|(a - ae_n) - (be_n - b)\| \\ &\geq \|a - ae_n\| - \|be_n - b\| \\ &\geq \beta - \|be_n - b\|. \end{aligned}$$

Letting $n \to \infty$, we see that $\|a + b\| \geq \beta$. This is true for all $b \in I$, so $\|a + I\| = \beta$. $\square$

8.6.5 Theorem *Let A be a C*-algebra and I be a closed ideal in A. The involution*

$$(a + I)^* := a^* + I$$

is well-defined and A/I is a C*-algebra with respect to this involution and the quotient norm.

Proof. We give a proof under the assumption that A is unital as the non-unital case requires but a minor modification.

That the involution is well-defined follows from Proposition 8.6.3. By Remark 8.2.2, it suffices to check that

$$\|a + I\|^2 \leq \|a^*a + I\|$$

for all $a \in A$. By Lemma 8.6.4,

$$\|a + I\|^2 = \inf\{\|a - ax\|^2 : x \in I, x = x^* \text{ and } \sigma(x) \subset [0, 1]\}.$$

Suppose $x \in I$ is such that $x = x^*$ and $\sigma(x) \subset [0, 1]$, then $\|1_A - x\| \leq 1$. Therefore

$$\|a + I\|^2 \leq \|a - ax\|^2 = \|(1_A - x)a^*a(1_A - x)\| \leq \|a^*a(1_A - x)\|.$$

Taking an infimum over all such x, we conclude that $\|a + I\|^2 \leq \|a^*a + I\|$. $\square$

8.6.6 Example If H is a Hilbert space, then we know that $\mathcal{K}(H)$ is a closed two-sided ideal in $\mathcal{B}(H)$. The corresponding quotient is called the *Calkin algebra* and is written as

$$\mathcal{Q}(\mathbf{H}) := \mathcal{B}(\mathbf{H})/\mathcal{K}(\mathbf{H}).$$

The Calkin algebra is the playground for a number of fascinating ideas in operator theory. For instance, the quest to understand normal elements in $\mathcal{Q}(\mathbf{H})$ resulted in the Brown–Douglas–Fillmore theory and the introduction of tools from algebraic topology into the study of operator algebras. We do not discuss these matters here, but the interested reader will find more on the subject in Davidson [10].

The First Isomorphism Theorem in the context of C*-algebras now takes the expected form. Perhaps the only novelty is one that we inherit from Proposition 8.5.12, where we proved that an injective *-homomorphism is isometric.

8.6.7 THEOREM (FIRST ISOMORPHISM THEOREM) *Let* $\Phi : \mathbf{A} \to \mathbf{B}$ *be a *-homomorphism. Then*

  (i) $\ker(\Phi)$ *is a closed, two-sided ideal of* $\mathbf{A}$.

  (ii) *If* $\pi : \mathbf{A} \to \mathbf{A}/\ker(\Phi)$ *denotes the quotient map, then there is an isometric *-homomorphism*

$$\widehat{\Phi} : \mathbf{A}/\ker(\Phi) \to \mathbf{B}$$

  such that $\widehat{\Phi} \circ \pi = \Phi$.

  (iii) $\mathrm{Range}(\Phi)$ *is a (closed) C*-subalgebra of* $\mathbf{B}$.

Proof. That $\ker(\Phi)$ is a closed, two-sided ideal is evident. Define $\widehat{\Phi} : \mathbf{A}/\ker(\Phi) \to \mathbf{B}$ by

$$a + \ker(\Phi) \mapsto \Phi(a).$$

Then it is clearly a *-homomorphism that must be injective by Theorem 1.1.13. By Proposition 8.5.12, $\widehat{\Phi}$ is isometric. Hence $\mathrm{Range}(\Phi) = \mathrm{Range}(\widehat{\Phi})$ must be a closed C*-subalgebra of $\mathbf{B}$ since it is the isometric image of a C*-algebra. $\qquad\square$

This brings us to the end of our initial exploration of C*-algebra theory. The main ideas in this chapter are the Gelfand–Naimark characterization of commutative C*-algebras and the continuous functional calculus for normal elements. These will play a crucial role in Chapter 10. For the moment, it is time to solve some problems.

8.7 Exercises

Adjoint of an Operator

8.7.1 DEFINITION Let $\mathbf{H}$ and $\mathbf{K}$ be Hilbert spaces. A function $u : \mathbf{H} \times \mathbf{K} \to \mathbb{C}$ is said to be a ***sesquilinear form*** if it satisfies the following conditions for all $x, y \in \mathbf{H}, z, w \in \mathbf{K}$ and $\alpha, \beta \in \mathbb{C}$.

(a) $u(\alpha x + \beta y, z) = \alpha u(x, z) + \beta u(y, z)$.

(b) $u(x, \alpha z + \beta w) = \overline{\alpha} u(x, z) + \overline{\beta} u(x, w)$.

Furthermore, u is said to be ***bounded*** if there exists $M > 0$ such that $|u(x, y)| \leq M\|x\|\|y\|$ for all $x \in \mathbf{H}, y \in \mathbf{K}$.

EXERCISE 8.1 (⚒) Let $\mathbf{H}$ and $\mathbf{K}$ be Hilbert spaces. If $u : \mathbf{H} \times \mathbf{K} \to \mathbb{C}$ is a bounded sesquilinear form, then prove that there is a bounded linear operator $S : \mathbf{K} \to \mathbf{H}$ such that

$$u(x, y) = \langle x, S(y) \rangle$$

for all $x \in \mathbf{H}, y \in \mathbf{K}$.

Hint: For $y \in \mathbf{K}$ fixed, the map $x \mapsto u(x, y)$ is a bounded linear functional.

EXERCISE 8.2 (⚒) Use Exercise 8.1 to give an alternate proof of Proposition 8.1.3.

EXERCISE 8.3 Let $\mathbf{E} = c_{00}$ equipped with the usual inner product. Define $T : \mathbf{E} \to \mathbf{E}$ by

$$T((x_j)) := \left(\sum_{k=1}^{\infty} \frac{x_k}{k}, 0, 0, \dots \right).$$

Prove that $T \in \mathcal{B}(\mathbf{E})$ and that there is no bounded operator $S \in \mathcal{B}(\mathbf{E})$ such that $\langle T(x), y \rangle = \langle x, S(y) \rangle$ for all $x, y \in \mathbf{E}$.

EXERCISE 8.4 (⚒) Prove Proposition 8.1.5.

EXERCISE 8.5 Let $M = (a_{i,j})_{i,j \in \mathbb{N}}$ be an infinite matrix whose entries are in $\mathbb{C}$. For $i, j \in \mathbb{N}$, define

$$\delta_i := \sum_{j=1}^{\infty} |a_{i,j}| \text{ and } \gamma_j := \sum_{i=1}^{\infty} |a_{i,j}|.$$

Assume that $\alpha := \sup_{j\in\mathbb{N}} \gamma_j < \infty$ and $\beta := \sup_{i\in\mathbb{N}} \delta_i < \infty$. For $x = (x_1, x_2, \ldots) \in \ell^2$, define $T(x) = (y_1, y_2, \ldots)$ where

$$y_i := \sum_{j=1}^{\infty} a_{i,j} x_j.$$

We know from Exercise 5.45 that T defines a bounded operator on ℓ^2. Prove that T^* is also an operator of the same form, whose matrix is the 'conjugate transpose' of M.

C*-Algebras

EXERCISE 8.6 On the Banach algebra $\ell^1(\mathbb{Z})$, define a map $\delta : \ell^1(\mathbb{Z}) \to \ell^1(\mathbb{Z})$ by

$$\delta(x)_n := \overline{x_{-n}}$$

for any $x = (x_n) \in \ell^1(\mathbb{Z})$. Prove that δ is an involution on $\ell^1(\mathbb{Z})$ that does not satisfy the C*-identity.

EXERCISE 8.7 If $\mathbf{H}$ is an infinite dimensional Hilbert space, then prove that $\mathcal{K}(\mathbf{H})$ is a non-unital C*-algebra.

EXERCISE 8.8 Let $\mathbf{A}$ be a C*-algebra and X be a locally compact, Hausdorff space. Define $C_0(X, \mathbf{A})$ as in Exercise 7.13 to be the space of all continuous, $\mathbf{A}$-valued functions on X which vanish at infinity. For $f \in C_0(X, \mathbf{A})$ define $f^* \in C_0(X, \mathbf{A})$ by

$$f^*(x) := f(x)^*.$$

Show that this is a well-defined involution on $C_0(X, \mathbf{A})$ and that $C_0(X, \mathbf{A})$ is a C*-algebra when equipped with this involution and the supremum norm.

EXERCISE 8.9 (⚒) Complete the proof of Proposition 8.2.8.

EXERCISE 8.10 (⚒) Complete the proof of Theorem 8.2.11.

EXERCISE 8.11 Let $\mathbf{A}$ be a non-unital C*-algebra. For any $a \in \mathbf{A}$, prove that $0 \in \sigma_{\mathbf{A}}(a)$ (see Definition 8.2.12).

Commutative C*-Algebras

EXERCISE 8.12 Let $\mathbf{A} = C^1[0,1]$ equipped with the norm

$$\|f\| := \|f\|_\infty + \|f'\|_\infty.$$

By Exercise 7.2, $\mathbf{A}$ is a Banach algebra. Let $\zeta : [0,1] \to \mathbb{C}$ be the identity function $\zeta(t) = t$.

(i) Show that $\mathbf{A}$ is generated by $\{1_{\mathbf{A}}, \zeta\}$.

(ii) For $t \in [0,1]$, define $\tau_t : \mathbf{A} \to \mathbb{C}$ by $f \mapsto f(t)$. Prove that $\tau_t \in \Omega(\mathbf{A})$.

(iii) Show that the map $\eta_{\mathbf{A}} : [0,1] \to \Omega(\mathbf{A})$ given by $t \mapsto \tau_t$ is a homeomorphism. Conclude that the Gelfand representation of Theorem 8.3.6 is not surjective.

EXERCISE 8.13 Let $\mathbf{A}$ be the set of all 2×2 complex matrices of the form

$$\begin{pmatrix} a & b \\ 0 & a \end{pmatrix}$$

for some $a, b \in \mathbb{C}$. Think of $\mathbf{A}$ as a subset of $\mathcal{B}(\mathbb{C}^2)$ equipped with the operator norm.

(i) Show that $\mathbf{A}$ is a unital commutative Banach algebra.

(ii) Determine $\Omega(\mathbf{A})$.

       Hint: If $\tau \in \Omega(\mathbf{A})$, then determine $\tau\left(\begin{pmatrix} 0 & 1 \\ 0 & 0 \end{pmatrix}\right)$.

(iii) Show that the Gelfand transform $\Gamma_{\mathbf{A}} : \mathbf{A} \to C(\Omega(\mathbf{A}))$ is not injective.

EXERCISE 8.14 Let $\mathbf{A} = L^1[0,1]$ with multiplication given by convolution (as in Example 7.1.2) and $f \in \mathbf{A}$ be the constant function $f = \mathbf{1}$.

(i) For each $n \in \mathbb{N}$, prove that

$$f^n(t) = \underbrace{f * f * \ldots * f}_{n \text{ times}}(t) = \frac{t^{n-1}}{(n-1)!}.$$

(ii) Prove that $r(f) = 0$.

(iii) Prove that $\mathbf{A}$ is generated by f.

(iv) Conclude that $\Omega(\mathbf{A}) = \varnothing$ (compare this result with Lemma 8.3.2).

EXERCISE 8.15 Let $\mathbf{A}$ be a unital commutative Banach algebra.

(i) For all $a, b \in \mathbf{A}$, prove that $\sigma(a + b) \subset \sigma(a) + \sigma(b)$ and $\sigma(ab) \subset \sigma(a)\sigma(b)$.

(ii) Give an example to show that this is not true for all Banach algebras.

EXERCISE 8.16 (✕) Let $\mathbf{A}$ be a C*-algebra and $a \in \mathbf{A}$. If $b, c \in \mathbf{A}$ are self-adjoint elements such that $a = b + ic$, then prove that

$$b = \frac{a + a^*}{2} \text{ and } c = \frac{i(a^* - a)}{2}.$$

EXERCISE 8.17 (✕) Let $\mathbf{A}$ be a non-unital, commutative Banach algebra.

(i) Prove that $\Omega(\mathbf{A}) \cup \{0\}$ is a compact Hausdorff space when equipped with the weak-* topology. Conclude that $\Omega(\mathbf{A})$ is locally compact and Hausdorff.

(ii) If $a \in \mathbf{A}$, prove that $\hat{a} \in C_0(\Omega(\mathbf{A}))$, where 0 is treated as the *point at infinity*.

(iii) Prove that the Gelfand representation $\Gamma_{\mathbf{A}} : \mathbf{A} \to C_0(\Omega(\mathbf{A}))$ given by $a \mapsto \hat{a}$ is a homomorphism of Banach algebras.

EXERCISE 8.18 (✕) Let $\mathbf{A}$ be a non-unital commutative C*-algebra and let $(\mathbf{A}^u, \iota)$ denote its unitalization.

(i) For each $\tau \in \Omega(\mathbf{A})$, define $\tau^u : \mathbf{A}^u \to \mathbb{C}$ by the formula

$$\tau^u(\iota(a) + \lambda 1_{\mathbf{A}^u}) := \tau(a) + \lambda.$$

Prove that $\tau^u \in \Omega(\mathbf{A}^u)$.

(ii) Prove that the map $\tau \mapsto \tau^u$ induces a homeomorphism from $\Omega(\mathbf{A}) \cup \{0\}$ to $\Omega(\mathbf{A}^u)$ (when each is equipped with the corresponding weak-* topology).

(iii) Prove that the Gelfand representation $\Gamma_{\mathbf{A}} : \mathbf{A} \to C_0(\Omega(\mathbf{A}))$ is an isometric *-isomorphism.

EXERCISE 8.19 Let X and Y be compact Hausdorff spaces and let $g : Y \to X$ be a continuous function. Define $\Phi : C(X) \to C(Y)$ by

$$\Phi(f) = f \circ g.$$

(i) Prove that Φ is surjective if and only if g is injective.

(ii) Prove that Φ is injective if and only if g is surjective.

Hint: You will need both Urysohn's Lemma and Tietze's Extension Theorem.

Spectral Permanence Theorem

EXERCISE 8.20 (⚒) Let $\mathbf{A}$ be a unital Banach algebra and let $a \in \mathbf{A}$. Let $\mathbf{B}$ be the Banach subalgebra of $\mathbf{A}$ generated by $\{1_{\mathbf{A}}, a\}$. Prove that

$$\mathbf{B} = \overline{\{p(a) : p(x) \in \mathbb{C}[x]\}}.$$

Conclude that $\mathbf{B}$ is commutative.

EXERCISE 8.21 (⚒) Suppose that a unital Banach algebra $\mathbf{A}$ is generated by the set $\{1_{\mathbf{A}}, a\}$ for some $a \in \mathbf{A}$. Then the map $\widehat{a} : \Omega(\mathbf{A}) \to \sigma(a)$ given by

$$\widehat{a}(\tau) := \tau(a)$$

is well-defined by Exercise 8.20 and Proposition 8.3.3. Prove that $\widehat{a}$ is a homeomorphism.

EXERCISE 8.22 Let K be a compact subset of $\mathbb{C}$ and let X be a non-empty hole in K. Let $\lambda \in X$ and let $f \in C(K)$ be the function

$$f(z) := \frac{1}{z - \lambda}.$$

Prove that f cannot be approximated uniformly by polynomials in $C(K)$.

EXERCISE 8.23 Suppose that a unital Banach algebra $\mathbf{A}$ is generated by the set $\{1_{\mathbf{A}}, a\}$ for some $a \in \mathbf{A}$. Prove that $\sigma(a)$ cannot have any holes.

Hint: Use Exercise 8.21, Exercise 8.22 and the Gelfand representation.

EXERCISE 8.24 Let $\mathbf{A} = C(S^1)$ and $\mathbf{B}$ be the Banach subalgebra generated by $\{1_{\mathbf{A}}, \zeta\}$ where $\zeta(z) = z$. Let $\eta_{\mathbf{B}} : \mathbb{D} \to \Omega(\mathbf{B})$ be given by

$$\eta_{\mathbf{B}}(\lambda) := \tau_\lambda$$

as defined in Example 8.4.3. Prove that $\eta_{\mathbf{B}}$ is a homeomorphism.

EXERCISE 8.25 Let $A(\mathbb{D})$ denote the disc algebra from Exercise 7.1 and let $\mathbf{B}$ be the Banach subalgebra of $C(S^1)$ described in Example 8.4.3.

(i) Prove that the restriction map $\Psi : A(\mathbb{D}) \to C(S^1)$ given by

$$\Psi(f) := f|_{S^1}$$

is isometric.

(ii) Prove that $A(\mathbb{D}) \cong \mathbf{B}$.

Hint: If $f \in A(\mathbb{D})$, prove that f is the uniform limit of polynomials by considering $g(z) = f(rz)$ for some $0 < r < 1$.

EXERCISE 8.26 Let $\mathbf{A}$ be a unital Banach algebra and $\mathbf{B}$ be a commutative subalgebra of $\mathbf{A}$ that is maximal amongst all commutative subalgebras of $\mathbf{A}$ (in other words, if $\mathbf{C}$ is a commutative subalgebra such that $\mathbf{B} \subset \mathbf{C}$, then $\mathbf{B} = \mathbf{C}$).

(i) Prove that $\mathbf{B}$ is closed and that $1_{\mathbf{A}} \in \mathbf{B}$.

(ii) For any $b \in \mathbf{B}$, prove that $\sigma_{\mathbf{A}}(b) = \sigma_{\mathbf{B}}(b)$.

Continuous Functional Calculus

EXERCISE 8.27 Let $\mathbf{A}$ be a unital C*-algebra and $a \in \mathbf{A}$ be a normal element. If $f \in C(\sigma(a))$ and $g \in C(\sigma(f(a)))$, then prove that

$$(g \circ f)(a) = g(f(a)).$$

Hint: Let $\mathcal{A}$ be the collection of all functions $g \in C(\sigma(f(a)))$ that satisfy the conclusion and apply the Stone-Weierstrass Theorem to $\mathcal{A}$.

EXERCISE 8.28 Give a proof of Corollary 8.5.6 directly from Lemma 8.2.14 (without using the continuous functional calculus).

EXERCISE 8.29 Let $\mathbf{A}$ be a unital, commutative C*-algebra that is generated by a finite set $\{a_1, a_2, \ldots, a_n\}$. If

$$X := \{(\tau(a_1), \tau(a_2), \ldots, \tau(a_n)) : \tau \in \Omega(\mathbf{A})\},$$

then prove that there is an isometric $*$-isomorphism $\Theta : \mathbf{A} \to C(X)$ such that, for each $1 \leq k \leq n$, $\Theta(a_k) = \zeta_k$, where $\zeta_k : X \to \mathbb{C}$ is the function $\zeta_k(z_1, z_2, \ldots, z_n) := z_k$.

EXERCISE 8.30 (⚒) Let $\mathbf{H}$ be a Hilbert space and let $P \in \mathcal{B}(\mathbf{H})$ be an operator such that $P = P^* = P^2$. If $\mathbf{M} = \text{Range}(P)$, prove that $\mathbf{M}$ is a closed subpace of $\mathbf{H}$ and that P is the orthogonal projection of $\mathbf{H}$ onto $\mathbf{M}$.

EXERCISE 8.31 Let $\mathbf{H}$ be a Hilbert space and $T \in \mathcal{B}(\mathbf{H})$ be a normal operator. For any $f \in C(\sigma(T))$, prove that the operator $f(T)$ is a projection in $\mathcal{B}(\mathbf{H})$ if and only if f is the characteristic function of a set in $\sigma(T)$.

EXERCISE 8.32 Let $T \in \mathcal{B}(\mathbf{H})$ be a normal operator.

(i) If $\sigma(T)$ is finite, then prove that T is a linear combination of projections.

(ii) If $\sigma(T)$ is a singleton, then prove that T is a scalar multiple of the identity operator.

EXERCISE 8.33 (⚒) Let $T \in \mathcal{B}(\mathbf{H})$ be a normal operator. If $\lambda \in \sigma(T)$ is an isolated point of $\sigma(T)$, then prove that λ is an eigenvalue of T.

EXERCISE 8.34 Let $T \in \mathcal{B}(\mathbf{H})$ be a bounded operator and $P \in \mathcal{B}(\mathbf{H})$ be a projection. If $TP = PT$, then prove that $\mathbf{M} := \text{Range}(P)$ is invariant under T.

EXERCISE 8.35 If $T \in \mathcal{B}(\mathbf{H})$ is a normal operator such that $\sigma(T)$ is disconnected, then prove that T has a non-trivial invariant subspace.

Note: In fact, the assumption on $\sigma(T)$ is superfluous (see Exercise 10.30).

EXERCISE 8.36 Let $\mathbf{A}$ be a unital C*-algebra and $h \in \mathbf{A}$ be a self-adjoint element. Prove that

$$u := \exp(ih) = \sum_{k=0}^{\infty} \frac{(ih)^k}{k!}$$

is a unitary in $\mathbf{A}$. (See Exercise 7.18.)

EXERCISE 8.37 Let $\mathbf{A}$ be a unital C*-algebra and $u \in \mathbf{A}$ be a unitary such that $\sigma(u) \neq S^1$. Then prove that there is a self-adjoint element $h \in \mathbf{A}$ such that $u = \exp(ih)$.

EXERCISE 8.38 (⚒) Let $\mathbf{A}$ be a unital C*-algebra and let $U(\mathbf{A})$ denote the set of all unitary elements in $\mathbf{A}$.

(i) Prove that $U(\mathbf{A})$ is a topological group.

(ii) Let $U_0(\mathbf{A})$ denote the connected component of $1_{\mathbf{A}}$ in $U(\mathbf{A})$. If $h \in \mathbf{A}$ is a self-adjoint element, then prove that $\exp(ih) \in U_0(\mathbf{A})$.

(iii) Give an example of a C*-algebra $\mathbf{A}$ and a unitary $u \in U(\mathbf{A})$ that cannot be expressed in the form $\exp(ih)$ for any self-adjoint element $h \in \mathbf{A}$.

Hint: See Exercise 7.22.

Ideals and Quotients of C*-Algebras

EXERCISE 8.39 Let $A = C[0,1]$ and $I = C_0((0,1/2))$, then choose $e_n \in I$ such that $e_n(x) = 1$ if $1/n \le x \le 1/2 - 1/n$ and $0 \le e_n \le 1$. For any $f \in I$, prove that

$$\lim_{n\to\infty} \|f - fe_n\| \to 0.$$

EXERCISE 8.40 Let X be a compact Hausdorff space and $U \subset X$ be an open set. Consider $I := C_0(U)$ as an ideal in $C(X)$. Prove that there is a net $\{e_\lambda : \lambda \in J\}$ of functions in I such that $0 \le e_\lambda(x) \le 1$ for all $x \in X$, such that

$$\lim_{\lambda} \|fe_\lambda - f\| = 0$$

for all $f \in C_0(U)$.

Note: If you are uncomfortable with nets, choose X to be a metric space and prove that there is a sequence (e_n) satisfying these conditions.

EXERCISE 8.41 Let A be a C*-algebra, B be a C*-subalgebra of A and let I be a closed ideal in A.

(i) Define $\Phi : B/(B \cap I) \to A/I$ by

$$\Phi(b + B \cap I) := b + I.$$

Prove that Φ is a *-homomorphism and that $\text{Range}(\Phi) = (B+I)/I$.

(ii) Prove that $(B+I)$ is a closed C*-subalgebra of A.

8.7.2 DEFINITION Let H be a Hilbert space, let $\mathcal{Q}(H)$ denote the Calkin algebra from Example 8.6.6 and let $\pi : \mathcal{B}(H) \to \mathcal{Q}(H)$ denote the quotient map. For $T \in \mathcal{B}(H)$, the *essential spectrum* of T is

$$\sigma_e(T) := \sigma_{\mathcal{Q}(H)}(\pi(T)).$$

Note that $\sigma(T)$ denotes the spectrum of T in $\mathcal{B}(H)$.

EXERCISE 8.42 For each $T \in \mathcal{B}(H)$, prove that

$$\sigma_e(T) \subset \bigcap_{K \in \mathcal{K}(H)} \sigma(T+K).$$

Multiplier Algebras

8.7.3 DEFINITION Let $\mathbf{A}$ be a C*-algebra. A *double centralizer* of $\mathbf{A}$ is a pair (L, R) of bounded linear maps on $\mathbf{A}$ such that

$$R(a)b = aL(b).$$

for all $a, b \in \mathbf{A}$. We think of L and R as elements of $\mathcal{B}(\mathbf{A})$, which is equipped with the operator norm.

EXERCISE 8.43 If (L, R) is a double centralizer of a C*-algebra $\mathbf{A}$, then

(i) Prove that $\|L\| = \|R\|$.

(ii) Define $L^* : \mathbf{A} \to \mathbf{A}$ by $L^*(a) := L(a^*)^*$ and define $R^* : \mathbf{A} \to \mathbf{A}$ by $R^*(a) := R(a^*)^*$. Prove that (R^*, L^*) is a double centralizer of $\mathbf{A}$.

EXERCISE 8.44 Let $\mathbf{A}$ be a C*-algebra and $M(\mathbf{A})$ be the set of all double centralizers of $\mathbf{A}$. Define addition and scalar multiplication of such pairs (L, R) componentwise. Define multiplication by

$$(L_1, R_1) \cdot (L_2, R_2) := (L_1 L_2, R_2 R_1).$$

Define a norm on $M(\mathbf{A})$ by $\|(L, R)\| := \|L\| = \|R\|$. Finally, define an involution on $M(\mathbf{A})$ by

$$(L, R)^* := (R^*, L^*).$$

Prove that $M(\mathbf{A})$ is a unital C*-algebra with these operations. The algebra $M(\mathbf{A})$ is called the *multiplier algebra* of $\mathbf{A}$.

EXERCISE 8.45 Let $\mathbf{A}$ be a C*-algebra. For each $c \in \mathbf{A}$, define $L_c : \mathbf{A} \to \mathbf{A}$ by $L_c(a) = ca$ and define $R_c : \mathbf{A} \to \mathbf{A}$ by $R_c(a) = ac$.

(i) Prove that (L_c, R_c) is a double centralizer and that $\|(L_c, R_c)\| = \|c\|$.

(ii) Prove that the map $\Psi : \mathbf{A} \to M(\mathbf{A})$ given by $c \mapsto L_c$ is an isometric *-homomorphism.

Identifying $\mathbf{A}$ with $\Psi(\mathbf{A})$, we think of $\mathbf{A}$ as a subset of $M(\mathbf{A})$.

EXERCISE 8.46 Let $\mathbf{A}$ be a C*-algebra and $\mathbf{I}$ be a closed ideal in $\mathbf{A}$. For each $a \in \mathbf{A}$, define $L_a : \mathbf{I} \to \mathbf{I}$ by $L_a(b) := ab$ and define $R_a : \mathbf{I} \to \mathbf{I}$ by $R_a(b) := ba$.

(i) Prove that $(L_a, R_a) \in M(\mathbf{I})$ for each $a \in \mathbf{A}$.

(ii) Prove that the map $\Phi_{\mathbf{I}} : \mathbf{A} \to M(\mathbf{I})$ given by $a \mapsto L_a$ is a *-homomorphism.

DEFINITION 8.7.4 An ideal $\mathbf{I}$ of a C*-algebra $\mathbf{A}$ is said to be *essential* if, for any $a \in \mathbf{A}$, $ab = 0$ for all $b \in \mathbf{I}$ implies that $a = 0$.

EXERCISE 8.47

(i) Let $\mathbf{A}$ be a C*-algebra and think of $\mathbf{A}$ as a subset of $M(\mathbf{A})$ by Exercise 8.45. Prove that $\mathbf{A}$ is an essential ideal of $M(\mathbf{A})$.

(ii) Let $\mathbf{I}$ be an essential ideal of a C*-algebra $\mathbf{A}$. Prove that the map $\Phi_{\mathbf{I}} : \mathbf{A} \to M(\mathbf{I})$ from Exercise 8.46 is injective.

EXERCISE 8.48 Let X be a locally compact Hausdorff space. Let $C_b(X)$ be the C*-algebra of bounded continuous functions on X and think of $C_0(X)$ as a subset of $C_b(X)$ and $M(C_0(X))$.

(i) Prove that $C_0(X)$ is an essential ideal in $C_b(X)$. Conclude that there is an isometric $*$-homomorphism

$$\Phi : C_b(X) \to M(C_0(X))$$

such that $\Phi(g) = g$ for all $g \in C_0(X)$.

(ii) Let $(L, R) \in M(C_0(X))$ and let $x \in X$ be fixed. If $f_1, f_2 \in C_0(X)$ are two functions such that $f_1(x) \neq 0$ and $f_2(x) \neq 0$, then prove that

$$\frac{L(f_1)(x)}{f_1(x)} = \frac{L(f_2)(x)}{f_2(x)}.$$

(iii) Fix $(L, R) \in M(C_0(X))$. For each $x \in X$, choose a function $f_x \in C_0(X)$ such that $f_x(x) \neq 0$. Define $h : X \to \mathbb{C}$ by

$$h(x) := \frac{L(f_x)(x)}{f_x(x)}.$$

Prove that h is a well-defined, $h \in C_b(X)$ and that $\Phi(h) = (L, R)$.

Hint: You will need Urysohn's Lemma (Theorem 7.2.3).

Conclude that $\Phi : C_b(X) \to M(C_0(X))$ is an isometric $*$-isomorphism.

Additional Reading

- When C*-algebras were first defined by Gelfand and Naimark in 1943 [21], the axioms they proposed were somewhat different from the ones that are in use today. It then took the work of many mathematicians for the definitions to reach their final form (which they did around 1960). For a 'behind-the-scenes' look at the development of C*-algebra theory (and the Gelfand–Naimark Theorems in particular), have a look at this very interesting book by Doran and Belfi.

Doran, Robert S. and Belfi, Victor A. (1986). *Characterizations of C*-algebras*. Vol. 101. Monographs and Textbooks in Pure and Applied Mathematics. xi+426. New York: Marcel Dekker, Inc. ISBN: 0-8247-7569-4.

- When Gelfand first studied normed rings, one of his prime interests was the study of abstract Harmonic analysis. Today, the two subjects are intimately related through the study of group C*-algebras and crossed products. Perhaps the most inspiring place to find the roots of these ideas is this delightful book from 1946.

Gelfand, Israel Moiseevich, Raikov, Dmitri Abramovich and Šilov, Georgiy Evgenievich (1946). Commutative normed rings. *Akademiya Nauk SSSR i Moskovskoe Matematicheskoe Obshchestvo. Uspekhi Matematicheskikh Nauk*, 1.2(12), 48–146. ISSN: 0042-1316.

Chapter 9

Measure and Integration

9.1 Positive Linear Functionals on C*-Algebras

In our quest to understand operators on a Hilbert space, we now take a detour. The purpose of this chapter is to prove a deep relationship between measures and linear functionals. The main theorem in this context is the Riesz–Markov–Kakutani Theorem, one of the most important theorems in modern analysis (this theorem is also referred to as the Riesz Representation Theorem, but we choose to use the longer name so as to prevent any confusion with Theorem 3.2.2). We will then use this result to identify the dual space of $C(X)$, the space of continuous functions on a compact metric space X.

We begin by studying *positivity* in the context of C*-algebras; a notion that was baked-in to the theory by the founding fathers of the subject.

9.1.1 DEFINITION An element a in a C*-algebra $\mathbf{A}$ is said to be *positive* if it is self-adjoint and $\sigma(a) \subset [0, \infty)$. We write $\mathbf{A}_+$ for the set of all positive elements in $\mathbf{A}$ and write $a \geq 0$ if $a \in \mathbf{A}_+$.

The next lemma tells us where to look for positive elements and allows us to construct some simple examples.

9.1.2 LEMMA If $a \in \mathbf{A}$ is positive, then there exists $b \in \mathbf{A}$ such that $b^*b = a$.

Proof. Suppose first that $\mathbf{A}$ is unital and let $f \mapsto f(a)$ denote the continuous functional calculus. Since $\sigma(a) \subset [0, \infty)$, $g(t) := t^{1/2}$ defines a continuous function on $\sigma(a)$ and $b := g(a)$ satisfies the required condition.

If $\mathbf{A}$ is non-unital, then we may apply the same argument in the unitalization $\mathbf{A}^u$. Since $g(0) = 0$, Remark 8.5.4 tells us that $b \in \mathbf{A}$. □

9.1.3 EXAMPLE

(i) Let $\mathbf{A} = C(X)$ denote the space of continuous functions on a compact Hausdorff space X. For any $f \in \mathbf{A}$, $\sigma(f) = \{f(x) : x \in X\}$. Therefore $f \in \mathbf{A}_+$ if and only if $f(x) \geq 0$ for all $x \in X$.

(ii) Let $\mathbf{A} = \mathcal{B}(\mathbf{H})$ denote the algebra of bounded operators on a Hilbert space. If $T \in \mathcal{B}(\mathbf{H})$ is positive, then there exists $S \in \mathcal{B}(\mathbf{H})$ such that $T = S^*S$. Therefore for any $x \in \mathbf{H}$,

$$\langle T(x), x \rangle = \|S(x)\|^2 \geq 0.$$

In fact, this property characterizes positive elements in $\mathcal{B}(\mathbf{H})$. To see this, suppose $T \in \mathcal{B}(\mathbf{H})$ is such that $\langle T(x), x \rangle \geq 0$ for each $x \in \mathbf{H}$. Then by Proposition 8.1.9, T is self-adjoint. Hence

$$\sigma(T) \subset \mathbb{R}.$$

Now suppose $\lambda \in \mathbb{R}$ is negative, then

$$\begin{aligned}
\|(T - \lambda I)x\|^2 &= \langle (T - \lambda I)x, Tx \rangle - \lambda \langle (T - \lambda I)x, x \rangle \\
&= \|Tx\|^2 - \lambda \langle x, Tx \rangle - \lambda \langle Tx, x \rangle + \lambda^2 \|x\|^2 \\
&\geq \lambda^2 \|x\|^2.
\end{aligned}$$

Hence $(T - \lambda I)$ is bounded below and $\lambda \notin \sigma_{ap}(T)$. However, T is normal so $\sigma_{ap}(T) = \sigma(T)$ by Corollary 8.1.16. We conclude that $\lambda \notin \sigma(T)$ and thus

$$\sigma(T) \subset [0, \infty).$$

Hence T is positive.

(iii) If $\mathbf{A}$ is a C*-algebra and $a \in \mathbf{A}$ is self-adjoint, then $\sigma(a) \subset \mathbb{R}$. By the Spectral Mapping Theorem, we conclude that $\sigma(a^2) \subset [0, \infty)$. Since a^2 is also self-adjoint, $a^2 \in \mathbf{A}_+$.

The main theorem we are after is the converse of Lemma 9.1.2 and the next two lemmas help us get there.

9.1.4 LEMMA *Let $\mathbf{A}$ be a C*-algebra.*

(i) *If $a, b \in \mathbf{A}_+$, then $a + b \in \mathbf{A}_+$.*

(ii) *$\mathbf{A}_+$ is a closed in $\mathbf{A}$.*

Proof. Once again, we assume without loss of generality that $\mathbf{A}$ is unital.

(i) Note that $c \in \mathbf{A}_+$ if and only if $\lambda c \in \mathbf{A}_+$ for all $\lambda > 0$. Therefore we may assume without loss of generality that $\|a\| \leq 1$ and $\|b\| \leq 1$. We set $z := \frac{a+b}{2}$ and prove that $z \in \mathbf{A}_+$. Note that z is self-adjoint, so that $\sigma(z) \subset \mathbb{R}$ by Proposition 8.4.10. Since $a \in \mathbf{A}_+$ and $\|a\| \leq 1$, it follows by the Spectral Mapping Theorem that $\sigma(1 - a) \subset [-1, 1]$. Hence $\|1 - a\| = r(1 - a) \leq 1$. Similarly, $\|1 - b\| \leq 1$. Therefore

$$\|1 - z\| = \left\| \frac{(1 - a) + (1 - b)}{2} \right\| \leq 1.$$

Hence if $t \in \sigma(z)$, then $|1 - t| \leq 1$. Since $t \in \mathbb{R}$, this implies $t \geq 0$. Thus $\sigma(z) \subset [0, \infty)$, proving that z is a positive element. Therefore $(a + b) \in \mathbf{A}_+$ as well.

(ii) Suppose (a_n) is a sequence in $\mathbf{A}_+$ such that $a_n \to a$. Since each a_n is self-adjoint, so is a and $\sigma(a) \subset \mathbb{R}$. Since (a_n) is bounded, we may assume that $\|a_n\| \leq 1$ for all $n \in \mathbb{N}$. As in part (i), it follows that $\|1 - a_n\| \leq 1$ for all $n \in \mathbb{N}$ and thus $\|1 - a\| \leq 1$. As before, we conclude that $\sigma(a) \subset [0, \infty)$. □

For the next lemma, we need the following result from Exercise 7.30. If $a, b \in \mathbf{A}$ are any two elements of a Banach algebra, then

$$\sigma(ab) \cup \{0\} = \sigma(ba) \cup \{0\}. \tag{9.1}$$

9.1.5 LEMMA *Let $a \in \mathbf{A}$ such that $-a^*a \in \mathbf{A}_+$. Then $a = 0$.*

Proof.

(i) Suppose first that a is self-adjoint. In that case, $-a^2 \in \mathbf{A}_+$. However, $\sigma(a) \subset \mathbb{R}$, so by the continuous functional calculus, $\sigma(a^2) \subset [0, \infty)$. We conclude that

$$\sigma(a^2) = \{0\}$$

Since a^2 is self-adjoint,

$$\|a\|^2 = \|a^2\| = r(a^2) = 0$$

and thus $a = 0$.

(ii) Now for the general case, suppose $-a^*a \in \mathbf{A}_+$, then $\sigma(a^*a) \subset (-\infty, 0]$. By Equation 9.1,

$$\sigma(aa^*) \subset (-\infty, 0]$$

as well. Hence $-aa^* \in \mathbf{A}_+$, so $-(a^*a + aa^*) \in \mathbf{A}_+$ by Lemma 9.1.4. Therefore

$$\sigma(a^*a + aa^*) \subset (-\infty, 0].$$

Write $a = b + ic$ where b and c self-adjoint. Then

$$a^*a + aa^* = (b - ic)(b + ic) + (b + ic)(b - ic) = 2b^2 + 2c^2.$$

Hence $-(b^2 + c^2) \in \mathbf{A}_+$. By Example 9.1.3, $c^2 \in \mathbf{A}_+$ and therefore $-b^2 \in \mathbf{A}_+$ by Lemma 9.1.4. By part (i), $b = 0$. Similarly, $c = 0$ and this forces $a = 0$. $\square$

9.1.6 THEOREM *For $a \in \mathbf{A}$, the following are equivalent:*

(i) *a is positive, in the sense that $a = a^*$ and $\sigma(a) \subset [0, \infty)$.*

(ii) *There exists $b \in \mathbf{A}$ such that $a = b^*b$.*

(iii) *There is a self-adjoint element $x \in \mathbf{A}$ such that $a = x^2$.*

(iv) *There is a unique positive element $z \in \mathbf{A}$ such that $a = z^2$.*

In particular,

$$\mathbf{A}_+ = \{b^*b : b \in \mathbf{A}\}.$$

Proof. We will assume that $\mathbf{A}$ is unital, so as to have access to the continuous functional calculus. The case when $\mathbf{A}$ is non-unital is left as an exercise for you to try (with the help of Remark 8.5.4). Note that $(i) \Rightarrow (ii)$ was proved in Lemma 9.1.2, while both $(iii) \Rightarrow (i)$ and $(iv) \Rightarrow (i)$ follow from the Spectral Mapping Theorem. We now prove the remaining implications.

$(ii) \Rightarrow (iii)$: If $a = b^*b$, then $a = a^*$ and $\sigma(a) \subset \mathbb{R}$. Define two functions $f, g \in C(\sigma(a))$ by

$$f(t) := \begin{cases} \sqrt{t} & : \text{if } t \geq 0 \\ 0 & : \text{if } t < 0 \end{cases} \quad \text{and} \quad g(t) := \begin{cases} 0 & : \text{if } t \geq 0 \\ \sqrt{-t} & : \text{if } t < 0 \end{cases}.$$

Let $x := f(a)$ and $y := g(a)$. Then x and y are self-adjoint and

$$a = x^2 - y^2$$

by the functional calculus. Furthermore $f(t)g(t) = 0$ for all $t \in \sigma(a)$, so $xy = yx = 0$. Therefore

$$(by)^*(by) = yb^*by = yay = yx^2y - y^4 = -y^4.$$

By Lemma 9.1.5, it follows that $y^4 = 0$. This implies that $y = (y^4)^{1/4} = 0$. Hence $a = x^2$.

$(ii) \Rightarrow (iv)$: In the proof of $(ii) \Rightarrow (iii)$, the element $x = f(a)$ is positive by the Spectral Mapping Theorem (since $f(t) \geq 0$ for all $t \in \sigma(a)$). Now if $z \in \mathbf{A}_+$ is another positive element such that $z^2 = a$, then z must commute with a and must therefore commute with any polynomial in a. Since x is obtained as a limit of such polynomials, it follows that $xz = zx$. Now let $\mathbf{B}$ denote the C*-subalgebra of $\mathbf{A}$ generated by $\{1, z, x\}$. Then $\mathbf{B}$ is commutative, unital and must contain a. Let

$$\Gamma : \mathbf{B} \to C(\Omega(\mathbf{B}))$$

be the Gelfand–Naimark isomorphism. Then $\Gamma(x)$ and $\Gamma(z)$ are both positive functions in $C(\Omega(\mathbf{B}))$ with the property that $\Gamma(x)^2 = \Gamma(z)^2 = \Gamma(a)$. However, any function in $C(\Omega(\mathbf{B}))$ has exactly one positive square root. Therefore $\Gamma(x) = \Gamma(z)$ and thus $x = z$. Hence a has a unique positive square root in $\mathbf{A}$. $\square$

The next definition will be useful to us in the study of operators on Hilbert spaces and we record it here for convenience.

9.1.7 DEFINITION Let $\mathbf{A}$ be a C*-algebra.

(i) If $c \in \mathbf{A}_+$, then there is a unique element $b \in \mathbf{A}_+$ such that $b^2 = c$. This element is called the *square root* of c and is denoted by $\sqrt{c}$.

(ii) If $a \in \mathbf{A}$, then $a^*a \in \mathbf{A}_+$ is a positive element. Therefore we may define $|a| := \sqrt{a^*a}$.

Let us now take a closer look at the set of self-adjoint elements in a C*-algebra. The notion of positivity allows us to define a partial order on this set, which gives it added structure.

9.1.8 DEFINITION For a C*-algebra $\mathbf{A}$, we write $\mathbf{A}_{sa}$ for the set of all self-adjoint elements in $\mathbf{A}$. For two elements $a, b \in \mathbf{A}_{sa}$, we write $a \leq b$ if $(b - a) \in \mathbf{A}_+$.

Notice that this relation is clearly reflexive. It is transitive because of Lemma 9.1.4 and it is anti-symmetric because of Theorem 9.1.6 (do verify this). Therefore $\mathbf{A}_{sa}$ is a partially ordered set. It is also a real vector space and the vector space operations respect the partial order in the sense that

(a) $a \leq b$ implies that $a + c \leq b + c$ for any $c \in \mathbf{A}_{sa}$.

(b) $a \leq b$ implies that $\lambda a \leq \lambda b$ for all $\lambda \geq 0$.

This turns $\mathbf{A}_{sa}$ into a ***partially ordered real vector space*** and it informs our study of C*-algebras to a great extent. In a sense, $\mathbf{A}_{sa}$ plays a role in $\mathbf{A}$ analogous to the role that $\mathbb{R}$ plays in $\mathbb{C}$. Furthermore, the set $\mathbf{A}_+$ plays a role analogous to that of $\mathbb{R}_+$. The next lemma is one such example.

9.1.9 LEMMA *Let a be a self-adjoint element in a C*-algebra $\mathbf{A}$. Then there exist positive elements a_+ and a_- such that $a = a_+ - a_-$ and $a_+ a_- = a_- a_+ = 0$. Furthermore, $\|a_+\| \leq \|a\|$ and $\|a_-\| \leq \|a\|$.*

Proof.   Assume first that $\mathbf{A}$ is unital. Since $\sigma(a) \subset \mathbb{R}$, we consider the continuous functional calculus applied to the functions

$$f(t) = \frac{|t| + t}{2} \text{ and } g(t) = \frac{|t| - t}{2}.$$

If $a_+ = f(a)$ and $a_- = g(a)$, then both a_+ and a_- are positive by the Spectral Mapping Theorem and they satisfy the required conditions.

Now if $\mathbf{A}$ is non-unital, then $0 \in \sigma(a)$. Since $f(0) = g(0) = 0$, it follows that both a_+ and a_- are elements of $\mathbf{A}$ by Remark 8.5.4. $\square$

9.1.10 DEFINITION A linear map $\theta : \mathbf{A} \to \mathbf{B}$ between two C*-algebras is said to be ***positive*** if $\theta(\mathbf{A}_+) \subset \mathbf{B}_+$.

Notice that if θ is positive, then $\theta(\mathbf{A}_{sa}) \subset \mathbf{B}_{sa}$ by Lemma 9.1.9 and the restricted map $\theta : \mathbf{A}_{sa} \to \mathbf{B}_{sa}$ is *order preserving* in the sense that $\theta(a) \leq \theta(b)$ in $\mathbf{B}_{sa}$ whenever $a \leq b$ in $\mathbf{A}_{sa}$.

9.1.11 EXAMPLE

(i) If $\Phi : \mathbf{A} \to \mathbf{B}$ is a $*$-homomorphism, then $\Phi(a^*a) = \Phi(a)^*\Phi(a)$ is positive in $\mathbf{B}$ for any $a \in \mathbf{A}$. By Theorem 9.1.6, $\Phi(\mathbf{A}_+) \subset \mathbf{B}_+$. Therefore, every $*$-homomorphism is positive.

(ii) Let X be a compact Hausdorff space and $\mathbf{A} = C(X)$. If μ is any positive, finite, Borel measure on X, then the map $\tau : \mathbf{A} \to \mathbb{C}$ given by

$$\tau(f) := \int_X f d\mu$$

is a positive linear functional on $\mathbf{A}$. Note that such a linear functional is not, in general, multiplicative.

(iii) Let $\mathbf{H}$ be a Hilbert space and $\mathbf{A} = \mathcal{B}(\mathbf{H})$. If $x \in \mathbf{H}$, then the map $\tau : \mathbf{A} \to \mathbb{C}$ given by

$$\tau(T) := \langle T(x), x \rangle$$

is a positive linear functional (by Example 9.1.3).

(iv) If θ_1 and θ_2 are two positive linear maps between two C*-algebras and $\lambda \in \mathbb{R}_+$, then $\lambda\theta_1 + \theta_2$ is also positive. This follows from Lemma 9.1.4.

(v) As a special case, the trace map $\tau : M_n(\mathbb{C}) \to \mathbb{C}$ given by

$$\tau(T) = \sum_{i=1}^{n} \langle T(e_i), e_i \rangle$$

is a positive linear functional on $M_n(\mathbb{C})$.

9.1.12 PROPOSITION *A positive linear functional on a C*-algebra is bounded.*

Proof. Let $\tau : \mathbf{A} \to \mathbb{C}$ be a positive linear functional on a C*-algebra $\mathbf{A}$. Let $S = \{a \in \mathbf{A}_+ : \|a\| \leq 1\}$ denote the positive part of the unit ball of $\mathbf{A}$.

(i) We first show that τ is bounded on S. Suppose not, then there is a sequence (a_n) in S such that $\tau(a_n) \geq 2^n$ for each $n \in \mathbb{N}$. Then the series

$$a := \sum_{n=1}^{\infty} \frac{a_n}{2^n}$$

converges in $\mathbf{A}$ (since it converges absolutely). Also, a is a limit of the partial sums of the series, so $a \in \mathbf{A}_+$ by Lemma 9.1.4. Since τ is order preserving,

$$\tau(a) \geq \tau\left(\sum_{n=1}^{N} \frac{a_n}{2^n}\right) \geq N$$

for each $N \in \mathbb{N}$. This is clearly impossible, so τ must be bounded on S.

(ii) Now suppose $M := \sup\{\tau(a) : a \in S\}$, then $M < \infty$. If $a \in \mathbf{A}$ with $\|a\| \leq 1$, we may write $a = b + ic$ where b and c are self-adjoint elements with $\|b\| \leq 1$ and $\|c\| \leq 1$. By Lemma 9.1.9,

$$a = b_+ - b_+ + ic_+ - ic_-$$

and b_+, b_-, c_+ and c_- are all in S. Therefore $|\tau(a)| \leq 4M$ and we conclude that τ is bounded. $\qquad\square$

We now collect some important properties of positive linear functionals. To begin with, we need the definition of a sesquilinear form (see Definition 8.7.1). If $\mathbf{A}$ is a complex vector space, a function $u : \mathbf{A} \times \mathbf{A} \to \mathbb{C}$ is said to be a *sesquilinear form* if it is linear in the first variable and conjugate-linear in the second variable. Furthermore, u is said to be *bounded* if there is $M > 0$ such that $|u(a,b)| \leq M\|a\|\|b\|$ for all $a, b \in \mathbf{A}$.

9.1.13 PROPOSITION *Let $\tau : \mathbf{A} \to \mathbb{C}$ be a positive linear functional on a C^*-algebra $\mathbf{A}$.*

(i) *For any $a \in \mathbf{A}$, $\tau(a^*) = \overline{\tau(a)}$.*

(ii) *The map $(a,b) \mapsto \tau(b^*a)$ is a bounded sesquilinear form on $\mathbf{A}$.*

(iii) *If $a, b \in \mathbf{A}$, then*

$$|\tau(b^*a)| \leq \tau(a^*a)^{1/2}\tau(b^*b)^{1/2}.$$

Proof.

(i) If $x \in \mathbf{A}$ is self-adjoint, then $x = x_+ - x_-$ for two positive elements $x_+, x_- \in$ $\mathbf{A}$. Therefore $\tau(x) = \tau(x_+) - \tau(x_-) \in \mathbb{R}$. Now if $a \in \mathbf{A}$, then we write $a = b + ic$ for two self-adjoint elements $b, c \in \mathbf{A}$. Then $a^* = b - ic$. Since $\tau(b)$ and $\tau(c)$ are both real numbers, it follows that

$$\tau(a^*) = \tau(b) - i\tau(c) = \overline{\tau(b) + i\tau(c)} = \overline{\tau(a)}.$$

(ii) Consider $u : \mathbf{A} \times \mathbf{A} \to \mathbb{C}$ by $u(a, b) := \tau(b^*a)$. This map is linear in the first variable because τ is linear. It is conjugate-linear in the second variable because $\tau(b^*a) = \overline{\tau(a^*b)}$ by part (i). That u is bounded follows from the fact that τ is bounded.

(iii) This is merely the Cauchy–Schwarz Inequality whose proof goes through *mutatis mutandis* for sesquilinear forms. $\qquad\square$

The next theorem gives us a rather simple way to identify positive linear functionals and it allows us to prove a very useful extension theorem.

9.1.14 THEOREM *Let $\mathbf{A}$ be a unital C*-algebra. A linear functional $\tau : \mathbf{A} \to \mathbb{C}$ is positive if and only if $\tau(1_\mathbf{A}) = \|\tau\|$.*

Proof.

(i) Suppose τ is positive and let $a \in \mathbf{A}$ with $\|a\| \leq 1$. By the Cauchy–Schwarz Inequality from Proposition 9.1.13,

$$|\tau(a)|^2 = |\tau(1_\mathbf{A}^*a)|^2 \leq \tau(a^*a)\tau(1_\mathbf{A}).$$

Now $a^*a \in \mathbf{A}_+$ and $\|a^*a\| \leq 1$. Therefore $0 \leq a^*a \leq 1_\mathbf{A}$. Hence

$$|\tau(a)|^2 \leq \tau(1_\mathbf{A})^2.$$

This is true for all $a \in \mathbf{A}$ with $\|a\| \leq 1$, so $\|\tau\| \leq \tau(1_\mathbf{A})$. The reverse inequality is obvious because $\|1_\mathbf{A}\| = 1$.

(ii) Now suppose $\|\tau\| = \tau(1_\mathbf{A})$. We wish to prove that τ is positive. We assume (by scaling if need be) that $\tau(1_\mathbf{A}) = 1$ and we break the proof into two steps.

(a) If $a \in \mathbf{A}_{sa}$, we claim that $\tau(a) \in \mathbb{R}$. To see this, first assume without loss of generality that $\|a\| \leq 1$ and write $\tau(a) = \alpha + i\beta$ for some $\alpha, \beta \in \mathbb{R}$.

Replacing a by $-a$ if need be, we may also assume that $\beta \leq 0$. Now for any $n \in \mathbb{N}$,

$$\|a - in\|^2 = \|(a - in)^*(a - in)\| = \|(a + in)(a - in)\|$$
$$= \|a^2 + n^2\| \leq \|a^2\| + n^2$$
$$\leq 1 + n^2.$$

Since $\tau(1_\mathbf{A}) = 1$,

$$|\alpha + i\beta - in|^2 = |\tau(a - in)|^2 \leq \|\tau\| \|a - in\|^2 \leq 1 + n^2.$$

If $\beta < 0$, this cannot hold for all $n \in \mathbb{N}$ and thus it must happen that $\beta = 0$. Therefore $\tau(a) \in \mathbb{R}$ for all $a \in \mathbf{A}_{sa}$.

(b) Now suppose $a \in \mathbf{A}_+$, then we wish to prove that $\tau(a) \geq 0$. Once again, we may assume that $\|a\| \leq 1$. As in the proof of Lemma 9.1.4, $\|1 - a\| \leq 1$ as well. Therefore

$$|1 - \tau(a)| = |\tau(1_\mathbf{A}) - \tau(a)| \leq \|1_\mathbf{A} - a\| \leq 1.$$

Since $\tau(a) \in \mathbb{R}$, it must happen that $\tau(a) \geq 0$. Thus τ is positive. $\qquad\square$

9.1.15 COROLLARY *Let $\mathbf{A}$ be a unital C*-algebra and $\mathbf{B} \subset \mathbf{A}$ be a C*-subalgebra of $\mathbf{A}$ such that $1_\mathbf{A} \in \mathbf{B}$. If $\tau : \mathbf{B} \to \mathbb{C}$ is a positive linear functional, then there is a positive linear functional $\widetilde{\tau} : \mathbf{A} \to \mathbb{C}$ such that $\widetilde{\tau}|_\mathbf{B} = \tau$ and $\|\tau\| = \|\widetilde{\tau}\|$.*

Proof. Let $\widetilde{\tau} : \mathbf{A} \to \mathbb{C}$ be a norm-preserving extension of τ given by the Hahn–Banach Theorem. Then

$$\|\widetilde{\tau}\| = \|\tau\| = \tau(1_\mathbf{A}) = \widetilde{\tau}(1_\mathbf{A}).$$

By Theorem 9.1.14, $\widetilde{\tau}$ is positive as well. $\qquad\square$

9.2   The Riesz–Markov–Kakutani Theorem

Given a compact Hausdorff space X and a finite, positive Borel measure μ, we may define a positive linear functional $\tau : C(X) \to \mathbb{C}$ by the formula

$$\tau(f) = \int_X f \, d\mu.$$

The goal of this section is to prove that *every* positive linear functional on $C(X)$ is of this form. While the statement of the theorem is true in more generality, we will prove a somewhat weaker version of the theorem. The reason is that we would like to avoid certain technicalities and for our purposes, this will suffice. Here is the theorem we wish to prove.

9.2.1 THEOREM (RIESZ, 1909, MARKOV, 1938, AND KAKUTANI, 1941) *Let X be a compact Hausdorff space and $\tau : C(X) \to \mathbb{C}$ be a positive linear functional. Then there is a unique finite, positive Baire measure μ on X such that*

$$\tau(f) = \int_X f d\mu \tag{9.2}$$

for all $f \in C(X)$.

The definition of a Baire measure is given below.

9.2.2 DEFINITION Let X be a compact Hausdorff space.

(i) The **Baire σ-algebra** $\mathfrak{A}_X$ is the smallest σ-algebra which makes every $f \in C(X)$ measurable. (i.e. $\mathfrak{A}_X$ is generated by sets of the form $f^{-1}(F)$, where $F \subset \mathbb{C}$ is closed and $f \in C(X)$). Members of $\mathfrak{A}_X$ are called **Baire sets**.

(ii) We denote the Borel σ-algebra (generated by all closed sets) by $\mathfrak{B}_X$. Note that

$$\mathfrak{A}_X \subset \mathfrak{B}_X.$$

The inclusion is strict in general. However, if X is a metric space, then $\mathfrak{A}_X = \mathfrak{B}_X$, because if $F \subset X$ is closed, then $F = f^{-1}(\{0\})$ where $f \in C(X)$ is the function $x \mapsto d(x, F)$.

(iii) A finite, positive measure μ is called a **Baire measure** if its domain contains $\mathfrak{A}_X$. Note that if μ is a Baire measure, then any continuous function is $\mathfrak{A}_X$-measurable. Hence the positive linear functional $\tau : C(X) \to \mathbb{C}$ defined by Equation 9.2 makes sense.

We now give an alternative description of $\mathfrak{A}_X$ that will be useful to us going forward. Recall that a G_δ set is the intersection of countably many open sets.

9.2.3 LEMMA $\mathfrak{A}_X$ *is the smallest σ-algebra containing all compact G_δ sets in X.*

Proof. Let K be a compact G_δ subset of X and let $\{U_1, U_2, \dots\}$ be a countable family of open sets such that $K = \bigcap_{n=1}^{\infty} U_n$. Furthermore, we may choose these sets so that $U_{n+1} \subset U_n$ for each $n \in \mathbb{N}$. Urysohn's Lemma then ensures that there is a continuous function $f_n : X \to [0,1]$ such that $f_n \equiv 1$ on K and $f_n \equiv 0$ on $X \setminus U_n$. Let $f : X \to \mathbb{C}$ be given by

$$f = \sum_{n=1}^{\infty} \frac{f_n}{2^n}.$$

Then $f \in C(X), f \equiv 1$ on K and $f(x) < 1$ whenever $x \notin K$. Hence if $g_n := f^n$, then (g_n) is a sequence of continuous functions such that $g_n \to \chi_K$ pointwise (Compare this with the proof of Proposition 2.3.12). Thus χ_K is the pointwise limit of a sequence of $\mathfrak{A}_X$-measurable functions and is thus $\mathfrak{A}_X$-measurable itself. In other words, $K \in \mathfrak{A}_X$. If $\mathfrak{A}_0$ denotes the smallest σ-algebra containing all compact G_δ sets, then we have just proved that $\mathfrak{A}_0 \subset \mathfrak{A}_X$.

Conversely, if $F \subset \mathbb{C}$ is a closed set, then F is a G_δ because

$$F = \bigcap_{n=1}^{\infty} \left\{ t \in \mathbb{C} : \inf_{y \in F} |t - y| < \frac{1}{n} \right\}.$$

Therefore if $f \in C(X)$, then $f^{-1}(F)$ is a G_δ that is closed in X and hence compact. In other words, $f^{-1}(F) \in \mathfrak{A}_0$. Hence $\mathfrak{A}_X \subset \mathfrak{A}_0$ as well. $\square$

We begin our proof of Theorem 9.2.1 by proving the uniqueness of the measure. For the proof, we need one fact you may have seen before (typically during the proof of the Fubini–Tonelli Theorem). It is equivalent to the Monotone Class Theorem and you will find a proof of it in this form in Billingsley [4, Theorem 3.2].

9.2.4 THEOREM (DYNKIN $\pi - \lambda$ THEOREM) *Let $\mathcal{P}$ and $\mathcal{L}$ be two collections of subsets of a set X. Assume that*

(a) *$\mathcal{P}$ is closed under finite intersections: If $E, F \in \mathcal{P}$, then $E \cap F \in \mathcal{P}$.*

(b) *$\varnothing \in \mathcal{L}$.*

(c) *If $E \in \mathcal{L}$, then $X \setminus E \in \mathcal{L}$.*

(d) *$\mathcal{L}$ is closed under countable disjoint unions: If $\{E_1, E_2, \ldots\}$ is a countable family of mutually disjoints sets in $\mathcal{L}$, then $\bigcup_{n=1}^{\infty} E_n \in \mathcal{L}$.*

(e) *$\mathcal{P} \subset \mathcal{L}$.*

Then the σ-algebra generated by $\mathcal{P}$ is contained in $\mathcal{L}$.

A collection satisfying condition (a) is called a π-system, while one satisfying conditions $(b), (c)$ and (d) is called a λ-system, which explains the name of the theorem.

9.2.5 LEMMA *Let X be a compact Hausdorff space and let μ and ν be two finite, positive, Baire measures such that*

$$\int_X f d\mu = \int_X f d\nu$$

for all $f \in C(X)$. Then $\mu \equiv \nu$ on $\mathfrak{A}_X$.

Proof. With a view to applying the Dynkin $\pi - \lambda$ Theorem, set

$$\mathcal{P} := \{K \subset X : K \text{ is a compact } G_\delta \text{ set}\} \text{ and}$$

$$\mathcal{L} := \{E \in \mathfrak{A}_X : \mu(E) = \nu(E)\}.$$

Then

(a) $\mathcal{P}$ is clearly closed under finite intersections.

(b) Clearly, $\varnothing \in \mathcal{L}$.

(c) Since $\mathbf{1} \in C(X)$, $\mu(X) = \nu(X) < \infty$. Therefore $\mathcal{L}$ is also closed under taking complements.

(d) Since both μ and ν are countably additive, it follows that $\mathcal{L}$ is closed under taking countable disjoint unions as well.

(e) Let K be a compact G_δ subset of X. Then by the proof of Lemma 9.2.3, there is a sequence (g_n) in $C(X)$ such that $g_n \to \chi_K$ pointwise and $\|g_n\| \leq 1$ for all $n \in \mathbb{N}$. Since μ and ν are both finite measures, the Dominated Convergence Theorem implies that

$$\mu(K) = \lim_{n \to \infty} \int_X g_n d\mu = \lim_{n \to \infty} \int_X g_n d\nu = \nu(K).$$

Hence $\mathcal{P} \subset \mathcal{L}$.

By the Dynkin $\pi - \lambda$ Theorem, $\mathcal{L}$ must contain the σ-algebra generated by $\mathcal{P}$. Lemma 9.2.3 now completes the proof. $\square$

We now turn to the existence of the measure. The idea of the proof is to first prove it for a special class of spaces and then extend it to all compact Hausdorff spaces using the results of the previous section.

9.2.6 DEFINITION A topological space X is said to be ***extremally disconnected*** if the closure of every open set is open.

Note that every discrete space is extremally disconnected. We now examine the space of continuous functions on an extremally disconnected space. As is customary, we will use the term *cl-open* to denote sets that are both closed and open. Before we begin, recall the following notion you would be familiar with.

9.2.7 Definition A collection $\mathcal{U}$ of subsets of a set X is called an *algebra* if it satisfies the following conditions.

(a) If $E \in \mathcal{U}$, then $X \setminus E \in \mathcal{U}$.

(b) If $E, F \in \mathcal{U}$, then $E \cup F \in \mathcal{U}$.

9.2.8 Lemma *Let X be an extremally disconnected, compact Hausdorff space. Then*

(i) *The collection $\mathcal{U}$ of all cl-open sets in X is an algebra.*

(ii) *The set $\mathcal{G} := \mathrm{span}\{\chi_E : E \in \mathcal{U}\}$ is dense in $C(X)$*

Proof. Part (i) is entirely obvious. For Part (ii), we verify that $\mathcal{G}$ satisfies the hypotheses of the Stone–Weierstrass Theorem (Section A.1).

(i) Clearly, $\mathcal{G}$ contains the constant function $\mathbf{1}$.

(ii) If $x, y \in X$ are distinct, then there exist open sets $U, V \subset Y$ such that $x \in U, y \in V$ and $U \cap V = \emptyset$. Then $E := \overline{U}$ is a cl-open set and $y \notin E$. Hence $\chi_E \in \mathcal{G}$ separates x from y.

(iii) If $f \in \mathcal{G}$, then $\overline{f} \in \mathcal{G}$ since each characteristic function is real-valued.

Thus $\mathcal{G}$ is dense in $C(X)$.   □

For the proof of the next result, we need the Carathéodory Extension Theorem. If you have not seen it before, you will find a detailed proof in Folland [18, Theorem 1.14].

9.2.9 Theorem (Carathéodory) *Let $\mathcal{U}$ be an algebra of subsets of a set X and let $\mu_0 : \mathcal{U} \to [0, \infty]$ be a function satisfying the following conditions.*

(a) $\mu_0(\emptyset) = 0$.

(b) *If $\{E_1, E_2, \ldots\}$ is a sequence of mutually disjoint sets in $\mathcal{U}$ such that $E := \bigcup_{n=1}^{\infty} E_n \in \mathcal{U}$, then $\mu_0(E) = \sum_{n=1}^{\infty} \mu_0(E_n)$.*

(c) $\mu_0(X) < \infty$.

Then there exists a unique finite measure μ defined on the σ-algebra generated by $\mathcal{U}$ such that $\mu|_{\mathcal{U}} = \mu_0$.

9.2.10 THEOREM *Let X be an extremally disconnected, compact Hausdorff space. If τ : $C(X) \to \mathbb{C}$ is a positive linear functional, then there is a unique finite, positive Baire measure μ on X such that*

$$\tau(f) = \int_X f d\mu$$

for all $f \in C(X)$.

Proof.   Uniqueness was proved in Lemma 9.2.5, so we only prove existence.

(i) Let $\mathcal{U}$ be the algebra of cl-open subsets of X as in Lemma 9.2.8. Define μ_0 : $\mathcal{U} \to [0, \infty)$ by

$$\mu_0(E) := \tau(\chi_E).$$

Then μ_0 is well-defined, $\mu_0(\varnothing) = 0$ and μ_0 is finitely additive (because τ is linear). Let $\{E_1, E_2, \dots, \}$ be a countable collection of mutually disjoint sets in $\mathcal{U}$ such that $E = \bigcup_{n=1}^{\infty} E_n \in \mathcal{U}$. Since all the sets are open and compact, all but finitely many E_i must be empty. Hence

$$\mu_0(E) = \sum_{n=1}^{\infty} \mu_0(E_n)$$

since μ_0 is finitely additive. Finally, $\mu_0(X) = \tau(\mathbf{1}) < \infty$. By Carathéodory's Theorem, μ_0 extends to a finite, positive measure $\mu : \mathfrak{M}(\mathcal{U}) \to [0, \infty)$, where $\mathfrak{M}(\mathcal{U})$ denotes the σ-algebra generated by $\mathcal{U}$.

(ii) We claim that $\mathfrak{A}_X \subset \mathfrak{M}(\mathcal{U})$. If $f \in C(X)$, we wish to prove that $f^{-1}(F) \in \mathfrak{M}(\mathcal{U})$ for any closed set $F \subset \mathbb{C}$. Since the set of all $\mathfrak{M}(\mathcal{U})$-measurable functions forms a vector space, we may assume that f is real-valued and that $f \geq 0$. Furthermore, since $\mathfrak{M}(\mathcal{U})$ is a σ-algebra, it suffices to show that

$$f^{-1}(-\infty, c] \in \mathfrak{M}(\mathcal{U})$$

for each $c \in \mathbb{R}$. So fix $c \in \mathbb{R}$ and let $E := f^{-1}(-\infty, c]$. For each $n \in \mathbb{N}$, set $E_n := \{x \in X : f(x) < c + 1/n\}$. Then E_n is open and $\overline{E_n} \in \mathcal{U}$ (because X is extremally disconnected). Since f is continuous, $\overline{E_n} \subset f^{-1}(-\infty, c + 1/n]$. Therefore

$$E = \bigcap_{n=1}^{\infty} E_n \subset \bigcap_{n=1}^{\infty} \overline{E_n} \subset \bigcap_{n=1}^{\infty} f^{-1}(-\infty, c + 1/n] \subset E.$$

We conclude that $E = \bigcap_{n=1}^{\infty} \overline{E_n}$ and thus $E \in \mathfrak{M}(\mathcal{U})$. This proves that $\mathfrak{A}_X \subset \mathfrak{M}(\mathcal{U})$.

(iii) We may now define $\tau_\mu : C(X) \to \mathbb{C}$ by

$$\tau_\mu(f) := \int_X f \, d\mu$$

Then τ_μ is a positive linear functional on $C(X)$. For each $E \in \mathcal{U}$,

$$\tau_\mu(\chi_E) = \mu(E) = \mu_0(E) = \tau(\chi_E).$$

By linearity, τ_μ and τ agree on $\mathcal{G}$ (as defined in Lemma 9.2.8). Since $\mathcal{G}$ is dense in $C(X)$ and both τ_μ and τ are continuous (by Proposition 9.1.12), they must agree on all of $C(X)$. Hence $\tau(f) = \int_X f \, d\mu$ for all $f \in C(X)$. $\qquad\square$

We now need to transition from extremally disconnected spaces to arbitrary compact Hausdorff spaces. In that quest, the Stone–Čech compactification plays a crucial role. We now discuss the construction and defining properties of this object.

Let S be a locally compact Hausdorff space. Then $C_b(S)$ is a unital commutative C*-algebra. By the Gelfand–Naimark Theorem,

$$C_b(S) \cong C(\Omega(C_b(S)))$$

and $\Omega(C_b(S))$ is a compact Hausdorff space when equipped with the weak-$*$ topology. For each $s \in S$, define $\tau_s : C_b(S) \to \mathbb{C}$ by $\tau_s(f) := f(s)$. Then $\tau_s \in \Omega(C_b(S))$, so we may define $\eta : S \to \Omega(C_b(S))$ by $s \mapsto \tau_s$.

9.2.11 LEMMA *Let S be a locally compact Hausdorff space. Then $\eta : S \to \eta(S)$ is a homeomorphism and $\eta(S)$ is dense in $\Omega(C_b(S))$.*

Proof.

(i) Suppose $s_1, s_2 \in S$ are two distinct points, then by Urysohn's Lemma (Theorem 7.2.3), there is a continuous function $f \in C_b(S)$ such that $f(s_1) \neq f(s_2)$. Therefore $\tau_{s_1} \neq \tau_{s_2}$ and η is injective.

(ii) For each $f \in C_b(S)$, let $\widehat{f} : \Omega(C_b(S)) \to \mathbb{C}$ be the Gelfand transform of f given by $\widehat{f}(\tau) = \tau(f)$. Since $\Omega(C_b(S))$ is equipped with the weak-$*$ topology, η is continuous if and only if $\widehat{f} \circ \eta : S \to \mathbb{C}$ is continuous for each $f \in C_b(S)$ (Proposition 6.3.3). However, $\widehat{f} \circ \eta = f$ is continuous by definition. Therefore η is continuous.

(iii) If $F \subset S$ is closed, then we prove that $\eta(F)$ is closed in $\eta(S)$. Let $\tau_{s_0} \in \eta(S) \setminus \eta(F)$. Then since η is injective, $s_0 \notin F$. Now $\{s_0\}$ is compact, F is closed and S is locally compact and Hausdorff. Therefore Urysohn's Lemma gives us a function $f \in C_b(S)$ such that $f(s_0) = 1$ and $f|_F = 0$. Consider the set

$$U := \{\tau \in \eta(S) : |\tau(f) - \tau_{s_0}(f)| < 1/2\}.$$

Then U is open in the subspace topology of $\eta(S)$, it contains τ_{s_0} and $U \cap \eta(F) = \varnothing$. Hence $\eta(F)$ is closed and $\eta : S \to \eta(S)$ is a homeomorphism.

(iv) Suppose $\eta(S)$ is not dense in $\Omega(C_b(S))$. Then by Urysohn's Lemma, there would exist a non-zero function $f \in C(\Omega(C_b(S)))$ such that $f(\eta(s)) = 0$ for all $s \in S$. However, if $f(\eta(s)) = 0$ for all $s \in S$, then $f \equiv 0$, so no such function can exist. Hence $\eta(S)$ is dense in $\Omega(C_b(S))$.

□

The next result is the universal property satisfied by the pair $(\Omega(C_b(S)), \eta)$.

9.2.12 PROPOSITION *Let S be a locally compact Hausdorff space and let $\eta : S \to \Omega(C_b(S))$ be as above. If K is a compact Hausdorff space and $g : S \to K$ is a continuous function, then there is a unique continuous function $G : \Omega(C_b(S)) \to K$ such that $G \circ \eta = g$.*

Proof. The uniqueness of the function follows from the fact that $\eta(S)$ is dense in $\Omega(C_b(S))$, so we only prove existence. Given $g : S \to K$, we obtain a $*$-homomorphism $\Phi : C(K) \to C_b(S)$ given by $\Phi(f) := f \circ g$.

If $\Gamma : C_b(S) \to C(\Omega(C_b(S)))$ denotes the Gelfand–Naimark isomorphism, we define $\widetilde{\Phi} : C(K) \to C(\Omega(C_b(S)))$ by $\widetilde{\Phi} = \Gamma \circ \Phi$. By Remark 8.3.14, there is a continuous function $G : \Omega(C_b(S)) \to K$ such that $\widetilde{\Phi}(f) = f \circ G$ for all $f \in C(K)$. Now for any $f \in C(K)$, one may check that

$$f \circ G \circ \eta = f \circ g.$$

Since $C(K)$ separates points of K, it follows that $G \circ \eta = g$.

□

Notice that Proposition 9.2.12 characterizes the pair $(\Omega(C_b(S)), \eta)$, which ensures that there is only one such pair upto homeomorphism. This allows us to make the following definition.

9.2.13 DEFINITION Given a locally compact Hausdorff space S, the pair $(\Omega(C_b(S)), \eta)$ is called the **Stone–Čech compactification** of S. We usually omit the function and simply refer to the space $\Omega(C_b(S))$ by the symbol βS.

Let us now return to our problem and see how this object helps us.

9.2.14 LEMMA *If S is extremally disconnected, then βS is extremally disconnected.*

Proof. For simplicity, we identify S with $\eta(S)$ and assume that $S \subset \beta S$. Let $U \subset \beta S$ be an open set, then we wish to prove that $\overline{U}$ is open. Assume without loss of generality that $U \neq \emptyset$. Since S is dense in βS, $U \cap S \neq \emptyset$. Since S is extremally disconnected,

$$V := \overline{U \cap S} \cap S$$

is open in S. Furthermore, one may check that $\overline{U} = \overline{V}$ (do verify this!). Therefore it now suffices to show that $\overline{V}$ is open in βS. Since V is cl-open in S, $\chi_V : S \to \{0,1\}$ defines a continuous function. By Proposition 9.2.12, there is a continuous function $f : \beta S \to \{0,1\}$ such that $f|_S = \chi_V$. Then $f^{-1}(\{1\})$ is a closed set in βS containing V, so that $\overline{V} \subset f^{-1}(\{1\})$.

   Similarly, $\overline{S \setminus V} \subset f^{-1}(\{0\})$. However, f takes values in $\{0,1\}$ and $\beta S = \overline{S} = \overline{V} \cup \overline{S \setminus V}$. Therefore $\overline{V} = f^{-1}(\{1\})$ is open in βS.   □

9.2.15 LEMMA *Let X be a compact Hausdorff space. Then there is an extremally disconnected compact Hausdorff space Y and a continuous surjective map $p : Y \to X$.*

Proof. Let X be a compact Hausdorff space. Let S denote the set X equipped with the discrete topology. Then S is locally compact and extremally disconnected. Therefore $Y := \beta S$ is compact, Hausdorff and extremally disconnected. Let $f \in C(X)$, then f is bounded, so the induced map

$$\tilde{f} : S \to \mathbb{C}$$

is both continuous and bounded. This gives a $*$-homomorphism $\Phi : C(X) \to C_b(S)$ given by $f \mapsto \tilde{f}$. Since $C_b(S) \cong C(Y)$, Remark 8.3.14 implies that there is a continuous function $p : Y \to X$ such that

$$\Phi(f) = f \circ p$$

for all $f \in C(X)$. We now claim that p is surjective. Suppose not, then there would exist $x_0 \in X$ which does not belong to the range of p. Since Y is compact, $p(Y)$ is compact. By Urysohn's Lemma, there would exist $f \in C(X)$ such that $f(x_0) = 1$ and $f(p(y)) = 0$ for all $y \in Y$. This would imply that $\Phi(f) = 0$. Since Φ is clearly injective, $f = 0$. This contradiction proves that p is indeed surjective.   □

We now have all the ingredients to complete the proof.

Proof of Theorem 9.2.1. Uniqueness was proved in Lemma 9.2.5, so we only prove existence.

(i) Let $\tau : C(X) \to \mathbb{C}$ be a positive linear functional. Let Y be the extremally disconnected, compact Hausdorff space and $p : Y \to X$ the surjective map as in Lemma 9.2.15. Note that the $*$-homomorphism $\Phi : C(X) \to C(Y)$ given by

$$f \mapsto f \circ p$$

is an injective $*$-homomorphism and $\Phi(1) = 1$. By Theorem 8.6.7, $\mathbf{B} :=$ $\Phi(C(X))$ is a C*-subalgebra of $C(Y)$ that contains the unit of $C(Y)$.

(ii) Since $\mathbf{B}$ is isometrically isomorphic to $C(X)$, there is a positive linear functional $\rho : \mathbf{B} \to \mathbb{C}$ such that $\rho \circ \Phi = \tau$. By Corollary 9.1.15, there is a positive linear functional $\widetilde{\tau} : C(Y) \to \mathbb{C}$ such that $\widetilde{\tau}|_{\mathbf{B}} = \rho$. Therefore

$$\widetilde{\tau}(f \circ p) = \tau(f)$$

for each $f \in C(X)$. By Theorem 9.2.10, there is a finite, positive Baire measure ν on Y such that

$$\widetilde{\tau}(g) = \int_Y g \, d\nu$$

for all $g \in C(Y)$.

(iii) We write $\mathfrak{A}_Y$ and $\mathfrak{A}_X$ for the Baire σ-algebras on Y and X respectively. Since p is continuous, $p^{-1}(U)$ is open in Y whenever U is open in X. Since both spaces are compact, $p^{-1}(F)$ is a compact G_δ set in Y whenever F is a compact G_δ set in X. Since the set $\{p^{-1}(E) : E \in \mathfrak{A}_X\}$ is a σ-algebra, it follows that $p^{-1}(E) \in \mathfrak{A}_Y$ whenever $E \in \mathfrak{A}_X$. Therefore we may define $\mu : \mathfrak{A}_X \to [0, \infty)$ by

$$\mu(E) := \nu(p^{-1}(E))$$

Then μ is a finite, positive, Baire measure on X.

(iv) If $E \in \mathfrak{A}_X$ and $f = \chi_E$, then

$$\int_X f \, d\mu = \mu(E) = \nu(p^{-1}(E)) = \int_Y \chi_{p^{-1}(E)} \, d\nu = \int_Y (f \circ p) \, d\nu.$$

By linearity, this equation also holds for simple functions. By the Dominated Convergence Theorem, it must hold for any bounded $\mathfrak{A}_X$-measurable function as well. Therefore for any $f \in C(X)$,

$$\tau(f) = \tilde{\tau}(f \circ p) = \int_Y (f \circ p) d\nu = \int_X f d\mu.$$

□

In our statement of Theorem 9.2.1, the measure we have constructed is a Baire measure. It is a fact (that we do not prove here) that every Baire measure extends to a unique Borel measure. Therefore every positive linear functional on $C(X)$ may be represented by a unique Borel measure on X. As mentioned earlier, the two notions coincide for metric spaces, so we have proved

9.2.16 THEOREM   *Let X be a compact metric space and $\tau : C(X) \to \mathbb{C}$ be a positive linear functional. Then there is a unique finite, positive Borel measure μ on X such that*

$$\tau(f) = \int_X f d\mu$$

for all $f \in C(X)$.

Note that there are compact Hausdorff spaces for which the σ-algebras $\mathfrak{A}_X$ and $\mathfrak{B}_X$ *do not* coincide (see Exercise 9.14). For such spaces, one does need to extend Baire measures to Borel measures to use the full force of the Riesz–Markov–Kakutani Theorem. If you are interested in doing so, do have a look at Halmos [25].

9.3   Complex Measures

Given a compact Hausdorff space, we have described positive linear functionals on $C(X)$ in terms of measures. In order to extend this description to an arbitrary bounded linear functional, we now need to introduce complex measures.

9.3.1 DEFINITION   Let $(X, \mathfrak{M})$ be a measurable space.

(i) A function $\mu : \mathfrak{M} \to \mathbb{K}$ is said to be *countably additive* if

$$\mu \left(\bigcup_{n=1}^{\infty} E_n \right) = \sum_{n=1}^{\infty} \mu(E_n)$$

whenever $\{E_1, E_2, \ldots\}$ is a sequence of mutually disjoint sets in $\mathfrak{M}$.

(ii) A *real measure* (or a *signed measure*) is a countably additive function $\mu : \mathfrak{M} \to \mathbb{R}$.

(iii) A *complex measure* is a countably additive function $\mu : \mathfrak{M} \to \mathbb{C}$.

Every real measure is a complex measure. By definition, a real measure cannot take the values $\pm\infty$. Therefore a positive measure μ is a real measure only if it is finite. Now some authors allow real measures to take the values $+\infty$ or $-\infty$ (but not both). However, we are primarily interested in finite Baire measures, so this narrower definition is more useful.

The next example is the fundamental example of a complex measure. As it turns out, *every* complex measure arises in this way.

9.3.2 EXAMPLE Let ν be a positive measure on $(X, \mathfrak{M})$ and let $h : X \to \mathbb{K}$ be an integrable function. Define $\mu : \mathfrak{M} \to \mathbb{K}$ by

$$\mu(E) := \int_E h \, d\nu.$$

Then μ is a complex measure if $\mathbb{K} = \mathbb{C}$ and a real measure if $\mathbb{K} = \mathbb{R}$.

Proof. To prove countable additivity, suppose $\{E_1, E_2, \ldots\}$ form a partition of a set $E \in \mathfrak{M}$, then

$$\chi_E(x) = \sum_{n=1}^{\infty} \chi_{E_n}(x)$$

for each $x \in X$. Since

$$\left| \sum_{n=1}^{n} \chi_{E_n}(x) h(x) \right| \leq |h(x)|,$$

for each $n \in \mathbb{N}$, it follows by the Dominated Convergence Theorem that

$$\mu(E) = \int_X \left(\sum_{n=1}^{\infty} \chi_{E_n} h \right) d\nu = \sum_{n=1}^{\infty} \int_X \chi_{E_n} h \, d\nu = \sum_{n=1}^{\infty} \mu(E_n).$$

$\square$

Notice that every complex measure can be decomposed into a pair of real measures. If $\mu : \mathfrak{M} \to \mathbb{C}$ is a complex measure, then we may define $\operatorname{Re}(\mu) : \mathfrak{M} \to \mathbb{R}$ by $\operatorname{Re}(\mu)(E) := \operatorname{Re}(\mu(E))$ and $\operatorname{Im}(\mu) : \mathfrak{M} \to \mathbb{R}$ by $\operatorname{Im}(\mu)(E) := \operatorname{Im}(\mu(E))$. $\operatorname{Re}(\mu)$ and $\operatorname{Im}(\mu)$ are real measures and

$$\mu(E) = \operatorname{Re}(\mu)(E) + i \operatorname{Im}(\mu)(E)$$

for all $E \in \mathfrak{M}$. This decomposition will be used repeatedly in what follows.

9.3.3 LEMMA *Let μ be a complex measure on a measurable space $(X, \mathfrak{M})$. Then*

(i) $\mu(\varnothing) = 0$.

(ii) *If $\{E_1, E_2, \ldots\}$ is a sequence of mutually disjoint sets in $\mathfrak{M}$, then $\sum_{n=1}^{\infty} |\mu(E_n)| < \infty$.*

Proof.

(i) Consider the countable collection $\{\varnothing, \varnothing, \ldots\}$ of mutually disjoint sets. By countably additivity, $\mu(\varnothing) = \sum_{n=1}^{\infty} \mu(\varnothing)$. Since $\mu(\varnothing) \in \mathbb{C}$, it follows that $\mu(\varnothing) = 0$.

(ii) First assume μ is a real measure. In that case, write $\mathbb{N} = K_+ \sqcup K_-$, where $K_+ = \{i \in \mathbb{N} : \mu(E_i) > 0\}$ and $K_- = \{i \in \mathbb{N} : \mu(E_i) < 0\}$. If $F_+ = \bigcup_{i \in K_+} E_i$, then $\mu(F_+) \in \mathbb{R}$ and therefore

$$\sum_{i \in K_+} |\mu(E_i)| = \sum_{i \in K_+} \mu(E_i) = \mu(F_+) < \infty.$$

Similarly, if $F_- = \bigcup_{i \in K_-} E_i$, then $\mu(F_-) \in \mathbb{R}$ and thus

$$\sum_{i \in K_-} |\mu(E_i)| = - \sum_{i \in K_+} \mu(E_i) = -\mu(F_-) < \infty.$$

This proves that $\sum_{n=1}^{\infty} |\mu(E_n)| < \infty$.

Now if μ is a complex measure, then we write $\mu = \mathrm{Re}(\mu) + i \, \mathrm{Im}(\mu)$ and observe that

$$\sum_{n=1}^{\infty} |\mu(E_n)| \leq \sum_{n=1}^{\infty} \left(|\mathrm{Re}(\mu)(E_n)| + |\mathrm{Im}(\mu)(E_n)| \right) < \infty$$

since both $\mathrm{Re}(\mu)$ and $\mathrm{Im}(\mu)$ are real measures. $\qquad\qquad\square$

While the definition of a complex measure should not come as a surprise, it is as yet unclear how to define the *integral* of a function with respect to a complex measure. In order to do so, we need to construct a positive measure associated to the complex measure and use that to define the integral.

9.3.4 DEFINITION Let μ be a complex measure on a measurable space $(X, \mathfrak{M})$. The *total variation* of μ is the function $|\mu| : \mathfrak{M} \to [0, \infty]$ defined by

$$|\mu|(E) = \sup \left\{ \sum_{i=1}^{n} |\mu(E_i)| \right\}$$

where the supremum is taken over all finite collections $\{E_1, E_2, \ldots, E_n\} \subset \mathfrak{M}$ of mutually disjoint sets satisfying $\bigcup_{i=1}^{n} E_i \subset E$.

9.3.5 THEOREM *The total variation of a complex measure is a positive measure.*

Proof. Let μ be a complex measure on $(X, \mathfrak{M})$. Since $\mu(\varnothing) = 0$, it follows that $|\mu|(\varnothing) = 0$ by definition. Therefore it suffices to prove countable additivity. Let $\{E_1, E_2, \ldots\}$ be a countable collection of mutually disjoint sets in $\mathfrak{M}$ and let $E := \bigcup_{n=1}^{\infty} E_n$.

(i) Fix $n \in \mathbb{N}$. For each $i \in \{1, 2, \ldots, n\}$, let $\{E_{i,1}, E_{i,2}, \ldots, E_{i,n_i}\}$ be a finite collection of mutually disjoint sets satisfying $\bigcup_{j=1}^{n_i} E_{i,j} \subset E_i$. Then $\{E_{i,j} : 1 \leq i \leq n, 1 \leq j \leq n_i\}$ forms a finite collection of mutually disjoint sets whose union is contained in E. By definition of $|\mu|$, it follows that

$$\sum_{i=1}^{n} \sum_{j=1}^{n_i} |\mu(E_{i,j})| \leq |\mu|(E).$$

Taking a supremum over all such finite collections $\{E_{i,j} : 1 \leq j \leq n_i\}$ gives $\sum_{i=1}^{n} |\mu|(E_i) \leq |\mu|(E)$. Since this holds for each $n \in \mathbb{N}$, it follows that

$$\sum_{i=1}^{\infty} |\mu|(E_i) \leq |\mu|(E)$$

(ii) To prove the reverse inequality, let $\{F_1, F_2, \ldots, F_k\}$ be a finite collection of mutually disjoint sets in $\mathfrak{M}$ such that $\bigcup_{i=1}^{k} F_i \subset E$. Then for each $n \in \mathbb{N}$, $\{F_1 \cap E_n, F_2 \cap E_n, \ldots, F_k \cap E_n\}$ is a finite collection of mutually disjoint sets in $\mathfrak{M}$ such that $\bigcup_{i=1}^{k} (F_i \cap E_n) \subset E_n$. So by definition of $|\mu|$,

$$\sum_{n=1}^{\infty} |\mu|(E_n) \geq \sum_{n=1}^{\infty} \sum_{j=1}^{k} |\mu(F_j \cap E_n)| = \sum_{j=1}^{k} \sum_{n=1}^{\infty} |\mu(F_j \cap E_n)|,$$

where the second equality follows from Proposition 1.2.13. However,

$$\sum_{j=1}^{k} \sum_{n=1}^{\infty} |\mu(F_j \cap E_n)| \geq \sum_{j=1}^{k} \left| \sum_{n=1}^{\infty} \mu(F_j \cap E_n) \right| = \sum_{j=1}^{k} |\mu(F_j)|,$$

where the second equality follows from the countable additivity of μ. Combining this series of inequalities, we see that

$$\sum_{n=1}^{\infty} |\mu|(E_n) \geq \sum_{j=1}^{k} |\mu(F_j)|.$$

This is true for any such collection $\{F_1, F_2, \ldots, F_k\}$ and thus $\sum_{n=1}^{\infty} |\mu|(E_n) \geq |\mu|(E)$.

$\square$

The next proposition is a collection of results that you would have seen for positive measures. The proofs for a complex measure are identical and are left as an exercise (Exercise 9.25). Note that property (iv) given below is only true for *finite* positive measures. The reason that it works for all complex measures is that complex measures are necessarily finite.

9.3.6 PROPOSITION *Let μ be a complex measure on a measurable space $(X, \mathfrak{M})$.*

(i) *If $E, F \in \mathfrak{M}$ with $F \subset E$, then $\mu(E \setminus F) = \mu(E) - \mu(F)$.*

(ii) *If $E, F \in \mathfrak{M}$, then $\mu(E \cup F) = \mu(E) + \mu(F) - \mu(E \cap F)$.*

(iii) *If $\{E_1, E_2, \ldots\}$ is a countable collection in $\mathfrak{M}$ such that $E_1 \subset E_2 \subset \ldots$, then*

$$\mu\left(\bigcup_{n=1}^{\infty} E_n\right) = \lim_{n \to \infty} \mu(E_n).$$

(iv) *If $\{E_1, E_2, \ldots\}$ is a countable collection in $\mathfrak{M}$ such that $E_1 \supset E_2 \supset \ldots$, then*

$$\mu\left(\bigcap_{n=1}^{\infty} E_n\right) = \lim_{n \to \infty} \mu(E_n).$$

The next proposition ties the room together. As we will see in the next section, the set of all complex measures has a natural vector space structure and this next result allows us to define a norm on it.

9.3.7 PROPOSITION *If μ is a complex measure on $(X, \mathfrak{M})$, then $|\mu|(X) < \infty$.*

Proof. Writing $\mu = \mathrm{Re}(\mu) + i\,\mathrm{Im}(\mu)$, we may assume that μ is a real measure (as we did in Lemma 9.3.3). In that case, suppose that $|\mu|(X) = \infty$. Seeking a contradiction, we construct a sequence $\{E_1, E_2, \ldots, \}$ of measurable sets as follows.

Let $E_1 := X$. Since $|\mu|(E_1) = \infty$, there exists a finite collection $\{F_1, F_2, \ldots, F_k\}$ of mutually disjoints sets in $\mathfrak{M}$ such that $F := \bigcup_{i=1}^{n} F_i \subset E_1$ and $|\mu(F)| \geq 2 + |\mu(E_1)|$. By Proposition 9.3.6,

$$|\mu(E_1 \setminus F)| = |\mu(E_1) - \mu(F)| \geq |\mu(F)| - |\mu(E_1)| \geq 2.$$

Since $|\mu|$ is a measure, we have $|\mu|(F) + |\mu|(E_1 \setminus F) = |\mu|(E_1) = \infty$. Therefore at least one of $|\mu|(F)$ or $|\mu|(E_1 \setminus F)$ is infinite. Hence there is a set $E_2 \subset E_1$ such that $|\mu|(E_2) = \infty$ and $|\mu(E_2)| \geq 2$.

Repeating this process inductively, we obtain a sequence $\{E_1, E_2, \ldots\}$ of sets in $\mathfrak{M}$ such that

(i) $E_1 \supset E_2 \supset \ldots$,

(ii) $|\mu|(E_n) = \infty$ for all $n \in \mathbb{N}$ and

(iii) $|\mu(E_n)| \geq n$ for all $n \in \mathbb{N}$.

If $E := \bigcap_{n=1}^{\infty} E_n$, then by Proposition 9.3.6, $\mu(E) = \lim_{n \to \infty} \mu(E_n)$. In particular, this limit must exist. However, this cannot happen if property (iii) listed above were to hold. Therefore it must happen that $|\mu|(X) < \infty$. $\qquad\square$

The next lemma is a simple fact whose proof we have used before without explicitly stating it (see Proposition 2.3.12).

9.3.8 LEMMA *Let ν be a positive measure on $(X, \mathfrak{M})$. Then the set of all simple functions in $L^1(\nu)$ is dense in $L^1(\nu)$.*

Proof. Fix $f \in L^1(\nu)$ and we wish to prove that f is a limit of simple functions. We may write f as a linear combination of positive functions and therefore we may assume that f is itself non-negative. In that case, there is a sequence (s_n) of simple functions such that $0 \leq s_1 \leq s_2 \leq \ldots$ and $\lim_{n \to \infty} s_n(x) = f(x)$ for each $x \in X$. Since $|s_n - f| \leq 2|f| \in L^1(\nu)$, the Dominated Convergence Theorem tells us that $s_n \to f$ in $L^1(\nu)$. $\qquad\square$

We now wish to describe the relationship between a complex measure and its total variation. The next lemma is a small step towards that.

9.3.9 LEMMA *Let ν be a positive measure on $(X, \mathfrak{M})$ and $h \in L^1(\nu)$. Define $\mu : \mathfrak{M} \to \mathbb{C}$ by*

$$\mu(E) := \int_E h \, d\nu.$$

Then for each $E \in \mathfrak{M}$,

$$|\mu|(E) = \int_E |h| dv.$$

Proof.   Fix $E \in \mathfrak{M}$.

(i) Suppose $\{E_1, E_2, \ldots, E_k\}$ is a finite collection in $\mathfrak{M}$ of mutually disjoint sets such that $\bigcup_{i=1}^{k} E_i \subset E$. Then

$$\sum_{i=1}^{k} |\mu(E_i)| = \sum_{i=1}^{k} \left| \int_{E_i} h dv \right| \leq \sum_{i=1}^{k} \int_{E_i} |h| dv \leq \int_E |h| dv.$$

Therefore $|\mu|(E) \leq \int_E |h| dv$.

(ii) To prove the reverse inequality, fix $\epsilon > 0$. By Lemma 9.3.8, there is a simple function $g \in L^1(v)$ such that $\|h - g\|_1 < \epsilon$. Then we may choose mutually disjoint sets $E_1, E_2, \ldots, E_k$ in $\mathfrak{M}$ such that $\bigcup_{i=1}^{k} E_i \subset E$ and constants $\alpha_1, \alpha_2, \ldots, \alpha_k$ such that

$$g|_E = \sum_{i=1}^{k} \alpha_i \chi_{E_i}.$$

Then

$$|\mu|(E) \geq \sum_{i=1}^{k} |\mu(E_i)| = \sum_{i=1}^{k} \left| \int_{E_i} h dv \right|$$

$$\geq \sum_{i=1}^{k} \left| \int_{E_i} g dv \right| - \sum_{i=1}^{k} \left| \int_{E_i} (g - h) dv \right|$$

$$\geq \sum_{i=1}^{k} |\alpha_i| v(E_i) - \sum_{i=1}^{k} \int_{E_i} |g - h| dv$$

$$= \int_E |g| dv - \int_E |g - h| dv$$

$$\geq \int_E |g| dv - \epsilon$$

$$\geq \int_E |h| dv - 2\epsilon.$$

Since this is true for all $\epsilon > 0$, it follows that $|\mu|(E) \geq \int_E |h| dv$.   □

In order to state our next result, we need the following notion which we first encountered in the context of the Lebesgue Differentiation Theorem in Chapter 4. The Radon–Nikodym Theorem, which we now describe, is a generalization of Lebesgue's Differentiation Theorem (and therefore of the Fundamental Theorem of Calculus) to the context of complex measures.

9.3.10 DEFINITION Let ν be a complex measure on a measurable space $(X, \mathfrak{M})$ and let μ be a positive measure on $(X, \mathfrak{M})$. We say that ν is **absolutely continuous** with respect to μ if, for any $E \in \mathfrak{M}$, $\nu(E) = 0$ whenever $\mu(E) = 0$. If this happens, we write $\nu \ll \mu$.

If μ is a positive measure and $h \in L^1(\mu)$, then the function $\nu : \mathfrak{M} \to \mathbb{C}$ given by

$$\nu(E) := \int_E h \, d\mu \tag{9.3}$$

is a complex measure which is clearly absolutely continuous with respect to μ. The Radon–Nikodym Theorem proves the converse: If $\nu \ll \mu$, then ν must be of the form in Equation 9.3 for some function $h \in L^1(\mu)$. Furthermore, this function is unique (almost everywhere with respect to μ). In other words, if h and g are two functions in $L^1(\mu)$ are such that $\int_E h \, d\mu = \int_E g \, d\mu$ for all $E \in \mathfrak{M}$, then $\mu(\{x \in X : h(x) \neq g(x)\}) = 0$. A proof of this fundamental fact is described in Section A.2. For now we simply use the theorem to give the *polar decomposition* of a complex measure.

9.3.11 PROPOSITION *Let μ be a complex measure on a measurable space $(X, \mathfrak{M})$. Then there is a measurable function $h : X \to S^1$ such that*

$$\mu(E) = \int_E h \, d|\mu| \tag{9.4}$$

for each $E \in \mathfrak{M}$. Furthermore, this function is unique a.e. with respect to $|\mu|$.

Proof. Clearly, $\mu \ll |\mu|$, therefore the existence and uniqueness of $h \in L^1(|\mu|)$ is guaranteed by the Radon–Nikodym Theorem. We must now verify that h takes values in S^1. To see this, observe that

$$|\mu|(E) = \int_E |h| \, d|\mu|$$

for all $E \in \mathfrak{M}$ by Lemma 9.3.9. Then if $n \in \mathbb{N}$, consider $E_n := \{x \in X : |h(x)| > 1 + 1/n\}$, then

$$|\mu|(E_n) = \int_{E_n} |h| \, d|\mu| > \left(1 + \frac{1}{n}\right) |\mu|(E_n).$$

Therefore $|\mu|(E_n) = 0$. This is true for each $n \in \mathbb{N}$, so $E := \{x \in X : |h(x)| > 1\}$ must have measure zero. Similarly, $F := \{x \in X : |h(x)| < 1\}$ also has measure zero and $|h(x)| = 1$ a.e. with respect to μ. Redefining h on a set of measure zero, we may thus assume that h takes values in S^1.   □

The function h constructed in Proposition 9.3.11 is called the Radon–Nikodym derivative of μ with respect to $|\mu|$ and is denoted by

$$h_\mu := \frac{d\mu}{d|\mu|}.$$

This now allows us to define the integral with respect to a complex measure.

9.3.12 DEFINITION   Let μ be a complex measure on a measurable space $(X, \mathfrak{M})$. A function $f : X \to \mathbb{K}$ is said to be integrable with respect to μ if $f \in L^1(|\mu|)$. In that case, we define

$$\int_X f d\mu := \int_X f h_\mu d|\mu|.$$

Note that this is well-defined and agrees with Equation 9.4 when applied to characteristic functions. Finally, observe that for any $f \in L^1(\mu)$,

$$\left| \int_X f d\mu \right| \le \int_X |f| d|\mu|.$$

This inequality will get used repeatedly below.

9.4   The Dual of $C(X)$

9.4.1 DEFINITION   Let X be a compact Hausdorff space and let $\mathfrak{A}_X$ denote the Baire σ-algebra on X.

(i) A complex measure μ is said to be a **Baire measure** if the domain of μ contains $\mathfrak{A}_X$. We write $\mathbb{M}_b(X)$ for the set of all complex Baire measures on X.

(ii) If $\lambda, \mu \in \mathbb{M}_b(X)$ and $\alpha \in \mathbb{C}$, we define $(\lambda + \mu)$ and $\alpha\mu$ by

$$(\lambda + \mu)(E) := \lambda(E) + \mu(E) \text{ and } (\alpha\mu)(E) := \alpha\mu(E).$$

Notice that $(\lambda + \mu) \in \mathbb{M}_b(X)$ thanks to Lemma 9.3.3. It is clear that $\mathbb{M}_b(X)$ is a vector space under these operations.

(iii) If $\mu \in \mathbb{M}_b(X)$, we define $\|\mu\| := |\mu|(X)$. It is easy to verify that $\mathbb{M}_b(X)$ is a normed linear space (Exercise 9.35).

(iv) If $\mu \in \mathbb{M}_b(X)$, then Definition 9.3.12 allows us to define $\varphi_\mu : C(X) \to \mathbb{C}$ by

$$\varphi_\mu(f) := \int_X f d\mu.$$

Since $|h_\mu| = 1$, we have

$$|\varphi_\mu(f)| = \left| \int_X f h_\mu d|\mu| \right| \leq \int_X |f| d|\mu| \leq \|f\|_\infty |\mu|(X).$$

Thus φ_μ is a bounded linear functional on $C(X)$ and $\|\varphi_\mu\| \leq \|\mu\|$. This gives us a function $\Delta : \mathbb{M}_b(X) \to C(X)^*$ given by

$$\Delta(\mu) := \varphi_\mu.$$

That Δ is linear is easy to see (Exercise 9.36). In order to investigate the other properties of Δ, we need the following fact.

9.4.2 PROPOSITION *Let μ be a finite, positive Baire measure on a compact Hausdorff space X. Then for any $E \in \mathfrak{A}_X$,*

$$\mu(E) = \sup\{\mu(K) : K \text{ compact}, K \subset E \text{ and } K \in \mathfrak{A}_X\} \tag{9.5}$$

and

$$\mu(E) = \inf\{\mu(U) : U \text{ open}, U \supset E \text{ and } U \in \mathfrak{A}_X\}. \tag{9.6}$$

Proof. Let $\mathcal{S}$ be the collection of all subsets in $\mathfrak{A}_X$ that satisfy both Equation 9.5 and Equation 9.6. Since μ is a finite measure, $E \in \mathcal{S}$ if and only if, for each $\epsilon > 0$, there is a compact set $K \in \mathfrak{A}_X$ and an open set $U \in \mathfrak{A}_X$ such that $K \subset E \subset U$ and

$$\mu(U \setminus K) < \epsilon.$$

We show that $\mathcal{S} = \mathfrak{A}_X$ in the following steps.

(i) We claim that $\mathcal{S}$ contains all compact G_δ sets. If K is a compact G_δ, then K clearly satisfies Equation 9.5. Also, we may write K as $\bigcap_{n=1}^\infty U_n$, where each U_n is an open set. For each $n \in \mathbb{N}$, Urysohn's Lemma ensures that there is a continuous function $f_n : X \to [0,1]$ such that $f \equiv 0$ on K and $f \equiv 1$ on $X \setminus U_n$. Define

$$V_n := \{x \in X : \max\{f_1(x), f_2(x), \ldots, f_n(x)\} < \frac{1}{2}\}.$$

Then V_n is an open set, $K \subset V_n \subset U_n$ and $V_n \in \mathfrak{A}_X$. It then follows that $K = \bigcap_{n=1}^{\infty} V_n$. Furthermore, $V_1 \supset V_2 \supset \ldots$. Therefore

$$\mu(K) = \lim_{n \to \infty} \mu(V_n),$$

which proves that K satisfies Equation 9.6 as well.

(ii) We claim that $\mathcal{S}$ is closed under taking complements. Suppose $E \in \mathcal{S}$ and $\epsilon > 0$ is fixed. Since $\mu(E) < \infty$, there exists compact $K \in \mathfrak{A}_X$ and open $U \in \mathfrak{A}_X$ such that $K \subset E \subset U$ and $\mu(U \setminus K) < \epsilon$. Then $V := X \setminus K$ and $L := X \setminus U$ are both Baire sets, V is open, L is compact and $L \subset X \setminus E \subset V$. Furthermore, $\mu(V \setminus L) = \mu(U \setminus K) < \epsilon$. This proves that $X \setminus E \in \mathcal{S}$ as well.

(iii) We now claim that $\mathcal{S}$ is closed under countable unions. Let $\{E_1, E_2, \ldots\}$ be a countable family of sets in $\mathcal{S}$ and let $E := \bigcup_{n=1}^{\infty} E_n$. Fix $\epsilon > 0$. Then for each $n \in \mathbb{N}$, there is a compact set $K_n \in \mathfrak{A}_X$ and an open set $U_n \in \mathfrak{A}_X$ such that $K_n \subset E_n \subset U_n$ and $\mu(U_n \setminus K_n) < \frac{\epsilon}{2^{n+1}}$. Define

$$U := \bigcup_{n=1}^{\infty} U_n \text{ and } L := \bigcup_{n=1}^{\infty} K_n.$$

Then U and L are both in $\mathfrak{A}_X$, U is open and

$$\mu(U) - \mu(L) = \mu(U \setminus L) \leq \sum_{n=1}^{\infty} (\mu(U_n) - \mu(K_n)) \leq \frac{\epsilon}{2}.$$

For $n \in \mathbb{N}$, let $L_n := K_1 \cup K_2 \cup \ldots \cup K_n$. Then $L_n \in \mathfrak{A}_X$ is compact and $L_n \subset E \subset U$. Furthermore, $L_1 \subset L_2 \subset \ldots$ and $L = \bigcup_{n=1}^{\infty} L_n$. Therefore

$$\mu(L) = \lim_{n \to \infty} \mu(L_n).$$

Hence there exists $N \in \mathbb{N}$ such that

$$\mu(U \setminus L_N) = \mu(U) - \mu(L_N) < \epsilon.$$

We have thus verified that $E \in \mathcal{S}$.

We conclude that $\mathcal{S}$ is a σ-algebra that contains all compact G_δ sets. By Lemma 9.2.3, $\mathcal{S} = \mathfrak{A}_X$. $\qquad\square$

The properties described in Equation 9.5 and Equation 9.6 are called *inner regularity* and *outer regularity* respectively. When both properties hold (as they do in this case), we say that the measure is **regular**. Regularity allows us to prove that continuous functions approximate measurable functions. This is a fundamental fact in measure theory (one of Littlewood's three principles) and we have seen it before in the guise of a density theorem (Proposition 2.3.12).

9.4.3 THEOREM (LUSIN, 1912) *Let μ be a finite, positive Baire measure on a compact Hausdorff space X and let $f : X \to \mathbb{C}$ be a Baire-measurable function. Then for any $\epsilon > 0$, there is a continuous function $g \in C(X)$ such that*

$$\mu(\{x \in X : f(x) \neq g(x)\}) < \epsilon.$$

Furthermore if f is bounded, we may choose g so that $\|g\|_\infty \leq \|f\|_\infty$.

Proof. Fix $\epsilon > 0$ and let $\{U_1, U_2, \ldots\}$ be a countable basis for the topology of $\mathbb{C}$. For each $n \in \mathbb{N}$, $f^{-1}(U_n) \in \mathfrak{A}_X$, so there is an open set $V_n \in \mathfrak{A}_X$ such that

$$\mu(V_n \setminus f^{-1}(U_n)) < \frac{\epsilon}{2^{n+1}}.$$

If $E := \bigcup_{n=1}^\infty V_n \setminus f^{-1}(U_n)$, then $\mu(E) < \epsilon/2$. Now choose a compact set $K \subset X \setminus E$ such that $\mu((X \setminus E) \setminus K) < \epsilon/2$ and let $h := f|_K$.

We claim that h is continuous. Fix $n \in \mathbb{N}$ and observe that $h^{-1}(U_n) = f^{-1}(U_n) \cap K \subset V_n \cap K$. Conversely, $V_n \cap K \subset V_n \cap (X \setminus E) \subset f^{-1}(U_n)$. Hence $h^{-1}(U_n) = V_n \cap K$. Since each such set is open in K and $\{U_n\}$ forms a basis for the topology of $\mathbb{C}$, it follows that h is continuous on K.

Since K is closed in X, Tietze's Extension Theorem ensures that there is a continuous function $g \in C(X)$ such that $g|_K = h$ (and $\|g\|_\infty = \|h\|_\infty \leq \|f\|_\infty$ if f is bounded). Now observe that $g(x) = f(x)$ for all $x \in K$ and therefore

$$\mu(\{x \in X : g(x) \neq f(x)\}) \leq \mu(X \setminus K) \leq \mu((X \setminus E) \setminus K) + \mu(E) < \epsilon.$$

$\qquad\square$

9.4.4 THEOREM *If X is a compact Hausdorff space, then the map $\Delta : \mathbb{M}_b(X) \to C(X)^*$ is isometric.*

Proof. Fix $\mu \in \mathbb{M}_b(X)$ and write $\varphi_\mu := \Delta(\mu)$ as before. We know that $\|\varphi_\mu\| \leq \|\mu\|$ from Definition 9.4.1. To prove the reverse inequality, fix $\epsilon > 0$ and let $h = h_\mu \in$

$L^1(|\mu|)$ be the Radon–Nikodym derivative of μ with respect to $|\mu|$. Then $\|h\|_\infty = 1$, so by Lusin's Theorem, there is a continuous function $g \in C(X)$ such that $\|g\|_\infty = 1$ and

$$E := \{x \in X : g(x) \neq \overline{h}(x)\}$$

is such that $|\mu|(E) < \epsilon/2$. Then

$$\|\mu\| = \int_X |h|^2 d|\mu| = \int_X \overline{h} d\mu$$
$$\leq \left| \int_X g d\mu \right| + \left| \int_X (g - \overline{h}) d\mu \right|$$
$$\leq |\varphi_\mu(g)| + 2|\mu|(E)$$
$$\leq \|\varphi_\mu\| \|g\|_\infty + \epsilon$$
$$\leq \|\varphi_\mu\| + \epsilon.$$

This is true for any $\epsilon > 0$ and therefore $\|\mu\| \leq \|\varphi_\mu\|$ as well. $\qquad\square$

Our goal, of course, is to prove that Δ is an isomorphism. To that end, we make the following definitions concerning linear functionals. We do this in the more general setting of C*-algebras for clarity. Recall that if $\mathbf{A}$ is a C*-algebra, then $\mathbf{A}_{sa}$ is the real vector space of all self-adjoint elements in $\mathbf{A}$.

9.4.5 DEFINITION Let $\mathbf{A}$ be a C*-algebra and $\tau : \mathbf{A} \to \mathbb{C}$ be a bounded linear functional.

(i) Define $\tau^* : \mathbf{A} \to \mathbb{C}$ by $\tau^*(a) := \overline{\tau(a^*)}$. Observe that τ^* is also a bounded linear functional.

(ii) A linear functional $\tau \in \mathbf{A}^*$ is said to be *self-adjoint* if $\tau = \tau^*$. We write $(\mathbf{A}^*)_{sa}$ for the set of all self-adjoint bounded linear functionals on $\mathbf{A}$.

9.4.6 REMARK

(i) If $\tau : \mathbf{A} \to \mathbb{C}$ is a positive linear functional, then $\tau(a^*) = \overline{\tau(a)}$ for all $a \in \mathbf{A}$ by Proposition 9.1.13. Therefore every positive linear functional is self-adjoint. More generally, any real linear combination of positive linear functionals is self-adjoint.

(ii) If $\tau \in \mathbf{A}^*$ is self-adjoint, then it induces a function $\widetilde{\tau} : \mathbf{A}_{sa} \to \mathbb{R}$ by restriction. Treating $\mathbf{A}_{sa}$ as a real vector space, $\widetilde{\tau}$ is a bounded linear functional. Furthermore,

$$\|\widetilde{\tau}\| = \sup\{|\widetilde{\tau}(b)| : b \in \mathbf{A}_{sa}, \|b\| \leq 1\} \leq \|\tau\|.$$

(iii) Conversely, every element $a \in \mathbf{A}$ can be expressed uniquely in the form $a = b + ic$ where b and c are self-adjoint. Therefore if $\rho : \mathbf{A}_{sa} \to \mathbb{R}$ is a bounded linear functional on $\mathbf{A}_{sa}$, then the formula

$$\tau(b + ic) := \rho(b) + i\rho(c)$$

determines a self-adjoint bounded linear functional on $\mathbf{A}$ such that $\tilde{\tau} = \rho$ (you need to verify that τ is indeed *complex* linear). Therefore the map

$$\tau \mapsto \tilde{\tau}$$

determines a real-linear isomorphism between $(\mathbf{A}^*)_{sa}$ and $(\mathbf{A}_{sa})^*$.

(iv) If $\tau \in \mathbf{A}^*$ is a bounded linear functional, then

$$\rho := \frac{\tau + \tau^*}{2} \text{ and } \eta := \frac{\tau - \tau^*}{2i}$$

are both self-adjoint and $\tau = \rho + i\eta$. Therefore every bounded linear functional is a linear combination of self-adjoint linear functionals.

The next result is an extension theorem due to Kantorovič [33], who first proved it in the context of partially ordered real vector spaces. In our case, if $\mathbf{A}$ is a C*-algebra, then $\mathbf{A}_{sa}$ is one such vector space. Furthermore, $\mathbf{A}_+$ is closed under addition (Lemma 9.1.4). Therefore we may speak of functions defined on $\mathbf{A}_+$ that are *additive*.

9.4.7 LEMMA (KANTOROVIČ, 1940) *Let $\mathbf{A}$ be a C*-algebra and $T : \mathbf{A}_+ \to [0, \infty)$ be such that $T(a + b) = T(a) + T(b)$ for all $a, b \in \mathbf{A}_+$. Then there is a unique positive linear functional $\tau : \mathbf{A} \to \mathbb{C}$ such that*

$$\tau|_{\mathbf{A}_+} = T.$$

Proof. We leave the uniqueness part of the proof as an exercise (Exercise 9.38) and we prove existence in the following steps.

(i) Given T as above, define $\rho : \mathbf{A}_{sa} \to \mathbb{R}$ by

$$\rho(a) := T(a_+) - T(a_-)$$

where a_+ and a_- are as defined in the proof of Lemma 9.1.9. We first claim that ρ is additive. If $c = a + b$ with $a, b \in \mathbf{A}_{sa}$, then $c_+ - c_- = a_+ - a_- + b_+ - b_-$, so $c_+ + a_- + b_- = c_- + a_+ + b_+$. Applying T to both sides of this equation and using the fact that T is additive, we conclude that

$$\rho(c) = T(c_+) - T(c_-) = T(a_+) - T(a_-) + T(b_+) - T(b_-) = \rho(a) + \rho(b).$$

(ii) Since T is additive, $T(ra) = rT(a)$ for any $a \in \mathbf{A}_+$ and $r \in \mathbb{Q}$. Also, if $a, b \in \mathbf{A}_+$ are such that $0 \leq b \leq a$, then $T(a - b) \geq 0$, so

$$0 \leq T(b) \leq T(b) + T(a - b) = T(a).$$

Hence T is an increasing function on $\mathbf{A}_+$. Now fix $\lambda \geq 0$ and $a \in \mathbf{A}_+$. Then there are sequences (r_n) and (t_n) of non-negative rational numbers such that (r_n) is increasing, (t_n) is decreasing and $\lim_{n \to \infty} r_n = \lambda = \lim_{n \to \infty} t_n$. Then for each $n \in \mathbb{N}$, we have

$$r_n T(a) = T(r_n a) \leq T(\lambda a) \leq T(t_n a) = t_n T(a).$$

Taking limits, we see that $T(\lambda a) = \lambda T(a)$.

(iii) We now claim that ρ is $\mathbb{R}$-linear. Fix $a \in \mathbf{A}_{sa}$ and $\lambda \geq 0$, then

$$\rho(\lambda a) = T(\lambda a_+) - T(\lambda a_-) = \lambda T(a_+) - \lambda T(a_-) = \lambda \rho(a).$$

If $\lambda < 0$, then note that $\rho(\lambda a) = -\rho((-\lambda)a)$ since ρ is additive by part (i). Since $(-\lambda) > 0$, the previous argument shows that

$$\rho(\lambda a) = -\rho((-\lambda)a) = -(-\lambda)\rho(a) = \lambda \rho(a).$$

Therefore ρ is $\mathbb{R}$-linear.

(iv) Now define $\tau : \mathbf{A} \to \mathbb{C}$ by the formula

$$\tau(a) := \rho\left(\frac{a + a^*}{2}\right) + i\rho\left(\frac{a - a^*}{2i}\right).$$

Then it is easy to see that τ is complex linear and that $\tau|_{\mathbf{A}_+} = T$. In particular, τ is positive.

$\square$

Now consider the special case when $\mathbf{A} = C(X)$ for some compact Hausdorff space X. Then $\mathbf{A}_{sa} = C(X, \mathbb{R})$, the space of all real-valued continuous functions on X, which we think of as a Banach space over $\mathbb{R}$.

9.4.8 LEMMA *If $\rho \in C(X)^*$ is self-adjoint, then there exist two positive linear functionals ρ_+ and ρ_- in $C(X)^*$ such that $\rho = \rho_+ - \rho_-$.*

Proof. If $f \in C(X)_+$ is a non-negative function and $0 \leq g \leq f$, then $\rho(g) \in \mathbb{R}$ satisfies $|\rho(g)| \leq \|\rho\| \|g\|_\infty \leq \|\rho\| \|f\|_\infty$. Therefore we may define $T : C(X)_+ \to [0, \infty)$ by the formula

$$T(f) := \sup\{\rho(g) : g \in C(X)_+ \text{ such that } 0 \leq g \leq f\}.$$

Then T is well-defined and $|T(f)| \leq \|\rho\| \|f\|_\infty$ for all $f \in C(X)_+$. In order to apply Lemma 9.4.7, we must verify that T is additive. To that end, fix $f_1, f_2 \in C(X)_+$.

(i) If $g_1, g_2 \in C(X)_+$ are such that $0 \leq g_i \leq f_i$, then $0 \leq g_1 + g_2 \leq f_1 + f_2$ holds. Therefore

$$T(f_1 + f_2) \geq \rho(g_1 + g_2) = \rho(g_1) + \rho(g_2).$$

Taking suprema independently, we conclude that $T(f_1 + f_2) \geq T(f_1) + T(f_2)$.

(ii) Now if $g \in C(X)_+$ is such that $0 \leq g \leq f_1 + f_2$, then $g_1 := \min\{g, f_1\}$ is a non-negative function such that $0 \leq g_1 \leq f_1$. Also, if $g_2 := g - g_1$, then $0 \leq g_2 \leq f_2$ must hold. Hence

$$\rho(g) = \rho(g_1) + \rho(g_2) \leq T(f_1) + T(f_2).$$

Taking a supremum over all such $g \in C(X)_+$, we conclude that $T(f_1 + f_2) \leq T(f_1) + T(f_2)$.

We have proved that T is additive and must therefore extend to a positive linear functional $\rho_+ : C(X) \to \mathbb{C}$. Now define $\rho_- : C(X) \to \mathbb{C}$ by

$$\rho_- := \rho_+ - \rho.$$

An easy verification shows that ρ_- is also positive and clearly $\rho = \rho_+ - \rho_-$. $\qquad\square$

Aside. You might be wondering if a result analogous to Lemma 9.4.8 holds for arbitrary C*-algebras. While that is true (see Murphy [43, Theorem 3.3.10]), the proof *does not* go through verbatim. The reason is that, during the course of the preceding proof, we needed to take the minimum of two functions in $C(X)_+$. This step requires that the set of all self-adjoint elements in the C*-algebra form a *lattice* (i.e. that any two such elements must have an infimum). While this is clearly true for commutative C*-algebras, it is not true for noncommutative C*-algebras. In fact, Sherman proved in 1951 [55] that this property *characterizes* commutative C*-algebras.

We now have all the tools to complete our discussion on the dual of $C(X)$.

9.4.9 THEOREM *If X is a compact Hausdorff space, then the map $\Delta : \mathbb{M}_b(X) \to C(X)^*$ is an isometric isomorphism.*

Proof. By Theorem 9.4.4, it suffices to prove that Δ is surjective. So fix $\tau \in C(X)^*$. By Remark 9.4.6 and Lemma 9.4.8, we may express τ in the form

$$\tau = (\rho_+ - \rho_-) + i(\eta_+ - \eta_-)$$

for four positive linear functionals ρ_+, ρ_-, η_+ and η_- on $C(X)$. By the Riesz–Markov–Kakutani Theorem, there exist four positive, Baire measures $\mu_+, \mu_-, \nu_+, \nu_- \in \mathbb{M}_b(X)$ such that

$$\rho_\pm(f) = \int_X f d\mu_\pm \text{ and } \eta_\pm(f) = \int_X f d\nu_\pm$$

for all $f \in C(X)$. Let $\lambda := (\mu_+ - \mu_-) + i(\eta_+ - \eta_-) \in \mathbb{M}_b(X)$. Since Δ is linear, it follows that $\Delta(\lambda) = \tau$. $\qquad\square$

Some important facts about Δ are worth mentioning here. Since the proofs are built in to the construction, we encourage you to try it as an exercise.

9.4.10 COROLLARY *Let X be a compact Hausdorff space and let $\mu \in \mathbb{M}_b(X)$. Then*

(i) *μ is a positive measure if and only if $\Delta(\mu)$ is a positive linear functional.*

(ii) *μ is a real measure if and only if $\Delta(\mu)$ is a self-adjoint linear functional.*

Notice that ℓ^∞ is naturally isomorphic to $C_b(\mathbb{N}) \cong C(\beta\mathbb{N})$, where $\beta\mathbb{N}$ denotes the Stone–Čech compactification of $\mathbb{N}$. Therefore Theorem 9.4.9 identifies $(\ell^\infty)^*$ with $\mathbb{M}_b(\beta\mathbb{N})$. Similarly, if we treat $L^\infty[0,1]$ as a commutative C*-algebra and $X := \Omega(L^\infty[0,1])$, then the dual of $L^\infty[0,1]$ may be identified with $\mathbb{M}_b(X)$. For what it is worth, let us record these facts.

9.4.11 COROLLARY *There are isometric isomorphisms*

$$\Delta : \mathbb{M}_b(\beta\mathbb{N}) \to (\ell^\infty)^* \text{ and}$$

$$\Delta : \mathbb{M}_b(\Omega(L^\infty[0,1])) \to L^\infty[0,1]^*.$$

Unfortunately, the spaces $\beta\mathbb{N}$ and $\Omega(L^\infty[0,1])$ are both quite unwieldy to work with. Therefore one usually works with a somewhat user-friendly description of $(\ell^\infty)^*$ and $(L^\infty[0,1])^*$ in terms of *finitely additive* set functions of bounded variation. The details of this are worked out in the exercises below.

9.5 Exercises

Positive Linear Functionals on C*-Algebras

EXERCISE 9.1 ($\maltese$) Let **A** be a C*-algebra and $a, b \in \mathbf{A}_{sa}$.

 (i) If $a \leq b$, then prove that $c^*ac \leq c^*bc$ for any $c \in \mathbf{A}$.

 (ii) If **A** is unital, then prove that $b \leq \|b\|1_{\mathbf{A}}$.

EXERCISE 9.2 Let **A** be a C*-algebra and $a, b \in \mathbf{A}_+$ be such that $0 \leq a \leq b$. Prove that

$$\|a\| \leq \|b\|.$$

Hint: First assume **A** is unital and use Exercise 9.1.

EXERCISE 9.3 In the C*-algebra $\mathbf{A} = M_2(\mathbb{C})$, give an example of two elements $a, b \in \mathbf{A}_+$ such that $a \leq b$ but $a^2 \not\leq b^2$.

EXERCISE 9.4 Let **A** be a unital C*-algebra.

 (i) If $a \in \mathbf{A}_{sa}$ is a self-adjoint element with $\|a\| \leq 1$, then prove that $(1_{\mathbf{A}} - a^2) \in \mathbf{A}_+$.

 (ii) If $a \in \mathbf{A}_{sa}$ as in part (i), then prove that the elements $u := a + i\sqrt{1_{\mathbf{A}} - a^2}$ and $v := a - i\sqrt{1_{\mathbf{A}} - a^2}$ are both unitaries in **A**.

 (iii) Conclude that every element in **A** can be expressed as a linear combination of unitaries in **A**.

EXERCISE 9.5 ($\maltese$) Let **A** be a unital C*-algebra and $a \in GL(\mathbf{A})$.

 (i) Prove that $|a| \in GL(\mathbf{A})$ (see Definition 9.1.7).

 (ii) Prove that $\omega(a) := a|a|^{-1}$ is a unitary.

EXERCISE 9.6 Let **A** be a unital C*-algebra and $a \in GL(\mathbf{A})$.

 (i) Prove that there is a scalar $\lambda > 0$ such that $|a| \geq \lambda 1_{\mathbf{A}}$.

 (ii) Prove that there is a continuous function $f : [0, 1] \to GL(\mathbf{A})$ such that $f(0) = a$ and $f(1) = \omega(a)$ (see Exercise 9.5).

EXERCISE 9.7

(i) Let X be a compact Hausdorff space and let $f \in C(X)$ be a real-valued function. If $g, h \in C(X)_+$ are two positive elements such that $f = g - h$ and $gh = 0$, then prove that $g = f_+$ and $h = f_-$, where f_+ and f_- are the elements constructed in Lemma 9.1.9.

(ii) Let $\mathbf{A}$ be a unital C*-algebra and $a \in \mathbf{A}$ be a self-adjoint element. Suppose $b, c \in \mathbf{A}_+$ are two positive elements in A such that $a = b - c$ and $bc = cb = 0$. Prove that $b = a_+$ and $c = a_-$.

EXERCISE 9.8 Let $\mathbf{A}$ be a unital C*-algebra and $\tau : \mathbf{A} \to \mathbb{C}$ be a positive linear functional. For any $a \in \mathbf{A}$, prove that $|\tau(a)|^2 \leq \|\tau\| \tau(a^*a)$.

EXERCISE 9.9 (✹) Let $\tau : \mathbf{A} \to \mathbb{C}$ be a positive linear functional on a C*-algebra $\mathbf{A}$. Define $N_\tau := \{a \in \mathbf{A} : \tau(a^*a) = 0\}$.

(i) If $a \in N_\tau$ and $x \in \mathbf{A}$, prove that $\tau(xa) = \tau(a^*x) = 0$.

(ii) Prove that N_τ is a closed left ideal of $\mathbf{A}$.

(iii) Give an example to show that N_τ is not necessarily a two-sided ideal.

The Riesz–Markov–Kakutani Theorem

EXERCISE 9.10 Let X be a compact Hausdorff space.

(i) If K is a compact set and U is an open set such that $K \subset U$, then prove that there is a compact set $C \in \mathfrak{A}_X$ such that $K \subset C \subset U$.

(ii) If U is an open F_σ set, then prove that $U \in \mathfrak{A}_X$.

EXERCISE 9.11 Let X be a compact Hausdorff space. For each countable family $\mathcal{F} \subset C(X)$, let $\mathfrak{M}_\mathcal{F}(X)$ denote the smallest σ-algebra on X that makes every member of $\mathcal{F}$ measurable. Prove that

$$\mathfrak{A}_X = \bigcup \mathfrak{M}_\mathcal{F}(X),$$

where the union is taken over all countable subsets $\mathcal{F} \subset C(X)$.

EXERCISE 9.12 Let X be a compact Hausdorff space and let $E \in \mathfrak{A}_X$. By Exercise 9.11, there is a countable collection $\mathcal{F} = \{f_1, f_2, \ldots\} \subset C(X)$ such that

$E \in \mathfrak{M}_{\mathcal{F}}(X)$. Define $Y := \prod_{n=1}^{\infty} \mathbb{C}$ equipped with the product topology and define $F : X \to Y$ by

$$F(x) := (f_n(x))_{n=1}^{\infty}.$$

(i) Prove that $\mathfrak{M}_{\mathcal{F}}(X) = \{F^{-1}(A) : A \in \mathfrak{B}_Y\}$.

(ii) Prove that $E = F^{-1}(F(E))$.

EXERCISE 9.13 (⚒) Let X be a compact Hausdorff space.

(i) Prove that every compact Baire set is a G_δ set.

(ii) Prove that every open Baire set is an F_σ set.

Hint: If $E \in \mathfrak{A}_X$ is compact, choose $F : X \to Y$ is as in Exercise 9.12 and prove that $F(E)$ is a G_δ set.

EXERCISE 9.14 The aim of this problem (and the next one) is to construct a compact Hausdorff space for which the Baire σ-algebra and the Borel σ-algebra do not coincide.

Let $I := [0,1]$, J be an uncountable set and $X := \prod_{\alpha \in J} I$ equipped with the product topology. Fix a point $x = (x_\alpha) \in X$.

(i) If V is a basic open set containing x, then prove that there is a finite set $F_V \subset J$ such that

$$\{y = (y_\alpha) \in X : y_\beta = x_\beta \text{ for all } \beta \in F_V\} \subset V.$$

(ii) If K is a G_δ set containing x, then prove that there is a countable set $F_0 \subset J$ such that

$$\{y = (y_\alpha) \in X : y_\beta = x_\beta \text{ for all } \beta \in F_0\} \subset K.$$

(iii) From Exercise 9.13, conclude that $\{x\}$ is a Borel set that is not a Baire set.

EXERCISE 9.15 Let $I := [0,1]$, J be an uncountable set and $X := \prod_{\alpha \in J} I$ equipped with the product topology. If $F \subset J$ is finite, define $f_F : X \to \mathbb{C}$ by

$$f_F(x) := \prod_{\alpha \in F} x_\alpha$$

(with the understanding that $f_\emptyset$ is the constant function $\mathbf{1}$). Let $\mathcal{A}$ denote the subalgebra of $C(X)$ generated by the set $\{f_F : F \subset J \text{ finite}\}$.

(i) Prove that $\mathcal{A}$ is dense in $C(X)$.

(ii) If $f \in C(X)$, prove that there is a countable set $F_0 \subset J$ such that if $x = (x_\alpha), y = (y_\alpha) \in X$ are two points such that $x_\alpha = y_\alpha$ for all $\alpha \in F_0$, then $f(x) = f(y)$ (in other words, $f(x)$ only depends on countably many components of x).

(iii) Let $\mathfrak{M}$ denote the set of all subsets of $A \in \mathfrak{A}_X$ such that $\chi_A(x)$ depends only on countably many components of x. Prove that $\mathfrak{M}$ is a σ-algebra and that $\mathfrak{M} = \mathfrak{A}_X$.

(iv) If $f : X \to \mathbb{C}$ is a function that is $\mathfrak{A}_X$-measurable, then prove that $f(x)$ depends only on countably many components of x.

(v) Prove that $\mathfrak{A}_X \neq \mathfrak{B}_X$.

EXERCISE 9.16 Let X be a compact Hausdorff space and let μ be a Borel measure on X such that

$$\mu(E) = \inf\{\mu(U) : U \text{ open and } E \subset U\}$$

for all $E \in \mathfrak{B}_X$. For each compact set K, prove that there is a compact G_δ set C such that $\mu(K) = \mu(C)$.

9.5.1 DEFINITION Let X be a compact Hausdorff space. A Borel measure μ on X is said to be **regular** if it satisfies Equation 9.5 and Equation 9.6 on the Borel σ-algebra.

EXERCISE 9.17 Let X be a compact Hausdorff space and let μ and ν are two finite, positive, regular Borel measures on X. If

$$\int_X f d\mu = \int_X f d\nu$$

for all $f \in C(X)$, then prove that $\mu = \nu$ on $\mathfrak{B}_X$.

Hint: Use Exercise 9.16 along with Lemma 9.2.5.

EXERCISE 9.18 Let X be a compact Hausdorff space and let μ be a finite, positive Baire measure on X. Prove that there is atmost one positive, regular Borel measure $\widetilde{\mu}$ on X such that

$$\widetilde{\mu}|_{\mathfrak{A}_X} = \mu.$$

Note: In fact, one can prove that there is *exactly* one regular Borel measure that extends μ. Therefore there is a bijection between the set of all finite, positive Baire measures on X and the set of all finite, positive, regular Borel measures on X.

EXERCISE 9.19 Let X be a compact Hausdorff space and μ be a finite, positive, regular Borel measure on X. Define $\tau : C(X) \to \mathbb{C}$ by

$$\tau(f) := \int_X f d\mu.$$

Let U denote the union of all open sets that have measure zero and let $K := X \setminus U$.

(i) Prove that $\mu(U) = 0$.

(ii) For any $x \in K$ and any open set V containing x, prove that $\mu(V) > 0$.

(iii) If $N_\tau := \{f \in C(X) : \tau(|f|^2) = 0\}$, then prove that $N_\tau = C_0(U)$ (see Exercise 9.9).

9.5.2 **DEFINITION** The compact set K described above is called the **support** of μ and is denoted by $\mathrm{supp}(\mu)$.

EXERCISE 9.20 Let X be a compact Hausdorff space and μ be a finite, positive, regular Borel measure. For a point $x \in X$, prove that $x \in \mathrm{supp}(\mu)$ if and only if

$$\int_X f d\mu > 0$$

for any $f \in C(X)_+$ such that $f(x) > 0$.

EXERCISE 9.21 Let X be a compact Hausdorff space and μ be a finite, positive, regular Borel measure on X.

(i) For any open set $U \subset X$, prove that

$$\mu(U) = \sup \left\{ \int_X f d\mu : f \in C(X), 0 \leq f \leq 1 \text{ and } f|_{X \setminus U} \equiv 0 \right\}.$$

(ii) For any compact set $K \subset X$, prove that

$$\mu(K) = \inf \left\{ \int_X f d\mu : f \in C(X)_+, f \geq \chi_K \right\}.$$

EXERCISE 9.22 Let S be a locally compact normal space and A be a closed subset of S. Let $\eta : S \to \beta S$ denote the natural inclusion map. Prove that the closure of $\eta(A)$ in βS is homeomorphic to βA.

Hint: Prove that both sets satisfy Proposition 9.2.12.

EXERCISE 9.23 Let $\beta\mathbb{N}$ denote the Stone–Čech compactification of $\mathbb{N}$. As before, we think of $\mathbb{N}$ as a subset of $\beta\mathbb{N}$.

 (i) Prove that any subset of $\mathbb{N}$ is open in $\beta\mathbb{N}$.

 (ii) If A and B are two disjoint subsets of $\mathbb{N}$, then prove that $\overline{A}$ and $\overline{B}$ are disjoint in $\beta\mathbb{N}$.

EXERCISE 9.24

 (i) If (x_n) is a sequence in $\mathbb{N}$ that converges in $\beta\mathbb{N}$, then prove that (x_n) is eventually constant (i.e. there exists $N \in \mathbb{N}$ such that $x_j = x_N$ for all $j \geq N$).

 (ii) Prove that $\beta\mathbb{N}$ is not sequentially compact (and hence is not metrizable).

Note: The fact that $\beta\mathbb{N}$ is not metrizable also follows from Exercise 9.40 and the fact that ℓ^∞ is not separable.

Complex Measures

EXERCISE 9.25 (⚒) Prove Proposition 9.3.6.

EXERCISE 9.26 Let μ be a positive measure and ν be a complex measure on a measurable space $(X, \mathfrak{M})$. If $\nu \ll \mu$, then prove that $|\nu| \ll \mu$.

EXERCISE 9.27 Let μ_1, μ_2, μ_3 be three finite, positive measures on a measurable space $(X, \mathfrak{M})$.

 (i) If $\mu_1 \ll \mu_2$ and $\mu_2 \ll \mu_3$, then prove that $\mu_1 \ll \mu_3$.

 (ii) Give an example to show that $\mu_1 \ll \mu_2$ does not imply that $\mu_2 \ll \mu_1$.

EXERCISE 9.28 Give examples of two finite, positive, Borel measures μ and ν on $[0,1]$ such that $\mu \not\ll \nu$ and $\nu \not\ll \mu$.

EXERCISE 9.29 Let $X = \mathbb{N}$ and $\mathfrak{M} = 2^X$. Let $\lambda > 0$ be a fixed real number. Define two measures μ and ν on $(X, \mathfrak{M})$ by the formulae

$$\mu(E) = \begin{cases} |E| & : \text{ if E is finite,} \\ \infty & : \text{ otherwise.} \end{cases}$$

$$\nu(E) = \sum_{n \in E} e^{-\lambda} \frac{\lambda^n}{n!}.$$

Prove that $\mu \ll \nu$ and $\nu \ll \mu$.

EXERCISE 9.30 Let μ and ν be two positive measures on a measurable space $(X, \mathfrak{M})$. If $\nu \ll \mu$ and $\epsilon > 0$ is fixed, then prove that there is $\delta > 0$ such that

$$\nu(E) < \epsilon$$

whenever $E \in \mathfrak{M}$ is such that $\mu(E) < \delta$.

Hint: Suppose this were false, construct a sequence $E_1 \supset E_2 \supset \ldots$ of measurable sets such that $\mu(E_n) < 2^{-n}$ and $\nu(E_n) \geq \epsilon$ for all $n \in \mathbb{N}$.

EXERCISE 9.31 Let μ be a real measure on a measurable space $(X, \mathfrak{M})$ and let $h = \frac{d\mu}{d|\mu|}$ be the Radon–Nikodym derivative of μ with respect to $|\mu|$.

(i) Prove that h is real-valued almost everywhere with respect to $|\mu|$.

(ii) Let $A := \{x \in X : h(x) = 1\}$ and $B := \{x \in X : h(x) = -1\}$. Define two measures on $(X, \mathfrak{M})$ by the formulae

$$\mu^a(E) := \mu(E \cap A) \text{ and } \mu^b(E) := -\mu(E \cap B).$$

Prove that μ^a and μ^b are both positive measures such that $\mu = \mu^a - \mu^b$.

(iii) Suppose λ and ν are two positive measures on $(X, \mathfrak{M})$ such that $\mu = \lambda - \nu$, then prove that $\mu^a \leq \lambda$ and $\mu^b \leq \nu$.

EXERCISE 9.32 Let μ be a real measure on a measurable space $(X, \mathfrak{M})$ and let μ^a and μ^b be the measures defined in Exercise 9.31. Prove that

$$\mu^a = \frac{1}{2}(|\mu| + \mu) \text{ and } \mu^b = \frac{1}{2}(|\mu| - \mu).$$

Note: The measures μ^a and μ^b are referred to as μ^+ and μ^- respectively in Section A.2.

EXERCISE 9.33 Let $(X, \mathfrak{M})$ be a measurable space and let $\mu : \mathfrak{M} \to \mathbb{C}$ be a function satisfying the following conditions.

(a) μ is *finitely additive*: If $E, F \in \mathfrak{M}$ are disjoint, then $\mu(E \cup F) = \mu(E) + \mu(F)$.

(b) If $\{F_1, F_2, \ldots\}$ is a sequence in $\mathfrak{M}$ such that $F_1 \supset F_2 \supset \ldots$ such that

$$\bigcap_{n=1}^{\infty} F_n = \varnothing,$$

then $\lim_{n \to \infty} \mu(F_n) = 0$.

Prove that μ is countably additive (and thus a complex measure).

EXERCISE 9.34 Let $(X, \mathfrak{M})$ be a measurable space and let $\mathbb{M}_c(X)$ denote the space of all complex measures on X, treated as a normed linear space as in Definition 9.4.1. Prove that $\mathbb{M}_c(X)$ is a Banach space.

Hint: Use Exercise 9.33.

The Dual of $C(X)$

EXERCISE 9.35 (✗) Let $\mathbb{M}_b(X)$ denote the set of all complex Baire measures on a compact Hausdorff space X. Under the operations described in Definition 9.4.1, prove that $\mathbb{M}_b(X)$ is a normed linear space.

EXERCISE 9.36 (✗) Prove that the map $\Delta : \mathbb{M}_b(X) \to C(X)^*$ defined in Definition 9.4.1 is linear.

EXERCISE 9.37 Let $\mathbf{A}$ be a C*-algebra and $\rho : \mathbf{A}_{sa} \to \mathbb{R}$ be an $\mathbb{R}$-linear functional. Prove that the formula

$$\tau(a) := \rho\left(\frac{a + a^*}{2}\right) + i\rho\left(\frac{a - a^*}{2i}\right)$$

defines a $\mathbb{C}$-linear functional on $\mathbf{A}$.

EXERCISE 9.38 (✗) Prove that the positive linear functional constructed in Lemma 9.4.7 is unique.

EXERCISE 9.39 Let X be a compact Hausdorff space and $\rho \in C(X)^*$ be a self-adjoint linear functional. If ρ_+ are ρ_- are the two positive linear functionals constructed in Lemma 9.4.8 such that $\rho = \rho_+ - \rho_-$, then prove that $\|\rho\| = \|\rho_+\| + \|\rho_-\|$.

EXERCISE 9.40 (⚒) If X is a compact metric space, then prove that $C(X)$ is separable.

Hint: X has a countable dense set D and for each point $x \in D$, the function $f_x : X \to \mathbb{R}$ given by $y \mapsto d(x,y)$ is continuous.

9.41 EXERCISE (PROKHOROV, 1956) *Let X be a compact metric space and (μ_n) be a sequence of positive Baire measures on X such that $\mu_n(X) = 1$ for all $n \in \mathbb{N}$. Prove that there is a measure $\mu \in \mathbb{M}_b(X)$ and a subsequence (μ_{n_k}) of (μ_n) such that*

$$\lim_{k\to\infty} \int_X f d\mu_{n_k} = \int_X f d\mu$$

for all $f \in C(X)$.

Hint: Use Exercise 9.40 together with Helley's Theorem.

EXERCISE 9.42 Let $X := [0,1]$. For each $n \in \mathbb{N}$, define Baire measures μ_n by the formula

$$\mu_n(E) := m(E \cap [0,1/n]),$$

where m denotes the Lebesgue measure.

(i) Prove that there is a measure μ such that $\mu_n \xrightarrow{w^*} \mu$ (Here, we identify μ_n with $\Delta(\mu_n) \in C(X)^*$).

(ii) Prove that $\mu_n \ll m$ for all $n \in \mathbb{N}$, but $\mu \not\ll m$.

EXERCISE 9.43 Prove Corollary 9.4.10.

The Dual of ℓ^∞ and $L^\infty[0,1]$

EXERCISE 9.44 A function $\mu : 2^{\mathbb{N}} \to \mathbb{C}$ is said to be *finitely additive* if, whenever $E, F \in 2^{\mathbb{N}}$ are disjoint, then $\mu(E \cup F) = \mu(E) + \mu(F)$. Given such a function, define $|\mu| : 2^{\mathbb{N}} \to [0,\infty]$ by

$$|\mu|(E) = \sup\left\{\sum_{i=1}^n |\mu(E_i)|\right\}$$

where the supremum is taken over all finite collections $\{E_1, E_2, \ldots, E_n\} \subset 2^{\mathbb{N}}$ of mutually disjoint sets satisfying $\bigcup_{i=1}^n E_i \subset E$.

Prove that $|\mu|$ is also finitely additive.

9.5.3 DEFINITION

(i) A finitely additive function $\mu : 2^{\mathbb{N}} \to \mathbb{C}$ is said to have **bounded variation** if $|\mu|(\mathbb{N}) < \infty$. We write $\mathbb{M}_f(\mathbb{N})$ for the set of all such finitely additive functions of bounded variation.

(ii) For $\mu, \nu \in \mathbb{M}_f(\mathbb{N})$ and $\alpha \in \mathbb{C}$, define $(\mu + \nu)$ and $\alpha\mu$ by the formulae

$$(\mu + \nu)(E) := \mu(E) + \nu(E) \text{ and } (\alpha\mu)(E) := \alpha\mu(E).$$

(iii) For $\mu \in \mathbb{M}_f(\mathbb{N})$, define

$$\|\mu\| := |\mu|(\mathbb{N}).$$

EXERCISE 9.45 Under these operations, prove that $\mathbb{M}_f(\mathbb{N})$ is a normed linear space.

EXERCISE 9.46 Prove that the set of all simple functions in ℓ^∞ is dense in ℓ^∞.

EXERCISE 9.47 Let $\mu \in \mathbb{M}_f(\mathbb{N})$ and let $\mathcal{S}$ denote the set of all simple functions in ℓ^∞.

(i) Let $f \in \mathcal{S}$ be given by $f = \sum_{i=1}^n \alpha_i \chi_{E_i}$ where $\{E_1, E_2, \ldots, E_n\}$ are mutually disjoint sets such that $\bigcup_{i=1}^n E_i = \mathbb{N}$. Define $\psi : \mathcal{S} \to \mathbb{C}$ by

$$\psi(f) := \sum_{i=1}^n \alpha_i \mu(E_i).$$

Prove that ψ defines a bounded linear functional on $\mathcal{S}$.

(ii) Conclude that there is a unique bounded linear functional $\varphi_\mu \in (\ell^\infty)^*$ such that $\varphi_\mu|_{\mathcal{S}} = \psi$.

EXERCISE 9.48 Define $\Delta : \mathbb{M}_f(\mathbb{N}) \to (\ell^\infty)^*$ by

$$\Delta(\mu) := \varphi_\mu,$$

with φ_μ defined as in Exercise 9.47. Prove that Δ is linear and that $\|\Delta(\mu)\| \leq \|\mu\|$ for all $\mu \in \mathbb{M}_f(\mathbb{N})$.

EXERCISE 9.49 If $\varphi \in (\ell^\infty)^*$, define $\mu : 2^{\mathbb{N}} \to \mathbb{C}$ by

$$\mu(E) := \varphi(\chi_E).$$

(i) Prove that $\mu \in \mathbb{M}_f(\mathbb{N})$ and that $\|\mu\| \leq \|\varphi\|$.

(ii) Prove that the map

$$\Delta : \mathbb{M}_f(\mathbb{N}) \to (\ell^\infty)^*$$

is an isometric isomorphism. Conclude that there is an isometric isomorphism $\mathbb{M}_f(\mathbb{N}) \to \mathbb{M}_b(\beta\mathbb{N})$.

EXERCISE 9.50 Using (a modification of) the argument in Exercise 4.44, construct an element $\mu \in \mathbb{M}_f(\mathbb{N})$ that is not countably additive. If every element in $\mathbb{M}_b(\beta\mathbb{N})$ is countably additive, then why does this not violate the conclusion of Exercise 9.49?

Note: For an explanation of this problem, have a look at Carothers [7, Chapter 16].

9.5.4 DEFINITION Let $X = [0, 1]$, $\mathfrak{M}$ denote the σ-algebra of Lebesgue measurable subsets of $[0, 1]$ and let m denote the Lebesgue measure. A finitely additive function $\mu : \mathfrak{M} \to \mathbb{C}$ is said to be **absolutely continuous** with respect to m if

$$\mu(E) = 0$$

for any $E \in \mathfrak{M}$ with $m(E) = 0$. Define $\mathbb{M}_f([0, 1])$ to be the set of all such finitely additive functions that have bounded variation (as defined above) and are absolutely continuous with respect to m.

EXERCISE 9.51 Using the ideas of the previous problems, prove that there is an analogous isometric isomorphism

$$\Delta : \mathbb{M}_f([0, 1]) \to (L^\infty[0, 1])^*.$$

Note: If $\mu \in \mathbb{M}_f([0, 1])$ and $f \in L^\infty[0, 1]$, it is customary to write

$$\int_{[0,1]} f \, d\mu := \Delta(\mu)(f).$$

Additional Reading

- The notion of *positivity* may also be explored in the context of partially ordered real vector spaces (see the comments after Definition 9.1.8). A positive operator is then a linear transformation that preserves this partial order structure. These objects provide an interesting playground for functional analytic techniques and a good introduction to the subject is this book.

Aliprantis, Charalambos D. and Burkinshaw, Owen (2006). *Positive Operators*. Reprint of the 1985 original. xx+376. Dordrecht: Springer, ISBN: 978-1-4020-5007-7; 1-4020-5007-0. DOI: 10.1007/978-1-4020-5008-4.

- The proof of the Riesz–Markov–Kakutani Theorem presented in this chapter is due to Garling [19]. The more standard proof of the theorem (in full generality) can be found in the following book by Halmos. It also discusses the subtle differences between Baire and Borel measures, which would be useful should you choose to venture beyond the metrizable case.

Halmos, Paul R. (2013). *Measure Theory*. Vol. 18. Springer.

Chapter 10

Normal Operators on Hilbert Spaces

10.1 Compact Normal Operators

In the final chapter of the book, we will bring to bear all that we have learnt on the study of normal operators. In the process, we will prove a far reaching generalization of the Spectral Theorem for self-adjoint matrices that you would have first encountered in linear algebra. In fact, our first step in the process is to revisit that very result and prove it with more refined machinery.

For now let $\mathbf{H}$ denote a *finite dimensional* complex Hilbert space and let $T \in \mathcal{B}(\mathbf{H})$. If Λ is an ordered orthonormal basis for $\mathbf{H}$, we write $[T]_\Lambda$ for the matrix of T with respect to Λ. In other words, if $\Lambda = \{e_1, e_2, \ldots, e_n\}$, then the $(i,j)^{th}$ entry of $[T]_\Lambda$ is $\langle T(e_j), e_i \rangle$.

10.1.1 DEFINITION Let $\mathbf{H}$ be a finite dimensional Hilbert space. An operator $T \in \mathcal{B}(\mathbf{H})$ is said to be (unitarily) *diagonalizable* if $\mathbf{H}$ has an orthonormal basis consisting of eigenvectors of T.

Equivalently, T is diagonalizable if there is an orthonormal basis Λ of $\mathbf{H}$ such that $[T]_\Lambda$ is a diagonal matrix.

10.1.2 REMARK A word of warning: Our notion of diagonalizability is stronger than the usual notion from linear algebra. For instance, if $T : \mathbb{C}^2 \to \mathbb{C}^2$ is the operator $T(x,y) = (x + y, 2y)$, then the matrix of T in the standard basis is

$$A = \begin{pmatrix} 1 & 1 \\ 0 & 2 \end{pmatrix}.$$

445

This matrix is *similar* to a diagonal matrix. In other words, there is an invertible matrix P such that $P^{-1}AP$ is diagonal. However, the matrix P cannot be chosen to be a unitary, so T is not unitarily diagonalizable (why this is so is evident from Lemma 10.1.3 below).

Despite this, we will henceforth use the term *diagonalizable* to mean *unitarily diagonalizable*. The reason is that this condition is more easily captured in algebraic terms (without reference to the minimal or characteristic polynomial).

10.1.3 LEMMA *Let* **H** *be a finite dimensional Hilbert space. If* $T \in \mathcal{B}(\mathbf{H})$ *is diagonalizable, then* T *is normal.*

Proof. Let Λ be an orthonormal basis of **H** consisting of eigenvectors of T. For $v, w \in \Lambda$ and let $\lambda, \mu \in \mathbb{C}$ be such that $Tv = \lambda v$ and $Tw = \mu w$. If $v \neq w$, then

$$\langle T^*v, w \rangle = \langle v, Tw \rangle = \overline{\mu}\langle v, w \rangle = 0.$$

If $v = w$, then $\langle T^*v, w \rangle = \overline{\lambda}\|v\|^2 = \overline{\lambda}$. Hence $T^*v = \overline{\lambda}v$ for all $v \in \Lambda$. Therefore

$$TT^*(v) = |\lambda|^2 v = T^*T(v).$$

This is true for any $v \in \Lambda$ and therefore T is normal.   □

Our aim is to prove the converse of this lemma and the next two results help us get there. Notice that both these lemmas hold in infinite dimensional spaces as well.

10.1.4 LEMMA *Let* **H** *be a Hilbert space and* $T \in \mathcal{B}(\mathbf{H})$ *be a normal operator. If* $v \in \mathbf{H}$ *is an eigenvector of* T *corresponding to the eigenvalue* λ, *then* v *is an eigenvector of* T^* *corresponding to the eigenvalue* $\overline{\lambda}$.

Proof. If $Tv = \lambda v$, then $\|(T - \lambda I)v\| = 0$. Since T is normal, so is $(T - \lambda I)$. By Theorem 8.1.14, $\|Tv - \overline{\lambda}v\| = \|(T - \lambda I)^*v\| = \|(T - \lambda I)v\| = 0$.   □

The next lemma is entirely routine to verify and is left as an exercise (Exercise 10.1).

10.1.5 LEMMA *If* $T \in \mathcal{B}(\mathbf{H})$ *and* $\mathbf{M} < \mathbf{H}$ *is a subspace such that* $T(\mathbf{M}) \subset \mathbf{M}$, *then* $T^*(\mathbf{M}^{\perp}) \subset \mathbf{M}^{\perp}$.

We now arrive at the celebrated Spectral Theorem in the finite dimensional setting. Notice how simple the proof is given what we now know. Also notice how important it is that we are working with *complex* Hilbert spaces.

10.1.6 THEOREM (SPECTRAL THEOREM) *Let* **H** *be a finite dimensional Hilbert space and* $T \in \mathcal{B}(\mathbf{H})$ *be a normal operator. Then* T *is diagonalizable.*

Proof. We induct on $\dim(\mathbf{H})$. If $\dim(\mathbf{H}) = 1$, there is nothing to prove, so assume that $\dim(\mathbf{H}) \geq 2$ and that the result is true for any Hilbert space **K** with $\dim(\mathbf{K}) < \dim(\mathbf{H})$.

If $T \in \mathcal{B}(\mathbf{H})$ is normal, choose an eigenvector v of T (which exists by the Fundamental Theorem of Algebra). Also, assume that $\|v\| = 1$ and set $\mathbf{M} := \mathrm{span}(\{v\})$. Then $T(\mathbf{M}) \subset \mathbf{M}$, so $T^*(\mathbf{M}^\perp) \subset \mathbf{M}^\perp$ by Lemma 10.1.5. By Lemma 10.1.4, $T^*(\mathbf{M}) \subset \mathbf{M}$ as well and thus $T(\mathbf{M}^\perp) \subset \mathbf{M}^\perp$.

If $\mathbf{K} := \mathbf{M}^\perp$, then both T and T^* now restrict to operators on **K**, which are also normal on **K**. By induction, **K** has an orthonormal basis $\Lambda_{\mathbf{K}}$ consisting of eigenvectors of T. It is now easy to verify that $\Lambda_{\mathbf{H}} := \Lambda_{\mathbf{K}} \cup \{v\}$ is an orthonormal basis for **H** consisting of eigenvectors of T. $\square$

Let us take a look at the proof a little more closely, with a view towards generalizing it to infinite dimensional spaces. The first step ensures that the operator has an eigenvector. The second step relied on Lemma 10.1.5 and the third step was the induction step. Naturally, not every operator on an infinite dimensional space has eigenvectors, but we do know one class of operators for which that step would work.

10.1.7 DEFINITION Let **H** be a (possibly infinite dimensional) Hilbert space and $T \in \mathcal{K}(\mathbf{H})$ be a compact operator. We say that T is **diagonalizable** if **H** has an orthonormal basis consisting of eigenvectors of T.

We now have all the tools to prove the Spectral Theorem for compact operators. In principle, we wish to apply the same argument as above. However, we replace the induction argument by Zorn's Lemma since we are working with infinite dimensional spaces.

10.1.8 THEOREM (SPECTRAL THEOREM FOR COMPACT OPERATORS) *Let* **H** *be a Hilbert space. If* $T \in \mathcal{K}(\mathbf{H})$, *then* T *is diagonalizable if and only if it is normal.*

Proof. If T is diagonalizable, then the proof of Lemma 10.1.3 goes through verbatim to show that T must be normal. Now suppose T is a compact normal operator and that T is non-zero.

(i) By Corollary 8.5.6, $\|T\| = r(T)$. Since $\sigma(T)$ is compact, there exists $\lambda \in \sigma(T)$ such that $|\lambda| = \|T\|$. Since $T \neq 0, \lambda$ is non-zero. Since T is compact,

Theorem 7.5.7 ensures that λ is an eigenvalue. Hence there is a unit vector v such that $Tv = \lambda v$.

(ii) In order to apply Zorn's Lemma, we set

$\mathcal{F} := \{\Lambda \subset \mathbf{H} : \Lambda$ is an orthonormal set consisting of eigenvectors of $T\}$.

Then $\mathcal{F}$ is non-empty since $\{v\} \in \mathcal{F}$. Also, $\mathcal{F}$ is partially ordered by inclusion: $\Lambda_1 \leq \Lambda_2$ if $\Lambda_1 \subset \Lambda_2$. Now suppose $\mathcal{C}$ is a totally ordered subset of $\mathcal{F}$, then

$$\Lambda_0 := \bigcup_{\Lambda \in \mathcal{C}} \Lambda$$

is clearly an upper bound for $\mathcal{C}$ in $\mathcal{F}$. Therefore $\mathcal{F}$ has a maximal element, which we denote by $\Lambda_{\mathbf{H}}$.

(iii) We claim that $\Lambda_{\mathbf{H}}$ is an orthonormal basis of $\mathbf{H}$. To prove this, consider $\mathbf{M} := \mathrm{span}(\Lambda_{\mathbf{H}})$. Since $T(\Lambda_{\mathbf{H}}) \subset \mathbf{M}$, it follows that $T(\mathbf{M}) \subset \mathbf{M}$. Hence $T^*(\mathbf{M}^\perp) \subset \mathbf{M}^\perp$. Furthermore, every member of $\Lambda_{\mathbf{H}}$ is also an eigenvector for T^* by Lemma 10.1.4. Therefore $T(\mathbf{M}^\perp) \subset \mathbf{M}^\perp$ as well. Hence both T and T^* restrict to operators on $\mathbf{K} := \mathbf{M}^\perp$.

Since T is compact and normal, so is $S := T|_{\mathbf{K}}$ (do verify this). Suppose $S \neq 0$, then the argument of part (i) applies and $\mathbf{K}$ contains an eigenvector v_0 of T of norm one. However, $v_0 \in \Lambda_{\mathbf{H}}^\perp$ which would imply that $\Lambda_{\mathbf{H}} \cup \{v_0\} \in \mathcal{F}$. This would contradict the maximality of $\Lambda_{\mathbf{H}}$. Therefore it must happen that $S = 0$.

However, if $S = 0$, then *every* vector of $\mathbf{K}$ is an eigenvector of S (and therefore of T). The only conclusion then is that $\mathbf{K} = \{0\}$. Hence $\mathbf{M}^\perp = \{0\}$ and $\mathbf{M}$ is dense in $\mathbf{H}$. By Lemma 3.3.3, $\Lambda_{\mathbf{H}}$ is an orthonormal basis of $\mathbf{H}$. $\qquad\square$

Now that we have settled the question for compact operators, how do we generalize this idea to arbitrary bounded operators? In this general setting, the notion of diagonalizability suffers from a fatal flaw: There exist normal operators which do not have any eigenvectors. For instance, if $\mathbf{H} = L^2[0,1]$ and $T \in \mathcal{B}(\mathbf{H})$ is the operator

$$T(f)(x) := x f(x),$$

then T has no eigenvalues (Example 7.4.12).

To find a meaningful analogue of diagonalizability, we must therefore re-interpret the notion of a diagonal matrix. As it turns out, the correct analogue is that of a multiplication operator.

10.2 Multiplication Operators

Before we discuss the general notion of diagonalizability, we will take a moment to study these multiplication operators and understand their basic properties.

> **Note.** In order to keep the measure theory and operator theory somewhat manageable, we will now restrict ourselves to *separable* Hilbert spaces and σ-finite measure spaces.

10.2.1 DEFINITION Let $(X, \mathfrak{M}, \mu)$ be a σ-finite measure space. Given a measurable, essentially bounded function $\phi : X \to \mathbb{C}$, define $M_\phi : L^2(X, \mu) \to L^2(X, \mu)$ by

$$M_\phi(f) := \varphi f.$$

For any $f \in L^2(X, \mu)$,

$$\int_X |\phi f|^2 d\mu \le \|\phi\|_\infty^2 \int_X |f|^2 d\mu.$$

Therefore M_ϕ is a well-defined, bounded linear operator with $\|M_\phi\| \le \|\phi\|_\infty$. This is called the *multiplication operator* associated to ϕ.

10.2.2 EXAMPLE

(i) If $X = \{1, 2, \ldots, n\}$ is a finite set and μ is the counting measure, then $\phi : X \to \mathbb{C}$ corresponds to a tuple $(\phi(1), \phi(2), \ldots, \phi(n)) \in \mathbb{C}^n$. Furthermore, there is a natural identification

$$L^2(X, \mu) = \mathbb{C}^n.$$

Under this identification, M_ϕ corresponds to the operator $T_\phi : \mathbb{C}^n \to \mathbb{C}^n$ defined by $T_\phi(x_1, x_2, \ldots, x_n) = (\phi(1)x_1, \phi(2)x_2, \ldots, \phi(n)x_n)$. If Λ denotes the standard orthonormal basis of $\mathbb{C}^n$, then $[T_\phi]_\Lambda$ is a diagonal matrix.

(ii) If X is a countable set and μ is the counting measure, then $\phi : X \to \mathbb{C}$ is essentially bounded if and only if it is bounded. In that case, once again, M_ϕ corresponds to the operator $T_\phi : \ell^2(X) \to \ell^2(X)$ defined by $T_\phi((x_j)) = (\phi(j)x_j)$. Once again, we think of such a multiplication operator as a diagonal matrix.

Notice that if ϕ_1 and ϕ_2 are two essentially bounded functions on $(X, \mathfrak{M}, \mu)$ and $\phi_1 = \phi_2$ a.e., then the associated multiplication operators M_{ϕ_1} and M_{ϕ_2} are equal. This allows us to define a function $\Psi : L^\infty(X, \mu) \to \mathcal{B}(L^2(X, \mu))$ given by

$$\Psi(\phi) := M_\phi.$$

Notice that Ψ is linear and that $\|\Psi(\phi)\| \leq \|\phi\|_\infty$. It is also clear that

$$M_{\phi_1} M_{\phi_2} = M_{\phi_1 \phi_2}$$

for any $\phi_1, \phi_2 \in L^\infty(X, \mu)$. Hence Ψ is a homomorphism of Banach algebras. Finally, if $\phi \in L^\infty(X, \mu)$, then the calculation in Example 8.1.4 shows that $M_\phi^* = M_{\overline{\phi}}$. Therefore Ψ is a $*$-homomorphism of C*-algebras. In fact,

10.2.3 PROPOSITION *The map* $\Psi : L^\infty(X, \mu) \to \mathcal{B}(L^2(X, \mu))$ *is an isometric* $*$-*homomorphism.*

Proof.   By the above discussion, it suffices to prove that Ψ is isometric. To that end, fix $\phi \in L^\infty(X, \mu)$ and we must prove that $\|M_\phi\| \geq \|\phi\|_\infty$. We may assume that ϕ is non-zero and fix $0 < c < \|\phi\|_\infty$. Then the set

$$A_c := \{x \in X : |\varphi(x)| > c\}$$

has positive measure. Since X is σ-finite, there is a set $E \subset A_c$ whose measure is both positive and finite. If $f = \chi_E$, then f is non-zero in $L^2(X, \mu)$ and $\|f\|_2 = \mu(E)^{1/2}$. Also,

$$\|M_\phi f\|_2^2 = \int_X |\phi \chi_E|^2 d\mu \geq c^2 \mu(E) = c^2 \|f\|_2^2.$$

Hence $\|M_\phi\| \geq c$. This is true for any such c and thus $\|M_\phi\| \geq \|\phi\|_\infty$.   □

A by-product of this discussion is an important one which is worth recording. Since $L^\infty(X, \mu)$ is commutative, each $\phi \in L^\infty(X, \mu)$ commutes with its adjoint. Therefore

10.2.4 COROLLARY   *Every multiplication operator is normal.*

We now wish to understand the spectrum of such an operator. By the Spectral Permanence Theorem and Proposition 10.2.3, this boils down to understanding the spectrum of a function $\phi \in L^\infty(X, \mu)$.

10.2.5 EXAMPLE

(i) If $X = \{1, 2, \ldots, n\}$ and μ is the counting measure, then $\phi \in \ell^\infty(X, \mu)$ is invertible if and only if $0 \notin \{\phi(1), \phi(2), \ldots, \phi(n)\}$. Hence it follows that $\sigma(\phi) = \phi(X)$.

(ii) Let $X = \mathbb{N}$ and μ be the counting measure on $(\mathbb{N}, 2^{\mathbb{N}})$. If $\phi \in \ell^\infty = \ell^\infty(\mathbb{N})$, then ϕ is invertible in ℓ^∞ if and only if $0 \notin \overline{\phi(\mathbb{N})}$ (by Exercise 7.16). From this, it follows that $\sigma(\phi) = \overline{\phi(\mathbb{N})}$.

From these examples, one would expect the spectrum of M_ϕ to be related to the range of ϕ. For arbitrary σ-finite measure spaces, one needs the notion of essential range.

10.2.6 DEFINITION
Let $(X, \mathfrak{M}, \mu)$ be a σ-finite measure space and $\phi : X \to \mathbb{C}$ be essentially bounded. The *essential range* of ϕ is the set

$$\text{ess-range}(\phi) := \{\lambda \in \mathbb{C} : \mu(\{x \in X : |\phi(x) - \lambda| < r\}) > 0 \text{ for all } r > 0\}.$$

In other words, $\lambda \in \text{ess-range}(\phi)$ if and only if $\mu(\phi^{-1}(B(\lambda, r))) > 0$ for all $r > 0$, where $B(\lambda, r) = \{z \in \mathbb{C} : |\lambda - z| < r\}$.

Notice that if $\phi_1 = \phi_2$ a.e., then $\text{ess-range}(\phi_1) = \text{ess-range}(\phi_2)$.

10.2.7 PROPOSITION
Let $(X, \mathfrak{M}, \mu)$ be a σ-finite measure space and $\phi \in L^\infty(X, \mu)$. Then $\sigma(M_\phi) = \text{ess-range}(\phi)$.

Proof. Suppose $\lambda \notin \text{ess-range}(\phi)$, then there exists $r > 0$ such that the set $E := \{x \in X : |\phi(x) - \lambda| < r\}$ has measure zero. Define $\widetilde{\phi} : X \to \mathbb{C}$ by

$$\widetilde{\phi}(x) = \begin{cases} \frac{1}{\phi(x) - \lambda} & : \text{if } x \notin E, \\ 0 & : \text{if } x \in E. \end{cases}$$

Then $\widetilde{\phi}$ is measurable and essentially bounded and $(\phi - \lambda\mathbf{1})\widetilde{\phi} = \mathbf{1}$ almost everywhere. Therefore

$$(M_\phi - \lambda I)M_{\widetilde{\phi}} = M_{(\phi - \lambda\mathbf{1})}M_{\widetilde{\phi}} = I = M_{\widetilde{\phi}}(M_\phi - \lambda I).$$

In other words, $(M_\phi - \lambda I)$ is invertible and $\lambda \notin \sigma(M_\phi)$.

Conversely, if $\lambda \in \text{ess-range}(\phi)$, then for each $n \in \mathbb{N}$, the set $A_n := \{x \in X : |\phi(x) - \lambda| < 1/n\}$ has positive measure. Since μ is σ-finite, there is a set $E_n \subset$

A_n with finite, positive measure. If $f_n := \mu(E_n)^{-1/2}\chi_{E_n}$, then $f_n \in L^2(X,\mu)$ and $\|f_n\|_2 = 1$. Furthermore,

$$\left|(M_\phi - \lambda I)(f_n)(x)\right| \le \frac{1}{n}|f_n(x)|$$

for all $x \in X$. Integrating, we see that $\|(M_\phi - \lambda I)(f_n)\| \le 1/n \to 0$ as $n \to \infty$. Therefore $(M_\phi - \lambda I)$ is not invertible and $\lambda \in \sigma(M_\phi)$. $\quad\square$

10.2.8 REMARK

(i) During the course of the proof of Proposition 10.2.7, we have proved that every $\lambda \in$ ess-range(ϕ) is, in fact, in the *approximate point* spectrum of M_ϕ. Therefore $\sigma_{ap}(M_\phi) = \sigma(M_\phi)$. This agrees with Corollary 8.1.16, where we proved the same result for any normal operator.

(ii) Also if $\lambda \notin$ ess-range(ϕ), we have proved that $(M_\phi - \lambda I)^{-1}$ is another multiplication operator. Indeed, the collection $\Psi(L^\infty(X,\mu)) = \{M_\phi : \phi \in L^\infty(X,\mu)\}$ is a C*-subalgebra of $\mathcal{B}(L^2(X,\mu))$ that contains the identity operator. Therefore this is a consequence of the Spectral Permanence Theorem: If $(M_\phi - \lambda I)$ is invertible, then its inverse must lie in $\Psi(L^\infty(X,\mu))$.

10.3   The Spectral Theorem

Armed with all this information about multiplication operators, let us now return to our discussion on the Spectral Theorem. We would like to show that every normal operator can be modelled as a multiplication operator. In other words, every normal operator is equivalent to a multiplication operator in the following sense.

10.3.1 DEFINITION   Let $\mathbf{H}$ and $\mathbf{K}$ be two Hilbert spaces and $T \in \mathcal{B}(\mathbf{H})$ and $S \in \mathcal{B}(\mathbf{K})$ be two operators. We say that S and T are *unitarily equivalent* if there is a unitary operator $U : \mathbf{H} \to \mathbf{K}$ such that $T = U^{-1}SU$. If this happens, we write $S \sim_u T$.

Note that unitary equivalence is an equivalence relation. Since unitaries are the isomorphisms between Hilbert spaces, two unitarily equivalent operators are essentially indistinguishable. To motivate our next definition, let us return to the notion of diagonalizability for compact operators. For a countable set equipped with the counting measure, we use the lowercase letters ℓ^p to denote the L^p spaces.

10.3.2 LEMMA *Let* $\mathbf{H}$ *be a separable Hilbert space and* $T \in \mathcal{K}(\mathbf{H})$. *Then* T *is diagonalizable (in the sense of Definition 10.1.7) if and only if* T *is unitarily equivalent to a multiplication operator.*

Proof. Suppose that $\mathbf{H}$ has an orthonormal basis Λ consisting of eigenvectors of T. The Fourier transform obtained from Theorem 3.4.6 gives us a unitary operator $U : \mathbf{H} \to \ell^2(\Lambda)$. For each $v \in \Lambda$, let $\lambda_v \in \mathbb{C}$ be such that $Tv = \lambda_v v$. Define $\phi : \Lambda \to \mathbb{C}$ by $\phi(v) = \lambda_v$. Then

$$\|\phi\|_\infty = \sup_{v \in \Lambda} |\lambda_v| \leq r(T) < \infty.$$

Therefore $\phi \in \ell^\infty(\Lambda)$, so let $M_\phi \in \mathcal{B}(\ell^2(\Lambda))$ be the associated multiplication operator. For any $v \in \Lambda$, let $\delta_v \in \ell^2(\Lambda)$ denote the associated element of the standard orthonormal basis defined by $\delta_v(w) = \delta_{v,w}$. Then $U(v) = \delta_v$ and $M_\phi(\delta_v) = \lambda_v \delta_v$ for all $v \in \Lambda$. Hence $T = U^{-1} M_\phi U$.

Conversely, suppose T is unitarily equivalent to a multiplication operator. Since every multiplication operator is normal, T is normal. By Theorem 10.1.8, T is diagonalizable. □

This result now allows us to make the following definition, which agrees with Definition 10.1.7 when the operator is compact.

10.3.3 DEFINITION Let $\mathbf{H}$ be a separable Hilbert space. An operator $T \in \mathcal{B}(\mathbf{H})$ is said to be *diagonalizable* if T is unitarily equivalent to a multiplication operator.

The next lemma (which follows directly from Corollary 10.2.4) is the 'easy direction' of the Spectral Theorem we are about to prove.

10.3.4 LEMMA *If* $T \in \mathcal{B}(\mathbf{H})$ *is diagonalizable, then it is normal.*

Our proof of the Spectral Theorem proceeds in two stages. We first prove it for a special class of normal operators and then extend it to the general case. For a normal operator $T \in \mathcal{B}(\mathbf{H})$, we will write $\mathcal{A}(T)$ for the C*-algebra generated by $\{T, I\}$ (see Example 8.2.6).

10.3.5 DEFINITION A vector $e \in \mathbf{H}$ is said to be *cyclic* with respect to T if the set $\mathcal{A}(T)e := \{Se : S \in \mathcal{A}(T)\}$ is dense in $\mathbf{H}$.

10.3.6 EXAMPLE

(i) If $\mathbf{H} = L^2[0,1]$ and $T \in \mathcal{B}(\mathbf{H})$ is the operator $T(f)(x) := xf(x)$, then the vector $e \in L^2[0,1]$ given by $e(x) = 1$ is a cyclic vector. The reason is that the set $\mathcal{A}(T)e$ contains all polynomials and the polynomials are dense in $\mathbf{H}$ (by the Weierstrass Approximation Theorem and Proposition 2.3.12).

(ii) Let $\mathbf{H} := \mathbb{C}^2$ and $T \in \mathcal{B}(\mathbf{H})$ be given by $T(x,y) = (x,0)$. Then for any $e \in \mathbf{H}, \mathcal{A}(T)e \subset \mathbb{C} \oplus \{0\}$, so T does not have a cyclic vector. However, if $\mathbf{H}_1 := \mathbb{C} \oplus \{0\}$ and $\mathbf{H}_2 := \{0\} \oplus \mathbb{C}$, then $T(\mathbf{H}_i) \subset \mathbf{H}_i$ for $i = 1,2$, and the operators $T|_{\mathbf{H}_i} \in \mathcal{B}(\mathbf{H}_i)$ both have cyclic vectors.

Indeed, we will soon see that this phenomenon holds in general: Every operator can be expressed as a direct sum of operators each of which has a cyclic vector.

We are now ready to prove the first step of the Spectral Theorem. At the heart of the proof lies the Riesz–Markov–Kakutani Theorem, which provides us a way to construct a measure space from thin air.

10.3.7 THEOREM (SPECTRAL THEOREM – SPECIAL CASE) *Let* $\mathbf{H}$ *be a separable Hilbert space and* $T \in \mathcal{B}(\mathbf{H})$ *be a normal operator with a cyclic vector. Then* T *is diagonalizable.*

Proof. Let $e \in \mathbf{H}$ be a cyclic vector. Let $X := \sigma(T)$, so that X is a compact metric space and let $f \mapsto f(T)$ denote the continuous functional calculus. Define $\tau : C(X) \to \mathbb{C}$ by

$$\tau(f) := \langle f(T)e, e \rangle.$$

Then τ is clearly a linear functional. Also, if $f \in C(X)$ is positive, then $f(T)$ is a positive operator. Therefore $\tau(f) \geq 0$ (see Example 9.1.3), so τ is a positive linear functional.

By the Riesz–Markov–Kakutani Theorem, there is a finite, positive Borel measure μ on X such that

$$\langle f(T)e, e \rangle = \int_X f d\mu$$

for all $f \in C(X)$. For any $f, g \in C(X)$,

$$\langle f(T)e, g(T)e \rangle = \langle g(T)^* f(T)e, e \rangle = \int_X \overline{g} f d\mu.$$

Define $V : C(X) \to \mathbf{H}$ by $V(f) := f(T)e$. If we think of $C(X)$ as a subset of $L^2(X, \mu)$, then V preserves the inner product on $C(X)$. Since $C(X)$ is dense in $L^2(X, \mu)$ by

Lusin's Theorem (Theorem 9.4.3), V extends to an isometry

$$U : L^2(X, \mu) \to \mathbf{H}.$$

Furthermore, Range(U) contains $\mathcal{A}(T)e$, which is dense in $\mathbf{H}$. Hence Range(U) = $\mathbf{H}$ and U is a unitary.

Now let $\phi \in C(X)$ be the function $\phi(z) = z$. Then for any $g \in C(X)$,

$$UM_\phi(g) = U(\phi g) = \phi(T)g(T)e = Tg(T)e = TU(g).$$

This holds for any $g \in C(X)$. Once again, since $C(X)$ is dense in $L^2(X, \mu)$, it follows that $T = UM_\phi U^{-1}$. $\square$

In order to transition from the special case to the general case, we need the following notions.

10.3.8 REMARK For each $n \in \mathbb{N}$, let $(\mathbf{H}_n, \langle \cdot, \cdot \rangle_{\mathbf{H}_n})$ be a Hilbert space. Let $\mathbf{H}$ denote the direct sum

$$\mathbf{H} := \bigoplus_{n=1}^{\infty} \mathbf{H}_n = \{(x_n) : x_i \in \mathbf{H}_i \text{ and } \sum_{i=1}^{\infty} \|x_i\|_{\mathbf{H}_i}^2 < \infty\}.$$

Then $\mathbf{H}$ is a Hilbert space with the inner product given by $\langle (x_n), (y_n) \rangle := \sum_{i=1}^{\infty} \langle x_i, y_i \rangle_{\mathbf{H}_i}$ (see Exercise 3.12). For each $n \in \mathbb{N}$, suppose $T_n \in \mathcal{B}(\mathbf{H}_n)$ and assume that $\sup\{\|T_n\| : n \in \mathbb{N}\} < \infty$. Define $T : \mathbf{H} \to \mathbf{H}$ by

$$T((x_n)) := (T_n(x_n)).$$

Then it is easy to see that T is a bounded linear operator with $\|T\| \le \sup_{n \in \mathbb{N}} \|T_n\|$. We denote this operator by $T = \bigoplus_{n=1}^{\infty} T_n$.

10.3.9 LEMMA *Let $\{\mathbf{H}_n : n \in \mathbb{N}\}$ be a countable family of separable Hilbert spaces and $T_n \in \mathcal{B}(\mathbf{H}_n)$ be such that $\sup_{n \in \mathbb{N}} \|T_n\| < \infty$. If each T_n is diagonalizable, then so is $\bigoplus_{n=1}^{\infty} T_n$.*

Proof.

(i) Let $\mathbf{H} := \bigoplus_{n=1}^{\infty} \mathbf{H}_n$. Since each $\mathbf{H}_n$ is separable, one can show that $\mathbf{H}$ is separable as well (Exercise 10.11).

(ii) For each $n \in \mathbb{N}$, there is a σ-finite measure space $(X_n, \mathfrak{M}_n, \mu_n)$, a function $\phi_n \in L^\infty(X_n, \mu_n)$ and a unitary $U_n : \mathbf{H}_n \to L^2(X_n, \mu_n)$ such that $T_n = U_n^{-1} M_{\phi_n} U_n$. Now define

$$X := \bigsqcup_{n=1}^\infty X_n$$

and $\mathfrak{M} := \{E \subset X : E \cap X_n \in \mathfrak{M}_n \text{ for all } n \in \mathbb{N}\}$. Define $\mu : \mathfrak{M} \to [0, \infty]$ by

$$\mu(E) := \sum_{n=1}^\infty \mu_n(E \cap X_n).$$

It is easy to see that $(X, \mathfrak{M}, \mu)$ is a measure space. Since each μ_n is σ-finite, one may verify that μ is also σ-finite (Exercise 10.12).

(iii) If $f : X \to \mathbb{C}$ is an $\mathfrak{M}$-measurable function, then $f_n := f|_{X_n}: X_n \to \mathbb{C}$ is $\mathfrak{M}_n$-measurable and

$$\int_X f d\mu = \sum_{n=1}^\infty \int_{X_n} f_n d\mu_n.$$

Hence the map $f \mapsto (f_n)$ establishes a unitary map $W : L^2(X, \mu) \to \bigoplus_{n=1}^\infty L^2(X_n, \mu_n)$. Since each $U_n : \mathbf{H}_n \to L^2(X_n, \mu_n)$ is a unitary, we may define a unitary $U : \mathbf{H} \to L^2(X, \mu)$ by

$$U := W^{-1} \circ \left(\bigoplus_{n=1}^\infty U_n \right).$$

(iv) Let $\phi_n \in L^\infty(X_n, \mu_n)$ be as in part (i). By Proposition 10.2.3, $\|\phi_n\|_\infty = \|M_{\phi_n}\| = \|T_n\|$. Hence $\sup_{n\in\mathbb{N}} \|\phi_n\|_\infty < \infty$. We may now define $\phi : X \to \mathbb{C}$ such that $\phi|_{X_n} = \phi_n$ for each $n \in \mathbb{N}$. Then $\phi \in L^\infty(X, \mu)$ and a short calculation proves that

$$U^{-1} M_\phi U = T.$$

□

The next lemma is the final piece of the puzzle that allows us to reduce the general case of the Spectral Theorem to the special case. Before we begin, recall that if $T \in \mathcal{B}(\mathbf{H})$ is a bounded operator, then a subspace $\mathbf{M} < \mathbf{H}$ is said to be *reducing* for T if $\mathbf{M}$ is invariant for both T and T^*. Notice that if T is normal, then a subspace $\mathbf{M}$ is reducing for T if and only if $S(\mathbf{M}) \subset \mathbf{M}$ for all $S \in \mathcal{A}(T)$. In other words, $\mathbf{M}$ is $\mathcal{A}(T)$-invariant.

$\boxed{10.3.10}$ LEMMA *Let $\mathbf{H}$ be a separable Hilbert space and $T \in \mathcal{B}(\mathbf{H})$ be a normal operator. Then there is a finite or countably infinite family $\{\mathbf{H}_1, \mathbf{H}_2, \ldots\}$ of subspaces of $\mathbf{H}$ satisfying the following conditions.*

(a) $\mathbf{H} \cong \bigoplus_{n=1}^{\infty} \mathbf{H}_n$.

(b) Each $\mathbf{H}_n$ is reducing for T and

(c) $T|_{\mathbf{H}_n}$ has a cyclic vector for each $n \in \mathbb{N}$.

Proof. For each $e \in \mathbf{H}$, set $\mathbf{H}_e := \overline{\mathcal{A}(T)e}$. This is a closed reducing subspace for T such that $T|_{\mathbf{H}_e}$ has a cyclic vector. Define

$$\mathcal{F} := \{\Lambda \subset \mathbf{H} : \|e\| = 1 \text{ and } \mathbf{H}_e \perp \mathbf{H}_f \text{ for all } e, f \in \Lambda\}.$$

Note that $\mathcal{F}$ is non-empty (because singleton sets are in $\mathcal{F}$) and it is partially ordered by inclusion. Furthermore, if $\mathcal{C}$ is a totally ordered subset of $\mathcal{F}$, then define

$$\Lambda_0 := \bigcup_{\Lambda \in \mathcal{C}} \Lambda.$$

Then since $\mathcal{C}$ is totally ordered, it is easy to see that $\Lambda_0 \in \mathcal{F}$. Thus every totally ordered subset of $\mathcal{F}$ has an upper bound. By Zorn's Lemma, $\mathcal{F}$ has a maximal element, which we denote by $\Lambda_{\mathbf{H}}$.

By definition, $\Lambda_{\mathbf{H}}$ is an orthonormal set. Since $\mathbf{H}$ is separable, $\Lambda_{\mathbf{H}}$ must be either finite or countably infinite. For convenience, we assume $\Lambda_{\mathbf{H}}$ is countably infinite, write $\Lambda_{\mathbf{H}} = \{e_1, e_2, \ldots\}$ and set $\mathbf{H}_n := \mathbf{H}_{e_n}$. Clearly, each $\mathbf{H}_n$ satisfies condition (b) and (c). To verify condition (a), it suffices to show that

$$\mathbf{K} := \mathrm{span}\{\mathcal{A}(T)e_n : n \in \mathbb{N}\}$$

is dense in $\mathbf{H}$. Suppose there were a unit vector $x \in \mathbf{K}^{\perp}$. Then if $e \in \Lambda_{\mathbf{H}}$ and $S_1, S_2 \in \mathcal{A}(T)$,

$$\langle S_1(e), S_2(x) \rangle = \langle S_2^* S_1(e), x \rangle = 0$$

since $S_2^* S_1 \in \mathcal{A}(T)$. Hence $\mathbf{H}_x \perp \mathbf{H}_e$ for any $e \in \Lambda_{\mathbf{H}}$. Therefore the set $\Lambda_{\mathbf{H}} \cup \{x\}$ would belong to $\mathcal{F}$, which would contradict the maximality of $\Lambda_{\mathbf{H}}$. Hence $\mathbf{K}^{\perp} = \{0\}$ and thus $\mathbf{K}$ is dense in $\mathbf{H}$. This proves that condition (a) is also satisfied. $\square$

10.3.11 THEOREM (SPECTRAL THEOREM – GENERAL CASE) *Let* **H** *be a separable Hilbert space. An operator* $T \in \mathcal{B}(\mathbf{H})$ *is normal if and only if it is diagonalizable.*

Proof. If T is diagonalizable, then T is normal by Lemma 10.3.4. Conversely, if T is normal, then we may choose T-reducing subspaces $\{\mathbf{H}_1, \mathbf{H}_2, \ldots\}$ such that $\mathbf{H} \cong \bigoplus_{n=1}^{\infty} \mathbf{H}_n$ and $T|_{\mathbf{H}_n}$ has a cyclic vector for each $n \in \mathbb{N}$. Clearly, T may be identified with the operator

$$\bigoplus_{n=1}^{\infty} T|_{\mathbf{H}_n}.$$

By the special case of the Spectral Theorem, $T|_{\mathbf{H}_n}$ is diagonalizable for each $n \in \mathbb{N}$. By Lemma 10.3.9, T is diagonalizable as well. $\qquad\square$

10.3.12 REMARK Some remarks are in order before we move on.

 (i) The version of the Spectral Theorem we have proved is not the only one that goes by this name. We will see another formulation of the theorem below, which many authors favour. We have chosen to present this version first because it is intuitive and directly generalizes the finite dimensional case.

 (ii) However, this version does have the drawback that the measure space produced in Theorem 10.3.11 is not unique. In subsequent sections, we will prove a version of the theorem that comes with a uniqueness clause.

 (iii) We have proved the Spectral Theorem for separable Hilbert spaces. The proof for non-separable spaces is not much more difficult. There are, however, some measure-theoretic and operator-theoretic nuances one has to take care of first (for instance, the direct sum of operators over an arbitrary index set). For the sake of brevity, we have avoided those arguments. The interested reader should be able to fill in the details in the general setting.

10.4   Borel Functional Calculus

Given a normal operator $T \in \mathcal{B}(\mathbf{H})$, the continuous functional calculus assigns to every function $f \in C(\sigma(T))$ an operator $f(T) \in \mathcal{B}(\mathbf{H})$. Now we wish to expand this definition beyond the context of continuous functions; to the class of *Borel* functions. This is a much larger class of functions that includes, for instance, characteristic functions of Borel sets. This broader functional calculus will give us a deeper understanding of normal operators and an alternate description of the Spectral Theorem.

> **Note.** Throughout the remainder of this chapter, X will denote a compact metric space.

10.4.1 DEFINITION A function $f : X \to \mathbb{C}$ is called a **Borel** function if it is measurable with respect to the Borel σ-algebra $\mathfrak{B}_X$. We write $B_\infty(X)$ for the space of all bounded Borel functions on X.

10.4.2 REMARK

(i) When equipped with the supremum norm, $B_\infty(X)$ is clearly a Banach space. Furthermore, under the usual pointwise multiplication and adjoint operation, $B_\infty(X)$ is a commutative C*-algebra.

(ii) Clearly, $C(X) \subset B_\infty(X)$. However, $B_\infty(X)$ is typically much larger than $C(X)$. For instance, if X is connected, then $C(X)$ has no non-trivial projections. However, for each $E \in \mathfrak{B}_X$, $\chi_E \in B_\infty(X)$ is a projection.

Given $T \in \mathcal{B}(\mathbf{H})$, our goal is to make sense of $f(T)$ for $f \in B_\infty(\sigma(T))$ in a way that extends the continuous functional calculus. We begin by recasting the continuous functional calculus in more abstract language.

10.4.3 DEFINITION Let $\mathbf{A}$ be a C*-algebra.

(i) A *representation* of $\mathbf{A}$ on a Hilbert space $\mathbf{H}$ is a $*$-homomorphism $\pi : \mathbf{A} \to \mathcal{B}(\mathbf{H})$. We often denote such a representation by the symbol $(\pi, \mathbf{H})$.

(ii) A representation $(\pi, \mathbf{H})$ is said to be *cyclic* if there is a vector $e \in \mathbf{H}$ such that $\pi(\mathbf{A})e := \{\pi(a)e : a \in \mathbf{A}\}$ is dense in $\mathbf{H}$. In this case, e is said to be a cyclic vector of $(\pi, \mathbf{H})$ and we denote such a representation by the triple $(\pi, \mathbf{H}, e)$.

(iii) A representation $(\pi, \mathbf{H})$ is said to be *non-degenerate* if the set $\pi(\mathbf{A})\mathbf{H} := \{\pi(a)x : a \in \mathbf{A}, x \in \mathbf{H}\}$ is dense in $\mathbf{H}$.

10.4.4 EXAMPLE

(i) Clearly, any cyclic representation is non-degenerate.

(ii) If $T \in \mathcal{B}(\mathbf{H})$ is normal, then the continuous functional calculus gives a representation $\Theta : C(\sigma(T)) \to \mathcal{B}(\mathbf{H})$. Since $\Theta(\mathbf{1}) = I$, the representation is non-degenerate. However, $(\Theta, \mathbf{H})$ is cyclic if and only if T has a cyclic vector.

(iii) Let X be a compact Hausdorff space and μ be a finite, positive Borel measure on X. Let $\mathbf{A} := L^\infty(X, \mu)$ and $\mathbf{H} := L^2(X, \mu)$. Then the map $\Psi : \mathbf{A} \to \mathcal{B}(\mathbf{H})$ given by $\varphi \mapsto M_\varphi$ is a representation of $\mathbf{A}$ on $\mathbf{H}$. Once again, since $\Psi(\mathbf{1}) = I$, Ψ is non-degenerate. In fact, since μ is a finite measure, $e := \mathbf{1}$ belongs to $\mathbf{H}$. Now the collection $\{\Psi(a)e : a \in \mathbf{A}\}$ contains $C(X)$. By Lusin's Theorem, this collection is dense in $\mathbf{H}$, so e is a cyclic vector for $(\Psi, \mathbf{H})$.

(iv) If $(\pi, \mathbf{H})$ is any representation of a C*-algebra $\mathbf{A}$ and $\mathbf{M} < \mathbf{H}$ is a subspace that is invariant under $\pi(\mathbf{A})$, then there is a natural representation $\mathbf{A} \to \mathcal{B}(\mathbf{M})$ given by $a \mapsto \pi(a)|_\mathbf{M}$. This representation is denoted by $(\pi|_\mathbf{M}, \mathbf{M})$. Since $\mathbf{A}$ is a C*-algebra, $\mathbf{M}^\perp$ is also invariant under $\pi(\mathbf{A})$ and we get a corresponding representation $(\pi|_{\mathbf{M}^\perp}, \mathbf{M}^\perp)$ as well.

(v) If $(\pi, \mathbf{H})$ is any representation of a C*-algebra $\mathbf{A}$, then $\mathbf{M} := \overline{\pi(\mathbf{A})\mathbf{H}}$ is a subspace that is invariant under $\pi(\mathbf{A})$. Furthermore, the representation $(\pi|_\mathbf{M}, \mathbf{M})$ is clearly non-degenerate and the representation $(\pi|_{\mathbf{M}^\perp}, \mathbf{M}^\perp)$ is the zero representation.

Returning to our original problem, we now think of the continuous functional calculus associated to a normal operator $T \in \mathcal{B}(\mathbf{H})$ as a representation of $\Theta : C(\sigma(T)) \to \mathcal{B}(\mathbf{H})$. We then wish to extend it *in a specific way* to a representation $\widehat{\Theta} : B_\infty(\sigma(T)) \to \mathcal{B}(\mathbf{H})$. To make sense of this, we must introduce some new notions of convergence on $\mathcal{B}(\mathbf{H})$. As in Chapter 6, these notions will lead to new topologies on $\mathcal{B}(\mathbf{H})$, which we will explore in the next section.

10.4.5 DEFINITION Let $\mathbf{H}$ be a Hilbert space, $(T_n) \subset \mathcal{B}(\mathbf{H})$ be a sequence and $T \in \mathcal{B}(\mathbf{H})$.

(i) We say that (T_n) *converges strongly* to T if $\lim_{n \to \infty} T_n(x) = T(x)$ for each $x \in \mathbf{H}$. If this happens, we write $T_n \overset{s}{\to} T$.

(ii) We say that (T_n) *converges weakly* to T if $\lim_{n \to \infty} \langle T_n(x), y \rangle = \langle T(x), y \rangle$ for all $x, y \in \mathbf{H}$. If this happens, we write $T_n \overset{w}{\to} T$.

If (T_n) converges to T in the operator norm of $\mathcal{B}(\mathbf{H})$, we simply write $T_n \to T$. While this notation does clash with that of Chapter 6, the context in which it is used should provide sufficient clarity.

10.4.6 EXAMPLE

(i) Clearly, if $T_n \to T$ in the norm, then $T_n \overset{s}{\to} T$. Furthermore, if $T_n \overset{s}{\to} T$, then $T_n \overset{w}{\to} T$ by the Cauchy–Schwarz Inequality.

(ii) Let $S \in \mathcal{B}(\ell^2)$ denote the left-shift operator $S((x_n)) := (x_2, x_3, \ldots)$ and let $T_n := S^n$. Then $T_n((x_j)) = (x_{n+1}, x_{n+2}, \ldots)$. Then

    (a) $\|T_n\| = 1$ for all $n \in \mathbb{N}$ (do verify this). Hence $T_n \not\to 0$ in the norm.

    (b) If $x \in \ell^2$ and $\epsilon > 0$, there exists $N \in \mathbb{N}$ such that $\sum_{j=N}^{\infty} |x_j|^2 < \epsilon$. Hence $\|T_n(x)\|_2^2 < \epsilon$ for all $n \geq N$. Therefore $T_n \xrightarrow{s} 0$.

(iii) Let $S \in \mathcal{B}(\ell^2)$ again denote the left-shift operator and let $T_n := (S^*)^n$ so that $T_n((x_j)) := (0, 0, \ldots, 0, x_1, x_2, \ldots)$ where the 0 appears n times. Then

    (a) Each T_n is an isometry, so (T_n) does not converge strongly to 0.

    (b) If $x = (x_n), y = (y_n) \in \ell^2$ and $\epsilon > 0$, there exists $N \in \mathbb{N}$ such that $\sum_{j=N}^{\infty} |y_j|^2 < \epsilon^2$. By the Cauchy-Schwarz Inequality, it then follows that

$$|\langle T_n(x), y \rangle| \leq \epsilon \|x\|$$

for all $n \geq N$. Hence $T_n \xrightarrow{w} 0$.

Now consider $B_\infty(X)$ for a compact metric space X. If $(f_n) \subset B_\infty(X)$ is a uniformly bounded sequence and $f : X \to \mathbb{C}$ is a function such that $\lim_{n \to \infty} f_n(z) = f(z)$ each $z \in X$, then it follows that $f \in B_\infty(X)$. In other words, $B_\infty(X)$ is closed under **bounded pointwise** convergence, which is weaker than norm convergence. For brevity, we will henceforth write $f_n \xrightarrow{bp} f$ if this happens. We now introduce those representations that respect this mode of convergence in $B_\infty(X)$.

<u>10.4.7 PROPOSITION</u> *Let X be a compact metric space and $\widehat{\pi} : B_\infty(X) \to \mathcal{B}(\mathbf{H})$ be a representation. Then the following conditions are equivalent.*

(a) For any sequence $(f_n) \subset B_\infty(X)$ and $f \in B_\infty(X)$, if $f_n \xrightarrow{bp} f$, then $\widehat{\pi}(f_n) \xrightarrow{s} \widehat{\pi}(f)$.

(b) For any sequence $(f_n) \subset B_\infty(X)$ and $f \in B_\infty(X)$, if $f_n \xrightarrow{bp} f$, then $\widehat{\pi}(f_n) \xrightarrow{w} \widehat{\pi}(f)$.

If either of these conditions holds, then we say that $\widehat{\pi}$ is a σ-normal representation.

Proof.   It suffices to prove that (b) implies (a), so suppose $\widehat{\pi}$ satisfies condition (b). Let $(f_n) \subset B_\infty(X)$ and $f \in B_\infty(X)$ be such that such that $f_n \xrightarrow{bp} f$. To prove that $\widehat{\pi}(f_n) \xrightarrow{s} \widehat{\pi}(f)$, we may repace f_n by $(f_n - f)$ and assume that $f = 0$.

If $g_n(z) := |f_n(z)|^2$, then $g_n \xrightarrow{bp} 0$. By hypothesis, $\pi(g_n) \xrightarrow{w} 0$. Hence if $x \in \mathbf{H}$, then

$$\lim_{n \to \infty} \|\widehat{\pi}(f_n)x\|^2 = \lim_{n \to \infty} \langle \widehat{\pi}(f_n)^* \widehat{\pi}(f_n)x, x \rangle = \lim_{n \to \infty} \langle \widehat{\pi}(g_n)x, x \rangle = 0.$$

This happens for each $x \in \mathbf{H}$, so $\widehat{\pi}(f_n) \xrightarrow{s} 0$. $\square$

The next example is the prototype of a σ-normal representation.

10.4.8 EXAMPLE Suppose μ is a finite, positive Borel measure on a compact metric space X and let $\mathbf{H} := L^2(X, \mu)$. For each $f \in B_\infty(X)$, let M_f denote the associated multiplication operator defined exactly as before. Then define a representation $\widehat{\pi} : B_\infty(X) \to \mathcal{B}(\mathbf{H})$ by

$$\widehat{\pi}(f) := M_f.$$

We claim that $\widehat{\pi}$ is σ-normal. To see this, suppose $(f_n) \subset B_\infty(X)$ is such that $f_n \xrightarrow{bp} f$. Once again, we assume without loss of generality that $f = 0$. If $g \in L^2(X, \mu)$, then $\lim_{n \to \infty} f_n(x)g(x) = 0$ for each $x \in X$. Furthermore, there exists $M > 0$ such that $\|f_n\|_\infty \leq M$ for all $n \in \mathbb{N}$. Therefore $|f_n g|^2 \leq M|g|^2 \in L^1(X, \mu)$ almost everywhere. By the Dominated Convergence Theorem, it follows that

$$\lim_{n \to \infty} \|\widehat{\pi}(f_n)g\|_2^2 = \lim_{n \to \infty} \int_X |f_n g|^2 d\mu = 0.$$

Hence $\widehat{\pi}(f_n) \xrightarrow{s} 0$ and $\widehat{\pi}$ is a σ-normal representation.

This example also tells us to think of a σ-normal representation as a way of encoding the Dominated Convergence Theorem into the definition of a representation. In fact, we may just as well encode the Monotone Convergence Theorem. For a sequence $(f_n) \subset B_\infty(X)$ and $f \in B_\infty(X)$, we write $f_n \nearrow f$ if each f_n is non-negative, $f_n(x) \leq f_{n+1}(x)$ for all $n \in \mathbb{N}$ and $f(x) = \lim_{n \to \infty} f_n(x)$ for each $x \in X$.

10.4.9 LEMMA *A non-degenerate representation $\widehat{\pi} : B_\infty(X) \to \mathcal{B}(\mathbf{H})$ is σ-normal if and only if $\widehat{\pi}(f_n) \xrightarrow{w} \widehat{\pi}(f)$ whenever $f_n \nearrow f$ in $B_\infty(X)$.*

Proof. We begin by proving that $\widehat{\pi}(\mathbf{1}) = I$ by appealing to the fact that $\widehat{\pi}$ is non-degenerate. For any $x \in \mathbf{H}$ and any $\epsilon > 0$, there exists $f \in B_\infty(X)$ and $y \in \mathbf{H}$ such that $\|\widehat{\pi}(f)y - x\| < \epsilon$. Since $\widehat{\pi}(\mathbf{1})\widehat{\pi}(f) = \widehat{\pi}(f)$ and $\|\widehat{\pi}(\mathbf{1})\| \leq 1$, it follows that

$$\|\widehat{\pi}(f)y - \widehat{\pi}(\mathbf{1})x\| < \epsilon.$$

Therefore $\|x - \widehat{\pi}(\mathbf{1})x\| < 2\epsilon$. This is true for each $\epsilon > 0$ and each $x \in \mathbf{H}$, so we conclude that $\widehat{\pi}(\mathbf{1}) = I$.

Now if $\widehat{\pi}$ is a σ-normal representation and $f_n \nearrow f$ in $B_\infty(X)$, then $f_n \xrightarrow{bp} f$ and thus $\widehat{\pi}(f_n) \xrightarrow{w} \widehat{\pi}(f)$.

Conversely, suppose $\widehat{\pi}$ satisfies the given 'monotone convergence' property and suppose $(f_n) \subset B_\infty(X)$ is a sequence such that $f_n \xrightarrow{bp} 0$. Then there exists $\alpha > 0$ such that $\|f_n\|_\infty^2 \le \alpha$. For each $k \in \mathbb{N}$, define $g_k \in B_\infty(X)$ by

$$g_k := \inf_{n \ge k}(\alpha - |f_n|^2).$$

Then each g_k is non-negative and $g_k \nearrow \alpha\mathbf{1}$ in $B_\infty(X)$. By hypothesis,

$$\widehat{\pi}(g_k) \xrightarrow{w} \alpha I.$$

However, $g_k \le \alpha\mathbf{1} - |f_k|^2$. Therefore $\widehat{\pi}(g_k) \le \alpha I - \widehat{\pi}(|f_k|^2)$ in $\mathcal{B}(\mathbf{H})$ since $\widehat{\pi}$ is a $*$-homomorphism. If $x \in \mathbf{H}$,

$$
\begin{aligned}
\|\widehat{\pi}(f_k)x\|^2 &= \langle \widehat{\pi}(f_k)x, \widehat{\pi}(f_k)x \rangle \\
&= \langle \widehat{\pi}(|f_k|^2)x, x \rangle \\
&\le \alpha\|x\|^2 - \langle \widehat{\pi}(g_k)x, x \rangle \\
&\to 0 \text{ as } k \to \infty.
\end{aligned}
$$

Therefore $\widehat{\pi}(f_k) \xrightarrow{s} 0$ and $\widehat{\pi}$ is σ-normal. $\qquad\square$

The conclusion from both these results is that a σ-normal representation is one that has a strong measure-theoretic flavour. We will see this highlighted in a couple of sections, when we use such a representation to build an 'operator-valued measure'. Before that though, we have to prove the following theorem.

10.4.10 THEOREM *Let X be a compact metric space and $\pi : C(X) \to \mathcal{B}(\mathbf{H})$ be a non-degenerate representation. Then there is a unique σ-normal representation $\widehat{\pi} : B_\infty(X) \to \mathcal{B}(\mathbf{H})$ such that $\widehat{\pi}|_{C(X)} = \pi$.*

Our journey is long, so we begin with a couple of lemmas. Also, if $V \subset \mathbb{C}$, we define $\mathrm{diam}(V) := \sup\{|x - y| : x, y \in V\}$.

10.4.11 LEMMA *Let $\mathcal{S}(X)$ denote the set of all simple, Borel-measurable functions on X. Then $\mathcal{S}(X)$ is dense in $B_\infty(X)$.*

Proof. Let $f \in B_\infty(X)$ and $\epsilon > 0$. Then $f(X) \subset \mathbb{C}$ is bounded, so there exist finitely many disjoint Borel sets $\{V_1, V_2, \ldots, V_n\} \subset \mathbb{C}$ such that $\mathrm{diam}(V_i) < \epsilon$ for all

$1 \leq i \leq n$ and

$$f(X) \subset \bigcup_{i=1}^{n} V_i.$$

For each $1 \leq i \leq n$, we may assume without loss of generality that $f^{-1}(V_i) \neq \varnothing$ and choose $\alpha_i \in f(X) \cap V_i$. Define $g := \sum_{i=1}^{n} \alpha_i \chi_{f^{-1}(V_i)}$. Then $g \in \mathcal{S}(X)$ and it is easy to see that $\|g - f\|_\infty < \epsilon$. $\square$

The next lemma tells us that a σ-normal representation is essentially built up from a family of complex measures which are compatible in a certain sense.

10.4.12 LEMMA *Let $\widehat{\pi}: B_\infty(X) \to \mathcal{B}(\mathbf{H})$ be a non-degenerate σ-normal representation. For each $x, y \in \mathbf{H}$, define $\mu_{x,y} : \mathfrak{B}_X \to \mathbb{C}$ by*

$$\mu_{x,y}(E) := \langle \widehat{\pi}(\chi_E)(x), y \rangle.$$

Then $\mu_{x,y}$ is a complex measure on X such that

$$\int_X f d\mu_{x,y} = \langle \widehat{\pi}(f)x, y \rangle \tag{10.1}$$

for all $f \in B_\infty(X)$.

Proof.

(i) Clearly, $\mu_{x,y}(\varnothing) = 0$, so we wish to prove countable additivity. Since $\widehat{\pi}$ is linear, it is clear that $\mu_{x,y}$ is *finitely* additive. Now choose a countable collection $\{E_1, E_2, \ldots\} \subset \mathfrak{B}_X$ of disjoint sets and let $E = \bigcup_{n=1}^{\infty} E_n$. If $F_n := \bigcup_{j=1}^{n} E_j$, then

$$\mu_{x,y}(F_n) = \sum_{j=1}^{n} \mu_{x,y}(E_j).$$

Moreover, $\chi_{F_n} \nearrow \chi_E$ in $B_\infty(X)$. Since $\widehat{\pi}$ is a σ-normal representation,

$$\mu_{x,y}(E) = \langle \widehat{\pi}(\chi_E)x, y \rangle = \lim_{n \to \infty} \langle \widehat{\pi}(\chi_{F_n})x, y \rangle = \lim_{n \to \infty} \mu_{x,y}(F_n) = \sum_{j=1}^{\infty} \mu_{x,y}(E_j).$$

(ii) By part (i), Equation 10.1 holds whenever f is the characteristic function of a set $E \in \mathfrak{B}_X$. By linearity, it also holds whenever f is a simple function. Now if $f \in B_\infty(X)$, there is a sequence $(f_n) \subset \mathcal{S}(X)$ such that $f_n \to f$ uniformly on X. In particular, $f_n \xrightarrow{bp} f$. Since $\widehat{\pi}$ is σ-normal,

$$\langle \widehat{\pi}(f)x, y \rangle = \lim_{n \to \infty} \langle \widehat{\pi}(f_n)x, y \rangle = \lim_{n \to \infty} \int_X f_n d\mu_{x,y} = \int_X f d\mu_{x,y}.$$

where the last equality holds by the Dominated Convergence Theorem. $\square$

We are now in a position to prove the uniqueness part of Theorem 10.4.10.

10.4.13 PROPOSITION *Let X be a compact metric space and let $\pi : C(X) \to \mathcal{B}(\mathbf{H})$ be a non-degenerate representation. Suppose $\pi_1 : B_\infty(X) \to \mathcal{B}(\mathbf{H})$ and $\pi_2 : B_\infty(X) \to \mathcal{B}(\mathbf{H})$ are two σ-normal representations that extend π. Then $\pi_1 = \pi_2$.*

Proof. We know that $\pi_1(f) = \pi_2(f)$ for each $f \in C(X)$. For $x, y \in \mathbf{H}$ fixed, define complex Borel measures $\mu_{x,y}$ and $\lambda_{x,y}$ by the formulae

$$\mu_{x,y}(E) := \langle \pi_1(\chi_E)x, y \rangle \text{ and } \lambda_{x,y}(E) := \langle \pi_2(\chi_E)x, y \rangle.$$

By Lemma 10.4.12,

$$\int_X f d\mu_{x,y} = \int_X f d\lambda_{x,y}$$

for all $f \in C(X)$. In other words, both $\mu_{x,y}$ and $\lambda_{x,y}$ define the same linear functional on $C(X)$. By Theorem 9.4.4, it follows that $\mu_{x,y}(E) = \lambda_{x,y}(E)$ for each $E \in \mathcal{B}_X$. Hence

$$\langle \pi_1(g)x, y \rangle = \langle \pi_2(g)x, y \rangle$$

must hold for every simple function $g \in B_\infty(X)$. Once again, since $\mathcal{S}(X)$ is dense in $B_\infty(X)$ and both π_1 and π_2 are σ-normal representations, it follows that

$$\langle \pi_1(g)x, y \rangle = \langle \pi_2(g)x, y \rangle$$

for each $g \in B_\infty(X)$. Hence $\pi_1(g) = \pi_2(g)$ for each $g \in B_\infty(X)$. $\square$

10.4.14 REMARK We record a useful fact we have just observed. If μ and ν are two complex Borel measures on X such that

$$\int_X f d\mu = \int_X f d\nu$$

for all $f \in C(X)$, then this equation also holds for all $f \in B_\infty(X)$.

The next lemma is a straight-forward consequence of Lusin's Theorem and we omit the proof (do try Exercise 10.16). With this lemma in hand, we are ready to prove Theorem 10.4.10.

10.4.15 LEMMA *Let $\{\mu_1, \mu_2, \ldots, \mu_n\}$ be a finite collection of complex Borel measures on X. If $f \in B_\infty(X)$ and $\epsilon > 0$, there exists $g \in C(X)$ such that*

$$\int_X |f - g| d|\mu_i| < \epsilon$$

for all $1 \le i \le n$.

Proof of Theorem 10.4.10. Let $\pi : C(X) \to \mathcal{B}(\mathbf{H})$ be a non-degenerate representation. We wish to construct a σ-normal representation $\widehat{\pi} : B_\infty(X) \to \mathcal{B}(\mathbf{H})$ that extends π. The proof is long, so we break it into steps.

(i) Fix $x, y \in \mathbf{H}$ and consider $\tau_{x,y} : C(X) \to \mathbb{C}$ given by

$$\tau_{x,y}(f) := \langle \pi(f)x, y \rangle.$$

This is clearly a linear functional. Since $\|\pi(f)\| \le \|f\|_\infty$, it is bounded and $\|\tau_{x,y}\| \le \|x\|\|y\|$. By Theorem 9.4.9, there is a complex Borel measure $\mu_{x,y}$ on X such that

$$\int_X f d\mu_{x,y} = \langle \pi(f)x, y \rangle$$

for all $f \in C(X)$. Furthermore, $\|\mu_{x,y}\| \le \|x\|\|y\|$.

(ii) Since $\mu_{x,y}$ is a Borel measure, we may integrate bounded Borel functions with respect to it. So fix $f \in B_\infty(X)$ and define $\eta_f : \mathbf{H} \times \mathbf{H} \to \mathbb{C}$ by

$$\eta_f(x,y) := \int_X f d\mu_{x,y}.$$

We claim that η_f is a sesquilinear form. To see this, fix $x_1, x_2, y \in \mathbf{H}$ and $\epsilon > 0$. Then by Lemma 10.4.15, there exists $g \in C(X)$ such that

$$\int_X |f - g| d|\mu| < \epsilon$$

for each $\mu \in \{\mu_{x_1,y}, \mu_{x_2,y}, \mu_{x_1+x_2,y}\}$. Hence

$$|\eta_f(x_1 + x_2, y) - \eta_f(x_1, y) - \eta_f(x_2, y)| \le 3\epsilon + |\eta_g(x_1 + x_2, y)$$
$$- \eta_g(x_1, y) - \eta_g(x_2, y)|.$$

However, $\eta_g(x,y) = \langle \pi(g)x, y \rangle$ and so the last term is zero. Hence

$$|\eta_f(x_1 + x_2, y) - \eta_f(x_1, y) - \eta_f(x_2, y)| \leq 3\epsilon.$$

This is true for all $\epsilon > 0$ and thus $\eta_f(x_1 + x_2, y) = \eta_f(x_1, y) + \eta_f(x_2, y)$. Similarly arguments may now be used to verify the other axioms of a sesquilinear form. Since $\|\mu_{x,y}\| \leq \|x\|\|y\|$, it follows that $|\eta_f(x,y)| \leq \|f\|_\infty \|x\|\|y\|$. Hence η_f is a bounded sesquilinear form on $\mathbf{H}$.

(iii) By Exercise 8.1, there exists $T_f \in \mathcal{B}(\mathbf{H})$ such that

$$\eta_f(x,y) = \int_X f d\mu_{x,y} = \langle T_f(x), y \rangle$$

for all $x, y \in \mathbf{H}$. Also, $\|T_f\| \leq \|f\|_\infty$. Therefore we may define $\widehat{\pi} : B_\infty(X) \to \mathcal{B}(\mathbf{H})$ by

$$\widehat{\pi}(f) := T_f.$$

Firstly, if $f \in C(X)$, then for any $x, y \in \mathbf{H}$,

$$\langle \widehat{\pi}(f)x, y \rangle = \int_X f d\mu_{x,y} = \langle \pi(f)x, y \rangle.$$

Hence $\widehat{\pi}(f) = \pi(f)$ and thus $\widehat{\pi}$ is an extension of π.

(iv) We claim that $\widehat{\pi}$ is a representation. We begin by proving that $\widehat{\pi}$ is linear. Fix $f_1, f_2 \in B_\infty(X)$. Then

$$\eta_{f_1+f_2}(x,y) = \int_X (f_1 + f_2) d\mu_{x,y} = \int_X f_1 d\mu_{x,y} + \int_X f_2 d\mu_{x,y} = \eta_{f_1}(x,y) + \eta_{f_2}(x,y).$$

Hence

$$\langle T_{f_1+f_2}(x), y \rangle = \langle T_{f_1}(x), y \rangle + \langle T_{f_2}(x), y \rangle.$$

Therefore $\widehat{\pi}(f_1 + f_2) = \widehat{\pi}(f_1) + \widehat{\pi}(f_2)$. Similarly, $\widehat{\pi}(\alpha f) = \alpha \widehat{\pi}(f)$ for all $\alpha \in \mathbf{C}$ and thus $\widehat{\pi}$ is linear.

(v) We now show that $\widehat{\pi}(\overline{f}) = \widehat{\pi}(f)^*$ for all $f \in B_\infty(X)$.

(a) To begin with, if $f \in C(X)$ is a non-negative function, $\pi(f)$ is a positive operator in $\mathcal{B}(\mathbf{H})$. Therefore for $x \in \mathbf{H}$, we have

$$\int_X f d\mu_{x,x} = \langle \pi(f)x, x \rangle \geq 0.$$

Hence $\mu_{x,x}$ is a positive measure (by Corollary 9.4.10).

(b) If $f \in B_\infty(X)$ is a real-valued function, then

$$\langle T_f(x), x \rangle = \int_X f d\mu_{x,x} \in \mathbb{R}$$

for each $x \in H$. This property characterizes self-adjoint operators (Proposition 8.1.9). Hence $\widehat{\pi}(f) = \widehat{\pi}(f)^*$ for every real-valued function $f \in B_\infty(X)$.

(c) Finally, for an arbitrary function $f \in B_\infty(X)$, we write f in the form $f = g + ih$, where g and h are real-valued functions in $B_\infty(X)$. Then $\overline{f} = g - ih$, and $\widehat{\pi}(g)$ and $\widehat{\pi}(h)$ are both self-adjoint operators. Therefore

$$\widehat{\pi}(\overline{f}) = \widehat{\pi}(g - ih) = \widehat{\pi}(g) - i\widehat{\pi}(h) = (\widehat{\pi}(g) + i\widehat{\pi}(h))^* = \widehat{\pi}(f)^*.$$

(vi) We now claim that $\widehat{\pi}(fg) = \widehat{\pi}(f)\widehat{\pi}(g)$ for all $f, g \in B_\infty(X)$.

(a) To begin with, fix $f, h \in C(X)$ and note that

$$\widehat{\pi}(fh) = \widehat{\pi}(f)\widehat{\pi}(h)$$

holds since $\widehat{\pi}$ is an extension of π. Therefore if $x, y \in H$, then

$$\int_X fh d\mu_{x,y} = \langle \pi(fh)x, y \rangle = \langle \pi(f)\pi(h)x, y \rangle = \int_X f d\mu_{\pi(h)x,y} = \int_X f d\mu_{\widehat{\pi}(h)x,y}.$$

By Remark 10.4.14,

$$\int_X fh d\mu_{x,y} = \int_X f d\mu_{\widehat{\pi}(h)x,y}$$

for all $f \in B_\infty(X)$. In other words, for each $f \in B_\infty(X)$ and $h \in C(X)$,

$$\langle \widehat{\pi}(fh)x, y \rangle = \langle \widehat{\pi}(f)\widehat{\pi}(h)x, y \rangle$$

and thus $\widehat{\pi}(fh) = \widehat{\pi}(f)\widehat{\pi}(h)$.

(b) Now for $f \in B_\infty(X), g \in C(X)$ and $x, y \in \mathbf{H}$ fixed,

$$\int_X gf d\mu_{x,y} = \int_X fg d\mu_{x,y} = \langle \widehat{\pi}(fg)x, y \rangle$$

$$= \langle \widehat{\pi}(f)\widehat{\pi}(g)x, y \rangle \quad \text{(by part (a))}$$

$$= \langle \widehat{\pi}(g)x, \widehat{\pi}(f)^* y \rangle$$

$$= \langle \widehat{\pi}(g)x, \widehat{\pi}(\overline{f})y \rangle \quad \text{(by step (v))}$$

$$= \int_X g d\mu_{x, \widehat{\pi}(\overline{f})y}.$$

This equation holds for all $g \in C(X)$. Once again, by Remark 10.4.14, it follows that

$$\int_X gf d\mu_{x,y} = \int_X g d\mu_{x, \widehat{\pi}(\overline{f})y}$$

for all $g \in B_\infty(X)$. In other words, for all $f, g \in B_\infty(X)$,

$$\langle \widehat{\pi}(fg)x, y \rangle = \langle \widehat{\pi}(gf)x, y \rangle = \langle \widehat{\pi}(g)x, \widehat{\pi}(\overline{f})y \rangle$$

$$= \langle \widehat{\pi}(g)x, \widehat{\pi}(f)^* y \rangle = \langle \widehat{\pi}(f)\widehat{\pi}(g)x, y \rangle.$$

Hence $\widehat{\pi}(fg) = \widehat{\pi}(f)\widehat{\pi}(g)$.

(vii) We now show that $\widehat{\pi}$ is σ-normal. Suppose $(f_n) \subset B_\infty(X)$ is such that $f_n \xrightarrow{bp} 0$. Then for any $x, y \in \mathbf{H}$, the Dominated Convergence Theorem implies that

$$\lim_{n \to \infty} \langle \widehat{\pi}(f_n)x, y \rangle = \lim_{n \to \infty} \int_X f_n d\mu_{x,y} = 0.$$

Hence $\widehat{\pi}(f_n) \xrightarrow{w} 0$. $\qquad\square$

The most important application of Theorem 10.4.10 is, of course, the following corollary.

10.4.16 COROLLARY *Let* $T \in \mathcal{B}(\mathbf{H})$ *be a normal operator. Then there is a unique* σ-*normal representation* $\widehat{\Theta} : B_\infty(\sigma(T)) \to \mathcal{B}(\mathbf{H})$ *such that* $\widehat{\Theta}(f) = f(T)$ *for all* $f \in C(\sigma(T))$.

For $f \in B_\infty(\sigma(T))$, we once again write $f(T) := \widehat{\Theta}(f)$. The map $f \mapsto f(T)$ is called the *Borel Functional Calculus* of T.

10.5 Von Neumann Algebras

Given a normal operator $T \in \mathcal{B}(\mathbf{H})$, write $\mathcal{V}(T)$ for the range of the Borel functional calculus,

$$\mathcal{V}(T) = \{f(T) : f \in B_\infty(\sigma(T))\}.$$

Notice that $\mathcal{V}(T)$ is a C*-algebra that contains $\mathcal{A}(T)$, the C*-algebra generated by $\{T, I\}$. However, $\mathcal{V}(T)$ might be strictly larger than $\mathcal{A}(T)$. For instance, if $E \in \mathfrak{B}_{\sigma(T)}$, then $\chi_E(T) \in \mathcal{V}(T)$ is a projection. Moreover, since simple functions are dense in $B_\infty(\sigma(T))$,

$$\mathcal{V}(T) = \overline{\text{span}\{\chi_E(T) : E \in \mathfrak{B}_{\sigma(T)}\}}.$$

However, if $\sigma(T)$ is connected, then $\mathcal{A}(T) \cong C(\sigma(T))$ does not have any non-trivial projections! This remarkable contrast between $\mathcal{V}(T)$ and $\mathcal{A}(T)$ leads us to the notion of a von Neumann algebra, which we now explore.

To begin with, we must first introduce some new topologies on $\mathcal{B}(\mathbf{H})$. These topologies are the natural home for the notions of strong and weak convergence from Definition 10.4.5. These are also locally convex topologies, which gives us access to the Hahn–Banach Separation Theorems.

Recall from Chapter 6 the notion of a weak topology on a set: Let X be a set and Y a topological space. Given a collection $\mathcal{G}$ of functions from X to Y, there is a unique smallest topology $\tau_\mathcal{G}$ on X that makes each member of $\mathcal{G}$ continuous. This is called the weak topology defined by $\mathcal{G}$.

10.5.1 DEFINITION Let $\mathbf{H}$ be a Hilbert space.

(i) For each $x \in \mathbf{H}$, define $p_x : \mathcal{B}(\mathbf{H}) \to \mathbf{H}$ by $p_x(T) := T(x)$. The weak topology defined by the family $\{p_x : x \in \mathbf{H}\}$ is called the **strong operator topology** (SOT) on $\mathcal{B}(\mathbf{H})$.

(ii) For each pair $x, y \in \mathbf{H}$, define $q_{x,y} : \mathcal{B}(\mathbf{H}) \to \mathbb{C}$ by $q_{x,y}(T) := \langle T(x), y \rangle$. The weak topology defined by the family $\{q_{x,y} : x, y \in \mathbf{H}\}$ is called the **weak operator topology** (WOT) on $\mathcal{B}(\mathbf{H})$.

For convenience, we write σ_N, σ_S and σ_W for the norm, strong and weak operator topologies respectively.

10.5.2 REMARK Let $\mathbf{H}$ be a Hilbert space.

(i) Note that each p_x is continuous with respect to the norm topology because $|p_x(T) - p_x(S)| \leq \|T - S\| \|x\|$. Similarly, each $q_{x,y}$ is continuous with respect to the strong topology by the Cauchy–Schwarz Inequality. Hence $\sigma_W \subset \sigma_S \subset \sigma_N$.

(ii) A subset of $\mathcal{B}(\mathbf{H})$ is said to be SOT-open if it belongs to σ_S and WOT-open if it belongs to σ_W. We define 'SOT-closed' and 'WOT-closed' similarly. Hence

$$\text{WOT-open} \Rightarrow \text{SOT-open} \Rightarrow \text{norm–open and}$$

$$\text{WOT-closed} \Rightarrow \text{SOT-closed} \Rightarrow \text{norm–closed.}$$

For a subset $A \subset \mathcal{B}(\mathbf{H})$, we write $\overline{A}^{SOT}$ to denote the SOT-closure of A (the intersection of all SOT-closed sets containing A). The terms $\overline{A}^{WOT}$ and $\overline{A}^{\|\cdot\|}$ are defined analogously.

(iii) We now identify the basic open sets in σ_S (Compare this with Remark 6.3.5).

(a) If $x \in \mathbf{H}$ and $\epsilon > 0$, then

$$\{T \in \mathcal{B}(\mathbf{H}) : \|T(x)\| < \epsilon\}$$

is a sub-basic open neighbourhood of 0.

(b) More generally, if $x_1, x_2, \ldots, x_n \in \mathbf{H}$ and $\epsilon > 0$, the set

$$\{T \in \mathcal{B}(\mathbf{H}) : \|T(x_i)\| < \epsilon \text{ for all } 1 \leq i \leq n\}$$

is a basic open neighbourhood of 0, so that every SOT-open set containing 0 must contain one such set.

(c) If $T_0 \in \mathcal{B}(\mathbf{H})$, then every SOT-open set containing T_0 must contain a set of the form

$$\{T \in \mathcal{B}(\mathbf{H}) : \|T(x_i) - T_0(x_i)\| < \epsilon \text{ for all } 1 \leq i \leq n\}$$

for some finite set $\{x_1, x_2, \ldots, x_n\} \subset \mathbf{H}$ and some $\epsilon > 0$.

(iv) We may describe the basic open sets in σ_W similarly. If $T_0 \in \mathcal{B}(\mathbf{H})$, then any WOT-open set containing T_0 must contain a set of the form

$$\{T \in \mathcal{B}(\mathbf{H}) : |\langle T(x_i), y_i \rangle - \langle T_0(x_i), y_i \rangle| < \epsilon \text{ for all } 1 \leq i \leq n\}$$

for some finite set $\{x_1, x_2, \ldots, x_n, y_1, y_2, \ldots, y_n\} \subset \mathbf{H}$ and some $\epsilon > 0$.

(v) For a sequence $(T_n) \subset \mathcal{B}(\mathbf{H})$ and $T \in \mathcal{B}(\mathbf{H})$, we have $T_n \xrightarrow{s} T$ if and only if $T_n \to T$ with respect to σ_S. Similarly, $T_n \xrightarrow{w} T$ if and only if $T_n \to T$ with respect to σ_W. The proofs of these facts are easily verified (Exercise 10.19). Hence Example 10.4.6 shows that $\sigma_W \neq \sigma_S$ and that $\sigma_S \neq \sigma_N$ in general.

Recall (Definition 6.7.4) that a locally convex topological vector space is one with a family of seminorms which define basic open neighbourhoods.

10.5.3 PROPOSITION   *Both the weak and strong operator topologies give* $\mathcal{B}(\mathbf{H})$ *the structure of a locally convex topological vector space.*

Proof.   We prove the result for σ_S as the proof for σ_W is analogous (Exercise 10.17).

If $S, T \in \mathcal{B}(\mathbf{H})$ are distinct, then there exists $x_0 \in \mathbf{H}$ such that $T(x_0) \neq S(x_0)$. If $\epsilon := \|T(x_0) - S(x_0)\|/3$, then the sets $U_S := \{R \in \mathcal{B}(\mathbf{H}) : \|R(x_0) - S(x_0)\| < \epsilon\}$ and $U_T := \{R \in \mathcal{B}(\mathbf{H}) : \|R(x_0) - T(x_0)\| < \epsilon\}$ are both SOT-open neighbourhoods of S and T, respectively, such that $U_S \cap U_T = \emptyset$. Hence σ_S is Hausdorff.

Let $a : \mathcal{B}(\mathbf{H}) \times \mathcal{B}(\mathbf{H}) \to \mathcal{B}(\mathbf{H})$ denote the addition map. Let $U \subset \mathcal{B}(\mathbf{H})$ be an open set and $(S_0, T_0) \in a^{-1}(U)$. Then $S_0 + T_0 \in U$, so there is a finite set $\{x_1, x_2, \dots, x_n\} \subset \mathbf{H}$ and $\epsilon > 0$ such that

$$\{R \in \mathcal{B}(\mathbf{H}) : \|R(x_i) - (S_0 + T_0)(x_i)\| < \epsilon \text{ for all } 1 \leq i \leq n\} \subset U.$$

Let $V := \{S \in \mathcal{B}(\mathbf{H}) : \|S(x_i) - S_0(x_i)\| < \epsilon/2 \text{ for all } 1 \leq i \leq n\}$ and $W := \{T \in \mathcal{B}(\mathbf{H}) : \|T(x_i) - T_0(x_i)\| < \epsilon/2 \text{ for all } 1 \leq i \leq n\}$, then it is clear that $V \times W$ is an open neighbourhood of (S_0, T_0) in the product topology $\sigma_S \times \sigma_S$ and

$$V \times W \subset a^{-1}(U).$$

Thus $a^{-1}(U)$ is open, proving that a is continuous. Similarly, one can show that the scalar multiplication map $s : \mathbb{C} \times \mathcal{B}(\mathbf{H}) \to \mathcal{B}(\mathbf{H})$ is also continuous. Hence, $(\mathcal{B}(\mathbf{H}), \sigma_S)$ is a topological vector space.

For each finite set $F := \{x_1, x_2, \dots, x_k\} \subset \mathbf{H}$, define $p_F : \mathcal{B}(\mathbf{H}) \to \mathbb{R}$ by

$$p_F(T) := \max\{\|T(x_i)\| : 1 \leq i \leq k\}.$$

Then the collection $\Gamma_S := \{p_F : F \subset \mathbf{H} \text{ finite}\}$ is a family of seminorms such that every open set containing 0 contains a set of the form

$$U_p^n := \left\{T \in \mathcal{B}(\mathbf{H}) : p(T) < \frac{1}{n}\right\}$$

for some $p \in \Gamma_S$ and $n \in \mathbb{N}$. Hence σ_S is locally convex. $\square$

The next proposition identifies all linear functionals on $\mathcal{B}(\mathbf{H})$ that are SOT-continuous. What is most remarkable, though, is that these are precisely those linear functionals that are WOT-continuous! The significance of this will be seen shortly, but it suffices to say that this will have useful consequences when taken together with the Hahn–Banach Separation Theorems.

10.5.4 PROPOSITION *Let* $\mathbf{H}$ *be a Hilbert space and* $\tau : \mathcal{B}(\mathbf{H}) \to \mathbb{C}$ *be a linear functional. Then the following statements are equivalent.*

(i) τ *is WOT-continuous.*

(ii) τ *is SOT-continuous.*

(iii) *There exist vectors* $\{x_1, x_2, \ldots, x_n, y_1, y_2, \ldots, y_n\} \subset \mathbf{H}$ *such that*

$$\tau(T) = \sum_{i=1}^{n} \langle T(x_i), y_i \rangle \tag{10.2}$$

for all $T \in \mathcal{B}(\mathbf{H})$.

Proof. Note that $(i) \Rightarrow (ii)$ and $(iii) \Rightarrow (i)$ are obvious by definition. Therefore we prove $(ii) \Rightarrow (iii)$.

Suppose $\tau : \mathcal{B}(\mathbf{H}) \to \mathbb{C}$ is SOT-continuous, then the set

$$V := \{T \in \mathcal{B}(\mathbf{H}) : |\tau(T)| < 1\}$$

is SOT-open. Let Γ_S be the collection of seminorms described in the proof of Proposition 10.5.3. Then there exists $p \in \Gamma_S$ and $n \in \mathbb{N}$ such that

$$U_p^n := \left\{T \in \mathcal{B}(\mathbf{H}) : p(T) < \frac{1}{n}\right\} \subset V.$$

Hence there exist finitely many vectors $\{x_1, x_2, \ldots, x_n\} \subset \mathbf{H}$ and $\epsilon > 0$ such that for any $S \in \mathcal{B}(\mathbf{H})$,

$$\max_{1 \leq i \leq k} \|S(x_i)\| < \epsilon \Rightarrow |\tau(S)| < 1.$$

For any $T \in \mathcal{B}(\mathbf{H})$, let $M_T := \max\{\|T(x_i)\| : 1 \leq i \leq k\}$ and set

$$S := \frac{\epsilon}{2} \frac{T}{M_T}.$$

Then $|\tau(S)| < 1$. Unwrapping this, we see that

$$|\tau(T)| \leq \frac{2}{\epsilon} \max_{1 \leq i \leq k} \|T(x_i)\|.$$

for each $T \in \mathcal{B}(\mathbf{H})$. Scaling τ if need be, we may assume that $\epsilon = 2$ and therefore

$$|\tau(T)| \leq \left(\sum_{i=1}^{k} \|T(x_i)\|^2 \right)^{1/2} \tag{10.3}$$

for each $T \in \mathcal{B}(\mathbf{H})$. Now let $\mathbf{K} := \bigoplus_{i=1}^{k} \mathbf{H}$, then $\mathbf{K}$ is a Hilbert space with the inner product given by

$$((u_i), (v_i))_{\mathbf{K}} := \sum_{i=1}^{k} \langle u_i, v_i \rangle.$$

Let $\mathbf{M}_0 := \mathrm{span}\{(T(x_1), T(x_2), \ldots, T(x_k)) : T \in \mathcal{B}(\mathbf{H})\} \subset \mathbf{K}$. Then we may define $\varphi : \mathbf{M}_0 \to \mathbb{C}$ by

$$\varphi(T(x_1), T(x_2), \ldots, T(x_k)) := \tau(T).$$

Equation 10.3 shows that this is well-defined and bounded (it is clearly linear). If $\mathbf{M} := \overline{\mathbf{M}_0}$, then φ extends to a bounded linear functional $\varphi : \mathbf{M} \to \mathbb{C}$. Since $\mathbf{M}$ is a Hilbert space, there exists $y = (y_1, y_2, \ldots, y_k) \in \mathbf{M}$ such that $\varphi(x) = (x, y)_{\mathbf{K}}$ for all $x \in \mathbf{M}$. Hence

$$\tau(T) = ((T(x_i)), (y_i))_{\mathbf{K}} = \sum_{i=1}^{k} \langle T(x_i), y_i \rangle.$$

for all $T \in \mathcal{B}(\mathbf{H})$. $\qquad\square$

Recall that the Hahn–Banach Separation Theorem views convex sets through the lens of closed hyperplanes. Therefore if two locally convex topologies have the same continuous linear functionals, then they must have the same closed convex sets.

10.5.5 THEOREM *Let $\mathbf{H}$ be a Hilbert space and C be a convex subset of $\mathcal{B}(\mathbf{H})$. Then C is WOT-closed if and only if it is SOT-closed.*

Proof. If C is WOT-closed, then it is clearly SOT-closed. Now suppose C is SOT-closed and convex. If $T_0 \in \mathcal{B}(\mathbf{H}) \setminus C$, then the Hahn–Banach Separation Theorem (Theorem 6.7.10) ensures that there is a SOT-continuous linear functional $\psi : \mathcal{B}(\mathbf{H}) \to \mathbb{C}$ such that

$$\mathrm{Re}(\psi)(T_0) < \beta := \inf_{S \in C} \mathrm{Re}(\psi)(S).$$

By Proposition 10.5.4, ψ is WOT-continuous. Therefore the set $\{T \in \mathcal{B}(\mathbf{H}) : \mathrm{Re}(\psi)(T) < \beta\}$ is WOT-open. This set contains T_0 and is disjoint from C. Therefore $T_0 \notin \overline{C}^{WOT}$. Thus we have proved that C is WOT-closed. $\qquad\square$

In particular, this theorem states that a subalgebra of $\mathcal{B}(\mathbf{H})$ is SOT-closed if and only if it is WOT-closed. This brings us to the definition of a von Neumann algebra. Notice that unlike a C*-algebra (which was defined abstractly), a von Neumann algebra is defined as a subalgebra of $\mathcal{B}(\mathbf{H})$. We will see later that every C*-algebra can also be realized as an algebra of operators; until then, however, one must live with this unaesthetic disparity. Moreover, a von Neumann algebra may also be defined abstractly (as the dual space of a Banach space). Some authors prefer to start with this abstract definition and arrive at the concrete one we give here (see, for instance, Sakai [53]).

10.5.6 DEFINITION Let $\mathbf{H}$ be a Hilbert space. A self-adjoint subalgebra $\mathbf{A} \subset \mathcal{B}(\mathbf{H})$ is called a *von Neumann algebra* if it contains the identity operator and is SOT-closed.

10.5.7 EXAMPLE

(i) $\mathcal{B}(\mathbf{H})$ is clearly a von Neumann algebra.

(ii) Let $\mathbf{A}$ denote the subalgebra of $M_n(\mathbb{C})$ consisting of diagonal matrices. Clearly, $\mathbf{A}$ is a self-adjoint subalgebra of $M_n(\mathbb{C}) \cong \mathcal{B}(\mathbb{C}^n)$. We claim that $\mathbf{A}$ is WOT-closed. Let $\Lambda = \{e_1, e_2, \ldots, e_n\}$ denote the standard orthonormal basis of $\mathbb{C}^n$ and let $\tau_{i,j} : \mathcal{B}(\mathbb{C}^n) \to \mathbb{C}$ denote the linear functional

$$T \mapsto \langle T(e_j), e_i \rangle.$$

Then observe that

$$\mathbf{A} = \bigcap_{i \neq j} \ker(\tau_{i,j}).$$

Each $\tau_{i,j}$ is clearly WOT-continuous and therefore $\mathbf{A}$ is WOT-closed. Hence $\mathbf{A}$ is a von Neumann algebra.

(iii) Suppose $\mathbf{A}$ is a self-adjoint subalgebra of $\mathcal{B}(\mathbf{H})$. Define the *commutant* of $\mathbf{A}$ to be

$$\mathbf{A}' := \{T \in \mathcal{B}(\mathbf{H}) : TS = ST \text{ for all } S \in \mathbf{A}\}.$$

Since $\mathbf{A}$ is self-adjoint, it is easy to verify that $\mathbf{A}'$ is also a self-adjoint subalgebra of $\mathcal{B}(\mathbf{H})$ that contains the identity operator. We claim that $\mathbf{A}'$

is also SOT-closed and thus a von Neumann algebra. To see this, suppose $T_0 \in \overline{\mathbf{A}'}^{SOT}$ and $S \in \mathbf{A}$ with $\|S\| \leq 1$. For any $x \in \mathbf{H}$ and any $\epsilon > 0$, the set

$$V := \{T \in \mathcal{B}(\mathbf{H}) : \|T(x) - T_0(x)\| < \epsilon \text{ and } \|TS(x) - T_0S(x)\| < \epsilon\}$$

is an SOT-open set containing T_0. Therefore there exists $T \in \mathbf{A}'$ such that $T \in V$. Then

$$\|T_0S(x) - ST_0(x)\| \leq \|T_0S(x) - TS(x)\| + \|ST(x) - ST_0(x)\| < 2\epsilon.$$

This is true for any $\epsilon > 0$ and thus $ST_0 = T_0S$. Hence $T_0 \in \mathbf{A}'$, proving that $\mathbf{A}'$ is a von Neumann algebra.

The next example is so important that it warrants a proposition all of its own. Given a σ-finite measure space $(X, \mathfrak{M}, \mu)$, there is an isometric $*$-homomorphism $\Psi : L^\infty(X, \mu) \to \mathcal{B}(L^2(X, \mu))$ whose range is the algebra of multiplication operators (Proposition 10.2.3). In what follows, we identify $L^\infty(X, \mu)$ with its image in $\mathcal{B}(L^2(X, \mu))$.

10.5.8 PROPOSITION *If $(X, \mathfrak{M}, \mu)$ is a σ-finite measure space, then $L^\infty(X, \mu)$ is a von Neumann algebra.*

Proof.   Let $\mathbf{A} := \{M_\phi : \phi \in L^\infty(X, \mu)\}$. Then $\mathbf{A}$ is a C*-algebra in $\mathcal{B}(L^2(X, \mu))$ that contains the identity operator. By Example 10.5.7, it suffices to prove that $\mathbf{A} = \mathbf{A}'$. Since $\mathbf{A}$ is commutative, $\mathbf{A} \subset \mathbf{A}'$. Conversely, suppose $T \in \mathbf{A}'$, then we wish to prove that $T \in \mathbf{A}$.

(i) Suppose first that μ is a finite measure. Let $\phi := T(\mathbf{1})$, where $\mathbf{1} \in L^2(X, \mu)$ denotes the constant function 1. Then $\phi \in L^2(X, \mu)$. For any $g \in L^\infty(X, \mu) \subset L^2(X, \mu)$, we have

$$g\phi = M_g(\phi) = M_g(T(\mathbf{1})) = T(M_g(\mathbf{1})) = T(g). \qquad (10.4)$$

Now choose $c > 0$ so that the set $E := \{x \in X : |\phi(x)| > c\}$ has positive (finite) measure. Set $g := \chi_E \in L^\infty(X, \mu)$ and note that

$$c^2\mu(E) \leq \int_E |\phi|^2 d\mu = \|g\phi\|_2^2 = \|T(g)\|^2 \leq \|T\|^2\|g\|_2^2 = \|T\|^2\mu(E).$$

Therefore $c \leq \|T\|$ must hold. This is true for any such $c > 0$ and thus $|\phi(x)| \leq \|T\|$ holds almost everywhere. Hence $\phi \in L^\infty(X, \mu)$. Then $M_\phi \in$

$\mathcal{B}(L^2(X, \mu))$ and by Equation 10.4,

$$M_\phi(g) = T(g)$$

for all $g \in L^\infty(X, \mu)$. However, $L^\infty(X, \mu)$ is dense in $L^2(X, \mu)$ and both operators are bounded. Thus $T = M_\phi \in \mathbf{A}$.

(ii) Now suppose μ is a σ-finite measure. Then there is a sequence (E_n) of disjoint measurable subsets of X, each of which has finite measure, such that

$$X = \bigsqcup_{n=1}^\infty E_n.$$

For $n \in \mathbb{N}$ fixed, define $\mathfrak{M}_n := \{E \cap E_n : E \in \mathfrak{M}\}$ and let $\mu_n : \mathfrak{M}_n \to [0, \infty]$ be given by restricting μ to $\mathfrak{M}_n$. Then $(E_n, \mathfrak{M}_n, \mu_n)$ is a finite measure space. Furthermore, for any $f \in L^2(X, \mu)$, $f\chi_{E_n} \in L^2(E_n, \mu_n)$ and the map $f \mapsto (f\chi_{E_n})$ gives an isometric isomorphism

$$L^2(X, \mu) \cong \bigoplus_{n=1}^\infty L^2(E_n, \mu_n).$$

Now since $T \in \mathbf{A}', TM_{\chi_{E_n}} = M_{\chi_{E_n}}T$ for each $n \in \mathbb{N}$. Therefore we may consider $T_n \in \mathcal{B}(L^2(E_n, \mu_n))$ given by restricting T. Under the isomorphism described above, it is clear that T decomposes as

$$T = \bigoplus_{n=1}^\infty T_n.$$

Now set $\mathbf{A}_n := \{M_\phi : \phi \in L^\infty(E_n, \mu_n)\} \subset \mathcal{B}(L^2(E_n, \mu_n))$. Then $T_n \in \mathbf{A}'_n$. By part (i), there exists $\phi_n \in L^\infty(E_n, \mu_n)$ such that $T_n = M_{\phi_n}$. Furthermore,

$$\|\phi_n\|_\infty \le \|T_n\| \le \|T\|$$

for each $n \in \mathbb{N}$. Thus the sequence (ϕ_n) is uniformly bounded and we may define $\phi \in L^\infty(X, \mu)$ such that $\phi|_{E_n} = \phi_n$ for each $n \in \mathbb{N}$. It is now easy to see that $T = M_\phi \in \mathbf{A}$.

We conclude that $\mathbf{A} = \mathbf{A}'$ and thus $\mathbf{A}$ is a von Neumann algebra. $\qquad \square$

John von Neumann (1903–1957) was a Hungarian-American mathematician whose prodigious mathematical talent was recognized at an early age. On his father's advice, though, von Neumann studied chemistry at the University of Berlin! His abiding interest in mathematics continued despite this and he received his doctorate in 1926, with a thesis on set theory. He simultaneously developed a deep interest in quantum mechanics and built a strong mathematical framework for the subject. Along the way, he realized the importance of certain operator algebras, which he referred to as 'rings of operators'. In 1957, Dixmier named them *von Neumann algebras* in his honour.

von Neumann's contributions to mathematics and physics are extraordinary both in depth and in breadth. Apart from quantum mechanics and set theory, he studied ergodic theory, game theory, computer science and even served as a consultant to the US armed forces.

During the proof of Proposition 10.5.8 we showed that the algebra $\mathbf{A} := L^\infty(X, \mu)$ satisfies $\mathbf{A} = \mathbf{A}'$. Therefore if $\mathbf{B}$ is a commutative self-adjoint algebra that contains $\mathbf{A}$, then

$$\mathbf{B} \subset \mathbf{B}' \subset \mathbf{A}' = \mathbf{A}.$$

Hence $\mathbf{A}$ is *maximal* amongst the collection of all commutative self-adjoint algebras in $\mathcal{B}(L^2(X, \mu))$. A subalgebra of $\mathcal{B}(\mathbf{H})$ that satisfies this property is called a maximal abelian self-adjoint algebra, abbreviated to *MASA*. It is not hard to see that a MASA is automatically a von Neumann algebra. Although we will not dwell on it, these are useful objects in von Neumann algebra theory.

Let us now return to the von Neumann algebra generated by a normal operator and relate it to the range of the Borel functional calculus.

10.5.9 REMARK

(i) Since an SOT-closed set is also norm–closed, it follows that any von Neumann algebra is a (unital) C*-algebra. Furthermore, the SOT-closure of a self-adjoint set is again self-adjoint. Therefore if $\mathbf{A} \subset \mathcal{B}(\mathbf{H})$ is a C*-algebra that contains the unit of $\mathcal{B}(\mathbf{H})$, then $\overline{\mathbf{A}}^{SOT}$ is a von Neumann algebra.

(ii) Furthermore, if $\{\mathbf{A}_\alpha : \alpha \in J\}$ is a family of von Neumann algebras on $\mathbf{H}$, then

$$\bigcap_{\alpha \in J} \mathbf{A}_\alpha$$

is also a von Neumann algebra. Therefore for any subset $\mathcal{S} \subset \mathcal{B}(\mathbf{H})$, there is a *smallest* von Neumann algebra containing $\mathcal{S}$. This is denoted by $W^*(\mathcal{S})$.

(iii) In particular, if $T \in \mathcal{B}(\mathbf{H})$ is a single operator, we write $W^*(T)$ for $W^*(\{T\})$. If T is normal and $\mathcal{A}(T)$ denotes the C*-algebra generated by $\{I, T\}$, then

$$W^*(T) = \overline{\mathcal{A}(T)}^{SOT} = \overline{\mathcal{A}(T)}^{WOT}.$$

Note that $\mathcal{A}(T) \cong C(\sigma(T))$ via the continuous functional calculus. The next result tells us that $W^*(T)$ is the natural home of the Borel functional calculus.

Recall that if $T \in \mathcal{B}(\mathbf{H})$ is normal, we write $\mathcal{V}(T) := \{f(T) : f \in B_\infty(\sigma(T))\}$.

10.5.10 THEOREM *Let* $\mathbf{H}$ *be a Hilbert space and* $T \in \mathcal{B}(\mathbf{H})$ *be a normal operator. Then* $\mathcal{V}(T) \subset W^*(T)$.

Proof. Let $\mathbf{A}$ be a von Neumann algebra containing T and let $f \in B_\infty(\sigma(T))$. We claim that $f(T) \in \mathbf{A}$. To prove this, we let U be a WOT-open set containing $f(T)$ and prove that $U \cap \mathbf{A} \neq \emptyset$. By Remark 10.5.2, there exist finitely many vectors $\{x_1, x_2, \ldots, x_n, y_1, y_2, \ldots, y_n\} \subset \mathbf{H}$ and $\epsilon > 0$ such that

$$V := \{S \in \mathcal{B}(\mathbf{H}) : |\langle S(x_i), y_i \rangle - \langle f(T)x_i, y_i \rangle| < \epsilon\} \subset U.$$

Let $X := \sigma(T)$. For each $1 \leq i \leq n$, by the proof of Theorem 10.4.10, there is a complex Borel measure μ_i on X such that

$$\int_X h d\mu_i = \langle h(T)x_i, y_i \rangle$$

for all $h \in B_\infty(X)$. By Lemma 10.4.15 applied to the collection $\{\mu_1, \mu_2, \ldots, \mu_n\}$, there exists $g \in C(X)$ such that

$$\int_X |f - g| d|\mu_i| < \epsilon$$

for all $1 \leq i \leq n$. Then $g(T) \in \mathcal{A}(T)$ and

$$|\langle g(T)(x_i), y_i \rangle - \langle f(T)x_i, y_i \rangle| \leq \int_X |f - g| d|\mu_i| < \epsilon$$

for all $1 \leq i \leq n$. However, since $\mathbf{A}$ is a von Neumann algebra containing T, it must contains $\mathcal{A}(T)$ and therefore $g(T) \in \mathbf{A}$. We conclude that $\mathbf{A} \cap U \neq \emptyset$ for any

WOT-open set U containing $f(T)$. Since $\mathbf{A}$ is WOT-closed, $f(T) \in \mathbf{A}$. This is true for any von Neumann algebra $\mathbf{A}$ containing T and thus $f(T) \in W^*(T)$.   □

In the context of Theorem 10.5.10, more can be said. In fact, it is true that $\mathcal{V}(T) = W^*(T)$. In other words, $\mathcal{V}(T)$ is itself WOT-closed. This is a remarkable fact and one that surely merits some attention. However, it will take us too far afield and does not have any immediate applications either. Therefore we omit the proof and refer you to Conway [9] for the details (Zhu [62] is another good reference). We end this section with a corollary that is interesting in and of itself.

10.5.11 COROLLARY   *A von Neumann algebra is the closed linear span of its projections.*

Proof.   Let $\mathbf{A}$ be a von Neumann algebra on a Hilbert space $\mathbf{H}$. Then every element in $\mathbf{A}$ can be expressed as a linear combination of self-adjoint elements. If $T \in \mathbf{A}$ is self-adjoint, then $\mathcal{V}(T) \subset \mathbf{A}$. As mentioned at the start of this section,

$$\mathcal{V}(T) = \overline{\operatorname{span}\{\chi_E(T) : E \in \mathfrak{B}_{\sigma(T)}\}}$$

Hence $\mathbf{A} = \overline{\operatorname{span}\{\chi_E(T) : T \in \mathbf{A} \text{ self-adjoint}, E \in \mathfrak{B}_{\sigma(T)}\}}$. Since each element of the form $\chi_E(T)$ is a projection, this proves the result.   □

The theory of von Neumann algebras is deep and we have barely scratched the surface. The important take-away from our discussion is that although von Neumann algebras are C*-algebras, they behave quite differently. They have a lot of projections and tend to have a more *measure-theoretic* flavour. At the end of this chapter, we will list some books that you could look into if you wish to learn more.

For now we return to our study of normal operators and the Spectral Theorem.

10.6   Spectral Measures

The Borel functional calculus will now allow us to recast the Spectral Theorem in a more *operator-theoretic* manner; one that does not rely on measure spaces. Moreover, the version of the theorem we now prove also carries a uniqueness clause which is useful in applications. Our earlier formulation of the Spectral Theorem (in terms of multiplication operators) is not unique in any natural way.

Before we begin, we must first take a closer look at projections in $\mathcal{B}(\mathbf{H})$. Recall that a projection is an operator $P \in \mathcal{B}(\mathbf{H})$ such that $P = P^2 = P^*$. If $P \in \mathcal{B}(\mathbf{H})$ is a projection, then $\operatorname{Range}(P)$ is a closed subspace of $\mathbf{H}$. Moreover, if $\mathbf{M}$ is a closed subspace of $\mathbf{H}$, then there is a unique projection $P \in \mathcal{B}(\mathbf{H})$ such that $\operatorname{Range}(P) = \mathbf{M}$

(Remark 8.5.8). Finally, if $P \in \mathcal{B}(\mathbf{H})$ is a projection, then $\sigma(P) \subset \{0, 1\}$ by Lemma 8.5.9. Therefore every projection is a positive operator in the sense of Definition 9.1.1.

10.6.1 DEFINITION Let P_1 and P_2 be two projections in $\mathcal{B}(\mathbf{H})$.

(i) Since both P_1 and P_2 are self-adjoint, we write $P_1 \leq P_2$ if $(P_2 - P_1)$ is a positive operator.

(ii) If $P_1 P_2 = P_2 P_1 = 0$, then we write $P_1 \perp P_2$.

10.6.2 LEMMA *Let P_1 and P_2 be two projections in $\mathcal{B}(\mathbf{H})$.*

(i) $P_1 \leq P_2$ *if and only if* $\text{Range}(P_1) \subset \text{Range}(P_2)$. *In that case,* $(P_2 - P_1)$ *is a projection.*

(ii) $P_1 \perp P_2$ *if and only if* $\text{Range}(P_1) \perp \text{Range}(P_2)$. *In that case,* $(P_2 + P_1)$ *is a projection.*

Proof.

(i) Suppose $\text{Range}(P_1) \subset \text{Range}(P_2)$. Let $\mathbf{M} := \text{Range}(P_1)^{\perp} \cap \text{Range}(P_2)$. Then $\mathbf{M}$ is a closed subspace and $\mathbf{M} \perp \text{Range}(P_1)$. Furthermore, since $\mathbf{H} = \text{Range}(P_1) + \text{Range}(P_1)^{\perp}$, it follows that $\text{Range}(P_2) = \text{Range}(P_1) + \mathbf{M}$. By Exercise 3.8, $P_2 = P_1 + P$, where P is a projection satisfying $\text{Range}(P) = \mathbf{M}$. In particular, $P_2 - P_1 = P \geq 0$.
Conversely, if $P_1 \leq P_2$ and $x \in \text{Range}(P_1)$, then write $x = y + z$ where $y \in \text{Range}(P_2)$ and $z \in \text{Range}(P_2)^{\perp}$. Then

$$0 \leq \langle (P_2 - P_1)(x), x \rangle = \langle y - x, x \rangle = -\langle z, x \rangle = -\|z\|^2.$$

Hence $z = 0$ and $x = y \in \text{Range}(P_2)$. Thus $\text{Range}(P_1) \subset \text{Range}(P_2)$.

(ii) If $P_1 P_2 = P_2 P_1 = 0$ and $x \in \text{Range}(P_1)$ and $y \in \text{Range}(P_2)$, then

$$\langle x, y \rangle = \langle P_1(x), P_2(y) \rangle = \langle P_2^* P_1(x), y \rangle = \langle P_2 P_1(x), y \rangle = 0.$$

Thus $\text{Range}(P_1) \perp \text{Range}(P_2)$.
Conversely, if $\text{Range}(P_1) \perp \text{Range}(P_2)$, then $\text{Range}(P_1) \subset \text{Range}(P_2)^{\perp} = \ker(P_2)$. Therefore $P_2 P_1 = 0$. Similarly $P_1 P_2 = 0$ as well. Now it is easy to verify that $P := P_1 + P_2$ is self-adjoint and satisfies $P^2 = P$. $\qquad\square$

The next result we are after is stated in somewhat more general terms than what we absolutely need. The reason is that it is widely applicable in this form and that the proof in the special case is not far different. Before we begin, note that if $T \in \mathcal{B}(\mathbf{H})$ and $x, y \in \mathbf{H}$, then one has a Polarization identity similar to that of Proposition 3.1.3.

$$\langle T(x), y \rangle = \frac{1}{4} \sum_{k=0}^{4} \langle T(x + i^k y), x + i^k y \rangle. \tag{10.5}$$

10.6.3 Theorem (Vigier) *Let $(T_n) \subset \mathcal{B}(\mathbf{H})$ be a uniformly bounded sequence of self-adjoint operators such that $T_n \leq T_{n+1}$ for each $n \in \mathbb{N}$. Then there exists a self-adjoint operator $T \in \mathcal{B}(\mathbf{H})$ such that $T_n \xrightarrow{s} T$.*

Proof. Let $M > 0$ be such that $\|T_n\| \leq M$ for all $n \in \mathbb{N}$. If $x \in \mathbf{H}$ is fixed, then $\langle T_n(x), x \rangle \leq \langle T_{n+1}(x), x \rangle$ by Example 9.1.3. Therefore the sequence $(\langle T_n(x), x \rangle)_{n=1}^{\infty}$ is an increasing sequence of real numbers that is bounded above by $M\|x\|^2$. Therefore this sequence converges. By the Polarization identity above, it follows that the sequence $(\langle T_n(x), y \rangle)_{n=1}^{\infty}$ converges for any $x, y \in \mathbf{H}$. We may now define $\eta : \mathbf{H} \times \mathbf{H} \to \mathbb{C}$ by

$$\eta(x, y) := \lim_{n \to \infty} \langle T_n(x), y \rangle.$$

It is easily verified that η is a sesquilinear form and that $|\eta(x, y)| \leq M\|x\|\|y\|$ for all $x, y \in \mathbf{H}$. By Exercise 8.1, there is an operator $T \in \mathcal{B}(\mathbf{H})$ such that

$$\langle T(x), y \rangle = \eta(x, y)$$

for all $x, y \in \mathbf{H}$. Since each T_n is self-adjoint, $\langle T_n(x), x \rangle \in \mathbb{R}$ for each $x \in \mathbf{H}$. Taking a limit, we see that $\langle T(x), x \rangle \in \mathbb{R}$ for each $x \in \mathbf{H}$. By Proposition 8.1.9, T is self-adjoint.

By definition, $T_n \xrightarrow{w} T$. To see that $T_n \xrightarrow{s} T$ requires one more step. Note that for any $n \in \mathbb{N}, \langle T_n(x), x \rangle \leq \langle T(x), x \rangle$. Thus $T_n \leq T$ and we may set $S_n := \sqrt{T - T_n}$ as in Definition 9.1.7. Now if $x \in \mathbf{H}$,

$$\begin{aligned}
\|T(x) - T_n(x)\|^2 &= \|S_n(S_n(x))\|^2 \\
&\leq \|S_n\|^2 \|S_n(x)\|^2 \\
&\leq \|T - T_n\| \langle S_n(x), S_n(x) \rangle \\
&\leq 2M \langle (T - T_n)(x), x \rangle.
\end{aligned}$$

This last term converges to 0 as $n \to \infty$ and thus $T_n \xrightarrow{s} T$. $\square$

10.6.4 COROLLARY *Let* $(P_n) \subset \mathcal{B}(\mathbf{H})$ *be a sequence of projections.*

(i) *Suppose* $P_n \leq P_{n+1}$ *for all* $n \in \mathbb{N}$. *Then there is a projection* $P \in \mathcal{B}(\mathbf{H})$ *such that* $P_n \xrightarrow{s} P$.

(ii) *Suppose* $P_i \perp P_j$ *whenever* $i \neq j$. *Then there is a projection* $P \in \mathcal{B}(\mathbf{H})$ *such that* $P(x) = \sum_{n=1}^{\infty} P_n(x)$ *for each* $x \in \mathbf{H}$.

Proof.

(i) By Vigier's Theorem, there exists $P \in \mathcal{B}(\mathbf{H})$ that is self-adjoint such that $P_n \xrightarrow{s} P$. Now if $x, y \in \mathbf{H}$,

$$\langle P(x), y \rangle = \lim_{n \to \infty} \langle P_n(x), y \rangle = \lim_{n \to \infty} \langle P_n(x), P_n(y) \rangle = \langle P(x), P(y) \rangle = \langle P^2(x), y \rangle.$$

Therefore $P = P^2$.

(ii) For each $n \in \mathbb{N}$, the operator $Q_n := P_1 + P_2 + \ldots + P_n$ is a projection by Lemma 10.6.2. Since P_n is positive, $Q_n \leq Q_{n+1}$ for each $n \in \mathbb{N}$. We may thus apply part (i) to the sequence (Q_n). $\qquad \square$

The preceding results have taught us that projections (which are intimately tied to closed subspaces of the ambient Hilbert space) interact favourably with each other with respect to the partial order '$\leq$' inherited from $\mathcal{B}(\mathbf{H})_{sa}$. The next definition takes this one step further by taking these set theoretic relationships and imbuing it with the structure of a measure.

10.6.5 DEFINITION Let $(X, \mathfrak{M})$ be a measurable space and $\mathbf{H}$ be a Hilbert space. A *spectral measure* (or a *resolution of the identity*) on X is a map $P : \mathfrak{M} \to \mathcal{B}(\mathbf{H})$ satisfying the following conditions.

(a) $P(E)$ is a projection in $\mathcal{B}(\mathbf{H})$ for each $E \in \mathfrak{M}$.

(b) $P(\emptyset) = 0$ and $P(X) = I$.

(c) P is countably additive: If $\{E_1, E_2, \ldots\}$ is a countable family of disjoint sets in $\mathfrak{M}$, then

$$P \left(\bigcup_{n=1}^{\infty} E_n \right) = \sum_{n=1}^{\infty} P(E_n)$$

where the series converges strongly (as in part (ii) of Corollary 10.6.4). In other words, for each $x \in \mathbf{H}$,

$$P \left(\bigcup_{n=1}^{\infty} E_n \right)(x) = \lim_{k \to \infty} \sum_{n=1}^{k} P(E_n)(x). \tag{10.6}$$

In particular, this limit exists.

Before we look at any examples, let us first prove some simple properties about spectral measures which are reminiscent of the properties of ordinary measures.

10.6.6 LEMMA *Let $P : \mathfrak{M} \to \mathcal{B}(\mathbf{H})$ be a spectral measure on a measurable space $(X, \mathfrak{M})$ and let $E, F \in \mathfrak{M}$.*

(i) *If $F \subset E$, then $P(F) \leq P(E)$.*

(ii) *$P(E \cap F) = P(E)P(F)$. In particular, if $E \cap F = \emptyset$, then $P(E) \perp P(F)$.*

Proof.   Notice that since $P(\emptyset) = 0$, P is finitely additive: If $E, F \in \mathfrak{M}$ are disjoint, then $P(E \cup F) = P(E) + P(F)$.

(i) If $F \subset E$, then $E = F \sqcup (E \setminus F)$. Therefore $P(E) = P(F) + P(E \setminus F)$. Since $P(E \setminus F) \geq 0$, it follows that $P(F) \leq P(E)$.

(ii)

    (a) We first prove the special case: If E and F are disjoint, then $F \subset X \setminus E$, so $P(F) \leq P(X \setminus E)$. Since $P(X) = I$,

$$I = P(E) + P(X \setminus E)$$

by finite additivity. Therefore $P(X \setminus E) = I - P(E)$. Hence $\mathrm{Range}(P(F)) \subset \mathrm{Range}(P(X \setminus E)) = \mathrm{Range}(P(E))^{\perp}$. By Lemma 10.6.2, $P(F) \perp P(E)$.

    (b) Now suppose E and F are any two sets in $\mathfrak{M}$, set $E' := E \setminus F$ and $F' := F \setminus E$. By the previous step, $P(E') \perp P(E \cap F)$. Since $E = E' \sqcup (E \cap F)$, we have

$$P(E) = P(E') + P(E \cap F).$$

Similarly, $P(F) = P(F') + P(E \cap F)$ and $P(F') \perp P(E \cap F)$. Since $E' \cap F' = \emptyset$, $P(E')P(F') = 0$ by part (a). Therefore

$$P(E)P(F) = (P(E') + P(E \cap F))(P(F') + P(E \cap F)) = P(E \cap F)^2 = P(E \cap F).$$
$$\square$$

As a by-product of this lemma, we see that the limit in Equation 10.6 exists. For if $\{E_n : n \in \mathbb{N}\} \subset \mathfrak{M}$ is a countable collection of mutually disjoint sets, then $(P(E_n))$ is a sequence of mutually orthogonal projections. Therefore the limit in Equation 10.6 exists by Corollary 10.6.4.

The following are some examples of spectral measures. More than anything else, these examples are meant to indicate the connection between spectral measures and the Spectral Theorem we have proved earlier.

10.6.7 EXAMPLE

(i) Let $\mathbf{H} = \mathbb{C}^n$ and $T \in \mathcal{B}(\mathbf{H})$ a normal operator. Write $X := \sigma(T) = \{\lambda_1, \lambda_2, \dots, \lambda_k\}$ and let $\mathfrak{M} := 2^X$ denote the power set of X. For each $1 \leq i \leq k$, let P_i denote the projection onto the closed subspace $\mathbf{M}_i := \ker(T - \lambda_i I)$. Since T is diagonalizable, $\mathbf{H} = \bigoplus_{i=1}^k \mathbf{M}_i$. Therefore $I = \sum_{i=1}^k P_i$. Define $P : \mathfrak{M} \to \mathcal{B}(\mathbf{H})$ by

$$P(E) := \sum_{i \in E} P_i.$$

Since the $\{P_i\}$ are mutually orthogonal, this is a projection-valued function on $\mathfrak{M}$ (by Lemma 10.6.2). By the earlier calculation, $P(X) = I$ and $P(\emptyset) = 0$. Finally, the function is finitely additive by definition and is therefore countably additive as well. Therefore P is a spectral measure.

(ii) Let $\mathbf{H}$ be an infinite dimensional, separable Hilbert space and let $T \in \mathcal{K}(\mathbf{H})$ be a compact normal operator. By Theorem 7.5.7, we may write

$$X := \sigma(T) = \{\lambda_0 = 0, \lambda_1, \lambda_2, \dots\}$$

where $\lim_{n \to \infty} \lambda_n = 0$. Once again, let P_i denote the projection onto $\mathbf{M}_i := \ker(T - \lambda_i I)$, with P_0 being the projection onto $\mathbf{M}_0 := \ker(T)$. If $\mathfrak{M} := 2^X$, define $P : \mathfrak{M} \to \mathcal{B}(\mathbf{H})$ by

$$P(E)(x) := \sum_{i \in E} P_i(x).$$

Notice that if E is infinite, then this sum is defined in the SOT-sense. We know that T is diagonalizable. Therefore the $\{P_i\}$ are all mutually orthogonal. For each $E \in \mathfrak{M}$, $P(E)$ is then a projection by Corollary 10.6.4. Also $\mathbf{H} = \bigoplus_{i=0}^{\infty} \mathbf{M}_i$. So $P(X) = I$. Finally, P is countably additive by Corollary 10.6.4. Hence P is a spectral measure.

(iii) Let X be a compact metric space and μ be a positive Borel measure on X. Let $\mathbf{H} := L^2(X, \mu)$ and define $P : \mathfrak{B}_X \to \mathcal{B}(\mathbf{H})$ by

$$P(E) := M_{\chi_E}.$$

Then $P(E)$ is a projection because χ_E is a projection in $L^\infty(X, \mu)$. Also, $P(X) = I$ and $P(\emptyset) = 0$. As for countable additivity, suppose $\{E_1, E_2, \dots\}$

is a countable collection of mutually disjoint sets, then for each $x \in X$,

$$\chi_E(x) = \sum_{i=1}^{\infty} \chi_{E_i}(x),$$

where $E = \bigcup_{i=1}^{\infty} E_i$. If $g \in \mathbf{H}$, then $\chi_E(x)g(x) = \sum_{i=1}^{\infty} \chi_{E_i}(x)g(x)$. By the Dominated Convergence Theorem (applied to the partial sums of the series), we see that

$$\left\| M_{\chi_E}(g) - \sum_{i=1}^{\infty} M_{\chi_{E_i}}(g) \right\|_2^2 = \int_X \left| \chi_E g - \sum_{i=1}^{\infty} \chi_{E_i} g \right|^2 d\mu = 0.$$

Hence $P(E)(g) = \sum_{i=1}^{\infty} P(E_i)(g)$ for each $g \in \mathbf{H}$. Thus P is a spectral measure on $(X, \mathfrak{B}_X)$.

The next lemma tells us that a spectral measure is itself made up of a family of measures, in the spirit of Lemma 10.4.12. Since the proof is also similar, it is left as an exercise (Exercise 10.28).

10.6.8 LEMMA *Let $P : \mathfrak{M} \to \mathcal{B}(\mathbf{H})$ be a spectral measure on a measurable space $(X, \mathfrak{M})$.*

(i) If $x, y \in \mathbf{H}$, then the map $P_{x,y}(E) := \langle P(E)x, y \rangle$ defines a complex measure on $(X, \mathfrak{M})$.

(ii) If $x \in \mathbf{H}$, the measure $P_{x,x}$ is positive.

With this, we arrive at the main theorem of this section. Together with the Borel functional calculus, it will give us the much-anticipated second formulation of the Spectral Theorem.

10.6.9 THEOREM *Let X be a compact metric space, $\mathbf{H}$ be a Hilbert space and $\widehat{\pi} : B_{\infty}(X) \to \mathcal{B}(\mathbf{H})$ be a non-degenerate σ-normal representation. Then there is a unique spectral measure $P : \mathfrak{B}_X \to \mathcal{B}(\mathbf{H})$ such that*

$$\int_X f \, dP_{x,y} = \langle \widehat{\pi}(f)x, y \rangle. \tag{10.7}$$

for any $f \in B_{\infty}(X)$ and any $x, y \in \mathbf{H}$.

Proof. We define $P : \mathfrak{B}_X \to \mathcal{B}(\mathbf{H})$ by $P(E) := \widehat{\pi}(\chi_E)$.

(i) We verify that P is a spectral measure:

(a) Since χ_E is a projection in $B_{\infty}(X)$, $P(E)$ is a projection for each $E \in \mathfrak{B}_X$. It is also clear that $P(\emptyset) = 0$.

(b) Since $\hat{\pi}$ is non-degenerate, $\hat{\pi}(1) = I$ by the proof of Lemma 10.4.9. Therefore $P(X) = I$.

(c) To verify countable additivity, suppose $\{E_1, E_2, \ldots\}$ is a countable collection of mutually disjoint sets. If $E = \bigcup_{i=1}^\infty E_i$, then set $F_n := \bigcup_{i=1}^n E_i$. Then $\chi_{F_n} = \sum_{i=1}^n \chi_{E_i}$ and $\chi_{F_n} \xrightarrow{bp} \chi_E$ in $B_\infty(X)$. Since $\hat{\pi}$ is σ-normal, it follows that

$$\hat{\pi}(\chi_{F_n}) \xrightarrow{s} \hat{\pi}(\chi_E).$$

Unwrapping this, we see that $P(E) = \sum_{i=1}^\infty P(E_n)$ where the series converges strongly.

(ii) Fix $x, y \in H$ and $f \in B_\infty(X)$. If $f = \chi_E$ for some $E \in \mathfrak{B}_X$, then

$$\int_X f dP_{x,y} = P_{x,y}(E) = \langle \hat{\pi}(f)x, y \rangle.$$

By linearity, Equation 10.7 holds whenever f is a simple function. As before, we appeal to the fact that $S(X)$ is dense in $B_\infty(X)$ to conclude that Equation 10.7 also holds for all $f \in B_\infty(X)$ (once again, one needs to use the fact that $\hat{\pi}$ is σ-normal).

(iii) As for uniqueness, suppose $Q : \mathfrak{B}_X \to \mathcal{B}(H)$ is another spectral measure satisfying Equation 10.7. Then for any $E \in \mathfrak{B}_X$ and any $x, y \in H$,

$$\langle Q(E)x, y \rangle = \int_X \chi_E dQ_{x,y} = \int_X \chi_E dP_{x,y} = \langle P(E)x, y \rangle.$$

Therefore $Q(E) = P(E)$ for all $E \in \mathfrak{B}_X$. □

10.6.10 DEFINITION Let X be a compact metric space, H a Hilbert space and $\hat{\pi} : B_\infty(X) \to \mathcal{B}(H)$ a σ-normal representation. Let $P : \mathfrak{B}_X \to \mathcal{B}(H)$ be the unique spectral measure constructed in Theorem 10.6.9. Then for each $f \in B_\infty(X)$, we define

$$\int_X f dP := \hat{\pi}(f).$$

This notation agrees with Equation 10.7 in the sense that, for each $x, y \in H$,

$$\left\langle \left(\int_X f dP \right)(x), y \right\rangle = \int_X f dP_{x,y}.$$

The next corollary are now immediate from Theorem 10.6.9 and Theorem 10.4.10.

10.6.11 COROLLARY *Let X be a compact metric space and $\pi : C(X) \to \mathcal{B}(\mathbf{H})$ be a non-degenerate representation. Then there is a unique spectral measure $P : \mathfrak{B}_X \to \mathcal{B}(\mathbf{H})$ such that*

$$\pi(f) = \int_X f dP$$

for each $f \in C(X)$.

Applying this to the continuous functional calculus gives us a second version of the Spectral Theorem.

10.6.12 COROLLARY (SPECTRAL THEOREM) *Let $T \in \mathcal{B}(\mathbf{H})$ be a normal operator. Then there exists a unique spectral measure P defined on the Borel σ-algebra of $\sigma(T)$ such that*

$$T = \int_{\sigma(T)} \zeta dP,$$

where $\zeta \in C(\sigma(T))$ is the function $\zeta(z) = z$.

We end the section with a description of this spectral measure in the all-important case of a multiplication operator.

10.6.13 EXAMPLE Let $(X, \mathfrak{M}, \mu)$ be a σ-finite measure space, $\mathbf{H} := L^2(X, \mu)$ and let $T \in \mathcal{B}(\mathbf{H})$ be the multiplication operator M_ϕ for some fixed $\phi \in L^\infty(X, \mu)$. We wish to determine the associated spectral measure.

Let $Y := \sigma(T) = \text{ess-range}(\phi)$ and let $\widehat{\pi} : B_\infty(Y) \to \mathcal{B}(\mathbf{H})$ be the map

$$\widehat{\pi}(f) := M_{f \circ \phi}.$$

Then it is easy to verify that $\widehat{\pi}$ is a representation of $B_\infty(Y)$. In fact, by the argument in Example 10.4.8, $\widehat{\pi}$ is a σ-normal representation. Now if $\zeta \in B_\infty(Y)$ is the identity function $\zeta(z) = z$, then $\widehat{\pi}(\zeta) = T$. By the uniqueness clause of Corollary 10.4.16, $\widehat{\pi}$ is the Borel functional calculus of T. Hence

$$f(T) = M_{f \circ \phi}$$

for each $f \in B_\infty(Y)$. Therefore the spectral measure $P : \mathfrak{B}_Y \to \mathcal{B}(\mathbf{H})$ associated to T is given by

$$P(E) = \chi_E(T) = M_{\chi_E \circ \phi} = M_{\chi_{\phi^{-1}(E)}}$$

for each $E \in \mathfrak{B}_Y$.

10.7 Applications of the Spectral Theorem

We have now proved four results concerning normal operators that merit repeating. Let $\mathbf{H}$ be a separable Hilbert space and $T \in \mathcal{B}(\mathbf{H})$ be a normal operator. Then

(i) there is an isometric $*$-homomorphism $\Theta : C(\sigma(T)) \to \mathcal{B}(\mathbf{H})$ such that $\Theta(p(z, \bar{z})) = p(T, T^*)$ for any polynomial $p(x, y) \in \mathbb{C}[x, y]$.

(ii) there is a σ-finite measure space $(X, \mathfrak{M}, \mu)$ and a function $\phi \in L^\infty(X, \mu)$ such that T is unitarily equivalent to the multiplication operator M_ϕ.

(iii) there is a unique σ-normal representation $\widehat{\Theta} : B_\infty(\sigma(T)) \to \mathcal{B}(\mathbf{H})$ such that $\widehat{\Theta}|_{C(\sigma(T))} = \Theta$.

(iv) there is a unique spectral measure $P : \mathfrak{B}_{\sigma(T)} \to \mathcal{B}(\mathbf{H})$ such that

$$T = \int_{\sigma(T)} \zeta \, dP,$$

where $\zeta \in C(\sigma(T))$ is the identity function $\zeta(z) = z$.

These theorems are immensely useful and we now give some applications.

Polar Decomposition

We begin with a simple and extremely useful application of the continuous functional calculus. Given an operator $T \in \mathcal{B}(\mathbf{H})$, the continuous functional calculus tells us that there is a positive operator $|T| \in \mathcal{B}(\mathbf{H})$ with the property that $|T|^2 = T^*T$ (see Definition 9.1.7). We will now use this operator to construct a decomposition of T that is reminiscent of the polar decomposition of a complex number into its modulus (a positive real number) and an exponential term (a complex number of unit modulus). The role of the modulus is played by $|T|$, while that of the exponential term is played by a partial isometry.

10.7.1 DEFINITION An operator $W \in \mathcal{B}(\mathbf{H})$ is called a *partial isometry* if

$$\|W(x)\| = \|x\|$$

for all $x \in \ker(W)^\perp$. In other words, W acts as an isometry on the orthogonal complement of its kernel. The space $\ker(W)^\perp$ is called the *initial space* of W and Range(W) is called the *final space* of W.

Note that both the initial and final spaces of a partial isometry are closed subspaces. As it turns out, the corresponding projection operators are realizable in terms of the partial isometry.

10.7.2 LEMMA *If W is a partial isometry, then W^*W and WW^* are projections onto the initial and final space of W, respectively.*

Proof. Let $P := W^*W$. Our first goal is to prove that P is indeed a projection.

(i) For $x \in \ker(W)^\perp$ and $y \in \ker(W)$, we have

$$\langle P(x), y \rangle = \langle W(x), W(y) \rangle = 0.$$

Hence $P(x) \in \ker(W)^\perp$.

(ii) Furthermore if $x \in \ker(W)^\perp$, then $\langle W(x), W(x) \rangle = \langle x, x \rangle$. By the polarization identity,

$$\langle W(x), W(y) \rangle = \langle x, y \rangle$$

for all $x, y \in \ker(W)^\perp$. Now fix $x \in \ker(W)^\perp$. For any $y \in \mathbf{H}$, write $y = y_1 + y_2$ where $y_1 \in \ker(W)$ and $y_2 \in \ker(W)^\perp$. Then

$$\langle P(x), y \rangle = \langle W(x), W(y) \rangle = \langle W(x), W(y_2) \rangle = \langle x, y_2 \rangle = \langle x, y \rangle.$$

Hence $P(x) = x$, so P is a projection.

(iii) By part (i), $\text{Range}(P) \subset \ker(W)^\perp$. Conversely, if $P(x) = x$, then for any $y \in \ker(W)$,

$$\langle x, y \rangle = \langle P(x), y \rangle = \langle W(x), W(y) \rangle = 0.$$

Therefore $x \in \ker(W)^\perp$. We conclude that P is a projection onto $\ker(W)^\perp$, the initial space of W.

The argument for $Q := WW^*$ is similar (try Exercise 10.32). $\qquad\square$

10.7.3 THEOREM (POLAR DECOMPOSITION) *For each $T \in \mathcal{B}(\mathbf{H})$, there is a partial isometry $W \in \mathcal{B}(\mathbf{H})$ such that $\ker(W) = \ker(|T|)$ and $T = W|T|$. Furthermore, if $T = VS$ with S positive and V a partial isometry such that $\ker(V) = \ker(S)$, then $S = |T|$ and $V = W$.*

This unique expression $T = W|T|$ is called the *polar decomposition* of T.

Proof.   For $x \in \mathbf{H}$,

$$\|Tx\|^2 = \langle Tx, Tx \rangle = \langle T^*Tx, x \rangle = \langle |T|^2 x, x \rangle = \langle |T|x, |T|x \rangle = \||T|x\|^2.$$

Hence we may define $W : \text{Range}(|T|) \to \text{Range}(T)$ by

$$W(|T|x) = Tx$$

and W is both well-defined and an isometry. Since $T^*T(x) = |T|(|T|x)$, it follows that $\text{Range}(T^*T) \subset \text{Range}(|T|)$. Moreover, it can be shown that $\text{Range}(|T|) \subset \overline{\text{Range}(T^*T)}$ (Exercise 10.33). Hence $\overline{\text{Range}(|T|)} = \overline{\text{Range}(T^*T)}$. Since $\overline{\text{Range}(T^*T)} = \ker(T)^\perp$ by Lemma 8.1.15, W extends to an isometry

$$W : \ker(T)^\perp \to \overline{\text{Range}(T)}.$$

Now we extend W to all of $\mathbf{H}$ by setting it to be zero on $\ker(T)$ and we get a partial isometry satisfying the required conditions.

As for uniqueness, suppose that $T = VS$ as in the statement of the theorem. By Lemma 10.7.2, V^*V is the projection Q onto the initial space of V, which is $\ker(V)^\perp = \ker(S)^\perp = \overline{\text{Range}(S)}$. Thus $T^*T = S^*V^*VS = SQS = S^2$. By the uniqueness of the positive square root (Theorem 9.1.6), it follows that $S = |T|$. Since

$$Tx = V|T|x = W|T|x,$$

it follows that V and W agree on $\text{Range}(|T|)$. However, $\text{Range}(|T|)$ is a dense subset of both their initial spaces and thus $V = W$ must hold. □

10.7.4 REMARK

(i) If $\mathbf{A}$ is a unital C*-algebra and $a \in \mathbf{A}$, then $|a|$ is a positive element in $\mathbf{A}$. However, there may not exist an element $w \in \mathbf{A}$ such that $a = w|a|$. For instance, suppose $\mathbf{A} = C[-1, 1]$ and $f \in \mathbf{A}$ is the identity function $f(t) = t$. If $w \in \mathbf{A}$ is such that $f = w|f|$, then it must happen that

$$w(t) = \begin{cases} -1 & : t \in [-1, 0) \text{ and} \\ 1 & : t \in (0, 1]. \end{cases}$$

Such a function cannot be continuous and therefore no such element can exist.

(ii) However, if **A** is a unital C*-algebra, and $a \in \mathbf{A}$ is *invertible*, then one may obtain a polar decomposition of the type $a = u|a|$, where u is a unitary. The proof is relatively straight-forward and was an exercise in Chapter 9 (Exercise 9.5).

(iii) Finally, if $\mathbf{A} \subset \mathcal{B}(\mathbf{H})$ is a von Neumann algebra and $T \in \mathbf{A}$ has polar decomposition $T = W|T|$ in $\mathcal{B}(\mathbf{H})$, then $W \in \mathbf{A}$. We do not prove this result here, as it will take us too far afield. The interested reader will find a proof Vol. 2 of Kadison and Ringrose [32, Theorem 6.1.3].

We will now use the polar decomposition to study compact operators.

Compact Operators

In Chapter 7, we had studied the spectrum of a compact operator in some detail. Later, in Theorem 10.1.8, we had proved that a compact normal operator is diagonalizable. We now wish to expand on those themes and say something about the *ideal* of compact operators. We do this by exploiting the second version of the Spectral Theorem (in terms of spectral measures). We begin with a lemma that helps us identify eigenvalues of a normal operator (which may remind you of Exercise 8.33).

10.7.5 LEMMA *Let $T \in \mathcal{B}(\mathbf{H})$ be a normal operator with spectral measure P. An element $\lambda \in \mathbb{C}$ is an eigenvalue of T if and only if $P(\{\lambda\}) \neq 0$. Furthermore, in that case, $P(\{\lambda\})$ is the projection onto $\ker(T - \lambda I)$.*

Proof.   Let $X := \sigma(T)$ and let $\widehat{\Theta} : B_\infty(X) \to \mathcal{B}(\mathbf{H})$ be the Borel functional calculus. If $\zeta \in B_\infty(X)$ is the identity function $\zeta(z) = z$, then $T = \widehat{\Theta}(\zeta)$ and if $E \subset X$ is a Borel set, then $P(E) = \widehat{\Theta}(\chi_E)$. Furthermore, for any $f \in B_\infty(X)$ and $x, y \in \mathbf{H}$,

$$\int_X f dP_{x,y} = \langle \widehat{\Theta}(f)x, y \rangle,$$

where $P_{x,y}$ is the complex Borel measure on X defined by $P_{x,y}(E) = \langle P(E)x, y \rangle$.

(i) If $Q = P(\{\lambda\})$, then

$$TQ = \widehat{\Theta}(\zeta \chi_{\{\lambda\}})) = \widehat{\Theta}(\lambda \chi_{\{\lambda\}}) = \lambda Q.$$

Thus if $Q \neq 0$, then any element of Range(Q) is an eigenvector with eigenvalue λ.

(ii) Conversely, suppose λ is an eigenvalue with eigenvector x, then we claim that $P(\{\lambda\})(x) = x$, which would imply that $P(\{\lambda\}) \neq 0$.

(a) Define $E_n := \{z \in \sigma(T) : |z - \lambda| \geq \frac{1}{n}\}$ and write $Q_n := P(E_n)$. Then

$$Q_n T = \widehat{\Theta}(\chi_{E_n} \zeta) = \widehat{\Theta}(\zeta \chi_{E_n}) = T Q_n.$$

Hence $(T - \lambda I)Q_n(x) = Q_n(T - \lambda I)x = 0$. Now recall that $P_{x,x}$ is a positive measure on $\mathcal{B}_X$ and we use this fact in the following calculation.

$$
\begin{aligned}
0 = \|(T - \lambda I)Q_n(x)\|^2 &= \langle (T - \lambda I)Q_n(x), (T - \lambda I)Q_n(x) \rangle \\
&= \langle Q_n^*(T - \lambda I)^*(T - \lambda I)Q_n(x), x \rangle \\
&= \langle \widehat{\Theta}(\chi_{E_n}\overline{(\zeta - \lambda \mathbf{1})}(\zeta - \lambda \mathbf{1})\chi_{E_n})x, x \rangle \\
&= \int_X \chi_{E_n}(z)|z - \lambda|^2 dP_{x,x}(z) \\
&\geq \frac{1}{n^2} \int_X \chi_{E_n} dP_{x,x} \\
&= \frac{1}{n^2} \langle Q_n(x), x \rangle \\
&= \frac{1}{n^2} \langle Q_n(x), Q_n(x) \rangle = \frac{1}{n^2} \|Q_n(x)\|^2.
\end{aligned}
$$

Hence $Q_n(x) = 0$ for all $n \in \mathbb{N}$.

(b) Now observe that $E_n \subset E_{n+1}$ and

$$E := \sigma(T) \setminus \{\lambda\} = \bigcup_{n=1}^{\infty} E_n.$$

Since $\widehat{\Theta}$ is a σ-representation, it follows that $P(E)(x) = \lim_{n \to \infty} Q_n(x) = 0$. Hence

$$x = P(X)(x) = P(\{\lambda\})(x) + P(E)(x) = P(\{\lambda\})(x).$$

This proves that $P(\{\lambda\}) \neq 0$.

(iii) Finally, we observe from the proof that $P(\{\lambda\})(x) = x$ if and only if $x \in \ker(T - \lambda I)$. Hence $P(\{\lambda\})$ is the projection onto $\ker(T - \lambda I)$. $\qquad \square$

The next lemma helps us identify normal compact operators using the associated spectral measure.

10.7.6 LEMMA *Let $T \in \mathcal{B}(\mathbf{H})$ be a normal operator with spectral measure P. Then T is compact if and only if*

$$Q_\epsilon := P(\{z \in \sigma(T) : |z| > \epsilon\})$$

is a finite rank projection for each $\epsilon > 0$.

Proof. Let $X := \sigma(T)$, $E_\epsilon := \{z \in X : |z| > \epsilon\}$ and $F_\epsilon := X \setminus E_\epsilon$.

(i) Assume that Q_ϵ has finite rank for each $\epsilon > 0$. Then observe that

$$T - TQ_\epsilon = \int_X \zeta dP - \int_X \zeta \chi_{E_\epsilon} dP = \int_X \zeta \chi_{F_\epsilon} dP = f(T),$$

where $f(z) = z\chi_{F_\epsilon}(z)$. Hence

$$\|T - TQ_\epsilon\| = \|f\|_\infty = \sup\{|z| : z \in F_\epsilon\} \leq \epsilon.$$

By hypothesis, $TQ_\epsilon \in \mathcal{K}(\mathbf{H})$ for each $\epsilon > 0$ and thus T is a limit of finite rank bounded operators. Since $\mathcal{K}(\mathbf{H})$ is closed, $T \in \mathcal{K}(\mathbf{H})$.

(ii) Conversely, suppose T is compact. Define

$$g(z) := \frac{1}{z}\chi_{E_\epsilon}(z).$$

Then $g \in B_\infty(X)$ and $\chi_{E_\epsilon}(z) = g(z)z$. Hence

$$Q_\epsilon = g(T)T \in \mathcal{K}(\mathbf{H}),$$

since $\mathcal{K}(\mathbf{H})$ is an ideal. Furthermore, Q_ϵ is a projection and must therefore have finite rank (by Exercise 7.40). $\square$

We now revisit the Spectral Theorem for compact normal operators and give it a fresh coat of paint. Recall from Theorem 7.5.7 that the spectrum of a compact operator is necessarily a countable set, whose non-zero elements are all eigenvalues. In Theorem 10.1.8, we proved that there is a basis consisting of eigenvectors. The next theorem merely recasts this in operator-theoretic terms.

10.7.7 THEOREM (SPECTRAL THEOREM FOR COMPACT OPERATORS) *Let T be a compact normal operator on a Hilbert space $\mathbf{H}$ and let P denote the associated spectral measure. By Theorem 7.5.7, we may write*

$$\sigma(T) \setminus \{0\} = \{\lambda_1, \lambda_2, \ldots\}$$

where each λ_k is an eigenvalue of T. If $Q_k := P(\{\lambda_k\})$, then

$$T = \sum_{k=1}^{\infty} \lambda_k Q_k,$$

where the series converges in the operator norm.

Proof. Let $X := \sigma(T)$ and let $\Theta : C(X) \to \mathcal{B}(\mathbf{H})$ denote the continuous functional calculus of T. For our proof, we assume that X is infinite as the finite case is similar (indeed, easier). By Theorem 7.5.7, (λ_k) is a sequence of non-zero complex numbers converging to 0. Also, each λ_k is an isolated point of X. For each $n \in \mathbb{N}$, we define

$$s_n := \sum_{k=1}^{n} \lambda_k \chi_{\{\lambda_k\}}.$$

Then $s_n \in C(X)$ and

$$\lim_{n \to \infty} \|\zeta - s_n\|_\infty \leq \lim_{n \to \infty} \sup_{k > n} |\lambda_k| = 0.$$

Therefore

$$T = \Theta(\zeta) = \lim_{n \to \infty} \Theta(s_n) = \sum_{k=1}^{\infty} \lambda_k \chi_{\{\lambda_k\}}(T) = \sum_{k=1}^{\infty} \lambda_k Q_k$$

and the sum converges in the norm topology (since we are dealing with the *continuous* functional calculus). $\qquad\square$

As promised, we now look at the ideal of compact operators. Our goal is to prove that $\mathcal{B}(\mathbf{H})$ has exactly one non-trivial, norm–closed ideal when $\mathbf{H}$ is separable and infinite dimensional. The next result is the first step towards that.

10.7.8 PROPOSITION *For any $T \in \mathcal{B}(\mathbf{H})$, $T \in \mathcal{K}(\mathbf{H})$ if and only if $T^*T \in \mathcal{K}(\mathbf{H})$.*

Proof. If $T \in \mathcal{K}(\mathbf{H})$ then $T^*T \in \mathcal{K}(\mathbf{H})$ since $\mathcal{K}(\mathbf{H})$ is an ideal. Conversely, if $S := T^*T \in \mathcal{K}(\mathbf{H})$, then $S^n \in \mathcal{K}(\mathbf{H})$ for all $n \geq 1$. Hence $p(S) \in \mathcal{K}(\mathbf{H})$ for any polynomial $p(z) \in \mathbb{C}[z]$ such that $p(0) = 0$. Since S is self-adjoint, $\sigma(S) \subset \mathbb{R}$. By the Weierstrass Approximation Theorem, we conclude that $f(S) \in \mathcal{K}(\mathbf{H})$ for any $f \in C(\sigma(S))$ such that $f(0) = 0$. In particular,

$$|T| = \sqrt{S} \in \mathcal{K}(\mathbf{H}).$$

By the polar decomposition, we may write T in the form $T = W|T|$ for some $W \in \mathcal{B}(\mathbf{H})$. Since $\mathcal{K}(\mathbf{H})$ is an ideal, it follows that $T \in \mathcal{K}(\mathbf{H})$. $\square$

10.7.9 LEMMA *Let* $\mathbf{H}$ *be a separable Hilbert space and* $\mathbf{J} \lhd \mathcal{B}(\mathbf{H})$ *be a two-sided ideal that contains a non-compact operator. Then* $\mathbf{J} = \mathcal{B}(\mathbf{H})$.

Proof. Let $T_0 \in \mathbf{J}$ be non-compact, then $T := T_0^* T_0 \in \mathbf{J}$ is normal. Furthermore, T is not compact by Proposition 10.7.8. Let P be the spectral measure associated to T. By Lemma 10.7.6, there exists $\epsilon > 0$ such that the projection

$$Q_\epsilon := P(\{z \in \sigma(T) : |z| > \epsilon\})$$

has infinite rank. Furthermore, by the proof of that lemma, there is an operator $S \in \mathcal{B}(\mathbf{H})$ such that

$$Q_\epsilon = ST.$$

Since $\mathbf{J}$ is an ideal, $Q_\epsilon \in \mathbf{J}$. Let $\mathbf{M} := \mathrm{Range}(Q_\epsilon)$, then $\mathbf{M}$ is a closed subspace of $\mathbf{H}$ and $\dim(\mathbf{M}) = \dim(\mathbf{H}) = \aleph_0$. Hence there is a unitary $V : \mathbf{H} \to \mathbf{M}$, which we treat as an isometry in $\mathcal{B}(\mathbf{H})$. By Lemma 10.7.2, $VV^* = Q_\epsilon$ and $V^*V = I$. Therefore

$$I = (V^*V)(V^*V) = V^* Q_\epsilon V.$$

Since $Q_\epsilon \in \mathbf{J}$, it follows that $I \in \mathbf{J}$ and thus $\mathbf{J} = \mathcal{B}(\mathbf{H})$. $\square$

In our quest to understand compact operators better, we first say a few words on finite rank operators. In Example 5.6.5, we had described a class of finite rank operator between two normed linear spaces. As it turns out, on Hilbert spaces, *all* finite rank operators admit such a description. The proof of this fact appeared as an exercise earlier (Exercise 3.31), but we now give a complete proof.

10.7.10 DEFINITION Let $\mathbf{H}$ be a Hilbert space and $x, y \in \mathbf{H}$. Define $S_{x,y} \in \mathcal{B}(\mathbf{H})$ by

$$S_{x,y}(z) := \langle z, x \rangle y.$$

Then $S_{x,y}$ is a rank one, bounded operator (which is therefore compact).

Recall that $\mathcal{F}(\mathbf{H})$ denotes the set of all finite rank operators in $\mathcal{B}(\mathbf{H})$.

10.7.11 LEMMA *If* $\mathbf{H}$ *is a Hilbert space, then* $\mathcal{F}(\mathbf{H}) = \text{span}\{S_{x,y} : x, y \in \mathbf{H}\}$.

Proof. If $T \in \mathcal{F}(\mathbf{H})$, then we choose a finite orthonormal basis $\{y_1, y_2, \ldots, y_n\}$ of Range(T). For any $z \in \mathbf{H}$, $T(z) = \sum_{i=1}^{n} \alpha_i y_i$, where

$$\alpha_i = \langle T(z), y_i \rangle = \langle z, T^*(y_i) \rangle.$$

Hence $T = \sum_{i=1}^{n} S_{x_i, y_i}$, where $x_i = T^*(y_i)$ for $1 \leq i \leq n$. $\square$

We conclude with a complete description of closed ideals in $\mathcal{B}(\mathbf{H})$. If $\mathbf{H}$ is finite dimensional, $\mathcal{B}(\mathbf{H}) \cong M_n(\mathbb{C})$ where $n = \dim(H)$. In that case, $\mathcal{B}(\mathbf{H})$ has no non-trivial ideals at all (see Example 7.1.4). In the infinite dimensional separable case, the following holds.

10.7.12 THEOREM *If* $\mathbf{H}$ *is a separable, infinite dimensional Hilbert space, then the only non-trivial closed, two-sided ideal of* $\mathcal{B}(\mathbf{H})$ *is* $\mathcal{K}(\mathbf{H})$.

Proof. Let $\mathbf{J} \neq \{0\}$ be a closed ideal in $\mathcal{B}(\mathbf{H})$, then by Lemma 10.7.9, it suffices to show that $\mathcal{K}(\mathbf{H}) \subset \mathbf{J}$. To that end, choose a non-zero operator $T \in \mathbf{J}$. Then there exists $x_0 \in \mathbf{H}$ such that $x_1 := T(x_0) \neq 0$. For any $y_0, y_1 \in \mathbf{H}$ of norm 1, consider two rank one operators,

$$A := S_{y_0, x_0} \text{ and } B := S_{x_1, y_1}.$$

Then for any $z \in \mathbf{H}$,

$$BTA(z) = BT(\langle z, y_0 \rangle x_0) = \langle z, y_0 \rangle BT(x_0)$$
$$= \langle z, y_0 \rangle B(x_1) = \langle z, y_0 \rangle y_1$$
$$= S_{y_0, y_1}(z).$$

Hence every rank one operator belongs to $\mathbf{J}$. By the previous lemma, $\mathcal{F}(\mathbf{H}) \subset \mathbf{J}$. By Corollary 5.6.11, $\mathcal{F}(\mathbf{H})$ is dense in $\mathcal{K}(\mathbf{H})$. Since $\mathbf{J}$ is closed, it follows that $\mathcal{K}(\mathbf{H}) \subset \mathbf{J}$. $\square$

Fuglede's Theorem

The next application of the Spectral Theorem is a commutativity theorem due to Fuglede. The proof that we present is due to Halmos [23], and is meant to highlight the fact that every normal operator is equivalent to a multiplication operator.

10.7.13 THEOREM (FUGLEDE, 1950) *Let* $\mathbf{H}$ *be a separable Hilbert space and* $S, T \in$ $\mathcal{B}(\mathbf{H})$ *are such that* $ST = TS$. *If* T *is normal, then* $ST^* = T^*S$.

Proof. By the Spectral Theorem, we may assume that $\mathbf{H} = L^2(X, \mu)$ for some σ-finite measure space $(X, \mathfrak{M}, \mu)$ and that $T = M_\phi$ for some $\phi \in L^\infty(X, \mu)$. We will prove the superficially stronger statement that $Sg(T) = g(T)S$ for each $g \in$ $B_\infty(\sigma(T))$. In fact, it is easy to verify that the conclusion of the theorem is equivalent to this statement.

Set $Y := \sigma(T) = \text{ess-range}(\phi)$. We know that $\mathcal{V}(T) = \overline{\text{span}\{\chi_E(T) : E \in \mathcal{B}_Y\}}$. Therefore it suffices to prove our result when g is the characteristic function of a Borel set. Let $P : \mathcal{B}_Y \to \mathcal{B}(\mathbf{H})$ denote the spectral measure associated to T. We thus need to show that the set

$$\mathfrak{N} := \{E \in \mathcal{B}_Y : P(E)S = SP(E)\}$$

is all of $\mathcal{B}_Y$. We prove this in the following steps. In what follows, for each $E \in \mathcal{B}_Y$, we write $V(E) := \text{Range}(P(E))$.

(i) Firstly, observe that $E \in \mathfrak{N}$ if and only if $V(E)$ is invariant under S. Furthermore, by Example 10.6.13, $P(E) = M_{\chi_{\phi^{-1}(E)}}$. Therefore $f \in V(E)$ if and only if f vanishes outside $\phi^{-1}(E)$, in the sense that $\mu(\text{supp}(f) \cap (X \setminus \phi^{-1}(E))) = 0$ (do verify this).

(ii) Let $D := \{z \in Y : |z| \leq 1\}$ and we wish to describe the elements of $V(D)$. We claim that $f \in V(D)$ if and only if

$$\sup\{\|T^n(f)\| : n \in \mathbb{N}\} < \infty.$$

(a) If $f \in V(D)$, then f vanishes outside $\phi^{-1}(D)$. Now $\phi^{-1}(D) = \{x \in X : |\phi(x)| \leq 1\}$. Therefore if $n \in \mathbb{N}$, then

$$\|T^n(f)\|^2 = \int_X |\phi^n f|^2 d\mu = \int_{\phi^{-1}(D)} |\phi^n|^2 |f|^2 d\mu \leq \int_X |f|^2 d\mu.$$

Hence $\sup\{\|T^n(f)\| : n \in \mathbb{N}\} \leq \|f\|_2 < \infty$.

(b) Conversely, if f does not vanish outside $\phi^{-1}(D)$, then there is a set $A \in \mathcal{B}_X$ of finite positive measure on which f is non-zero and $|\phi| > 1$. Then

$$\|T^n(f)\|^2 = \int_X |\phi^n f|^2 d\mu \geq \int_A |\phi^n|^2 |f|^2 d\mu \to \infty.$$

This proves the claim.

(iii) We now claim that $D \in \mathfrak{N}$. If $f \in V(D)$, then there is $M > 0$ such that $\|T^n(f)\| \leq M$ for all $n \in \mathbb{N}$. Then $\|T^n(S(f))\| = \|S(T^n(f))\| \leq \|S\|M$ for all $n \in \mathbb{N}$. By part (ii), we conclude that $S(f) \in V(D)$. Thus $V(D)$ is invariant under S and $D \in \mathfrak{N}$.

(iv) For any $\lambda \in \mathbb{C}$ and $r > 0$, consider $E := \{z \in X : |z - \lambda| \leq r\}$. Then

$$\phi^{-1}(E) = \left\{ x \in X : \left| \frac{\phi(x) - \lambda}{r} \right| \leq 1 \right\}.$$

Since $SM_\phi = M_\phi S$, it follows that $SM_{\widetilde{\phi}} = M_{\widetilde{\phi}} S$, where $\widetilde{\phi}(x) := (\phi(x) - \lambda)/r$. By part (iii), it follows that $V(E)$ is invariant under S. Therefore $E \in \mathfrak{N}$.

(v) If $E, F \in \mathfrak{N}$, then $E \cap F \in \mathfrak{N}$ because $P(E \cap F) = P(E)P(F)$ by Lemma 10.6.6.

(vi) If $E, F \in \mathfrak{N}$, then $E = (E \cap F) \sqcup (E \setminus F)$. Therefore $P(E \setminus F) = P(E) - P(E \cap F)$. Since $E \cap F \in \mathfrak{N}$, it follows that $E \setminus F \in \mathfrak{N}$ as well. In particular, if $E \in \mathfrak{N}$, then $Y \setminus E \in \mathfrak{N}$.

(vii) If $E, F \in \mathfrak{N}$, then $E \cup F = (E \setminus F) \sqcup F$. Hence $P(E \cup F) = P(E \setminus F) + P(F)$. By part (vi), $E \setminus F \in \mathfrak{N}$ and therefore $E \cup F \in \mathfrak{N}$. Hence $\mathfrak{N}$ is closed under finite unions.

(viii) Now suppose $\{E_1, E_2, \ldots\}$ is a sequence of sets in $\mathfrak{N}$. We wish to prove that $E := \bigcup_{n=1}^\infty E_n \in \mathfrak{N}$. For each $n \in \mathbb{N}$, set

$$F_n := E_n \setminus \left(\bigcup_{i=1}^{n-1} E_i \right).$$

Then $F_n \in \mathfrak{N}$ by step (vi) and (vii). Also, $E = \bigsqcup_{n=1}^\infty F_n$. Therefore

$$P(E) = \sum_{n=1}^\infty P(F_n),$$

where the series converges in the strong operator topology. Since $P(F_n)S = SP(F_n)$ for each $n \in \mathbb{N}$, it follows that $P(E)S = SP(E)$. In other words, $E \in \mathfrak{N}$ as well.

We conclude that $\mathfrak{N}$ is a σ-algebra that contains all closed discs (intersected with Y). Hence $\mathfrak{N} = \mathfrak{B}_Y$, which proves the result. $\qquad\square$

10.7.14 COROLLARY (PUTNAM, 1951) *Let $\mathbf{H}_1$ and $\mathbf{H}_2$ be two separable Hilbert spaces. For $i \in \{1, 2\}$, let $T_i \in \mathcal{B}(\mathbf{H}_i)$ be normal operators. If $S : \mathbf{H}_1 \to \mathbf{H}_2$ is a bounded operator such that $ST_1 = T_2 S$, then $ST_1^* = T_2^* S$.*

Proof. Let $\mathbf{H} := \mathbf{H}_1 \oplus \mathbf{H}_2$ and let $S', T' \in \mathcal{B}(\mathbf{H})$ be the operators given by the matrices

$$S' = \begin{pmatrix} 0 & 0 \\ S & 0 \end{pmatrix} \text{ and } T' = \begin{pmatrix} T_1 & 0 \\ 0 & T_2 \end{pmatrix}.$$

Then T' is normal and $S'T' = T'S'$. Now Fuglede's Theorem applies. □

Unitary Operators

Our final application of the Spectral Theorem is meant to showcase the Borel functional calculus. Before we study unitary operators, let us say a few things about unitary elements in an arbitrary (unital) C*-algebra. Let $\mathbf{A}$ be a unital C*-algebra and $h \in \mathbf{A}$ be a self-adjoint element. Then the element

$$u := \exp(ih)$$

is a unitary because $u^* = \exp(-ih) = u^{-1}$. However, not every unitary is necessarily of this form. To understand why, let us revisit some exercises from earlier chapters (specifically, Exercise 7.22 and Exercise 8.38).

Given a unital C*-algebra $\mathbf{A}$, we write $U(\mathbf{A})$ for the set of all unitary elements in $\mathbf{A}$. Note that $U(\mathbf{A})$ is a group. In fact, the multiplication is continuous (as $\mathbf{A}$ is a Banach algebra) and so is the inverse operation, since the latter coincides with the adjoint operation. Hence $U(\mathbf{A})$ is a *topological group*. We write $U_0(\mathbf{A})$ for the connected component of the identity in $U(\mathbf{A})$. Our first lemma gives us a hint as to why every unitary may not be an exponential.

10.7.15 Lemma *If h is a self-adjoint element in a unital C*-algebra $\mathbf{A}$, then $\exp(ih) \in U_0(\mathbf{A})$.*

Proof. For each $t \in [0,1]$, define $u_t := \exp(ith)$, then $u_0 = 1_{\mathbf{A}}$ and $u_1 = \exp(ih)$. Since $\|u_t\| \leq \exp(\|h\|)$ for all $t \in [0,1]$, it follows that the map $t \mapsto u_t$ is a continuous function from $[0,1]$ to $U(\mathbf{A})$ (do verify this). We have proved that $\exp(ih)$ is in the path component of $1_{\mathbf{A}}$ and is thus in $U_0(\mathbf{A})$. □

Our next example should now come as no surprise.

10.7.16 Example Let $\mathbf{A} = C(S^1)$ and let $u \in \mathbf{A}$ be the identity function $u(z) = z$. If there were $h \in \mathbf{A}$ such that $u = \exp(ih)$, then the proof of the previous lemma would show u to be a contractible loop. However, u is not contractible (it has non-zero winding number) and thus u cannot be expressed as an exponential.

It is interesting then that the following result should hold. Recall from Lemma 8.5.9 that the spectrum of a unitary element must lie in the circle S^1.

10.7.17 PROPOSITION *Let u be a unitary in a unital C^*-algebra $\mathbf{A}$ such that $\sigma(u) \neq S^1$. Then there exists $h \in \mathbf{A}_{sa}$ such that $u = \exp(ih)$.*

Proof. Since $\sigma(u) \neq S^1$, there exists $\theta \in \mathbb{R}$ such that $\exp(i\theta) \notin \sigma(u)$. Then there exists a real-valued function $f \in C(\sigma(u))$ such that

$$f(\exp(it)) = t$$

for all $t \in (\theta, \theta + 2\pi)$. It follows that $\exp(if(z)) = z$ for all $z \in \sigma(u)$. Applying the continuous functional calculus, the element $h := f(u)$ is a self-adjoint and satisfies $u = \exp(ih)$. $\qquad\square$

The previous result is restricted to the case $\sigma(u) \neq S^1$ only because we are forced to use the continuous functional calculus. If we remove this constraint and allow ourselves the use of the *Borel* functional calculus, this restriction also disappears! This brings us to the main result of this subsection.

10.7.18 THEOREM *Let $\mathbf{H}$ be a Hilbert space and $U \in \mathcal{B}(\mathbf{H})$ be a unitary operator. Then there is a self-adjoint operator $T \in \mathcal{B}(\mathbf{H})$ such that $U = \exp(iT)$.*

Proof. Define $f : S^1 \to \mathbb{R}$ by $f(e^{i\theta}) = \theta$, for $0 \leq \theta < 2\pi$. Then f is Borel measurable and bounded. Since $\sigma(U) \subset S^1$, we may define

$$T := f(U).$$

T is self-adjoint since f is real-valued. Since $\exp(i(f(z)) = z$ for all $z \in S^1$, it follows that $U = \exp(iT)$. $\qquad\square$

By Lemma 10.7.15, every unitary in $\mathcal{B}(\mathbf{H})$ may be connected to the identity operator by a path consisting of unitaries. We conclude that

10.7.19 COROLLARY *For any Hilbert space $\mathbf{H}$, the unitary group $U(\mathcal{B}(\mathbf{H}))$ is path connected.*

In fact, more is true. It is a theorem due to Kuiper [35] that $U(\mathcal{B}(\mathbf{H}))$ is contractible in the norm topology, provided $\mathbf{H}$ is an infinite dimensional Hilbert space (this is patently false if $\mathbf{H}$ is finite dimensional). Although we do not prove this here, it is a result that plays an important role in the study of fibre bundles and thus in K-theory.

This brings us to the end of this long discussion on the Spectral Theorem. We now end the book with some results for arbitrary C^*-algebras that take inspiration from the ideas we have explored in this chapter.

10.8   The Gelfand–Naimark Representation

We now revisit the first major theorem of this chapter, the Spectral Theorem in the special case where the operator had a cyclic vector (Theorem 10.3.7). Contained in the proof is the following idea.

10.8.1 REMARK   Let X be a compact metric space and let $\tau : C(X) \to \mathbb{C}$ be a positive linear functional. By the Riesz-Markov-Kakutani Theorem, there is a finite, positive, Borel measure μ on X such that

$$\tau(f) = \int_X f d\mu$$

for all $f \in C(X)$. Set $\mathbf{H} := L^2(X, \mu)$. For each $f \in C(X)$, define a multiplication operator $M_f \in \mathcal{B}(\mathbf{H})$ by

$$M_f(g) := fg.$$

The map $\pi : C(X) \to \mathcal{B}(\mathbf{H})$ defined by $f \mapsto M_f$ is now a representation of $C(X)$ on $\mathbf{H}$ (in the sense of Definition 10.4.3). If $e = \mathbf{1} \in \mathbf{H}$ denotes the constant function, then

$$\langle \pi(f)e, e \rangle = \int_X f d\mu = \tau(f). \tag{10.8}$$

In other words, the positive linear functional helps us to build a representation of $C(X)$ and that representation may be used to recover the positive linear functional by means of Equation 10.8.

In the Spectral Theorem, we started with a normal operator $T \in \mathcal{B}(\mathbf{H})$, took $X = \sigma(T)$ and applied this idea to realize T as a multiplication operator on $L^2(X, \mu)$. We now take a step back and analyze this process if $C(X)$ were replaced by an arbitrary C*-algebra. Naturally, we would not have access to the Riesz–Markov–Kakutani Theorem. However, as we will soon see, it is possible to construct a representation in this abstract setting as well. This procedure is called the Gelfand–Naimark–Segal (GNS) construction and it will lead to a foundational representation theorem for C*-algebras. For simplicity, we will focus on the case of *unital* C*-algebras, although many of the theorems below also hold in the non-unital case.

Let $\mathbf{A}$ be a unital C*-algebra. Recall from Definition 10.4.3 that a representation $(\pi, \mathbf{H})$ of $\mathbf{A}$ is a *-homomorphism $\pi : \mathbf{A} \to \mathcal{B}(\mathbf{H})$. Furthermore, $(\pi, \mathbf{H})$ is said to be cyclic if there is a vector $e \in \mathbf{H}$ such that $\pi(\mathbf{A})e$ is dense in $\mathbf{H}$. If such a vector exists, we simply say that the triple $(\pi, \mathbf{H}, e)$ is a cyclic representation.

10.8.2 DEFINITION Let $\mathbf{A}$ be a unital C*-algebra and let $\tau : \mathbf{A} \to \mathbb{C}$ be a positive linear functional.

(i) A *GNS representation* for τ is a cyclic representation $(\pi, \mathbf{H}, e)$ of $\mathbf{A}$ such that

$$\tau(a) = \langle \pi(a)e, e \rangle$$

for all $a \in \mathbf{A}$.

(ii) Two such GNS representations $(\pi, \mathbf{H}, e)$ and $(\pi', \mathbf{H}', e')$ are said to be *equivalent* if there is a unitary operator $U : \mathbf{H} \to \mathbf{H}'$ such that $U(e) = e'$, and $U\pi(a)U^{-1} = \pi'(a)$ for all $a \in \mathbf{A}$.

At the moment, we do not know if such a representation exists and our goal is to construct one. We begin with a proof of Exercise 9.9.

10.8.3 LEMMA *Let $\mathbf{A}$ be a unital C*-algebra and $\tau : \mathbf{A} \to \mathbb{C}$ be a positive linear functional. Define $N_\tau := \{a \in \mathbf{A} : \tau(a^*a) = 0\}$.*

(i) *If $a \in N_\tau$ and $x \in \mathbf{A}$, then $\tau(xa) = \tau(a^*x) = 0$.*

(ii) *N_τ is a closed left ideal of $\mathbf{A}$.*

(iii) *For any $a, b \in \mathbf{A}$, $\tau(b^*a^*ab) \leq \|a^*a\|\tau(b^*b)$.*

Proof.

(i) By the Cauchy–Schwarz Inequality (Proposition 9.1.13), we have

$$|\tau(xa)| \leq \tau(a^*a)^{1/2}\tau(xx^*)^{1/2} = 0.$$

Hence $\tau(xa) = 0$ for all $x \in \mathbf{A}$. In particular, $\tau(x^*a) = 0$. By Proposition 9.1.13, $\tau(a^*x) = \overline{\tau(x^*a)} = 0$ as well.

(ii) If $a, b \in N_\tau$, then

$$\tau((a+b)^*(a+b)) = \tau(a^*a + b^*a + a^*b + b^*b) = 0$$

by part (i). Hence N_τ is closed under addition. That it is closed under scalar multiplication is trivial to verify. Thus N_τ is a subspace of $\mathbf{A}$. Now suppose $a \in N_\tau$ and $b \in \mathbf{A}$, then taking $x = a^*b^*b$ in part (i), we see that

$$\tau((ba)^*(ba)) = \tau(a^*b^*ba) = 0.$$

Thus $ba \in N_\tau$ and N_τ is a left ideal in $\mathbf{A}$.

(iii) Suppose $b \in N_\tau$. Then $ab \in N_\tau$ by part (ii) and the inequality holds trivially. Therefore we fix $b \in \mathbf{A}$ such that $\tau(b^*b) > 0$ and look to prove the inequality. To do this, we define $\rho : \mathbf{A} \to \mathbb{C}$ by

$$\rho(c) := \frac{\tau(b^*cb)}{\tau(b^*b)}.$$

Then ρ is clearly a linear functional on $\mathbf{A}$. Also, if $c \in \mathbf{A}_+$ is positive, then there exists $d \in \mathbf{A}$ such that $c = d^*d$. In that case, $b^*cb = (db)^*db \in \mathbf{A}_+$. Hence $\rho(c) \geq 0$. Thus ρ is a positive linear functional. By Theorem 9.1.14,

$$\|\rho\| = \rho(1_\mathbf{A}) = 1.$$

Hence for any $a \in \mathbf{A}$, $\rho(a^*a) \leq \|a^*a\|$. This is precisely what we needed to prove. $\qquad\square$

10.8.4 LEMMA *Let $\mathbf{A}$ be a unital C*-algebra and $\tau : \mathbf{A} \to \mathbb{C}$ be a positive linear functional. Define $N_\tau := \{a \in \mathbf{A} : \tau(a^*a) = 0\}$ and let $\mathbf{K}_\tau := \mathbf{A}/N_\tau$. Then $\mathbf{K}_\tau$ is a vector space and the map $\langle \cdot, \cdot \rangle : \mathbf{K}_\tau \times \mathbf{K}_\tau \to \mathbb{C}$ given by*

$$\langle a + N_\tau, b + N_\tau \rangle := \tau(b^*a)$$

is a well-defined inner product on $\mathbf{K}_\tau$.

Proof. That $\mathbf{K}_\tau$ is a vector space follows from the fact that N_τ is a subspace of $\mathbf{A}$. We now verify the properties of $\langle \cdot, \cdot \rangle$.

(i) We first verify that it is well-defined. If $a, c \in \mathbf{A}$ are such that $a + N_\tau = c + N_\tau$, then $(a - c) \in N_\tau$. Thus for any $b \in \mathbf{A}$,

$$\langle a + N_\tau, b + N_\tau \rangle - \langle c + N_\tau, b + N_\tau \rangle = \tau(b^*(a - c)) = 0,$$

by Lemma 10.8.3. One may prove well-definedness in the second variable similarly.

(ii) It is entirely obvious that $\langle \cdot, \cdot \rangle$ is linear in the first variable.

(iii) If $a, b \in \mathbf{A}$, then by part (i) of Proposition 9.1.13, we have

$$\langle a + N_\tau, b + N_\tau \rangle = \tau(b^*a) = \overline{\tau(a^*b)} = \overline{\langle b + N_\tau, a + N_\tau \rangle}.$$

(iv) Finally, if $a \in \mathbf{A}$ is such that $\langle a + N_\tau, a + N_\tau \rangle = 0$, then it follows that $a \in N_\tau$. Thus $\langle \cdot, \cdot \rangle$ is positive definite and an inner product on $\mathbf{K}_\tau$. $\qquad\square$

10.8.5 THEOREM (GELFAND AND NAIMARK, 1943, AND SEGAL, 1947) *Let* **A** *be a unital C*-algebra and* $\tau : \mathbf{A} \to \mathbb{C}$ *be a positive linear functional. Then* τ *has a GNS representation and any two GNS representations for* τ *are equivalent to each other.*

Proof. We prove both existence and uniqueness below.

(i) Existence: Let $N_\tau := \{a \in \mathbf{A} : \tau(a^*a) = 0\}$ and let $\mathbf{K}_\tau := \mathbf{A}/N_\tau$ equipped with the inner product from Lemma 10.8.4.

(a) For each $a \in \mathbf{A}$, define $M_a : \mathbf{K}_\tau \to \mathbf{K}_\tau$ by

$$M_a(b + N_\tau) := ab + N_\tau.$$

Note that M_a is well-defined because N_τ is a left ideal of **A**. Furthermore, for any $b \in \mathbf{A}$, we have

$$\begin{aligned}
\|M_a(b + N_\tau)\|^2 &= \langle ab + N_\tau, ab + N_\tau \rangle \\
&= \tau(b^*a^*ab) \\
&\leq \|a^*a\|\tau(b^*b) \\
&= \|a\|^2\|b + N_\tau\|^2.
\end{aligned}$$

(Note that the penultimate inequality holds by part (iii) of Lemma 10.8.3). Hence M_a defines a bounded linear operator on $\mathbf{K}_\tau$ such that $\|M_a\| \leq \|a\|$.

(b) Let $\mathbf{H}_\tau$ denote the Hilbert space completion of $\mathbf{K}_\tau$ (see Exercise 3.13). Then M_a extends uniquely to a bounded linear operator on $\mathbf{H}_\tau$ (by Proposition 3.2.4). We denote this extension by M_a once again and define $\pi_\tau : \mathbf{A} \to \mathcal{B}(\mathbf{H}_\tau)$ by

$$\pi_\tau(a) := M_a.$$

If $a, b, c \in \mathbf{A}$, then

$$M_aM_b(c + N_\tau) = a(bc + N_\tau) = abc + N_\tau = M_{ab}(c + N_\tau).$$

In other words, M_aM_b and M_{ab} agree on a dense subset of $\mathbf{H}_\tau$ and are thus equal. Hence $\pi_\tau(ab) = \pi_\tau(a)\pi_\tau(b)$. The proof that π_τ is a linear map is also similar. Finally, for any $a, b, c \in \mathbf{A}$, we have

$$
\begin{aligned}
\langle M_a(b + N_\tau), c + N_\tau \rangle &= \langle ab + N_\tau, c + N_\tau \rangle \\
&= \tau(c^* ab) \\
&= \tau((a^* c)^* b) \\
&= \langle b + N_\tau, a^* c + N_\tau \rangle \\
&= \langle b + N_\tau, M_{a^*}(c + N_\tau) \rangle.
\end{aligned}
$$

As before, this proves that $(M_a)^* = M_{a^*}$. Thus π_τ is a $*$-homomorphism.

(c) If $e := 1_\mathbf{A} + N_\tau \in \mathbf{H}_\tau$, then observe that

$$
\pi_\tau(\mathbf{A})(e) = \{a + N_\tau : a \in \mathbf{A}\} = \mathbf{K}_\tau,
$$

which is dense in $\mathbf{H}_\tau$. Hence $(\pi_\tau, \mathbf{H}_\tau, e)$ is a cyclic representation.

(d) Finally, for any $a \in \mathbf{A}$,

$$
\langle \pi_\tau(a)e, e \rangle = \langle a + N_\tau, 1_\mathbf{A} + N_\tau \rangle = \tau(a).
$$

Thus $(\pi_\tau, \mathbf{H}_\tau, e)$ is a GNS representation for τ.

(ii) Uniqueness: Suppose $(\pi, \mathbf{H}, f)$ is another GNS representation for τ. Define $U : \mathbf{K}_\tau \to \mathbf{H}$ by

$$
U(a + N_\tau) := \pi(a)f.
$$

Then

(a) U is well-defined: If $a, b \in \mathbf{A}$ are such that $a + N_\tau = b + N_\tau$, then $c := (a - b) \in N_\tau$. Thus

$$
\|\pi(c)f\|^2 = \langle \pi(c)f, \pi(c)f \rangle = \langle \pi(c^* c)f, f \rangle = \tau(c^* c) = 0.
$$

(Note that the penultimate equality holds because $(\pi, \mathbf{H}, f)$ is a GNS representation for τ.) We conclude that $\pi(a)f = \pi(b)f$ and thus U is well-defined.

(b) That U is linear is trivial to check. Also, note that

$$\|U(a + N_\tau)\|^2 = \langle \pi(a)f, \pi(a)f \rangle$$
$$= \langle \pi(a^*a)f, f \rangle$$
$$= \tau(a^*a)$$
$$= \langle a + N_\tau, a + N_\tau \rangle.$$

Hence U is bounded and must extend to an isometric linear map $U :$ $\mathbf{H}_\tau \to \mathbf{H}$. Observe that U is also surjective since f is a cyclic vector for the representation $(\pi, \mathbf{H})$. Hence U is a unitary.

(c) Finally, if $a, b \in \mathbf{A}$,

$$U^{-1}\pi(a)U(b + N_\tau) = U^{-1}\pi(a)\pi(b)f$$
$$= U^{-1}\pi(ab)f$$
$$= ab + N_\tau = \pi_\tau(a)(b + N_\tau).$$

Thus $U^{-1}\pi(a)U = \pi_\tau(a)$ for all $a \in \mathbf{A}$. $\qquad\square$

For convenience, the GNS representation constructed in the first part of the proof will often be referred to as *the* GNS representation associated to the positive linear functional.

10.8.6 EXAMPLE Let μ be a positive Borel measure on a compact Hausdorff space X and let $\tau : C(X) \to \mathbb{C}$ be the positive linear functional

$$f \mapsto \int_X f \, d\mu.$$

Then one may verify that

$$N_\tau = \{f \in C(X) : f \equiv 0 \text{ a.e.} [\mu]\}.$$

Now $\mathbf{H}_\tau$ is the completion of $\mathbf{K}_\tau = C(X)/N_\tau$. A moment's thought shows that $\mathbf{H}_\tau \cong L^2(X, \mu)$. Furthermore, the GNS representation associated to τ is precisely the map $\pi : C(X) \to \mathcal{B}(L^2(X, \mu))$ given by

$$f \mapsto M_f,$$

where M_f is the multiplication operator associated to f as in Remark 10.8.1.

Before we go any further, we need the notion of a direct sum of representations.

10.8.7 REMARK Let $\{(\mathbf{H}_\lambda, \langle \cdot, \cdot \rangle_\lambda) : \lambda \in J\}$ be a (possibly uncountable) family of Hilbert spaces.

(i) We wish to define the Hilbert space direct sum exactly as in Exercise 3.14. Set

$$\mathbf{E} := \{(x_\lambda) : x_\lambda \in \mathbf{H}_\lambda \text{ and } x_\lambda \neq 0 \text{ for only finitely many } \lambda \in J\}.$$

Define addition and scalar multiplication pointwise and an inner product by

$$\langle (x_\lambda), (y_\lambda) \rangle := \sum_{\lambda \in J} \langle x_\lambda, y_\lambda \rangle_\lambda.$$

Note that this is well-defined because it is a finite sum. Then, the completion of $\mathbf{E}$ with respect to this inner product is called the Hilbert space direct sum of the $\{\mathbf{H}_\lambda : \lambda \in J\}$ and is denoted by

$$\mathbf{H} = \bigoplus_{\lambda \in J} \mathbf{H}_\lambda.$$

(ii) We now wish to define the direct sum of operators as in Remark 10.3.8. For each $\lambda \in J$, let $T_\lambda \in \mathcal{B}(\mathbf{H}_\lambda)$ be a bounded linear operator. If $\sup\{\|T_\lambda\| : \lambda \in J\} < \infty$, then we may define $T : \mathbf{E} \to \mathbf{E}$ by

$$T((x_\lambda)) := (T_\lambda(x_\lambda)).$$

Then T is a well-defined, bounded linear operator. By Proposition 3.2.4, T extends to a bounded linear operator $T \in \mathcal{B}(\mathbf{H})$, which is denoted by

$$T = \bigoplus_{\lambda \in J} T_\lambda.$$

(iii) Now suppose $\mathbf{A}$ is a C*-algebra. For each $\lambda \in J$, suppose $\pi_\lambda : \mathbf{A} \to \mathcal{B}(\mathbf{H}_\lambda)$ is a representation of $\mathbf{A}$. Then for each $a \in \mathbf{A}$, $\|\pi_\lambda(a)\| \leq \|a\|$. Hence we may define $\pi : \mathbf{A} \to \mathcal{B}(\mathbf{H})$ by

$$\pi(a) := \bigoplus_{\lambda \in J} \pi_\lambda(a).$$

It is easy to see that $(\pi, \mathbf{H})$ is a representation of $\mathbf{A}$. We denote this representation by

$$\pi = \bigoplus_{\lambda \in J} \pi_\lambda.$$

In order to prove the Gelfand–Naimark Theorem, we need an adequate supply of representations over which to take a direct sum. These are the GNS representations of a class of positive linear functionals which we now identify.

10.8.8 DEFINITION Let **A** be a C*-algebra. A *state* on **A** is a positive linear functional of norm 1. We write $S(\mathbf{A})$ for the set of all states on **A**.

Note that if **A** is unital, then

$$S(\mathbf{A}) = \{\tau \in \mathbf{A}^* : \|\tau\| = \tau(1_\mathbf{A}) = 1\}$$

by Theorem 9.1.14. The next proposition tells us that not only do states exist, they are quite plentiful.

10.8.9 PROPOSITION *Let **A** be a unital C*-algebra and $a \in \mathbf{A}$ be a self-adjoint element. Then there exists $\tau \in S(\mathbf{A})$ such that $|\tau(a)| = \|a\|$.*

Proof.   Let $\mathbf{B} := \mathcal{A}(a)$ denote the (commutative) C*-subalgebra of **A** generated by $\{a, 1_\mathbf{A}\}$. By the Gelfand–Naimark Theorem (Theorem 8.3.12), there is an isometric ∗-isomorphism

$$\Gamma : \mathbf{B} \to C(\Omega(\mathbf{B})),$$

given by $b \mapsto \widehat{b}$, where $\widehat{b} : \Omega(\mathbf{B}) \to \mathbb{C}$ is the function $\rho \mapsto \rho(b)$. In particular, we have

$$\|a\| = \sup\{|\rho(a)| : \rho \in \Omega(\mathbf{B})\}.$$

Since $\Omega(\mathbf{B})$ is compact, there exists $\rho \in \Omega(\mathbf{B})$ such that $|\rho(a)| = \|a\|$. Now if $b \in \mathbf{B}_+$, then there exists $c \in \mathbf{B}$ such that $b = c^*c$. Then $\rho(b) = |\rho(c)|^2 \geq 0$. Thus ρ is a positive linear functional on **B**.

By Corollary 9.1.15, there is a positive linear functional $\tau : \mathbf{A} \to \mathbb{C}$ such that $\tau|_\mathbf{B} = \rho$. In particular,

$$\|\tau\| = \tau(1_\mathbf{A}) = \rho(1_\mathbf{A}) = 1.$$

Thus $\tau \in S(\mathbf{A})$. Finally, $|\tau(a)| = |\rho(a)| = \|a\|$.   □

We are now in a position to prove the main result of this section. It is a representation theorem of fundamental importance that informs virtually all aspects of the subject. Recall that if **H** is a Hilbert space, then any norm–closed, self-adjoint subalgebra of $\mathcal{B}(\mathbf{H})$ is a C*-algebra. Such a C*-algebra is often referred to as a *concrete* C*-algebra. The theorem simply says that any *abstract* C*-algebra (as defined by Definition 8.2.1) may be realized as a concrete C*-algebra.

10.8.10 THEOREM (GELFAND AND NAIMARK, 1943) *For every C*-algebra* **A**, *there is a Hilbert space* **H** *such that* **A** *is isometrically isomorphic to a C*-subalgebra of* $\mathcal{B}(\mathbf{H})$.

Proof.   By embedding **A** in its unitalization if need be, we may assume that **A** is unital. For each $\tau \in S(\mathbf{A})$, let $(\pi_\tau, \mathbf{H}_\tau, e_\tau)$ be the GNS representation associated to τ (as constructed in Theorem 10.8.5). Let

$$\mathbf{H}_u := \bigoplus_{\tau \in S(\mathbf{A})} \mathbf{H}_\tau \text{ and } \pi_u := \bigoplus_{\tau \in S(\mathbf{A})} \pi_\tau$$

as defined above. We know that $(\pi_u, \mathbf{H}_u)$ is a representation of **A**. We claim that π_u is injective. To see this, suppose $a \in \mathbf{A}$ is such that $\pi_u(a) = 0$. Then $\pi_\tau(a) = 0$ for each $\tau \in S(\mathbf{A})$. We may choose $\tau \in S(\mathbf{A})$ such that $|\tau(a^*a)| = \|a^*a\|$ by Proposition 10.8.9. Then

$$\tau(a^*a) = \langle \pi_\tau(a^*a)e_\tau, e_\tau \rangle = \|\pi_\tau(a)e_\tau\|^2 = 0.$$

This implies that $\|a\|^2 = \|a^*a\| = |\tau(a^*a)| = 0$. Hence $a = 0$, proving that π_u is injective. By Proposition 8.5.12, π_u is isometric and thus sets up an isomorphism between **A** and $\pi_u(\mathbf{A})$. $\square$

Before we give any applications of this theorem, some comments are in order.

10.8.11 REMARK

  (i) The representation π_u constructed in the proof of the theorem is called the **universal** representation of the C*-algebra.

 (ii) Any representation $(\pi, \mathbf{H})$ of **A** that is injective is called a *faithful* representation. Therefore the content is the previous theorem is simply that **A** has at least one faithful representation.

(iii) In the proof of the theorem, we took a direct sum over *all* states of the C*-algebra. This is not, strictly speaking, necessary. By Exercise 10.43, one may choose a certain subcollection of states (known as *pure* states) instead.

 (iv) Finally, if **A** is a *separable* C*-algebra, then the Hilbert space chosen in Theorem 10.8.10 may be taken to be separable (see Exercise 10.47).

We end with a useful application. Let **A** be a C*-algebra and $n \in \mathbb{N}$. Let $M_n(\mathbf{A})$ denote the set of all $n \times n$ matrices with entries in **A**. Since **A** is an algebra, we may equip $M_n(\mathbf{A})$ with the usual algebraic operations (including matrix multiplication) to make it an algebra. For a matrix $(a_{i,j}) \in M_n(\mathbf{A})$, we define an adjoint by

$$(a_{i,j})^* = (a_{j,i}^*).$$

It is clear that this defines an involution on $M_n(\mathbf{A})$ and we would like to determine if there is a norm on $M_n(\mathbf{A})$ that satisfies the C*-identity. As it turns out, the easiest way to prove this uses Theorem 10.8.10.

Let $(\pi, \mathbf{H})$ be a faithful representation of $\mathbf{A}$. Define $\mathbf{H}^n$ to be the direct sum of n copies $\mathbf{H}$ and define $\Phi : M_n(\mathbf{A}) \to \mathcal{B}(\mathbf{H}^n)$ by the formula

$$\Phi \begin{pmatrix} a_{1,1} & a_{1,2} & \dots & a_{1,n} \\ a_{2,1} & a_{2,2} & \dots & a_{2,n} \\ \vdots & \vdots & \vdots & \vdots \\ a_{n,1} & a_{n,2} & \dots & a_{n,n} \end{pmatrix} \begin{pmatrix} x_1 \\ x_2 \\ \vdots \\ x_n \end{pmatrix} := \begin{pmatrix} \pi(a_{1,1})(x_1) + \pi(a_{1,2})(x_2) + \ldots + \pi(a_{1,n})(x_n) \\ \pi(a_{2,1})(x_1) + \pi(a_{2,2})(x_2) + \ldots + \pi(a_{2,n})(x_n) \\ \vdots \\ \pi(a_{n,1})(x_1) + \pi(a_{n,2})(x_2) + \ldots + \pi(a_{n,n})(x_n) \end{pmatrix}.$$

It is easy to show that Φ is an injective homomorphism of algebras that preserves the involution. Therefore for $a = (a_{i,j}) \in M_n(\mathbf{A})$, we may define

$$\|a\| := \|\Phi(a)\|_{\mathcal{B}(\mathbf{H}^n)}.$$

Then this defines a norm on $M_n(\mathbf{A})$ that satisfies the C*-identity. By Corollary 8.2.17, this norm does not depend on the choice of faithful representation π. Thus we obtain

10.8.12 COROLLARY *If* $\mathbf{A}$ *is a C*-algebra, then there is a unique norm on* $M_n(\mathbf{A})$ *that makes it a C*-algebra.*

This brings us to the end of the chapter and, with it, the book. I hope that you have had an enjoyable journey exploring functional analysis with me. I bid you farewell in customary fashion, with some problems that I hope you will enjoy.

10.9 Exercises

Compact Normal Operators

EXERCISE 10.1 (⚒) Prove Lemma 10.1.5.

EXERCISE 10.2 (SINGULAR VALUE DECOMPOSITION) *Let* $\mathbf{H}$ *be a separable Hilbert space and* $T \in \mathcal{K}(\mathbf{H})$ *be any compact operator. Prove that there exist orthonormal sequences* (e_n) *and* (f_k) *in* $\mathbf{H}$ *and a sequence* (μ_j) *of positive real numbers such that*

$$T = \sum_{k=1}^{\infty} \mu_k \langle \cdot, e_k \rangle f_k,$$

where this sequence converges in the operator norm on $\mathcal{B}(\mathbf{H})$. *Moreover, if* $\{\mu_k\}$ *is infinite, then* $\lim_{k \to \infty} \mu_k = 0$.

EXERCISE 10.3 Let $\mathbf{H}$ be a separable Hilbert space. If $T \in \mathcal{K}(\mathbf{H})$, then use Exercise 10.2 to prove that T is a norm limit of finite rank operators (This gives an alternate proof of Corollary 5.6.11).

Multiplication Operators

EXERCISE 10.4 Let X be a compact Hausdorff space and $\phi \in C(X)$. If μ is a positive Borel measure on X, then $\phi \in L^\infty(X, \mu)$ as well. If $\mathrm{supp}(\mu) = X$, then prove that ess-range$(\phi) = \phi(X)$.

EXERCISE 10.5 Let $(X, \mathfrak{M}, \mu)$ be a σ-finite measure space, $\phi \in L^\infty(X, \mu)$ and let $M_\phi \in \mathcal{B}(L^2(X, \mu))$ be the associated multiplication operator.

   (i) Prove that M_ϕ is injective if and only if $\{x \in X : \phi(x) = 0\}$ has measure zero.

   (ii) Prove that M_ϕ is surjective if and only if there exists $c > 0$ such that $\{x \in X : |\phi(x)| \le c\}$ has measure zero.

Hint: For part (ii), if no such c exists, then consider $E_k := \{x \in X : \frac{1}{2^{k+1}} \le |\phi(x)| \le \frac{1}{2^k}\}$. Prove that there is a subsequence (E_{n_k}) each of which has positive measure.

EXERCISE 10.6 Let $(X, \mathfrak{M}, \mu)$ be a σ-finite measure space, $\phi \in L^\infty(X, \mu)$ and let $M_\phi \in \mathcal{B}(L^2(X, \mu))$ be the associated multiplication operator. For any $\lambda \in \mathbb{C}$, prove that $\lambda \in \sigma_p(M_\phi)$ if and only if there is a set $E \in \mathfrak{M}$ such that $\mu(E) > 0$ and $E \subset \phi^{-1}(\{\lambda\})$.

10.9.1 DEFINITION Let $(X, \mathfrak{M}, \mu)$ be a measure space. A set $E \in \mathfrak{M}$ is said to be an *atom* if $\mu(E) > 0$ and E does not contain any measurable subsets of positive measure. A measure space $(X, \mathfrak{M}, \mu)$ is said to be *atomless* if it does not contain any atoms.

EXERCISE 10.7 Let $(X, \mathfrak{M}, \mu)$ be an atomless, σ-finite measure space and $\phi \in L^\infty(X, \mu)$. If the corresponding multiplication operator M_ϕ is compact, then prove that $\phi = 0$ a.e. $[\mu]$.

Hint: If $\epsilon > 0$ is such that $E := \{x \in X : |\phi(x)| \geq \epsilon\}$ has positive measure, then construct a norm bounded, orthonormal sequence $(f_n) \subset L^2(X, \mu)$ such that $\|M_\phi f_n\| \geq \epsilon$ for all $n \in \mathbb{N}$.

Spectral Theorem

EXERCISE 10.8 Let $\mathbf{H}$ be a separable Hilbert space and $S, T \in \mathcal{K}(\mathbf{H})$ be compact normal operator. Prove that S is unitarity equivalent to T if and only if $\sigma_p(T) = \sigma_p(S)$ and $\dim\ker(S - \lambda I) = \dim\ker(T - \lambda I)$ for all $\lambda \in \mathbb{C}$.

EXERCISE 10.9 (⚒) Let $\mathbf{H}$ be a separable Hilbert space and $T \in \mathcal{B}(\mathbf{H})$ be a normal operator with a cyclic vector $e \in \mathbf{H}$. Set $X := \sigma(T)$ and let μ be the positive Borel measure obtained in the Spectral Theorem (Theorem 10.3.7).

(i) If $f \in C(X)$ is such that $f(T)e = 0$, then prove that $f = 0$ in $C(X)$.

(ii) Let $\text{supp}(\mu)$ be the support of μ as defined in Definition 9.5.2. Prove that $\text{supp}(\mu) = X$.

EXERCISE 10.10 (⚒) Let (H, T, e, X, μ) be exactly as in Exercise 10.9.

(i) Prove that $\lambda \in \mathbb{C}$ is an eigenvalue of T if and only if $\mu(\{\lambda\}) > 0$.

(ii) Use part (i) to give an alternate proof of Exercise 8.33 (in this special case).

EXERCISE 10.11 (⚒) Let $\{\mathbf{H}_n : n \in \mathbb{N}\}$ be a countable family of separable Hilbert spaces. Prove that $\oplus_{n=1}^\infty \mathbf{H}_n$ is separable.

EXERCISE 10.12 (⚒) In the proof of Lemma 10.3.9, prove that $(X, \mathfrak{M}, \mu)$ is a σ-finite measure space.

EXERCISE 10.13 Let $\mathbf{H}$ be a separable Hilbert space and $T \in \mathcal{B}(\mathbf{H})$ be a normal operator. For any $\epsilon > 0$, prove that there is a normal operator $S \in \mathcal{B}(\mathbf{H})$ with finite spectrum such that $\|S - T\| < \epsilon$.

Hint: If $(X, \mathfrak{M}, \mu)$ is a σ-finite measure space and $\phi \in L^\infty(X, \mu)$ is a simple function, then prove that $\sigma(M_\phi)$ is finite.

Borel Functional Calculus

EXERCISE 10.14 Let X be a topological space and $\mathcal{A}$ be an algebra of complex-valued functions on X containing $\mathbf{1}$. Suppose that $\mathcal{A}$ is closed under bounded pointwise convergence (in the sense described before Proposition 10.4.7). Prove that the set $\mathfrak{N} := \{E \subset X : \chi_E \in \mathcal{A}\}$ is a σ-algebra.

EXERCISE 10.15 Let X be a compact metric space.

(i) Prove that $B_\infty(X)$ is the smallest algebra (of complex-valued functions on X) that contains $C(X)$ and is closed under bounded pointwise convergence.

Hint: For each closed set $F \subset X$, prove that χ_F is a pointwise limit of a sequence of uniformly bounded continuous functions. Then use Exercise 10.14.

(ii) Give an example to show that a bounded Borel function need not be a pointwise limit of a sequence of continuous functions.

EXERCISE 10.16 (⚒) Prove Lemma 10.4.15.

Hint: Apply Lusin's Theorem with $\nu := |\mu_1| + |\mu_2| + \ldots + |\mu_n|$.

Von Neumann Algebras

EXERCISE 10.17 (⚒) Prove Proposition 10.5.3 for the weak operator topology.

EXERCISE 10.18 Let $\mathbf{H}$ be a finite dimensional Hilbert space and $\{e_1, e_2, \ldots, e_n\}$ be a fixed orthonormal basis of $\mathbf{H}$.

(i) For each $T \in \mathcal{B}(\mathbf{H})$, define $\beta(T) := \max\{|\langle T(e_j), e_i\rangle| : 1 \leq i, j \leq n\}$. Prove that there is a constant $c > 0$ such that $\|T\| \leq c\beta(T)$ for each $T \in \mathcal{B}(\mathbf{H})$.

(ii) Prove that the weak operator topology, the strong operator topology and the norm topology coincide on $\mathcal{B}(\mathbf{H})$.

EXERCISE 10.19 (⚒) Let $\mathbf{H}$ be a Hilbert space and $(T_n) \subset \mathcal{B}(\mathbf{H})$.

(i) Prove that $T_n \xrightarrow{s} T$ (in the sense of Definition 10.4.5) if and only if $T_n \to T$ in the strong operator topology.

(ii) Prove that $T_n \xrightarrow{w} T$ if and only if $T_n \to T$ in the weak operator topology.

EXERCISE 10.20 Let $\mathbf{H}$ be a Hilbert space.

(i) Prove that the adjoint operation $* : \mathcal{B}(\mathbf{H}) \to \mathcal{B}(\mathbf{H})$ is continuous with respect to the weak operator topology.

(ii) If $\mathbf{H}$ is infinite dimensional, then prove that the adjoint operation is not continuous on $\mathcal{B}(\mathbf{H})$ with respect to the strong operator topology.

EXERCISE 10.21 Let $\mathbf{H}$ be a separable Hilbert space. Write $\mathcal{F}(\mathbf{H})$ for the set of all finite rank, bounded operators in $\mathcal{B}(\mathbf{H})$.

(i) Prove that $\mathcal{F}(\mathbf{H})$ is dense in $\mathcal{B}(\mathbf{H})$ with respect to the strong operator topology.

(ii) Prove that $(\mathcal{B}(\mathbf{H}), \sigma_S)$ is separable.

(iii) Prove parts (i) and (ii) for the weak operator topology.

EXERCISE 10.22 Let $\mathbf{H}$ be a separable Hilbert space and $\{x_n : n \in \mathbb{N}\}$ be a countable dense subset of the unit ball of $\mathbf{H}$. For $S, T \in \mathcal{B}(\mathbf{H})$, define

$$d_w(S, T) := \sum_{n,m=1}^{\infty} \frac{|\langle (S - T)x_n, x_m \rangle|}{2^{n+m}}.$$

(i) Prove that d_w is a metric on $\mathcal{B}(\mathbf{H})$.

(ii) If $D \subset \mathcal{B}(\mathbf{H})$ is a norm bounded set, then prove that the metric topology on D induced by d_w coincides with the weak operator topology.

EXERCISE 10.23 Let $\mathbf{H}$ be a separable Hilbert space and $\{x_n : n \in \mathbb{N}\}$ be a countable dense subset of the unit ball of $\mathbf{H}$. For $S, T \in \mathcal{B}(\mathbf{H})$, define

$$d_s(S, T) := \sum_{n=1}^{\infty} \frac{\|(S - T)(x_n)\|}{2^n}.$$

(i) Prove that d_s is a metric on $\mathcal{B}(\mathbf{H})$.

(ii) If $D \subset \mathcal{B}(\mathbf{H})$ is a norm bounded set, then prove that the metric topology on D induced by d_s coincides with the strong operator topology.

EXERCISE 10.24 Let $\mathbf{H}$ be a Hilbert space and let D denote the (norm) closed unit ball in $\mathcal{B}(\mathbf{H})$. For each $x, y \in \mathbf{H}$, let $D_{x,y} := \{z \in \mathbb{C} : |z| \leq \|x\|\|y\|\}$ and set

$$B := \prod_{x,y \in \mathbf{H}} D_{x,y},$$

equipped with the product topology. Define $\Theta : D \to B$ by

$$\Theta(T) := (\langle T(x), y \rangle)_{x,y \in \mathbf{H}}.$$

(i) If D is equipped with the weak operator topology, then prove that Θ induces a homeomorphism between D and $\Theta(D)$.

(ii) Prove D is compact in the weak operator topology.

Hint: Use ideas from the proof of the Banach–Alaoglu Theorem (Theorem 6.6.3), along with 8.1.

EXERCISE 10.25 Let $\mathbf{H}$ be an infinite dimensional Hilbert space and let D denote the (norm) closed unit ball in $\mathcal{B}(\mathbf{H})$.

(i) Prove that the strong and weak operator topologies on D do not coincide.

(ii) Prove that D is not compact in the strong operator topology.

Hint: Use Exercise 10.20 and Exercise 10.24.

EXERCISE 10.26 Let $(\pi, \mathbf{H})$ be a representation of a C*-algebra $\mathbf{A}$. Let $\mathbf{M} < \mathbf{H}$ be a closed subspace of $\mathbf{H}$ and let $P \in \mathcal{B}(\mathbf{H})$ be the unique projection operator such that $\text{Range}(P) = \mathbf{M}$. Prove that $\mathbf{M}$ is invariant under $\pi(\mathbf{A})$ if and only if $P \in \pi(\mathbf{A})'$ (the commutant of $\pi(\mathbf{A})$).

10.9.2 DEFINITION A representation $(\pi, \mathbf{H})$ of a C*-algebra $\mathbf{A}$ is said to be *irreducible* if the there is no non-trivial, closed subspaces of $\mathbf{H}$ that is invariant under $\pi(\mathbf{A})$.

EXERCISE 10.27 Let $(\pi, \mathbf{H})$ be a representation of a C*-algebra $\mathbf{A}$. Prove that $(\pi, \mathbf{H})$ is irreducible if and only if $\pi(\mathbf{A})' = \{\lambda I : \lambda \in \mathbf{C}\}$.

Hint: Use the fact that $\pi(\mathbf{A})'$ is a von Neumann algebra.

Spectral Measures

EXERCISE 10.28 (✗) Prove Lemma 10.6.8.

EXERCISE 10.29 Let X be a compact metric space and let $\pi : C(X) \to \mathcal{B}(\mathbf{H})$ be a non-degenerate representation. Let $P : \mathfrak{B}_X \to \mathcal{B}(\mathbf{H})$ be the associated spectral measure from Corollary 10.6.11. Prove that π is injective if and only if $P(E) \neq 0$ for all non-empty open sets $E \in \mathfrak{B}_X$.

EXERCISE 10.30 (✗) Let $\mathbf{H}$ be a separable Hilbert space and $T \in \mathcal{B}(\mathbf{H})$ be a normal operator. Let $P : \mathfrak{B}_{\sigma(T)} \to \mathcal{B}(\mathbf{H})$ be the associated spectral measure from Corollary 10.6.12. For each $E \in \mathfrak{B}_{\sigma(T)}$, prove that $P(E)T = TP(E)$. Conclude that if $\dim(H) > 1$, then T has a non-trivial invariant subspace (notice that this is a strengthening of Exercise 8.35).

EXERCISE 10.31 Suppose $\{T_1, T_2, \ldots, T_n\}$ is a finite set of normal operators on a Hilbert space $\mathbf{H}$ such that $T_iT_j = T_jT_i$ for all $1 \leq i, j \leq n$. Prove that there is a compact set $X \subset \mathbb{C}^n$ and a spectral measure $P : \mathfrak{B}_X \to \mathcal{B}(\mathbf{H})$ such that

$$T_k = \int_X \zeta_k dP,$$

where $\zeta_k \in C(X)$ is the function $\zeta_k(z_1, z_2, \ldots, z_n) := z_k$.

Applications of the Spectral Theorem

EXERCISE 10.32 (⚒) Complete the proof of Lemma 10.7.2.

EXERCISE 10.33 Let $T \in \mathcal{B}(\mathbf{H})$ be a self-adjoint operator and $f \in C(\sigma(T))$ be such that $f(0) = 0$. Prove that $\text{Range}(f(T)) \subset \overline{\text{Range}(T)}$.

EXERCISE 10.34 Let $\mathbf{A}$ be a unital C*-algebra and let $a \in \mathbf{A}$.

(i) For each $n \in \mathbb{N}$, prove that $(|a| + n^{-1}1_{\mathbf{A}}) \in GL(\mathbf{A})$.

(ii) If $d_n := a(|a| + n^{-1}1_{\mathbf{A}})^{-1/2}$, then prove that

$$\|d_n - d_m\| = \|f_{m,n}(|a|)\|,$$

where $f_{m,n} : \mathbb{R}_+ \to \mathbb{R}$ denotes the function $t \mapsto t[(m^{-1} + t)^{-1/2} - (n^{-1} + t)^{-1/2}]$.

(iii) Prove that (d_n) is a Cauchy sequence in $B_\infty A$.

(iv) If $d := \lim_{n\to\infty d_n}$, then prove that $a = d|a|^{1/2}$.

EXERCISE 10.35 Let $\mathbf{A}$ be a unital C*-algebra and $\mathbf{I}$ be an ideal of $\mathbf{A}$. For any $a \in \mathbf{A}$, prove that $a \in \mathbf{I}$ if and only if $a^*a \in \mathbf{I}$.

Hint: Use Exercise 10.34 along the lines of Proposition 10.7.8.

EXERCISE 10.36 Prove that two similar normal operators are unitarily equivalent.

Hint: Use Exercise 9.5 along with the Fuglede–Putnam Theorem.

EXERCISE 10.37 Let $T \in \mathcal{B}(\mathbf{H})$ be a normal operator and let P denote the spectral measure associated to T. Suppose that T satisfies the following conditions:

(a) $\sigma(T)$ is a countable set with 0 as the unique limit point.

(b) For each $\lambda \in \sigma(T) \setminus \{0\}$, the projection $P(\{\lambda\})$ has finite rank.

Then prove that T is a compact operator.

EXERCISE 10.38 Let $U(\mathbf{A})$ be the group of unitaries in a unital C*-algebra $\mathbf{A}$.

(i) If $u \in U(\mathbf{A})$ is such that $\|u - 1_{\mathbf{A}}\| < 2$, then prove that there exists $h \in \mathbf{A}_{sa}$ such that $u = \exp(ih)$.

(ii) If $u, v \in U(\mathbf{A})$ are such that $\|u - v\| < 2$, then prove that there exists $h \in \mathbf{A}_{sa}$ such that $u = v \exp(ih)$.

EXERCISE 10.39 Let $U(\mathbf{A})$ be the group of unitaries in a unital C*-algebra $\mathbf{A}$ and let $U_0(\mathbf{A})$ denote the connected component of $1_{\mathbf{A}}$ in $U(\mathbf{A})$. Prove that $u \in U_0(\mathbf{A})$ if and only if there exist finitely many self-adjoint elements $h_1, h_2, \ldots, h_n$ in $\mathbf{A}$ such that

$$u = \exp(ih_1) \exp(ih_2) \ldots \exp(ih_n).$$

Hint: Let G be the set of all elements of this form. Use Exercise 10.38 to prove that G is both open and closed in $U(\mathbf{A})$.

EXERCISE 10.40 Let $\mathbf{A}$ be a unital, commutative C*-algebra and let $U_0(\mathbf{A})$ be the connected component of the identity in $U(\mathbf{A})$. Prove that

$$U_0(\mathbf{A}) = \{\exp(ih) : h \in \mathbf{A}_{sa}\}.$$

The Gelfand–Naimark Representation

EXERCISE 10.41 If $(\pi, \mathbf{H}, e)$ is a cyclic representation of a unital C*-algebra $\mathbf{A}$, prove that there is a positive linear functional $\tau : \mathbf{A} \to \mathbb{C}$ such that $(\pi, \mathbf{H}, e)$ is a GNS representation of τ.

EXERCISE 10.42 Let $\mathbf{A}$ be a unital C*-algebra and let $S(\mathbf{A})$ denote the set of all states on $\mathbf{A}$. Prove that $S(\mathbf{A})$ is a convex set in $\mathbf{A}^*$ that is compact in the weak-$*$ topology.

10.9.3 DEFINITION Let $\mathbf{A}$ be a unital C*-algebra and let $S(\mathbf{A})$ denote the set of all states on $\mathbf{A}$. An extreme point of $S(\mathbf{A})$ is called a *pure state* of $\mathbf{A}$ (see Definition 6.8.1). We denote the set of all such pure states by $PS(\mathbf{A})$.

EXERCISE 10.43 (✗) In the proof of Proposition 10.8.9, prove that the state τ may be chosen to be a pure state.

EXERCISE 10.44 Let $\mathbf{A}$ be a unital C*-algebra and $a \in \mathbf{A}$ be a normal element. For each $\lambda \in \sigma(a)$, prove that there is a state $\tau \in S(\mathbf{A})$ such that $\tau(a) = \lambda$.

Hint: Use Proposition 8.3.3.

EXERCISE 10.45 Let $\mathbf{A}$ be a unital C*-algebra and $\mathbf{B}$ be a C*-subalgebra of $\mathbf{A}$ that contains $\mathbf{1_A}$. Suppose that for each pair $\tau_1, \tau_2 \in S(\mathbf{A})$, there exists $b \in \mathbf{B}$ such that $\tau_1(b) \neq \tau_2(b)$ (In other words, $\mathbf{B}$ separates the states of $\mathbf{A}$). Then prove that $\mathbf{B} = \mathbf{A}$.

EXERCISE 10.46 Let $\mathbf{A}$ be a unital, separable C*-algebra and let $\tau \in S(\mathbf{A})$. If $\mathbf{H}_\tau$ denotes the Hilbert space occuring in the GNS triple associated to τ, then prove that $\mathbf{H}_\tau$ is separable.

EXERCISE 10.47 (✗) Let $\mathbf{A}$ be a unital, separable C*-algebra. Prove that there is a separable Hilbert space $\mathbf{H}$ such that $\mathbf{A}$ is isometrically isomorphic to a C*-subalgebra of $\mathcal{B}(\mathbf{H})$.

Hint: Use Exercise 10.46.

EXERCISE 10.48 Let $\mathbf{A}$ be a C*-algebra and $a \in \mathbf{A}$ be a self-adjoint element. Prove that $a \in \mathbf{A}_+$ if and only if $\tau(a) \geq 0$ for each $\tau \in S(\mathbf{A})$.

EXERCISE 10.49 Let $\mathbf{A}$ be a C*-algebra and let $(\pi_u, \mathbf{H}_u)$ be the universal representation of $\mathbf{A}$. For each $\tau \in S(\mathbf{A})$, prove that there is a vector $x_\tau \in \mathbf{H}_u$ such that

$$\tau(a) = \langle \pi_u(a)x_\tau, x_\tau \rangle$$

for each $a \in \mathbf{A}$.

10.9.4 DEFINITION Let $\mathbf{A}$ be a C*-algebra. A state $\tau \in S(\mathbf{A})$ is said to be *faithful* if whenever $a \in \mathbf{A}_+$ is such that $\tau(a) = 0$, then $a = 0$ (Equivalently, if $N_\tau = \{0\}$).

EXERCISE 10.50 Let $\mathbf{A}$ be a unital C*-algebra and $\tau \in S(\mathbf{A})$ be a faithful state. Prove that the corresponding GNS representation $(\pi_\tau, \mathbf{H}_\tau, e_\tau)$ is faithful.

EXERCISE 10.51 Fix $n \in \mathbb{N}$ and let $\mathbf{A} := \mathcal{B}(\mathbb{C}^n) \cong M_n(\mathbb{C})$ treated as a C*-algebra (see Example 8.2.4). Let $\mathrm{tr} : \mathbf{A} \to \mathbb{C}$ denote the normalized trace, given by

$$\mathrm{tr}((a_{i,j})) := \frac{1}{n} \sum_{i=1}^{n} a_{i,i}.$$

(i) Prove that tr is a faithful state on $\mathbf{A}$.

(ii) For a fixed matrix $a \in \mathbf{A}$, define $\rho_a : \mathbf{A} \to \mathbb{C}$ by

$$\rho_a(b) := \mathrm{tr}(ab).$$

Prove that $\rho_a \in S(\mathbf{A})$ if and only if $\mathrm{tr}(a) = 1$ and $a \in \mathbf{A}_+$.

(iii) Prove that ρ_a is a pure state if and only if a is a projection onto a one-dimensional subspace of $\mathbb{C}^n$ (in which case, $\rho_a(b) = \langle b(x), x \rangle$ for some $x \in \mathrm{Range}(a)$).

(iv) Prove that ρ_a is a faithful state if and only if $\mathrm{tr}(a) = 1$ and $a \geq t 1_{\mathbf{A}}$ for some $t > 0$.

Additional Reading

- The saga of normal operators on Hilbert spaces is not complete, despite our best efforts. Together with some von Neumann algebra theory, one may extend the Borel functional calculus to an L^∞-functional calculus. This book by Zhu has a nice treatment of this material.

  Zhu, Ke He (1993). *An Introduction to Operator Algebras*. Studies in Advanced Mathematics. x+157. Boca Raton, FL: CRC Press. ISBN: 0-8493-7875-3.

- In Theorem 10.3.11, we had proved that any normal operator on a separable Hilbert space is equivalent to a direct sum of multiplication operators, each of which has a cyclic vector. As we mentioned earlier, the measure spaces obtained in this theorem need not be unique. However, it is possible to choose measure spaces in such a way that the resulting sequence of measures is unique in a certain sense. This is the Hahn-Hellinger Theorem and a detailed proof of it is given in this book by Sunder.

  Sunder, Viakalathur Shankar (1997). *Functional Analysis*. Birkhäuser Advanced Texts: Basler Lehrbücher. [Birkhäuser Advanced Texts: Basel Textbooks]. Spectral theory. x+241. Basel: Birkhäuser Verlag. ISBN: 3-7643-5892-0.

- We have briefly touched upon the subject of von Neumann algebras in this book. Perhaps the next stop for any reader interested in learning more should be a book devoted to Operator algebras and my personal choice would be this one by Kadison and Ringrose.

Kadison, Richard V. and Ringrose, John R. (1997). *Fundamentals of the Theory of Operator Algebras. Vol. I.* vol. 15. Graduate Studies in Mathematics. Elementary theory, Reprint of the 1983 original. American Mathematical Society, Providence, RI. xvi+398. ISBN: 0-8218-0819-2. DOI: 10.1090/gsm/015.

Appendices

A.1 The Stone–Weierstrass Theorem

Here is the statement of the Weierstrass Approximation Theorem that would be familiar to most of us, albeit stated in the language of normed linear spaces.

A.1.1 THEOREM (WEIERSTRASS, 1885) *Consider* $C[0,1]$ *equipped with the supremum norm and let* $\mathcal{A}$ *denote the set of all polynomials in one variable. Then* $\mathcal{A}$ *is dense in* $C[0,1]$.

The theorem we now wish to prove is a far-reaching generalization of Weierstrass' Approximation Theorem due to M.H. Stone. It is a theorem that is of fundamental importance to us and is used a few times in the book.

In order to make sense of the statement, we need one fact about spaces of continuous functions. Let X be a compact Hausdorff space and let $C(X)$ denote the space of all complex-valued, continuous functions on X. Given $f, g \in C(X)$, we may define the *product* of f and g by

$$(f \cdot g)(x) := f(x)g(x).$$

$C(X)$ is closed under this operation, so it the structure of a ring. What is more, this multiplication operation behaves well with respect to the existing vector space operations, in the sense that

$$(f + g) \cdot h = (f \cdot h) + (g \cdot h) \text{ and } \alpha(f \cdot g) = (\alpha f) \cdot g$$

for all $f, g, h \in C(X)$ and $\alpha \in \mathbb{C}$. This gives $C(X)$ the structure of an **algebra** (see Definition 7.1.1). Therefore we may now speak of a **subalgebra** of $C(X)$ - namely, a vector subspace of $C(X)$ that is closed under this multiplication operation. With this notion in place, we may now state the theorem.

A.1.2 THEOREM (STONE–WEIERSTRASS THEOREM (STONE, 1937)) *Let X be a compact Hausdorff space and let $C(X)$ be the algebra of continuous, complex-valued functions on X, equipped with the supremum norm. Let $\mathcal{A}$ be a subalgebra of $C(X)$ satisfying the following properties:*

(P1) $\mathcal{A}$ contains the constant function **1**.

(P2) For any pair of distinct points $x, y \in X$, there is a function $f \in \mathcal{A}$ such that $f(x) \neq f(y)$. In other words, $\mathcal{A}$ separates points of X.

(P3) If $f \in \mathcal{A}$, then $\overline{f} \in \mathcal{A}$, where

$$\overline{f}(x) := \overline{f(x)}.$$

Then $\mathcal{A}$ is dense in $C(X)$.

Before we jump into the proof, let us fix some notation: Let X and $C(X)$ be as above. In what follows, we will also have reason to discuss $C(X, \mathbb{R})$, the space of real-valued, continuous functions on X. We will use the symbol $C(X, \mathbb{K})$ when we wish to discuss both spaces simultaneously.

Let $\mathcal{A} \subset C(X, \mathbb{K})$ denote a subalgebra. If $\mathbb{K} = \mathbb{R}$, we require that $\mathcal{A}$ satisfy conditions (P1) and (P2) given in A.1.2 Theorem. If $\mathbb{K} = \mathbb{C}$, we require that $\mathcal{A}$ satisfy conditions (P1), (P2) and (P3).

A.1.3 EXAMPLE In the following examples, the subalgebra $\mathcal{A}$ described satisfies the three conditions (P1)–(P3).

(i) $\mathcal{A} = C(X, \mathbb{K})$ itself satisfies all these properties. Note that it separates points because of Urysohn's lemma (which we discussed in Remark 2.3.10 and Theorem 7.2.3).

(ii) Let $X = [0, 1]$ and $\mathcal{A}$ denotes the set of all polynomials in $C([0, 1], \mathbb{K})$.

(iii) Let $X = S^1 := \{z \in \mathbb{C} : |z| = 1\}$ and $\mathcal{A}$ denotes all polynomials in z and $\overline{z}$.

(iv) Let $X = [0, 1] \times [0, 1]$ and

$$\mathcal{A} = \left\{ (x, y) \mapsto \sum_{i=1}^{n} f_i(x) g_i(y) : f_i, g_j \in C[0, 1] \right\}.$$

In our quest to prove Theorem A.1.2, we may make one simplification right up front. Note that if $\mathcal{A}$ satisfies these conditions, then so does its closure, so we may as well assume that $\mathcal{A}$ is a *closed* subalgebra and prove that $\mathcal{A} = C(X, \mathbb{C})$.

A.1.4 LEMMA *Let $\mathcal{A} \subset C(X, \mathbb{R})$ be a closed subalgebra satisfying properties (P1) and (P2). Then for any $f, g \in \mathcal{A}$,*

(i) $|f| \in \mathcal{A}$ *and*

(ii) $\max\{f, g\}, \min\{f, g\} \in \mathcal{A}$.

Proof. For any $f, g \in C(X, \mathbb{R})$, we have

$$\max\{f, g\} = \frac{1}{2}\left(f + g + |f - g|\right) \text{ and } \min\{f, g\} = \frac{1}{2}\left(f + g - |f - g|\right).$$

Therefore it suffices to prove part (i).

If $f \in \mathcal{A}$, set $m := \|f\|_\infty$ and set $g(x) := \frac{|f(x)|}{m}$ for $x \in X$. Then $g(x) \in [0, 1]$ for each $x \in X$. Since $\mathcal{A}$ is a subspace of $C(X, \mathbb{R})$ it is enough to prove that $g \in \mathcal{A}$. By the Weierstrass Approximation Theorem, there is a sequence $(p_n)_{n=1}^\infty$ of polynomials such that (p_n) converges uniformly on $[0, 1]$ to the square root function. Hence

$$p_n\left(\frac{f^2}{m^2}\right) \longrightarrow \sqrt{\frac{f^2}{m^2}} = g$$

uniformly on $[0, 1]$. Since $\mathcal{A}$ is an algebra containing the constants,

$$p_n\left(\frac{f^2}{m^2}\right) \in \mathcal{A}$$

for all $n \in \mathbb{N}$. Since $\mathcal{A}$ is closed, $g \in \mathcal{A}$. $\square$

The previous lemma is the only point in the proof where we need the Weierstrass Approximation Theorem. In fact, one can explicitly construct a sequence of polynomials that converge uniformly to the square root function (by a use of Dini's Theorem). Therefore with a little more effort, one can make our proof completely independent of Weierstrass' Theorem and thus make the latter a proper corollary of Stone's result.

A.1.5 LEMMA *Let $\mathcal{A} \subset C(X, \mathbb{R})$ be a subalgebra satisfying properties (P1) and (P2). Then for any pair of real numbers α, β and any pair of distinct points $x, y \in X$, there is a function $g \in \mathcal{A}$ such that $g(x) = \alpha$ and $g(y) = \beta$.*

Proof.   Since $x \neq y$, we can choose $f \in \mathcal{A}$ such that $f(x) \neq f(y)$. Then the function g defined by

$$g(u) = \frac{\alpha(f(u) - f(y)) - \beta(f(x) - f(u))}{f(x) - f(y)}$$

is an element of $\mathcal{A}$ (since $\mathcal{A}$ is an algebra) and it satisfies the required properties. $\square$

The next result is the real version of the proof and contains the main idea - to use compactness of X together with Lemma A.1.4.

A.1.6 THEOREM (STONE–WEIERSTRASS THEOREM – REAL VERSION) *Let* $\mathcal{A} \subset C(X, \mathbb{R})$ *be a closed subalgebra satisfying properties (P1) and (P2). Then* $\mathcal{A} = C(X, \mathbb{R})$.

Proof.   Let $f \in C(X, \mathbb{R})$ and $\epsilon > 0$ be given. For any $x, y \in X$, by Lemma A.1.5, there is a function $g_{x,y} \in \mathcal{A}$ such that $g_{x,y}(x) = f(x)$ and $g_{x,y}(y) = f(y)$. Define

$$U_{x,y} := \{t \in X : g_{x,y}(t) < f(t) + \epsilon\} \text{ and}$$
$$V_{x,y} := \{t \in X : g_{x,y}(t) > f(t) - \epsilon\}.$$

Then $U_{x,y}$ and $V_{x,y}$ are open sets containing x and y, respectively. Fix $y \in X$. By the compactness of X, there is a finite set $\{x_1, x_2, \ldots, x_n\}$ such that $\{U_{x_i,y}\}_{i=1}^{n}$ covers X. Let

$$g_y := \min\{g_{x_i,y} : 1 \leq i \leq n\},$$

then $g_y \in \mathcal{A}$ (by Lemma A.1.4) and satisfies

$$g_y(t) < f(t) + \epsilon \text{ for all } t \in X \text{ and}$$
$$g_y(t) > f(t) - \epsilon \text{ for all } t \in V_y := \bigcap_{i=1}^{n} V_{x_i,y}.$$

We now select a finite subcover $\{V_{y_j}\}_{j=1}^{m}$ from $\{V_y : y \in X\}$ for X and define

$$g := \max\{g_{y_j} : 1 \leq j \leq m\}.$$

Then $g \in \mathcal{A}$ (by Lemma A.1.4) and satisfies

$$f(t) - \epsilon < g(t) < f(t) + \epsilon$$

for all $t \in X$. Hence to every $\epsilon > 0$, there is an element $g \in \mathcal{A}$ such that $\|f - g\|_\infty < \epsilon$. Since $\mathcal{A}$ is closed, we see that $f \in \mathcal{A}$. This is true for every f in $C(X, \mathbb{R})$ and the theorem is proved. $\square$

We are now ready to complete the proof.

Proof of Theorem A.1.2. Let $\mathcal{A} \subset C(X, \mathbb{C})$ be a subalgebra satisfying conditions (P1), (P2) and (P3). Define

$$\mathcal{D} := \{\mathrm{Re}(f) : f \in \mathcal{A}\} \subset C(X, \mathbb{R}).$$

Since $\mathcal{A}$ is closed under complex conjugation (condition (P3)) and

$$\mathrm{Re}(f) = \frac{f + \bar{f}}{2} \in \mathcal{A},$$

we see that $\mathcal{D} \subset \mathcal{A}$. Since $\mathcal{A}$ is a vector subspace, $\mathrm{Im}(f) = \mathrm{Re}(if) \in \mathcal{D}$ for all $f \in \mathcal{A}$. Thus $\mathcal{D}$ satisfies all the hypotheses of Theorem A.1.6. If $f \in C(X)$, then write $f = g + ih$, where g, h are real-valued. By Theorem A.1.6, $g, h \in \mathcal{D}$. Since $\mathcal{D} \subset \mathcal{A}, f = g + ih \in \mathcal{A}$. Therefore $\mathcal{A} = C(X, \mathbb{C})$. $\qquad\square$

A.2 The Radon–Nikodym Theorem

This section is meant to be read alongside Chapter 9. Specifically, we will use the definitions of real and complex measures and absolute continuity introduced there. We now wish to prove the following theorem.

A.2.1 THEOREM (RADON, 1913, AND NIKODYM, 1930) *Let μ be a finite, positive measure on a measurable space $(X, \mathfrak{M})$. If ν is a complex measure such that $\nu \ll \mu$, then there is a measurable function $h \in L^1(\mu)$ such that*

$$\nu(E) = \int_E h\, d\mu$$

for all $E \in \mathfrak{M}$. Furthermore, if $g \in L^1(\mu)$ is another function such that $\nu(E) = \int_E g\, d\mu$ for all $E \in \mathfrak{M}$, then $\mu(\{x \in X : g(x) \neq h(x)\}) = 0$. In other words, the function h is unique almost everywhere with respect to μ.

The function h constructed above is called the **Radon–Nikodym derivative** of ν with respect to μ and is denoted by $\frac{d\nu}{d\mu}$. We begin by proving the uniqueness of this function.

A.2.2 PROPOSITION *Let μ be a finite, positive measure on (X, μ). Suppose $h, g \in L^1(\mu)$ are two functions such that*

$$\int_E h\, d\mu = \int_E g\, d\mu$$

for all $E \in \mathfrak{M}$. Then $g = h$ almost everywhere.

Proof. Simply apply the proof of Lemma 5.4.3 verbatim to the function $f := h - g$. □

Our proof of the existence of the Radon-Nikodym derivative works by first proving the result when ν is a positive measure and then decomposing a complex measure as a linear combination of positive measures. The next lemma (due to Schep [54]) is the first step. For two positive measures μ and ν defined on a measurable space $(X, \mathfrak{M})$, we write $\nu \leq \mu$ if $\nu(E) \leq \mu(E)$ for all $E \in \mathfrak{M}$.

A.2.3 LEMMA *Let μ and ν be two finite, positive measures on $(X, \mathfrak{M})$ such that $0 \leq \nu \leq \mu$. Then there exists a measurable function $h : X \to [0,1]$ such that*

$$\nu(E) = \int_E h \, d\mu$$

for all $E \in \mathfrak{M}$.

Proof. Let

$$H := \left\{ f : X \to [0,1] \text{ measurable, such that } \int_E f \, d\mu \leq \nu(E) \text{ for all } E \in \mathfrak{M} \right\}.$$

Then H is non-empty because the zero function is in H.

(i) If $f_1, f_2 \in H$, then we claim that $g := \max\{f_1, f_2\} \in H$. To see this, let $A := \{x \in X : f_1(x) \geq f_2(x)\}$ and let $B := X \setminus A$. Then for any $E \in \mathfrak{M}$,

$$\int_E g \, d\mu = \int_{E \cap A} g \, d\mu + \int_{E \cap B} g \, d\mu$$

$$= \int_{E \cap A} f_1 \, d\mu + \int_{E \cap B} f_2 \, d\mu$$

$$\leq \nu(E \cap A) + \nu(E \cap B) = \nu(E).$$

Thus $g \in H$.

(ii) If $M := \sup\{\int_X f \, d\mu : f \in H\}$, then $0 \leq M \leq \nu(X) < \infty$ and there exists a sequence (f_n) of functions in H such that $\int_X f_n \, d\mu > M - \frac{1}{n}$ for all $n \in \mathbb{N}$. Replacing f_n by $\max\{f_1, f_2, \ldots, f_n\}$ if need be (by appealing to part (i)), we may assume that $f_n \leq f_{n+1}$ for all $n \in \mathbb{N}$. Now set $h := \lim_{n \to \infty} f_n$. Then h is measurable and

$$\int_X h \, d\mu \geq M$$

by the Monotone Convergence Theorem. However, $h \in H$ by Fatou's Lemma and therefore

$$\int_X h \, d\mu = M.$$

(iii) We now show that $v(E) = \int_E h d\mu$ for all $E \in \mathfrak{M}$. Since $h \in H$, $v(E) \geq \int_E h d\mu$. Seeking a contradiction, we assume that there is a set $E \in \mathfrak{M}$ such that

$$v(E) > \int_E h d\mu.$$

Let $E_1 := \{x \in E : h(x) = 1\}$ and $E_0 := E \setminus E_1$, then

$$v(E) = v(E_0) + v(E_1) > \int_E h d\mu = \int_{E_0} h d\mu + \mu(E_1) \geq \int_{E_0} h d\mu + v(E_1).$$

Therefore $v(E_0) > \int_{E_0} f d\mu$. For each $n \geq 2$, let $F_n := \{x \in E_0 : h(x) < 1 - \frac{1}{n}\}$. Then $F_2 \subset F_3 \subset \ldots$ and $E_0 = \bigcup_{n=1}^\infty F_n$. Therefore there exists $N \in \mathbb{N}$ such that

$$v(F_N) > \int_{F_N} h d\mu.$$

Since $h(x) < 1 - \frac{1}{N}$ for all $x \in F_N$, we may choose $\epsilon > 0$ small enough so that $g := h + \epsilon \chi_{F_N} \in H$. However, in that case,

$$\int_X g d\mu = M + \epsilon \mu(F_N) \geq M + \epsilon v(F_N) > M.$$

This contradicts the choice of M. Therefore it must happen that

$$v(E) = \int_E h d\mu$$

for all $E \in \mathfrak{M}$. $\square$

We now prove a special case of the Radon–Nikodym Theorem for positive measures, the proof of which is due to von Neumann.

A.2.4 PROPOSITION *Let μ and v be two finite, positive measures on a measurable space $(X, \mathfrak{M})$. If $v \ll \mu$, then there is a non-negative function $h \in L^1(\mu)$ such that*

$$v(E) = \int_E h d\mu$$

for all $E \in \mathfrak{M}$.

Proof. Let $\lambda := \mu + v$, then λ is a finite, positive measure and $0 \leq v \leq \lambda$. By Lemma A.2.3, there is a function $g : X \to [0, 1]$ such that

$$v(E) = \int_E g d\lambda$$

for all $E \in \mathfrak{M}$. Hence $\mu(E) = \int_E (1 - g) d\lambda$ for all $E \in \mathfrak{M}$.

(i) If $D := \{x \in X : g(x) = 1\}$, then $\mu(D) = \int_D (1 - g)d\mu = 0$. Since $\nu \ll \mu$, it follows that $\nu(D) = 0$ and therefore $\lambda(D) = 0$. This implies that $0 \le g < 1$ almost everywhere with respect to all three measures.

(ii) Now note that $\nu(E) = \int_E g d\nu + \int_E g d\mu$. Therefore

$$\int_X (1 - g)\chi_E d\nu = \int_E (1 - g)d\nu = \int_E g d\mu = \int_X g\chi_E d\mu.$$

Taking linear combinations, we see that

$$\int_X (1 - g)s d\nu = \int_X gs d\mu$$

for any non-negative simple function $s : X \to [0, \infty]$. If $f : X \to [0, \infty]$ is any non-negative function, then the Monotone Convergence Theorem ensures that

$$\int_X (1 - g)f d\nu = \int_X gf d\mu.$$

(iii) Taking $f := (1 + g + g^2 + \ldots + g^n)\chi_E$, we see that

$$\int_E (1 - g^{n+1})d\nu = \int_E g(1 + g + g^2 + \ldots + g^n)d\mu.$$

Since $0 \le g < 1$, the Monotone Convergence Theorem allows us to take a limit of this expression and arrive at

$$\nu(E) = \lim_{n \to \infty} \int_E (1 - g^{n+1})d\nu$$

$$= \lim_{n \to \infty} \int_E g(1 + g + g^2 + \ldots + g^n)d\mu$$

$$= \int_E g(1 - g)^{-1}d\mu.$$

This proves the result with $h = g(1 - g)^{-1}$.

$\square$

In order to extend Proposition A.2.4 to complex measures, we need to express complex measures in terms of positive measures. Recall that every complex measure μ can be expressed as $\text{Re}(\mu) + i\,\text{Im}(\mu)$ where $\text{Re}(\mu)$ and $\text{Im}(\mu)$ are real measures. All that remains therefore is the next notion.

A.2.5 DEFINITION Let ν be a real measure on a measurable space $(X, \mathfrak{M})$. Then the total variation $|\nu|$ is also a real measure such that

$$|\nu(E)| \le |\nu|(E)$$

for each $E \in \mathfrak{M}$. Therefore we define the *positive* and *negative variations* of ν by the formula

$$\nu^+ := \frac{1}{2}(|\nu| + \nu) \text{ and } \nu^- := \frac{1}{2}(|\nu| - \nu).$$

Then ν^+ and ν^- are both positive measures such that $\nu = \nu^+ - \nu^-$.

Proof of Theorem A.2.1. Let ν be a complex measure and μ be a finite, positive measure such that $\nu \ll \mu$. If $E \in \mathfrak{M}$ is such that $\mu(E) = 0$, then choose any finite collection $\{E_1, E_2, \ldots, E_n\}$ of measurable sets such that $\bigcup_{i=1}^{n} E_i \subset E$. For any $1 \le i \le n$, $\mu(E_i) = 0$ and therefore $\nu(E_i) = 0$. Hence $\text{Re}(\nu)(E_i) = \text{Im}(\nu)(E_i) = 0$. This is true for any such collection, so it follows by definition of the total variation that

$$|\text{Re}(\nu)|(E) = |\text{Im}(\nu)|(E) = 0.$$

We conclude that $|\text{Re}(\nu)| \ll \mu$ and $|\text{Im}(\nu)| \ll \mu$. Hence $\text{Re}(\nu) \ll \mu$ and $\text{Im}(\nu) \ll \mu$ which implies that

$$\text{Re}(\nu)^{\pm} \ll \mu \text{ and } \text{Im}(\nu)^{\pm} \ll \mu.$$

Thus ν can be expressed as

$$\nu = \text{Re}(\nu)^+ - \text{Re}(\nu)^- + i(\text{Im}(\nu)^+ - \text{Im}(\nu)^-),$$

where each measure appearing on the right hand side is a finite, positive measure that is absolutely continuous with respect to μ. We may now apply Proposition A.2.4 to each of these measures and take a linear combination to construct the required Radon–Nikodym derivative. $\qquad\qquad \square$

Bibliography

[1] Aliprantis, Charalambos D. and Owen Burkinshaw. *Positive Operators*. Reprint of the 1985 original. Springer, Dordrecht, 2006, pp. xx+376. ISBN: 978-1-4020-5007-7; 1-4020-5007-0. DOI: 10.1007/978-1-4020-5008-4.

[2] Aronszajn, H. Theory of reproducing kernels. *Transactions of the American Mathematical Society* 68 (1950), pp. 337–404. ISSN: 0002-9947. DOI: 10.2307/1990404.

[3] Banach, Stefan. *Théorie des Opérations Linéaires*. 1932.

[4] Billingsley, Patrick. *Probability and Measure*. Anniversary edition [of MR1324786], With a foreword by Steve Lalley and a brief biography of Billingsley by Steve Koppes. Wiley Series in Probability and Statistics. John Wiley & Sons Inc. Hoboken, NJ, 2012, pp. xviii+624. ISBN: 978-1-118-12237-2.

[5] Bourgain, J. l^∞/c_0 has no equivalent strictly convex norm. *Proceedings of the American Mathematical Society* 78.2 (1980), pp. 225–226. ISSN: 0002-9939. DOI: 10.2307/2042258.

[6] Brezis, Haim. *Functional Analysis, Sobolev Spaces and Partial Differential Equations*. Universitext, Springer, New York, 2011, pp. xiv+599. ISBN: 978-0-387-70913-0.

[7] Carothers, N. L. *A Short Course on Banach Space Theory*. London Mathematical Society Student Texts. Vol. 64. Cambridge University Press, Cambridge, 2005, pp. xii+184. ISBN: 0-521-84283-2; 0-521-60372-2.

[8] Clarkson, James A. Uniformly convex spaces. *Transactions of the American Mathematical Society* 40.3 (1936), pp. 396–414. ISSN: 0002-9947. DOI: 10.2307/1989630.

[9] Conway, John B. *A Course in Functional Analysis*. Graduate Texts in Mathematics. Vol. 96. Springer-Verlag, New York, 1990, pp. xvi+399. ISBN: 0-387-97245-5.

[10] Davidson, Kenneth R. *C*-Algebras by Example*. Fields Institute Monographs. Vol. 6. American Mathematical Society, Providence, RI, 1996, pp. xiv+309. ISBN: 0-8218-0599-1. DOI: 10.1090/fim/006.

[11] Day, Mahlon M. Reflexive Banach spaces not isomorphic to uniformly convex spaces. *Bulletin of the American Mathematical Society* 47 (1941), pp. 313–317. ISSN: 0002-9904. DOI: 10.1090/S0002-9904-1941-07451-3.

[12]   Diestel, Joseph. *Sequences and Series in Banach Spaces*. Graduate Texts in Mathematics. Vol. 92. Springer-Verlag, New York, 1984, pp. xii+261. ISBN: 0-387-90859-5. DOI: 10.1007/978-1-4612-5200-9.

[13]   Doran, Robert S. and Victor A. Belfi. *Characterizations of C*-Algebras*. Monographs and Textbooks in Pure and Applied Mathematics. Vol. 101. Marcel Dekker, Inc., New York, 1986, pp. xi+426. ISBN: 0-8247-7569-4.

[14]   Douglas, Ronald G. *Banach Algebra Techniques in Operator Theory*. Second edition. Graduate Texts in Mathematics. Vol. 179. Springer-Verlag, New York, 1998, pp. xvi+194. ISBN: 0-387-98377-5. DOI: 10.1007/978-1-4612-1656-8.

[15]   Dummit, David S. and Richard M. Foote. *Abstract Algebra*. Third edition. John Wiley & Sons Inc. Hoboken, NJ, 2004, pp. xii+932. ISBN: 0-471-43334-9.

[16]   Enflo, Per. A counterexample to the approximation problem in Banach spaces. *Acta Mathematica* 130 (1973), pp. 309–317. ISSN: 0001-5962. DOI: 10.1007/BF02392270.

[17]   Foguel, Shaul R. On a theorem by A. E. Taylor. *Proceedings of the American Mathematical Society* 9 (1958), p. 325. ISSN: 0002-9939. DOI: 10.2307/2033162.

[18]   Folland, Gerald B. *Real Analysis: Modern Techniques and Their Applications*. Pure and Applied Mathematics (New York). A Wiley-Interscience Publication. John Wiley & Sons, Inc., New York, 1984, pp. xiv+350. ISBN: 0-471-80958-6.

[19]   Garling, D. J. H. A "short" proof of the Riesz representation theorem. *Proceedings of the Cambridge Philosophical Society* 73 (1973), pp. 459–460. ISSN: 0008-1981. DOI: 10.1017/s0305004100077021.

[20]   Gelfand, I. Normierte Ringe. *Rec. Math. [Mat. Sbornik]*. N.S., 9 (51) (1941), pp. 3–24.

[21]   Gelfand, I. and M. Neumark. On the imbedding of normed rings into the ring of operators in Hilbert space. In *C*-algebras: 1943–1993 (San Antonio, TX, 1993)*. Contemporary Mathematics. Vol. 167. Corrected reprint of the 1943 original [MR 5, 147]. American Mathematical Society, Providence, RI, 1994, pp. 2–19. DOI: 10.1090/conm/167/1292007.

[22]   Gelfand, I. M., D. A. Raikov, and G. E. Šilov. Commutative normed rings. *Akademiya Nauk SSSR i Moskovskoe Matematicheskoe Obshchestvo. Uspekhi Matematicheskikh Nauk* 1.2(12) (1946), pp. 48–146. ISSN: 0042-1316.

[23]   Halmos, P. R. What does the spectral theorem say? *American Mathematical Monthly* 70 (1963), pp. 241–247. ISSN: 0002-9890. DOI: 10.2307/2313117.

[24]   Halmos, Paul R. *Finite-dimensional Vector Spaces*. Second edition. Undergraduate Texts in Mathematics. Springer-Verlag, New York-Heidelberg, 1974, pp. viii+200.

[25]   Halmos, Paul R. *Measure Theory*. Vol. 18. Springer, 2013.

[26]   Hochstadt, Harry. *Integral Equations*. Wiley Classics Library. Reprint of the 1973 original, A Wiley-Interscience Publication. John Wiley & Sons, Inc., New York, 1989, pp. x+282. ISBN: 0-471-50404-1.

[27]   Hoffman, Kenneth and Ray Kunze. *Linear Algebra*. Second edition. Prentice-Hall Inc., Englewood Cliffs, N.J., 1971, pp. viii+407.

[28] James, Robert C. A non-reflexive Banach space isometric with its second conjugate space. *Proceedings of the National Academy of Sciences of the United States of America* 37.3 (1951), p. 174.

[29] James, Robert C. Bases in Banach spaces. *The American Mathematical Monthly* 89.9 (1982), pp. 625–640.

[30] James, Robert C. Reflexivity and the sup of linear functionals. *Israel Journal of Mathematics* 13 (1972), 289–300 (1973). ISSN: 0021-2172. DOI: 10.1007/BF02762803.

[31] Kadison, Richard V. and John R. Ringrose. *Fundamentals of the Theory of Operator Algebras. Vol. 1: Elementary Theory.* Graduate Studies in Mathematics. Vol. 15. Reprint of the 1983 original. American Mathematical Society, Providence, RI, 1997, pp. xvi+398. ISBN: 0-8218-0819-2. DOI: 10.1090/gsm/015.

[32] Kadison, Richard V. and John R. Ringrose. *Fundamentals of the Theory of Operator Algebras. Vol. II: Advanced Theory.* Graduate Studies in Mathematics. Vol. 16. Corrected reprint of the 1986 original. American Mathematical Society, Providence, RI, 1997, i–Vxxii and 399–1074. ISBN: 0-8218-0820-6. DOI: 10.1090/gsm/016.01.

[33] Kantorovitch, L. Linear operations in semi-ordered spaces. *Rec. Math. [Mat. Sbornik]*, *N.S.*, 7 (49) (1940), pp. 209–284.

[34] Kesavan, S. *Functional Analysis.* Texts and Readings in Mathematics. Vol. 52. Second corrected reprint of the 2009 original [MR2475358]. Hindustan Book Agency, New Delhi, 2017, pp. xii+269. ISBN: 978-93-80250-62-5.

[35] Kuiper, Nicolaas H. The homotopy type of the unitary group of Hilbert space. *Topology. An International Journal of Mathematics* 3 (1965), pp. 19–30. ISSN: 0040-9383. DOI: 10.1016/0040-9383(65)90067-4.

[36] Leon, Steven J., Å ke Bjöorck, and Walter Gander. Gram-Schmidt orthogonalization: 100 years and more. *Numerical Linear Algebra with Applications* 20.3 (2013), pp. 492–532. ISSN: 1070-5325. DOI: 10.1002/nla.1839.

[37] Limaye, Balmohan V. *Functional Analysis.* Second edition. New Age International Publishers Limited, New Delhi, 1996, pp. x+612. ISBN: 81-224-0849-4.

[38] Lindenstrauss, J. and L. Tzafriri. On the complemented subspaces problem. *Israel Journal of Mathematics* 9 (1971), pp. 263–269. ISSN: 0021-2172. DOI: 10.1007/BF02771592.

[39] Lindenstrauss, Joram and Lior Tzafriri. *Classical Banach Spaces. I.* Ergebnisse der Mathematik und ihrer Grenzgebiete, Band 92. Sequence spaces. Springer-Verlag, Berlin-New York, 1977, pp. xiii+188. ISBN: 3-540-08072-4.

[40] Loomis, L. H. *Introduction to Abstract Harmonic Analysis.* Izdat. Inostran. Lit., Moscow, 1956, p. 251.

[41] Lorentz, G. G. A contribution to the theory of divergent sequences. *Acta Mathematica* 80, 167–190. ISSN: 0001-5962. DOI: 10.1007/BF02393648.

[42] Munkres, James R. *Topology.* Second edition of [MR0464128]. Prentice Hall Inc., Upper Saddle River, NJ, 2000, pp. xvi+537. ISBN: 0-13-181629-2.

[43] Murphy, Gerard J. *C*-Algebras and Operator Theory*. English edition. Boston, MA, etc.: Academic Press, Inc., 1990, pp. x + 286. ISBN: 0-12-511360-9.

[44] Nagumo, Von Mitio. Einige analytische Untersuchungen in linearen, metrischen Ringen. *Japanese Journal of Mathematics: Transactions and Abstracts* 13 (1936), pp. 61–80. The Mathematical Society of Japan.

[45] Paulsen, Vern I. and Mrinal Raghupathi. *An Introduction to the Theory of Reproducing Kernel Hilbert Spaces*. Cambridge Studies in Advanced Mathematics. Vol. 152. Cambridge University Press, Cambridge, 2016, pp. x+182. ISBN: 978-1-107-10409-9. DOI: 10.1017/CBO978131 6219232.

[46] Phelps, R. R. Uniqueness of Hahn-Banach extensions and unique best approximation. *Transactions of the American Mathematical Society* 95 (1960), pp. 238–255. ISSN: 0002-9947. DOI: 10.2307/1993289.

[47] Pietsch, Albrecht. *History of Banach Spaces and Linear Operators*. Birkhäuser Boston Inc. Boston, MA, 2007, pp. xxiv+855. ISBN: 978-0-8176-4367-6; 0-8176-4367-2.

[48] Raman-Sundström, Manya. A pedagogical history of compactness. *American Mathematical Monthly* 122.7 (2015), pp. 619–635. ISSN: 0002-9890. DOI: 10.4169/amer.math.monthly.122.7.619.

[49] Royden, H. L. *Real Analysis*. Third edition. Macmillan Publishing Company, New York, 1988, pp. xx+444. ISBN: 0-02-404151-3.

[50] Rudin, Walter. *Principles of Mathematical Analysis*. Third edition. International Series in Pure and Applied Mathematics. McGraw-Hill Book Co., New York–Auckland–Düsseldorf, 1976, pp. x+342.

[51] Rudin, Walter. *Real and Complex Analysis*. Third edition. McGraw-Hill Book Co., New York, 1987, pp. xiv+416. ISBN: 0-07-054234-1.

[52] Runde, Volker. *Lectures on Amenability*. Lecture Notes in Mathematics. Vol. 1774. Springer-Verlag, Berlin, 2002, pp. xiv+296. ISBN: 3-540-42852-6. DOI: 10.1007/b82937.

[53] Sakai, Shôichirô. *C*-Algebras and W*-Algebras*. Classics in Mathematics. Reprint of the 1971 edition. Springer-Verlag, Berlin, 1998, pp. xii+256. ISBN: 3-540-63633-1. DOI: 10.1007/978-3-642-61993-9.

[54] Schep, Anton R. And still one more proof of the Radon-Nikodym theorem. *American Mathematical Monthly* 110.6 (2003), pp. 536–538. ISSN: 0002-9890. DOI: 10.2307/3647910.

[55] Sherman, S. Order in operator algebras. *American Journal of Mathematics* 73 (1951), pp. 227–232. ISSN: 0002-9327. DOI: 10.2307/2372173.

[56] Shioji, Naoki. Simple proofs of the uniform convexity of L^p and the Riesz representation theorem for L^p. *American Mathematical Monthly* 125.8 (2018), pp. 733–738. ISSN: 0002-9890. DOI: 10.1080/00029890.2018.1496762.

[57] Stein, Elias M. and Rami Shakarchi. *Fourier Analysis: An Introduction*. Princeton Lectures in Analysis. Vol. 1. Princeton University Press, Princeton, NJ, 2003, pp. xvi+311. ISBN: 0-691-11384-X.

[58]   Sunder, V. S. *Functional Analysis: Spectral Theory*. Birkhäuser Advanced Texts: Basler
         Lehrbücher. [Birkhäuser Advanced Texts: Basel Textbooks]. Birkhäuser Verlag, Basel,
         1997, pp. x+241. ISBN: 3-7643-5892-0.

[59]   Swartz, Charles. The evolution of the uniform boundedness principle. *Mathematical
         Chronicle* 19 (1990), pp. 1–18. ISSN: 0581-1155.

[60]   Taylor, A. E. The extension of linear functionals. *Duke Mathematical Journal* 5 (1939),
         pp. 538–547. ISSN: 0012-7094.

[61]   Wheeden, Richard L. and Antoni Zygmund. *Measure and Integral: An Introduction
         to Real Analysis*. Second edition. Pure and Applied Mathematics (Boca Raton). CRC
         Press, Boca Raton, FL, 2015, pp. xvii+514. ISBN: 978-1-4987-0289-8.

[62]   Zhu, Ke He. *An Introduction to Operator Algebras*. Studies in Advanced Mathematics.
         CRC Press, Boca Raton, FL, 1993, pp. x+157. ISBN: 0-8493-7875-3.

Index